AF344485

IEE CONTROL ENGINEERING SERIES 66

Series Editors: Professor D. Atherton
Professor G. W. Irwin
Professor S. Spurgeon

Variable Structure Systems: from Principles to Implementation

Other volumes in print in this series:

Variable Structure Systems: from Principles to Implementation

Edited by Asif Sabanovic, Leonid Fridman
and Sarah Spurgeon

The Institution of Electrical Engineers

Published by: The Institution of Electrical Engineers, London,
United Kingdom

The Institution of Electrical Engineers,
Michael Faraday House,
Six Hills Way, Stevenage,
Herts., SG1 2AY, United Kingdom

www.iee.org

British Library Cataloguing in Publication Data

Sabanovic, A.
  Variable structure systems : from principles to implementation. –
  (IEE control engineering series ; 66)
  1. Sliding mode control
  I. Title II. Fridman, L. III. Spurgeon, Sarah K. IV. Institution of Electrical Engineers
  629.8'36

ISBN 0 86341 350 1

Typeset in India by Newgen Imaging Systems (P) Ltd., Chennai, India
Printed in England by MPG Books Ltd., Bodmin, Cornwall

Contents

3   Deterministic output noise effects in sliding mode observation   45
Alex S. Poznyak

Contributors

N. Barrie Jones
University of Leicester, Department of
Engineering, Control and
Instrumentation Research Group,
University Road, Leicester, LE1 7RH,
United Kingdom

Giorgio Bartolini
Department of Electrical and
Electronic Engineering (DIEE),
University of Cagliari, Italy

Domingo Biel Solé
Department of Electronic Engineering,
Universitat Politecnica de Catalunya

Igor Boiko
SNC-Lavalin, Control and
Instrumentation Department, 909 5th
Avenue SW, Calgary, Alberta,
T2P 3G5, Canada

B. Castillo-Toledo
Centro de Investigación y de Estudios
Avanzados del IPN, A. P. 31-438,
C.P. 44550, Guadalajara, Jal., México

Hao-Chi Chang
Department of Electrical and
Mechanical Engineering, The Ohio
State University, Columbus,
OH 43210, United States of America

Guanrong Chen
Department of Electronic Engineering,
City University of Hong Kong,
Kowloon, Hong Kong SAR,
People's Republic of China

Enric Fossas Colet
Institute of Industrial and Control
Engineering, Universitat Politecnica
de Catalunya

Kemalettin Erbatur
Sabanci University, Electrical
Engineering and Computer Science
Program Mechatronics Research
Group, Istanbul, Turkey

Leonid Fridman
Department of Postgraduate Study,
Engineering Faculty, National
Autonomous University of Mexico
DEP-FI, UNAM, A. P. 70-256,
C.P.04510, Mexico, D.F., Mexico

Keng Boon Goh
University of Leicester, Department of
Engineering, Control and
Instrumentation Research Group,
University Road, Leicester, LE1 7RH,
United Kingdom

Karel Jezernik
University of Maribor, FERI,
Smetanova 17, 22000 Maribor,
Slovenia

Arie Levant
Applied Mathematics Department,
Tel-Aviv University, Ramat-Aviv,
69978 Tel-Aviv, Israel

Alexander G. Loukianov
Centro de Investigación y de Estudios
Avanzados del IPN, A. P. 31-438,
C.P. 44550, Guadalajara, Jal., México

Čedomir Milosavljević
University of Niš, Faculty of
Electronic Engineering, Department
of Automatic Control, Beogradska 14,
18000 Niš, Serbia and Montenegro

Alessandro Pisano
Department of Electrical and
Electronic Engineering (DIEE),
University of Cagliari, Italy

Andrei Polyakov
Department of Applied Mathematics,
Voronezh State University,
Universitetskaja pl. 1,
Voronezh, 394693, Russia

Alex S. Poznyak
CINVESTAV-IPN,
Departamento de Control Automatico,
A.P. 14-740, C.P. 07300 Mexico D.F.,
Mexico

Elisabetta Punta
ISSIA – Institute of Intelligent
Systems for Automation, CNR –
National Research Council of Italy

J. Rivera
Centro de Investigación y de Estudios
Avanzados del IPN, A. P. 31-438,
C.P. 44550, Guadalajara, Jal.,
México

Asif Sabanovic
Sabanci University, Electrical
Engineering and Computer Science
Program Mechatronics Research
Group, Istanbul, Turkey

Ilya A. Shkolnikov
The University of Alabama in
Huntsville, Department of Electrical
and Computer Engineering,
Huntsville, AL 35899,
United States of America

Yuri B. Shtessel
The University of Alabama in
Huntsville, Department of Electrical
and Computer Engineering,
Huntsville, AL 35899,
United States of America

Hebertt Sira-Ramírez
CINVESTAV IPN, Av. IPN No. 2508,
Departamento Ing. Electrica, Secc.
Mecatronica, Colonia San Pedro
Zacatenco, AP 14740,
07300 Mexico D.F.,
Mexico

Sarah K. Spurgeon
University of Leicester, Department of
Engineering, Control and
Instrumentation Research Group,
University Road, Leicester, LE1 7RH,
United Kingdom

Vadim Strygin
Department of Applied Mathematics,
Voronezh State University,
Universitetskaja pl. 1, Voronezh,
394693, Russia

Elio Usai
Department of Electrical and
Electronic Engineering (DIEE),
University of Cagliari,
Italy

Vadim I. Utkin
Department of Electrical Engineering,
Ohio State University, Columbus,
Ohio, 43210-1272,
United States of America

Luis Iván Lugo Villeda
CINVESTAV IPN, Av. IPN No. 2508,
Departamento Ing. Electrica, Secc.
Mecatronica, Colonia San Pedro
Zacaterco, AP 14740,
07300 Mexico D.F., Mexico

Yildiray Yildiz
Sabanci University, Electrical
Engineering and Computer Science
Program Mechatronics Research
Group, Istanbul, Turkey

Xinghuo Yu
School of Electrical and Computer
Engineering, Royal Melbourne
Institute of Technology, Melbourne,
VIC 3001, Australia

Preface

In the formulation of any control problem there will typically be discrepancies between the actual system and the mathematical model available to the designer. This mismatch may be due to unmodelled dynamics, variation in system parameters or the approximation of complex, possibly nonlinear, system behaviour by a straightforward model. The engineer must ensure that controllers have the ability to produce the required performance despite such mismatches. This has led to an intense interest in the development of so-called *robust control* methods. One particular approach to robust controller design is the so-called *variable structure* control methodology.

Variable structure control systems (VSCS) are characterised by a suite of feedback control laws and a decision rule. The decision rule, termed the switching function, has as its input some measure of the current system behaviour and produces as an output the particular feedback controller that should be used at that instant in time. The well known *sliding mode control* methodology is a particular type of VSCS. In sliding mode control, VSCS are designed to drive and then constrain the system state to lie within a neighbourhood of the switching function. There are a number of advantages of this approach. First, the dynamic behaviour of the system may be tailored by the particular choice of switching function. Second, the closed-loop response become totally insensitive to a particular class of uncertainty in the system; this provides a very strong and inherent robustness to the resulting controllers. Finally, analysis of the discontinuous signals applied to the system can be used as a technique to model the signal activity required in order to achieve the ideal performance from the system.

The concept of a variable structure control system originated in the Soviet Union in the 1960s and the design paradigm now forms a mature and established approach for robust control and estimation. This book is divided into three sections which cover the essential background to variable structure control systems, current topics of research interest within the area and descriptions of a range of application studies, respectively.

The first section provides the necessary basic background to enable a graduate engineer to design a variable structure controller/estimator. Formulation of the desired system performance is emphasised and practical issues, such as discrete implementation and the problem of noise on the output measurements, are a focus.

The second section presents an overview of some topics that are of current interest in the area of variable structural control. Issues such as the control of inherently nonminimum-phase systems, the application of variable structure control techniques in the field of chaos and the exciting possibilities of higher order sliding regimes are considered. This section will provide essential reading for anyone involved in research in the area of robust and nonlinear control.

The final section presents a range of application studies in the area of variable structural control and will be interesting reading for anyone involved in the area of control applications. A diverse range of studies is considered from motion control to automobile control and from sliding mode applications in fuzzy and neutral network systems to sliding mode application in power electronics.

The Editors would like to thank all the contributors for their hard work and cooperation in the preparation of this manuscript. The helpfulness and professionalism of the editorial and production staff at the Institution of Electrical Engineers is gratefully acknowledged. In an increasingly globalised world this text is a testament to the benefits of international collaboration.

Professor Sarah Spurgeon
Leicester
April 2004

Part I

Sliding mode control theory

Chapter 1

Sliding mode control

Vadim I. Utkin

1.1 Introduction[1]

The sliding mode control approach is recognised as an efficient tool to design robust controllers for complex high-order nonlinear dynamic plant operating under uncertain conditions. The research in this area was initiated in the former Soviet Union about 40 years ago, and the sliding mode control methodology has subsequently received much more attention from the international control community within the last two decades.

The major advantage of sliding mode is low sensitivity to plant parameter variations and disturbances which eliminates the necessity of exact modelling. Sliding mode control enables the decoupling of the overall system motion into independent partial components of lower dimension and, as a result, reduces the complexity of feedback design. Sliding mode control implies that control actions are discontinuous state functions which may easily be implemented by conventional power converters with 'on-off' as the only admissible operation mode. Due to these properties, the intensity of the research at many scientific centres of industry and universities is maintained at a high level, and sliding mode control has been proved to be applicable to a wide range of problems in robotics, electric drives and generators, process control, vehicle and motion control.

1.2 The concept of a 'sliding mode'

The 'sliding mode' phenomenon may appear in dynamic systems governed by ordinary differential equations with discontinuous state functions in the right-hand sides. The conventional example of sliding mode – a second order relay system – can

---

[1] The source of Sections 1.1–1.7 and 1.9 is V. Utkin, Section 6.43.21.14 'Sliding Mode Control', Part C, Chapter 6.43, 'Control Systems, Robotics and Automation', published with permission from EOLSS Publishers Co Ltd.

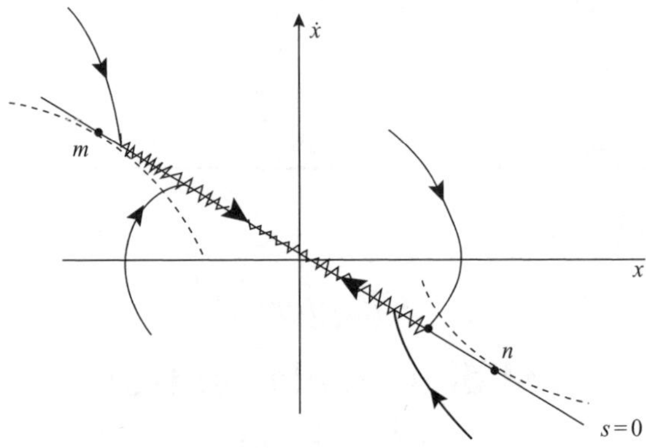

Figure 1.1   Sliding mode in a second relay system

be found in any text book on nonlinear control. The control input in the second order system

$$\ddot{x}+a_2\dot{x}+a_1x=u$$

$$u=-M\text{sign}(s), \qquad s=cx+\dot{x}, \quad a_1,a_2,M,c \text{ are const}$$

may take only two values, M and $-M$, and undergoes discontinuities on the straight line $s=0$ in the state plane $(x,\dot{x})$ (Fig. 1.1 for the case $a_1=a_2=0$). It follows from the analysis of the state plane that, in the neighbourhood segment mn on the switching line $s=0$, the trajectories run in opposite directions, which leads to the appearance of a sliding mode along this line. The equation of this line $\dot{x}+cx=0$ may be interpreted as the sliding mode equation. Note that the order of the equation is less than that of the original system and the sliding mode does not depend on the plant dynamics, and is determined by the parameter c only.

Sliding mode became the principle operation mode in so-called variable structure systems. A variable structure system consists of a set of continuous subsystems with a proper switching logic and, as a result, control actions are discontinuous functions of the system state, disturbances (if they are accessible for measurement), and reference inputs. The previous example of the relay system with state dependent amplitude of the control variable may serve as an illustration of a variable structure system: $u=-k|x|\text{sign}(s)$, k is constant.

Now the system with $a_1=0$ and $a_2<0$ consists of two unstable linear structures ($u=kx$ and $u=-kx$, Fig. 1.2) with $x=0$ and $s=0$ as switching lines. As it is clear from the system state plane, the state reaches the switching line $s=0$ for any initial conditions. Then, the sliding mode occurs on this line (Fig. 1.3) with the motion equation $\dot{x}+cx=0$, while the state vector decays exponentially. Similarly to the relay system, after the start of the sliding mode, the motion is governed by a reduced order equation which does not depend on the plant parameters. Now we demonstrate sliding modes in nonlinear affine systems of general form

$$\dot{x}=f(x,t)+B(x,t)u \tag{1.1}$$

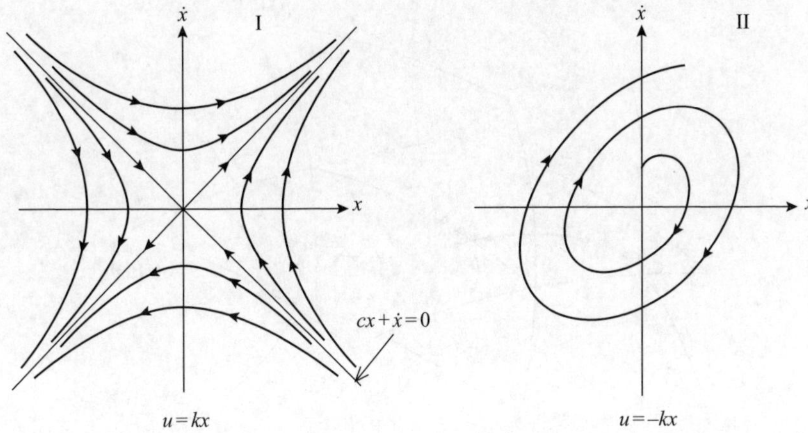

Figure 1.2   *State planes of two unstable structures*

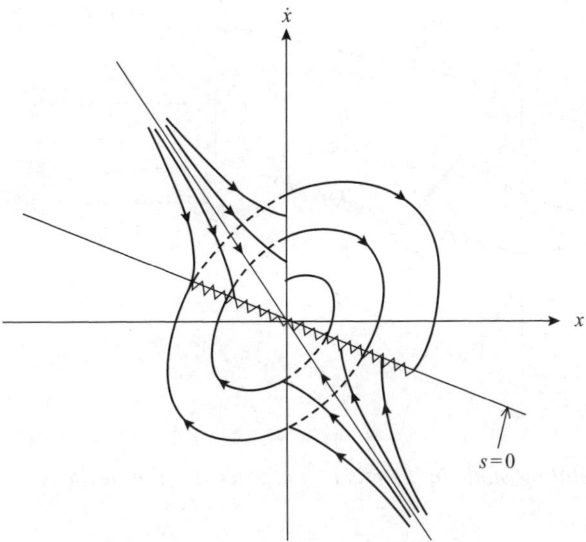

Figure 1.3   *State plane of variable structure system*

$$u_i = \begin{cases} u_i^+(x,t) & \text{if } s_i(x) > 0 \\ u_i^-(x,t) & \text{if } s_i(x) < 0 \end{cases} \qquad i = 1,\dots,m \qquad (1.2)$$

where $x \in R^n$ is a state vector, $u \in R^m$ is a control vector, $u_i^+(x,t)$, $u_i^-(x,t)$ and $s_i(x)$ are continuous functions of their arguments, $u_i^+(x,t) \neq u_i^-(x,t)$. The control is designed as a discontinuous function of the state such that each component undergoes discontinuities in some surface in the system state space.

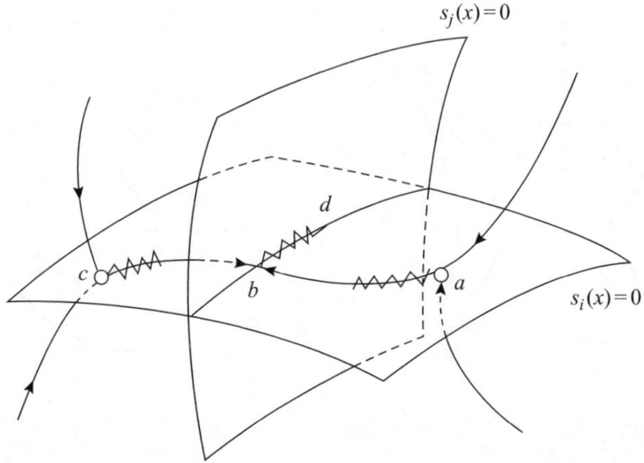

Figure 1.4   *Sliding mode in discontinuity surface and their intersection*

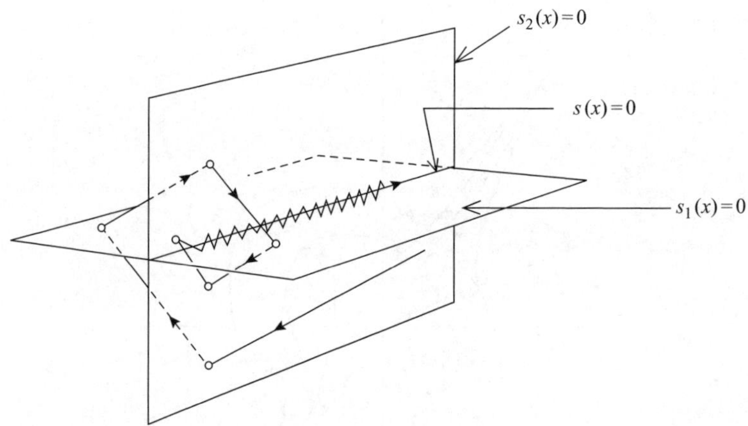

Figure 1.5   *Sliding mode in intersection of discontinuity surfaces*

Similar to the above example, state velocity vectors may be directed towards one of the surfaces and a sliding mode arises along it (arcs *ab* and *cb* in Fig. 1.4). It may arise also along the intersection of two surfaces (arc *bd*).

Figure 1.5 illustrates the sliding mode in the intersection even if it does not exist at each of the surfaces taken separately.

For the general case (1.1), a sliding mode may exist in the intersection of all discontinuity surfaces $s_i = 0$, or in the manifold

$$s(x)=0, \quad s^T(x)=[s_1(x),\ldots,s_m(x)] \tag{1.3}$$

of dimension $n - m$.

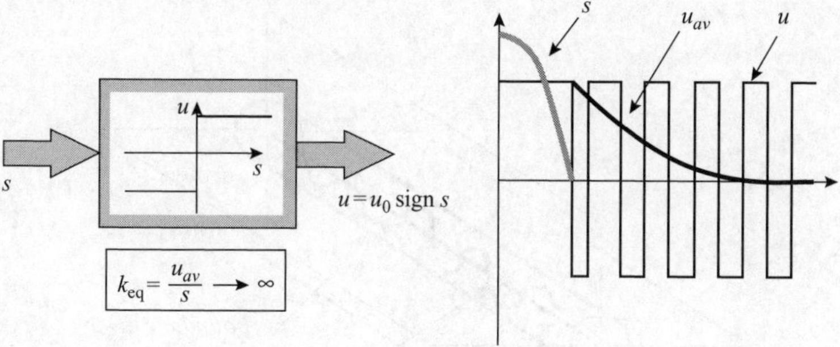

Figure 1.6   High gain implementation by sliding mode

Let us discuss the benefits of sliding modes, if they are enforced in the control system. First, in the sliding mode, the input s of the element implementing the discontinuous control is close to zero, while its output (speaking precisely, its average value u_{av}) takes finite values (Fig. 1.6).

Hence, the element implements high (theoretically infinite) gain, that is the conventional tool to reject disturbance and other uncertainties in the system behaviour. Unlike systems under a continuous control action, this property, called invariance, is attained using finite control actions. Second, since sliding mode trajectories belong to a manifold of a dimension lower than that of the original system, the order of the system is reduced as well. This enables a designer to simplify and decouple the design procedure. Both order reduction and invariance are transparent for the above two second-order systems.

1.3   Sliding mode equations

So far the arguments in favour of employing sliding modes in control systems have been discussed at the qualitative level. To justify them strictly, mathematical methods should be developed for describing this motion in the intersection of discontinuity surfaces and deriving the conditions for a sliding mode to exist.

The first problem means deriving differential equations of the sliding mode. Note that, for our second-order example, the equation of the switching line $\dot{x} + cx = 0$ was interpreted as the motion equation. But even for a time invariant second-order relay system

$$\dot{x}_1 = a_{11}x_1 + a_{12}x_2 + b_1 u$$

$$\dot{x}_2 = a_{21}x_1 + a_{22}x_2 + b_2 u, \qquad u = -M\operatorname{sign}(s), \qquad s = cx_1 + x_2;$$

$$M, a_{ij}, b_i, c \text{ are const}$$

the problem does not look trivial since in the sliding mode $s = 0$ is not a motion equation.

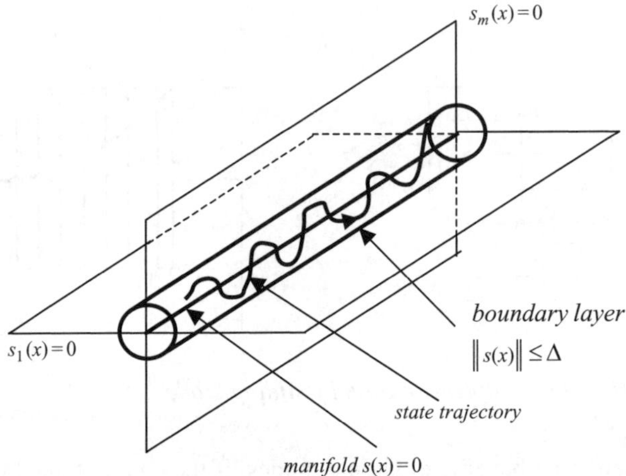

Figure 1.7   Boundary layer

The first problem arises due to discontinuities in the control, since the relevant motion equations do not satisfy the conventional theorems on existence-uniqueness of solutions. In situations when conventional methods are not applicable, the usual approach is to employ regularisation or replacing the initial problem by a closely similar one, for which familiar methods can be used. In particular, taking into account delay or hysteresis of a switching element, small time constants in an ideal model, replacing a discontinuous function by a continuous approximation are examples of regularisation since discontinuity points (if they exist) are isolated. The universal approach to regularisation consists of introducing a boundary layer $\|s\| < \Delta$, Δ – const around the manifold $s = 0$, where an ideal discontinuous control is replaced by a real one such that the state trajectories are not confined to this manifold but run arbitrarily inside the layer (Fig. 1.7).

The only assumption for this motion is that the solution exists in the conventional sense. If, with the width of the boundary layer Δ tending to zero, the limit of the solution exists, it is taken as a solution to the system with ideal sliding mode. Otherwise we have to recognise that the equations beyond discontinuity surfaces do not derive, unambiguously, equations in their intersection, or equations of the sliding mode.

The boundary layer regularisation enables substantiation of the so-called *Equivalent Control Method* intended for deriving sliding mode equations in the manifold $s = 0$ in system (1.1). Following this method, the sliding mode equation with a unique solution may be derived for the nonsingular matrix

$$G(x)B(x), \quad G(x) = \left\{ \frac{\partial s}{\partial x} \right\}, \quad \det(GB) \neq 0$$

First, the *equivalent control* should be found for the system (1.1) as the solution to the equation $\dot{s} = 0$ on the system trajectories (G and $(GB)^{-1}$ are assumed to exist):

$$\dot{s} = G\dot{x} = Gf + GBu_{eq} = 0, \qquad u_{eq} = -(GB)^{-1}Gf$$

Then the solution should be substituted into (1.1) for the control

$$\dot{x} = f - B(GB)^{-1}Gf \tag{1.4}$$

Equation (1.4) is the sliding mode equation with initial conditions $s(x(0), 0) = 0$.

Since $s(x) = 0$ in the sliding mode, m components of the state vector may be found as a function of the remaining $(n - m)$: $x_2 = s_0(x_1)$; $x_2, s_0 \in \Re^m$; $x_1 \in \Re^{n-m}$ and, correspondingly, the order of the sliding mode equation may be reduced by m:

$$\dot{x}_2 = f_1[x_1, t, s_0(x_1)], \qquad f_1 \in \Re^{n-m} \tag{1.5}$$

The idea of the equivalent control method may be easily explained with the help of geometric considerations. Sliding mode trajectories lie in the manifold $s = 0$ and the equivalent control u_{eq}, being a solution to the equation $\dot{s} = 0$, implies replacing the original discontinuous control by a continuous one such that the state velocity vector lies in the tangential manifold and as a result the state trajectories are in this manifold. It will be important for control design that the sliding mode equation is of reduced order, does not depend on the control and depends on the equation of the switching surfaces.

1.4   Existence conditions

The second mathematical problem in the analysis of sliding mode as a phenomenon is deriving the conditions for the sliding mode to exist. As with the second-order systems with scalar control studied in Section 1.2, the conditions may be obtained from geometrical considerations: the deviation from the switching surface s and its time derivative should have opposite signs in the vicinity of a discontinuity surface $s = 0$, or

$$\lim_{s \to +0} \dot{s} < 0 \quad \text{and} \quad \lim_{s \to -0} \dot{s} > 0 \tag{1.6}$$

Inequalities (1.6) with the condition $s\dot{s} < 0$ are referred to as *reaching conditions* – the condition for the state to reach the surface $s = 0$ after a finite time for arbitrary initial conditions. For the second-order relay system in Section 1.2, the domain of sliding mode on $s = 0$ or for $\dot{x} = -cx$ (sector mn on the switching line, Fig. 1.1) may be found analytically from these conditions:

$$\dot{s} = (-c^2 + a_2 c - a_1)x - M\mathrm{sign}(s) \quad \text{and} \quad |x| < \frac{M}{|-c^2 + a_2 c - a_1|}$$

As was demonstrated in the example in Fig. 1.5, for existence of a sliding mode in an intersection of a set of discontinuity surfaces $s_i(x) = 0$, $(i = 1, \ldots, m)$, it is not necessary to fulfil inequalities (1.6) for each of them. The trajectories should converge to the manifold $s^T = (s_1, \ldots, s_m) = 0$ and reach it after a finite time interval similarly to the systems with scalar control. The term 'converge' means that we deal with the problem of stability of the origin in an m-dimensional subspace $(s_1, \ldots, s_m)$, therefore the existence conditions may be formulated in terms of stability theory.

The non-traditional condition, *finite time convergence*, should take place. This last condition is important to distinguish the systems with sliding modes and the continuous system with state trajectories converging to some manifold asymptotically. For example, the state trajectories of the system $\ddot{x} - x = 0$ converge to the manifold $s = \dot{x} - x = 0$ asymptotically since $\dot{s} = -s$, however it would hardly be reasonable to call the motion in $s = 0$ a 'sliding mode'.

Further we will deal with the conditions for a sliding mode to exist for affine systems (1.1). To derive the existence conditions, the stability of the motion projection on the subspace s

$$\dot{s} = Gf + GBu \tag{1.7}$$

should be analysed.

The control (1.2) may be represented as $u(x,t) = u_0(x,t) + U(x,t)\mathrm{sign}(s)$, where $u_0(x) = (u^+(x,t) + u^-(x,t))/2$, $U(x)$ is a diagonal matrix with elements $U_i = (u_i^+(x,t) - u_i^-(x,t))/2$, $i = 1,\ldots,m$ and the discontinuous vector $\mathrm{sign}(s)$ is in the form of component-wise sign function $[\mathrm{sign}(s)]^T = [\mathrm{sign}(s_1),\ldots,\mathrm{sign}(s_m)]$.

Then the motion projection on subspace s is governed by

$$\dot{s} = d(x) - D(x)\mathrm{sign}(s) \tag{1.7'}$$

with $d = Gf + GBu_0$, $D = -GBU$.

To find the stability conditions of the origin $s = 0$ for the nonlinear system (1.7'), or the conditions for a sliding mode to exist, we will follow the standard methodology for stability analysis of nonlinear systems and try to find a Lyapunov function.

Definition 1.   The set $S(x)$ in the manifold $s(x) = 0$ is the domain of the sliding mode if for the motion governed by equation (1.7') the origin in the subspace s is asymptotically stable with finite convergence time for each x from $S(x)$.

Definition 2.   Manifold $s(x) = 0$ is referred to as *a sliding manifold* if a sliding mode exists at each point, or $S(x) = \{x : s(x) = 0\}$.

Theorem 1. $S(x)$ is a sliding manifold for the system with motion projection on subspace s governed by $\dot{s} = -D\mathrm{sign}(s)$ if the matrix $D + D^T > 0$ is positive definite.

Theorem 2. $S(x)$ is a sliding manifold for system (1.7') if $D(x) + D^T(x) > 0$, $\lambda_0 > d_0\sqrt{m}$, $\lambda_{\min}(x) > \lambda_0 > 0$, $\|d(x)\| < d_0$, $\lambda_{\min}$ is the minimal eigenvalue of matrix $(D + D^T)/2$, $\lambda_{\min} > 0$.

The statements of both the theorems may be proven using a sum of absolute values of s_i as a Lyapunov function $V = [\mathrm{sign}(s)]^T s > 0$. Similarly to the scalar case, the conditions of the theorems are the reaching conditions simultaneously if they hold for any state vector.

1.5 Design principles

The above mathematical results constitute the background of the design methods for sliding mode control involving two independent subproblems of lower dimensions:

- design of the desired dynamics for a system of the $(n-m)$th order by proper choice of a sliding manifold $s=0$;
- enforcing sliding motion in this manifold which is equivalent to a stability problem of the mth order system.

Since the principle operating mode is in the vicinity of the discontinuity points, the effects inherent in the systems with infinite feedback gains may be obtained with finite control actions. As a result sliding mode control is an efficient tool to control dynamic high-order nonlinear plants operating under uncertain conditions (e.g. unknown parameter variations and disturbances).

Formally the sliding mode is insensitive to 'uncertainties' in the systems satisfying the matching conditions

$$h(x,t) \in \text{range}(B)$$

where the vector $h(x,t)$ characterises all disturbance factors in a motion equation

$$\dot{x} = f(x,t) + B(x,t)u + h(x,t)$$

whose influence on the control process should be rejected. The matching condition means that the disturbance vector $h(x,t)$ may be represented as a linear combination of the columns of matrix B: $h(x,t) = Bh(x,t)$, $h(x,t) \in R^m$.

The design procedure may be illustrated easily for the systems represented in the *Regular Form*

$$\begin{aligned}
\dot{x}_1 &= f_1(x_1,x_2,t), & x_1 &\in R^{n-m} \\
\dot{x}_2 &= f_2(x_1,x_2,t) + B_2(x_1,x_2,t)u, & x_2 &\in R^m, \quad \det(B_2) \neq 0
\end{aligned} \tag{1.8}$$

The state subvector x_2 is handled as a fictitious control in the first equation of (1.8) and selected as a function of x_1 to provide the desired dynamics in the first subsystem (the design problem in the system of the $(n-m)$th order with m-dimensional control):

$$x_2 = -s_0(x_1)$$

Then the discontinuous control should be designed to enforce a sliding mode in the manifold

$$s(x_1,x_2) = x_2 + s_0(x_1) = 0 \tag{1.9}$$

(the design problem of the mth order with m-dimensional control).

After a finite time interval, a sliding mode in the manifold (1.9) starts and the system will exhibit the desired behaviour governed by $\dot{x}_1 = f_1[x_1, -s_0(x_1), t]$.

Note that the motion is of a reduced order and depends neither on the function $f_2(x_1,x_2,t)$ nor on function $B_2(x_1,x_2,t)$ in the second equation of the original system (1.8).

As shown in Section 1.3, sliding mode will start at manifold $s = 0$ if the matrix $GB + (GB)^T$ is positive-definite and the control is of the form $u = -M(x,t)\text{sign}(s)$ (component-wise) with a function $M(x,t)$ chosen to satisfy the inequality $M(x,t) > \lambda^{-1}\|Gf\|$, where λ is the lower bound of the eigenvalues of the matrix $GB + (GB)^T$.

To demonstrate a method of enforcing a sliding mode in the manifold $s = 0$ for an arbitrary nonsingular matrix GB, write the motion projection equation on subspace s in the form $\dot{s} = GB(u - u_{eq})$. Recall that the equivalent control u_{eq} is the value of the control such that the time derivative of the vector s is equal to zero.

Let $V = \frac{1}{2}s^T s$ be a Lyapunov function candidate and the control be of the form $u = -M(x,t)\text{sign}(s^*)$, $s^* = (GB)^T s$ to guarantee asymptotic stability of the motion projection on subspace s. Then

$$\dot{V} = \left[-s^{*T}\text{sign}(s^*) + s^{*T}\frac{u_{eq}}{M(x,t)}\right] M(x,t)$$

and $\dot{V}$ is negative definite for $M(x,t) > \|u_{eq}\|$.

This means that a sliding mode is enforced in the manifold $s^* = 0$ that is equivalent to its existence in the manifold $s = 0$ selected at the first step of the design procedure. It is important that the conditions for the sliding mode to exist are inequalities. Therefore an upper estimate of the disturbances is needed rather than precise information on their values.

Example

To demonstrate the sliding mode control design methodology consider the conventional problem of linear control theory: *eigenvalue placement* in a linear time invariant multidimensional system

$$\dot{x} = Ax + Bu$$

where x and u are n- and m-dimensional state and control vectors, respectively, A and B are constant matrices, $\text{rank}(B) = m$. The system is assumed to be controllable.

For any controllable system there exists a linear feedback $u = Fx$ (F being a constant matrix) such that the eigenvalues of the feedback system, i.e. of matrix $A + BF$, take the desired values and, as a result, the system exhibits desired dynamic properties.

Now we will show that the eigenvalue placement task may be solved in the framework of the sliding mode control technique dealing with a reduced order system. The core idea is to utilise the methods of linear control theory for reduced order equations and to employ one of the methods of enforcing sliding modes with desired dynamics.

As demonstrated in this section, the design becomes simpler for systems represented in the regular form. Reducing system equations to the regular form will be performed as a preliminary step in the design procedures. Since $\text{rank}(B) = m$, matrix B may be partitioned (after reordering the state vector components) as

$$B = \begin{bmatrix} B_1 \\ B_2 \end{bmatrix}$$

where $B_1 \in \mathfrak{R}^{(n-m)\times m}$, $B_2 \in \mathfrak{R}^{m\times m}$ with $\det B_2 \neq 0$. The nonsingular coordinate transformation

$$\begin{bmatrix} x_1 \\ x_2 \end{bmatrix} = Tx, \qquad T = \begin{bmatrix} I_{n-m} & -B_1 B_2^{-1} \\ 0 & B_2^{-1} \end{bmatrix} s$$

reduces the system equations to the regular form

$$\dot{x}_1 = A_{11}x_1 + A_{12}x_2$$

$$\dot{x}_2 = A_{21}x_1 + A_{22}x_2 + u$$

where $x_1 \in \mathfrak{R}^{(n-m)}$, $x_2 \in \mathfrak{R}^m$ and A_{ij} are constant matrices for $i, j = 1, 2$.

It follows from controllability of (A, B) that the pair (A_{11}, A_{12}) is controllable as well. Handling x_2 as an m-dimensional intermediate control in the controllable $(n-m)$-dimensional first subsystem all $(n-m)$ eigenvalues may be assigned arbitrarily by a proper choice of matrix C in $x_2 = -Cx_1$.

To provide the desired dependence between components x_1 and x_2 of the state vector, a sliding mode should be enforced in the manifold $s = x_2 + Cx_1 = 0$, where $s^T = (s_1, \ldots, s_m)$ is the difference between the real values of x_2 and its desired value $-Cx_1$.

After commencement of the sliding mode, the motion is governed by a reduced order system with the desired eigenvalues $\dot{x}_1 = (A_{11}x_1 - A_{12}C)x_1$.

For a piece-wise linear discontinuous control $u = -(\alpha|x| + \delta)\mathrm{sign}(s)$, with $|x| = \sum_{i=1}^{n} |x_i|$, $\mathrm{sign}(s)^T = [\mathrm{sign}(s_1), \ldots, \mathrm{sign}(s_m)]$; α and δ being constant positive values, calculate the time derivative of the positive definite function $V = \frac{1}{2}s^T s$

$$\dot{V} = s^T[(CA_{11} + A_{21})x_1 + (CA_{12} + A_{22})x_2] - (\alpha|x| + \delta)|s|$$

$$\leq |s||(CA_{11} + A_{21})x_1 + (CA_{12} + A_{22})x_2| - (\alpha|x| + \delta)|s|$$

It is evident that there exist values of α such that for any δ, the time derivative $\dot{V}$ is negative, which validates convergence of the state vector to the manifold $s = 0$ and existence of a sliding mode with the desired dynamics. The time interval preceding the sliding motion may be decreased by increasing the parameters α and δ in the control.

1.6   Discrete-time sliding mode control

Once a continuous-time dynamic system is in the sliding mode, its state trajectory is confined to a manifold in the state space. Generally speaking this method of system order reduction may be implemented by discontinuous control only, switching at infinite frequency. Most modern control systems are based on discrete-time microprocessor implementation. Since the switching frequency cannot exceed that of sampling, the ideal sliding mode cannot be implemented and discontinuities in the control result in oscillations at finite frequency referred to as *chattering*.

To develop the new concept *discrete-time sliding mode*, the motion equation should be replaced by the discrete-time equation

$$x_{k+1} = F(x_k, u_k), \qquad u_k = u(x_k), \quad x_i \in \mathfrak{R}^n, \quad u \in \mathfrak{R}^m \tag{1.10}$$

Similarly to continuous-time systems, the motion with state trajectories in a manifold

$$s(x) = 0, \qquad s \in R^m$$

and finite time needed to reach the manifold may occur in discrete-time system as well. The fundamental difference is that the control should be a continuous function of the state. The discrete-time sliding mode control with bounded control actions $\|u\| \leq u_0$ is of form

$$u_k = \begin{cases} u_{keq} & \text{if } \|u_{keq}\| \leq u_0 \\ \dfrac{u_{keq}}{\|u_{keq}\|} u_0 & \text{if } \|u_{keq}\| > u_0 \end{cases}$$

where the equivalent control u_{keq} is the solution to the algebraic equation $s[F(x_k, u_k)] = 0$ with respect to the control u_k. The equivalent control is a continuous state (but not time!) function. For example, in linear time-invariant discrete-time systems $x_{k+1} = Ax_k + Bu_k$ with a linear sliding manifold $s_k = Cx_k = 0$, the equivalent control is the control is the linear state function $u_{keq} = -(CB)^{-1}CAx_k$. For linear plants with unknown parameters in matrix A, the control u_{keq} cannot be found, and the modified version below should be applied

$$u_k = \begin{cases} -(CB)^{-1}s_k & \text{if } \|(CB)^{-1}s_k\| \leq u_0 \\ -\dfrac{u_0(CB)^{-1}s_k}{\|(CB)^{-1}s_k\|} & \text{if } \|(CB)^{-1}s_k\| > u_0 \end{cases} \tag{1.11}$$

For both versions, the control system is free of chattering and the motion equation is of a reduced order. The accuracy of the systems operating under uncertain conditions is of the order of a sampling interval.

1.7 Chattering problem

The subject of this section is of great importance whenever we intend to establish the bridge between the recommendations of the theory and real applications. Bearing in mind that the control has a high-frequency component, we should analyse the robustness or the problem of correspondence between an ideal sliding mode and real-life processes in the presence of unmodelled dynamics. Neglected small time constants (μ_1 and μ_2 in Fig. 1.8 with a linear plant) in plant models, sensors, and actuators lead to discrepancy in the dynamics (z_1 and z_2 are the state vectors of the unmodelled dynamics).

In accordance with singular perturbation theory, in systems with continuous control a fast component of the motion decays rapidly and a slow one depends on the small time constants continuously. In discontinuous control systems, the solution depends on the small parameters continuously as well. But unlike continuous systems,

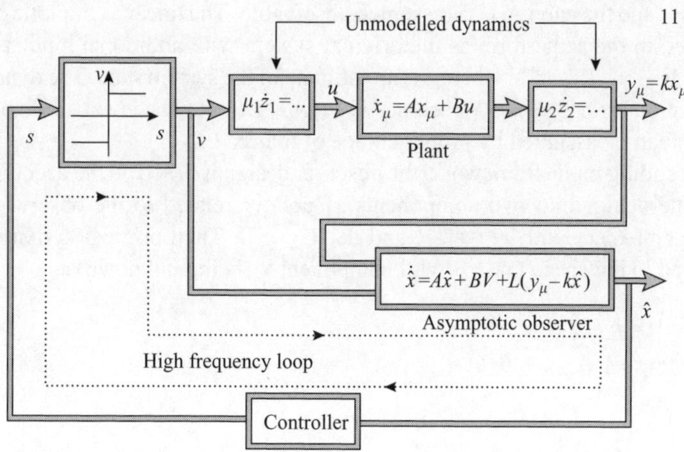

Figure 1.8   Chattering suppression in systems with observers

the switching of the control excites the unmodelled dynamics, which leads to oscillations in the state vector at a high frequency. The oscillations, usually referred to as *chattering*, are known to result in low control accuracy, high heat losses in electrical power circuits, and high wear of moving mechanical parts. These phenomena have been considered as serious obstacles for the application of sliding mode control in many papers and discussions. A recent study and practical experience showed that the chattering caused by unmodelled dynamics may be eliminated in systems with asymptotic observers, also known as Luenberger observers (Fig. 1.8). In spite of the presence of unmodelled dynamics, ideal sliding arises, and it is described by a singularly perturbed differential equation with solutions free of a high-frequency component and close to those of the ideal system.

As shown in Fig. 1.8 an asymptotic observer serves as a bypass for the high-frequency component, therefore the unmodelled dynamics are not excited. Preservation of sliding modes in systems with asymptotic observers predetermined successful application of the sliding mode control.

Another way to reduce chattering implies replacing the discontinuous control by its continuous approximation in a boundary layer. This may result in chattering as well as the presence of unmodelled fast dynamics if the gain in the boundary layer is too high. Since the values of the time constants, neglected in the ideal model, are unknown, the designer should be oriented towards the worst case and reduce the gain such that the unmodelled dynamics are not excited. As a result the disturbance rejection properties of discontinuous (or high gain) control are not utilised to the full extent.

1.8   Sliding mode observers

The idea underlying observer design may be illustrated for a linear time-invariant system $\dot{x} = Ax + Bu$, $x \in \Re^n$, $u \in \Re^m$ with measurable observed vector $y = Cx$, $y \in \Re^l$, rank$(C) = l$ and the pair (A, C) is assumed observable. The linear asymptotic observer is designed in the same form as the original system with additional input $\dot{\hat{x}} = A\hat{x} + Bu + L(C\hat{x} - y)$, $L \in \Re^{n \times l}$ where $\hat{x}$ is an estimate of the system state. The dynamics of the estimation error $\varepsilon_x = \hat{x} - x$ becomes $\dot{\varepsilon}_x = (A + LC)\varepsilon_x$. The eigenvalues of matrix $(A + LC)$ can be assigned by proper choice of matrix L.

In the sliding mode framework, the observer design is based on the decomposition of the state vector into two components x_1 and x_2 related to the observed vector as $y = C_1 x_1 + C_2 x_2$ with $x_1 \in \Re^{n-l}$ and $\det(C_2) \neq 0$. Then original system may be represented in the space (x_1, y) (with component x_1 being unknown) as

$$\dot{x}_1 = A_{11}x_1 + A_{12}y + B_1 u$$

$$\dot{y} = A_{21}x_1 + A_{22}y + B_2 u \qquad (1.12)$$

$$x_2 = C_2^{-1}(y - C_1 x_1)$$

The observer is described by differential equations

$$\dot{\hat{x}}_1 = A_{11}\hat{x}_1 + A_{12}\hat{y} + B_1 u + L_1 v$$

$$\dot{\hat{y}} = A_{21}\hat{x}_1 + A_{22}\hat{y} + B_2 u - v \qquad (1.13)$$

where $\hat{x}_1$ and $\hat{y}$ are estimates of the system state and input $v = M\,\mathrm{sign}(\hat{y} - y)$, $M > 0$, $M = \mathrm{const}$. If discontinuous vector function v is selected such that sliding mode is enforced in the manifold $\varepsilon_y = \hat{y} - y = 0$ then solution to the equation $\dot{\varepsilon}_y = 0$ leads to $v_{eq} = A_{21}(\hat{x}_1 - x_1)$. Now the dynamics estimation error can be determined from $\dot{\varepsilon}_{x1} = A_{11}\varepsilon_{x1} + A_{12}\varepsilon_y + L_1 v_{eq}$. By substituting the value for v_{eq} and $\varepsilon_y = 0$ one can derive estimation error dynamics as

$$\dot{\varepsilon}_{x1} = (A_{11} + L_1 A_{21})\varepsilon_{x1} \qquad (1.14)$$

The convergence can be provided by proper choice of the matrix L_1 and then x_2 can be found from $\hat{x}_2 = C_2^{-1}(y - C_1\hat{x}_1)$. It is easy to see that the sliding mode observer is equivalent to a reduced-order observer.

1.9   Conclusion

The chapter has outlined the basic concepts of sliding mode control theory, mathematical background and design principles oriented to control of high-order nonlinear dynamic plants operating under uncertain conditions.

An assessment of the scientific arsenal accumulated in the sliding mode control theory within several decades is beyond the objective of the chapter. Therefore we confine ourselves to mentioning research areas in the framework of the sliding mode control approach: control of infinite-dimensional systems, control of systems with delay, sliding mode observers, parameter and disturbance estimators, adaptive control and Lyapunov function based design methods. The list may be complemented by application oriented research: control of different types of electric machines; manipulators and mobile robots; motion control; control of flexible mechanical structures;

and automotive engines. Detailed information on the many theoretical and application results may be found in the books list in the next section and published in English at different stages of the development of sliding mode control where authors have summarised the vast amount of material accumulated by the time of their publication. This book will follow with a review of more of the basics of sliding mode control and estimation. Insight into some new topics where sliding modes are proving particularly fruitful are then presented. A range of recent application studies are then developed.

1.10   Further Reading

EDWARDS, C. and SPURGEON, S.: 'Sliding mode control: theory and applications' (Taylor and Francis, London, 1999)

FILIPPOV, A.: 'Differential equations with discontinuous right-hand sides' (Kluwer, Holland, 1988)

ITKIS, U.: 'Control systems of variable structure' (Wiley, New York, 1976)

SLOTINE, J.-J. E. and LI, W.: 'Applied nonlinear control' (Prentice Hall, Englewood Cliffs, New Jersey, 1991)

UTKIN, V.: 'Sliding modes and their applications in variable structure systems' (Mir Publ., Moscow, 1978, Translation of the book published by Nauka, Moscow, 1974 (in Russian))

UTKIN, V.: 'Sliding modes in control and optimization' (Springer Verlag, Berlin, 1992)

UTKIN, V., GULDNER, J., and SHI, J. X.: 'Sliding mode control in electro-mechanical systems' (Taylor and Francis, London, 1999)

YOUNG, K.-K. D. (Ed.): 'Variable structure control for robotics and aerospace application' (Elsevier Science Publishers B.V., Amsterdam, 1993)

YOUNG, K. D. and OZGUNER, U. (Eds): 'Variable structure systems, sliding mode and nonlinear control' (Springer Verlag, Berlin, 1999)

ZINOBER, A. S. (Ed.): 'Deterministic non-linear control' (Peter Peregrinus, UK, 1990)

ZINOBER, A. S. (Ed.): 'Variable structure and Lyapunov control' (Springer Verlag, London, 1993)

Chapter 2

Sliding mode regulator design

Vadim I. Utkin, Alexander G. Loukianov,
B. Castillo-Toledo and J. Rivera

2.1 Introduction

The regulator problem, in the classical sense, consists of designing a continuous state or error feedback controller such that the output of a system tracks a reference signal possibly in the presence of a disturbance signal. In the linear setting a complete solution of the problem was presented [1], based on the existence of a solution for a set of algebraic matrix equations. In the nonlinear framework, it was shown [2] that the solution can be posed in terms of the solution of a set of nonlinear differential equations, which represents a generalisation of the Francis conditions. This set of equations became known as the Francis-Isidori-Byrnes (FIB) equations. Basically, the regulator solution can be viewed as finding a steady-state surface on which the output tracking error map is zero, and which can be made attractive and invariant by feedback.

An alternative approach to deal with this problem is the use of the sliding mode technique to decompose and simplify the regulator design procedure and impose robustness properties [3, 4]. The underlying idea is to design a sliding surface on which the dynamics of the system are constrained to evolve by means of a discontinuous control law, instead of designing a continuous stabilising feedback, as in the case of the classical regulator problem. The sliding manifold contains the steady-state surface, and the dynamics of the system tend asymptotically, along the sliding manifold, to the steady-state behaviour.

In the full information case, static state feedback sliding mode regulator design has been investigated [4–6]. To overcome the limiting requirement of full information knowledge, a dynamic discontinuous error feedback strategy has been designed for linear systems [7], and for a class of nonlinear systems [8]. Considering that the state of the exosystem is accessible, a dynamic error feedback regulator has been proposed for a class of nonlinear systems with unitary relative degree [9].

In this chapter we address a number of issues for a general case of linear and nonlinear systems, including also a class of dynamic systems presented in the so-called Regular and Block Controllable forms.

To formalise the ideas, we briefly recall the basic facts on regulation theory. Consider a nonlinear system

$$\dot{x} = f(x) + g(x)u + d(x)w \tag{2.1}$$

$$y = h(x) \tag{2.2}$$

with state x, defined on a neighbourhood X of the origin of $\Re^n$, and $u \in \Re^m$, $y \in \Re^p$. The vector $f(x)$, the columns of $g(x)$ and $d(x)$ are smooth vector fields of class $C_{[t,\infty)}^{\infty}$, and in addition, it is assumed that $f(0) = h(0) = 0$. The output tracking error is defined as the difference between the output of the system, y, and a reference signal, $q(w)$, i.e.,

$$e = y - q(w) \tag{2.3}$$

where the reference signal, $q(w)$, is generated by a given external system described by

$$\dot{w} = s(w), \qquad s(0) = 0 \tag{2.4}$$

with state w, defined on a neighbourhood W of the origin of $\Re^s$. This system is characterised by the following assumption:

H1.   *The Jacobian matrix $S = [\partial s / \partial w]_{(0)}$ at the equilibrium point $w = 0$ has all eigenvalues on the imaginary axis.*

It is assumed also that only the components of the error e are available for measurement. It has been shown that the control action to (2.1) can be provided by an *error feedback* dynamic system [2]:

$$\dot{\xi} = \eta(\xi, e) \tag{2.5}$$

$$u = \theta(\xi) \tag{2.6}$$

and the solvability of the *Error Feedback Regulator Problem* (EFRP) under assumption H1, can be stated in terms of the existence of a pair of mappings $x = \pi(w)$ and $\xi = \rho(w)$, with $\pi(0) = 0$ and $\rho(0) = 0$, that solve the partial differential equation

(FIB equations)

$$f(\pi(w)) + g(\pi(w))\theta(\rho(w)) + d(\pi(w))w = \frac{\partial \pi(w)}{\partial w}s(w)$$

$$\eta(\rho(w), 0) = \frac{\partial \rho(w)}{\partial w}s(w)$$

$$h(\pi(w)) - q(w) = 0$$

The controller (2.6) can be chosen as

$$\theta(\xi) = c(\xi_1) + K(\xi_1 - \pi(\xi_2)) \tag{2.7}$$

where $\xi = col(\xi_1, \xi_2)$, and K is a matrix that places the eigenvalues of the linear approximation of the closed-loop system (2.1) and (2.7) at the equilibrium point $x = 0$, namely $(A + BK)$ in C^- where $A = [\partial f/\partial x]_{(0)}$, $B = g(0)$. If the mapping $c(\cdot)$ is chosen as a solution of

$$f(\pi(w)) + g(\pi(w))c(w) + d(\pi(w))w = \frac{\partial \pi(w)}{\partial w}s(w) \tag{2.8}$$

$$h(\pi(w)) - q(w) = 0 \tag{2.9}$$

then $\rho(w) = col(\pi(w), w)$ and $c(w) = \theta(\rho(w))$, i.e., the EFRP solution can be obtained from the solution of the *State Feedback Regulator Problem*, provided some additional assumption on the detectability of the composite system (2.1)–(2.4), is made as in Reference 2.

The linear solution may be derived by considering the linear approximation of the system (2.1)–(2.4) at the equilibrium point $(x, w) = (0, 0)$:

$$\dot{x} = Ax + Bu + Dw \tag{2.10}$$

$$e = Cx - Qw \tag{2.11}$$

$$\dot{w} = Sw \tag{2.12}$$

where $D = d(0)$, $C = [\partial h/\partial x]_{(0)}$ and $Q = [\partial q/\partial w]_{(0)}$. In this case, the conditions (2.8) and (2.9) take the form of the *Sylvester* matrix equation

$$A\Pi + B\Gamma + D = \Pi S \tag{2.13}$$

$$C\Pi - Q = 0 \tag{2.14}$$

where $\Gamma = [\partial c/\partial w]_{(0)}$ and $\Pi = [\partial \pi/\partial w]_{(0)}$ are the linear approximation of the control (2.7). In fact, the conditions (2.13) and (2.14) are added by the following trivially necessary conditions:

H2. *The pair $\{A, B\}$ is stabilisable and*

H3. *The pair $\left\{ [C \quad Q], \begin{bmatrix} A & D \\ 0 & S \end{bmatrix} \right\}$ is detectable.*

In the following we present the regulator problem from a sliding mode viewpoint. We define the problem and give the conditions for the existence of a solution.

2.2   Error feedback sliding mode control problem

Analogously to EFRP, the *Error Feedback Sliding Mode Regulation Problem* (EFSMRP) is defined as the problem of finding a dynamic discontinuous controller

$$\dot{\xi} = \eta(\xi, u, e) \tag{2.15}$$

$$u_i(\xi) = \begin{cases} u_i^+(\xi) & \text{if } \sigma_i(\xi) > 0 \\ u_i^-(\xi) & \text{if } \sigma_i(\xi) < 0 \end{cases} \qquad i = 1, \ldots, m \tag{2.16}$$

where $u = (u_1, \ldots, u_m)^T$; $u_i^+(\xi)$, $u_i^-(\xi)$ and the sliding manifold

$$\sigma(\xi) = 0, \qquad \sigma = (\sigma_1, \ldots, \sigma_m)^T \tag{2.17}$$

are chosen to induce asymptotic convergence of the state vector to the manifold (2.17), such that the following conditions hold:

- (SMS_{ef}) (Sliding Mode Stability). The state of the closed-loop system formed from closing the loop in the system (2.1) and (2.2), with the controllers (2.15)–(2.17), converges to the manifold (2.17) in a finite time;
- (S_{ef}). The equilibrium $(x, \xi) = (0, 0)$ of the sliding mode dynamics

$$\dot{x} = f(x, u_{eq}, 0)\,|_{\sigma(\xi)=0}$$

$$\dot{\xi} = \eta(\xi, u_{eq}, e)$$

  is asymptotically stable, where u_{eq} is the equivalent control defined as a solution of $\dot{\sigma} = 0$;
- (R_{ef}). There exists a neighbourhood $V \subset X \times \Xi \times W$ of $(0, 0, 0)$ such that, for each initial condition $(x_0, \xi_0, w_0) \in V$, the output tracking error (2.3) goes asymptotically to zero, i.e., $\lim_{t \to \infty} e(t) = 0$.

In the following, for both the linear and nonlinear cases, a solution for this problem will be presented.

2.3   Discontinuous dynamic regulator for multivariable linear systems

In this section, the EFSMRP solvability conditions will be derived for linear systems in the general form (2.10)–(2.12), and then a sliding regulator will be designed for linear systems presented in Regular and Block Controllable forms.

2.3.1   Solvability conditions

Consider the linear system (2.10)–(2.12). For this system, we define the steady-state error as $z = x - \Pi w$ where Π is a matrix to be defined later and thus rewrite the original equations as

$$\dot{\zeta} = \bar{A}\zeta + \bar{B}u \tag{2.18}$$

$$e = \bar{C}\zeta \tag{2.19}$$

where

$$\zeta = \begin{pmatrix} z \\ w \end{pmatrix}, \quad \bar{A} = \begin{pmatrix} A & A\Pi - \Pi S + D \\ 0 & S \end{pmatrix}, \quad \bar{B} = \begin{pmatrix} B \\ 0 \end{pmatrix}$$

$$\bar{C} = (C \quad (C\Pi - Q)), \quad \mathrm{rank}(B) = m$$

Then the system (2.15) can be designed in this case as an observer for ζ. For asymptotic stabilisation of the closed-loop system via error feedback the following assumption is introduced:

H4.   *The pair $\{\bar{C}, \bar{A}\}$ is detectable.*

Under this assumption, the system (2.15) with state $\xi = (\hat{z}, \hat{w})^T$ is designed as the observer:

$$\dot{\xi} = \bar{A}\xi + \bar{B}u + L(e - \hat{e}), \qquad \hat{e} = \bar{C}\xi \tag{2.20}$$

where ξ is the estimate of $\zeta = (z, w)^T$, and the matrix $L = (L_1, \ L_2)^T$ is chosen to stabilise the error dynamics:

$$\dot{\epsilon} = (\bar{A} - L\bar{C})\epsilon \tag{2.21}$$

where $\epsilon = \zeta - \xi = (\epsilon_1, \ \epsilon_2)^T$.

Once the observer is designed, a sliding manifold $\hat{\sigma}(\xi) = 0$ has to be chosen to satisfy the stability conditions. To this end, we choose

$$\hat{\sigma}(\xi) = (\Sigma \quad 0)\xi = \Sigma\hat{z} = 0 \tag{2.22}$$

where an appropriately chosen design matrix Σ will determine the dynamic response of the system on (2.22). To investigate the stability on this sliding manifold, we first prove the following lemma.

Lemma 1.   *Let the operator P be defined as $P = (I_n - B(\Sigma B)^{-1}\Sigma)$. Then the relation*

$$P(A\Pi - \Pi S + D) = 0 \tag{2.23}$$

is true if and only if there are matrices Π and Λ, such that

$$A\Pi - \Pi S + D = B\Lambda \tag{2.24}$$

Proof.   The operator P is a projection operator along the space of the rank of B over the Σ null space, i.e.,

$$PB = (I_n - B(\Sigma B)^{-1}\Sigma)B = 0$$

$$Pz = z \qquad \forall z \in \aleph, \ \aleph = \{z \in \Re^n \mid \Sigma z = 0\}$$

Thus, if condition (2.24) holds, then it follows that $P(A\Pi - \Pi S + D) = PB\Sigma = 0$. Conversely, if condition (2.23) is satisfied, then $(A\Pi - \Pi S + D)$ must be in the image of B, i.e., $A\Pi - \Pi S + D = B\Lambda$ for some matrix Λ.

From this result, a condition for a solution of the discontinuous regulator problem can be deduced.

Proposition 1. *Suppose that assumptions H1, H2 and H4 hold, and there exists a matrix Π that solves the linear equations*

$$A\Pi - \Pi S + D = B\Lambda \tag{2.25}$$

$$C\Pi - Q = 0 \tag{2.26}$$

for some matrix Λ. Then the EFSMRP for the linear system is solvable.

Proof. Choose the control u as

$$u = -k(\Sigma B)^{-1} \text{sign}(\hat\sigma), \qquad \hat\sigma = \Sigma\hat z, \quad k > 0$$

Using the Lyapunov function $V = \frac{1}{2}\hat\sigma^T\hat\sigma$, we obtain from the derivative of V taken along the trajectories of (2.20) the condition:

$$k > \| (\Sigma B)u_{eq} \|$$

that guarantees the (SMS_{ef}) condition. The equivalent control u_{eq} is calculated from $\dot{\hat\sigma} = 0$ as

$$\begin{aligned}
u_{eq} &= -(\Sigma B)^{-1}\Sigma\left[A\hat z + (A\Pi - \Pi S + D)\hat w + L_1 C\epsilon_1\right] \\
&= -(\Sigma B)^{-1}\Sigma[Az + (A\Pi - \Pi S + D)w - (A - L_1 C)\epsilon_1 \\
&\quad - ((A\Pi - \Pi S + D) - L_1(C\Pi - Q))\epsilon_2]
\end{aligned} \tag{2.27}$$

The reduced order sliding mode dynamics on $\hat\sigma = \sigma - \Sigma\epsilon_1 = 0$ are obtained by replacing (2.27) in (2.18), to yield:

$$\dot\zeta = \tilde A\zeta + E\epsilon, \quad \Sigma z - \Sigma\epsilon_1 = 0 \tag{2.28}$$

$$\dot\epsilon = (\bar A - L\bar C)\epsilon \tag{2.29}$$

$$e = (C \quad (C\Pi - Q))\zeta \tag{2.30}$$

where

$$\tilde A = \begin{pmatrix} PA & R \\ 0 & S \end{pmatrix}$$

$$E = \begin{pmatrix} (I_n - P)(A - L_1 C) & (I_n - P)((A\Pi - \Pi S + D) - L_1(C - \Pi Q)) \\ 0 & 0 \end{pmatrix}$$

with P already defined in Lemma 1, and $R = P(A\Pi - \Pi S + D)$. Using the condition (2.25) and Lemma 1 it yields that $R = 0$. Then, assuming that the observer estimation error decays rapidly by appropriate choice of L_1 and L_2 (under assumption H4), from (2.28) we have

$$\dot z = PAz|_{\Sigma z = 0} \tag{2.31}$$

Since the matrix Σ in (2.31) by assumption H2 can be chosen such that (ΣB) is invertible, and the $(n-m)$ eigenvalues of PA are arbitrarily placed in C^- [10], then $z(t) \to 0$ as $t \to \infty$, satisfying condition (S_{ef}). Now, if the tracking error equation (2.30) satisfies condition (2.26), then, $e(t) \to 0$ as $t \to \infty$, satisfying condition (R_{ef}).

Comparing the conditions (2.13) and (2.25), we note that the steady-state matrices Π and Γ for the state x and control u, respectively, in equation (2.13) have to be calculated. On the contrary, in the second case (2.25) only the matrix Π needs to be calculated such that the perturbation satisfies the matching condition [11]. The structure of equation (2.25) can be put in evidence using the decomposition of a linear system to Regular form.

2.3.2 Sliding regulator for linear systems in Regular form

In order to show the explicit form of condition (2.25) and sliding dynamics (2.31), the linear system (2.10) is first converted into Regular form [12]:

$$\begin{pmatrix} \dot{x}_1 \\ \dot{x}_2 \end{pmatrix} = \begin{pmatrix} A_{11} & A_{12} \\ A_{21} & A_{22} \end{pmatrix} \begin{pmatrix} x_1 \\ x_2 \end{pmatrix} + \begin{pmatrix} 0 \\ B_2 \end{pmatrix} u + \begin{pmatrix} D_1 \\ D_2 \end{pmatrix} w \tag{2.32}$$

$$e = C_1 x_1 + C_2 x_2 - Qw$$

where $x_1 \in \Re^{n-m}$, $x_2 \in \Re^m$, $\mathrm{rank}(B_2) = m$. Defining $z_1 = x_1 - \Pi_1 w$ and $z_2 = x_2 - \Pi_2 w$ with Π_1 and Π_2 constant matrices of proper dimension, the system (2.32) in the new variables z_1 and z_2 obeys the following dynamics:

$$\begin{pmatrix} \dot{z}_1 \\ \dot{z}_2 \end{pmatrix} = \begin{pmatrix} A_{11} & A_{12} \\ A_{21} & A_{22} \end{pmatrix} \begin{pmatrix} z_1 \\ z_2 \end{pmatrix} + \begin{pmatrix} 0 \\ B_2 \end{pmatrix} u + \begin{pmatrix} R_1 \\ R_2 \end{pmatrix} w \tag{2.33}$$

$$e = C_1 z_1 + C_2 z_2 + (C_1 \Pi_1 + C_2 \Pi_2 - Q)w$$

with $R_1 = A_{11}\Pi_1 + A_{12}\Pi_2 - \Pi_1 S + D_1$ and $R_2 = A_{21}\Pi_1 + A_{22}\Pi_2 - \Pi_2 S + D_2$.

Now, the system (2.15) with state $\xi = (\hat{z}_1, \hat{z}_2, \hat{w})^T$ is designed to have the following form:

$$\dot{\xi} = A'\xi + B'u + L'(e - \hat{e}), \qquad \hat{e} = C'\xi$$

where ξ is the estimate of

$$\zeta = (z_1, z_2, w)^T, \quad A' = \begin{pmatrix} A_{11} & A_{12} & R_1 \\ A_{21} & A_{22} & R_2 \\ 0 & 0 & S \end{pmatrix}, \quad B' = \begin{pmatrix} 0 \\ B_2 \\ 0 \end{pmatrix}, \quad L' = \begin{pmatrix} L_1 \\ L_2 \\ L_3 \end{pmatrix}$$

and

$$C' = (C_1 \quad C_2 \quad (C_1 \Pi_1 + C_2 \Pi_2 - Q))$$

The observer gain matrix L' is chosen to stabilise the observer error state $\varepsilon = \zeta - \xi = (\varepsilon_1, \varepsilon_2, \varepsilon_3)^T$, a dynamics of which are governed by

$$\dot{\varepsilon} = (A' - L'C')\varepsilon \tag{2.34}$$

The following assumption is thus necessary to guarantee the stability of the system (2.34).

H5.   *The pair $\{C', A'\}$ is detectable.*

Proposition 2.   *Suppose that assumptions H1, H2 and H5 hold, and there exist matrices Π_1 and Π_2 which solve the linear equations*

$$A_{11}\Pi_1 + A_{12}\Pi_2 - \Pi_1 S + D_1 = 0 \tag{2.35}$$

$$C_1\Pi_1 + C_2\Pi_2 - Q = 0 \tag{2.36}$$

Then the EFSMRP for linear system in the Regular form is solvable.

Proof.   We first specify the sliding surface (2.22) in terms of the estimated states as

$$\hat{\sigma} = \hat{z}_2 - \Sigma_1\hat{z}_1 = z_2 - \Sigma_1 z_1 - (\varepsilon_2 - \Sigma_1\varepsilon_1) = 0$$

where $\Sigma_1 \in \Re^{m \times (n-m)}$. The proposed sliding control law is given as $u = -kB_2^{-1}\,\text{sign}\,(\hat{\sigma})$. Then the requirement (SMS_{ef}) is fulfilled if $k > \|B_2 u_{eq}\|$, where u_{eq} is calculated from $\dot{\hat{\sigma}} = 0$ and has the following form:

$$\begin{aligned}
u_{eq} &= -B_2^{-1}[-\Sigma_1(A_{11}\hat{z}_1 + A_{12}\hat{z}_2 + R_1\hat{w} + L_1 C'\varepsilon) + A_{21}\hat{z}_1 + A_{22}\hat{z}_2 \\
&\quad + R_2\hat{w} + L_2 C'\varepsilon] \\
&= -B_2^{-1}[-\Sigma_1(A_{11}z_1 + A_{12}z_2 + R_1 w + (L_1 C' - G_1)\varepsilon) + A_{21}z_1 + A_{22}z_2 \\
&\quad + R_2 w + (L_2 C' - G_2)\varepsilon]
\end{aligned}$$

with $G_1 = (A_{11}\ \ A_{12}\ \ R_1)$ and $G_2 = (A_{21}\ \ A_{22}\ \ R_2)$. By condition (2.35) it follows that $R_1 = 0$ in (2.33), therefore, the reduced order sliding mode equation can be obtained as

$$\dot{z}_1 = A_{11}z_1 + A_{12}z_2, \quad z_2 = \Sigma_1 z_1 + (\varepsilon_2 - \Sigma_1\varepsilon_1) \tag{2.37}$$

$$\dot{w} = Sw$$

$$\dot{\varepsilon} = (A' - L'C')\varepsilon$$

$$e = (C_1 - C_2\Sigma_1)z_1 + (C_1\Pi_1 + C_2\Pi_2 - Q)w \tag{2.38}$$

It is known [12] that if the pair $\{A, B\}$ is controllable (stabilisable) then the pair $\{A_{11}, A_{12}\}$ is controllable (stabilisable) as well. Therefore there exists a matrix Σ_1 such that the matrix $(A_{11} + A_{12}\Sigma_1)$ in (2.37) is stable and hence $z_1(t)$ asymptotically tends to zero, satisfying condition (S_{ef}). In consequence, thanks to condition (2.36) the output tracking error $e(t)$ in (2.38) tends to zero too and condition (R_{ef}) is satisfied.

Note that the conditions (2.25) and (2.26) are modified as (2.35) and (2.36), respectively. On the other hand, the equation (2.35) as well as the system (2.37) can be further decomposed if the system (2.10) or (2.32) is represented in Block Controllable form.

2.3.3   *Block Controllable form with disturbances*

In this section a discontinuous regulator is proposed using the Block Control technique [13]. The underlying idea is to first reduce system (2.10) to a Block Controllable form (BC-form) in the presence of perturbations by means of a nonsingular transformation, and then, using the Block Control technique, to design a sliding surface on which the unperturbed part of the dynamics of the system is stable. Finally, the condition for the solution of the corresponding EFSMRP is derived.

The essential feature of the proposed method is the transformation of (2.10) into BC-form consisting of r blocks of the form:

$$\dot{x}_1 = A_{11}x_1 + B_1 x_2 + D_1 w$$

$$\dot{x}_i = \sum_{j=1}^{i} A_{ij}x_j + B_i x_{i+1} + D_i w, \qquad i = 2,\ldots,r-1$$

$$\dot{x}_r = \sum_{k=1}^{r} A_{rk}x_k + B_r u + D_r w \tag{2.39}$$

$$e = \sum_{k=1}^{r} M_k x_k - Q w$$

where the transformed vector $\bar{x}$ is decomposed as $\bar{x} = (x_1,\ldots,x_r)^T$, and $x_i \in \Re^{n_i}$, $i = 1,\ldots,r$. In the ith block, the vector x_{i+1} is regarded as a fictitious control vector, where $\mathrm{rank}(B_i) = n_i$. The integers $(n_1,n_2,\ldots,n_r)$ characterise the structure of the system (2.39) by the condition $n_1 \leq n_2 \leq \cdots \leq n_r \leq m$ with $\sum_{i=1}^{r} n_i = n$. It was shown that a necessary condition to transform the system (2.10) into BC-form (2.39), is that the pair $\{A, B\}$ must be controllable [13].

Introducing the steady-state $\Gamma_i w$ for the state vector x_i, we define the steady-state error z_i as

$$z_i = x_i - \Gamma_i w, \qquad i = 1,\ldots,r \tag{2.40}$$

Then, the states in (2.40) obtained from the evolution of (2.39) are of the following form:

$$\dot{z}_1 = A_{11}z_1 + B_1 z_2 + R_1 w$$

$$\dot{z}_i = \sum_{j=1}^{i} A_{ij}z_j + B_i z_{i+1} + R_i w, \qquad i = 2,\ldots,r-1 \tag{2.41}$$

$$\dot{z}_r = \sum_{k=1}^{r} A_{rk}z_k + B_r u + R_r w$$

$$e = \sum_{k=1}^{r} M_k z_k + \left(\sum_{k=1}^{r} M_k \Gamma_k - Q \right) w \tag{2.42}$$

where

$$R_1 = A_{11}\Gamma_1 + B_1\Gamma_2 + D_1 - \Gamma_1 S \tag{2.43}$$

$$R_i = \sum_{j=1}^{i} A_{ij}\Gamma_j + B_i\Gamma_{i+1} + D_i - \Gamma_i S, \qquad i=2,\ldots,r \tag{2.44}$$

The system (2.15) with state $\xi = (\hat{z}_1,\ldots,\hat{z}_r,\hat{w})^T$ is designed as follows:

$$\dot{\hat{z}}_1 = A_{11}\hat{z}_1 + B_1\hat{z}_2 + L_1(e - \hat{e})$$

$$\dot{\hat{z}}_i = \sum_{j=1}^{i} A_{ij}\hat{z}_j + B_i\hat{z}_{i+1} + L_i(e - \hat{e}), \qquad i=2,\ldots,r-1$$

$$\dot{\hat{z}}_r = \sum_{k=1}^{r} A_{rk}\hat{z}_k + B_r u + R_r\hat{w} + L_r(e - \hat{e}) \tag{2.45}$$

$$\dot{\hat{w}} = S\hat{w} + L_{r+1}(e - \hat{e})$$

$$\hat{e} = \sum_{k=1}^{r} M_k\hat{z}_k + \left(\sum_{k=1}^{r} M_k\Gamma_k - Q\right)\hat{w}$$

where $\xi = (\hat{z}_1,\ldots,\hat{z}_r,\hat{w})^T$ is the estimate of $\zeta = (z_1,\ldots,z_r,w)^T$, and $\tilde{L} = (L_1,\ldots,L_{r+1})^T$ is the observer gains matrix. Assuming that $R_i = 0$, $i=1,\ldots,r-1$, (2.43) and (2.44), then the observer error state $\epsilon = \zeta - \xi = (\epsilon_1,\ldots,\epsilon_{r+1})^T$ obeys the following dynamics:

$$\dot{\epsilon} = (\tilde{A} - \tilde{L}\tilde{C})\epsilon \tag{2.46}$$

with

$$\tilde{A} = \begin{pmatrix} A_{11} & B_1 & 0 & \cdots & 0 & 0 \\ A_{21} & A_{22} & B_2 & \cdots & 0 & 0 \\ \vdots & & & \vdots & & \\ A_{r1} & A_{r2} & A_{r3} & \cdots & A_{rr} & R_r \\ 0 & 0 & 0 & \cdots & 0 & S \end{pmatrix}$$

and

$$\tilde{C} = \begin{pmatrix} M_1 & \cdots & M_r & \left(\sum_{k=1}^{r} M_k\Gamma_k - Q\right) \end{pmatrix}$$

Similar to the previous case we assume that

H6.   *The pair $\{\tilde{C}, \tilde{A}\}$ is detectable.*

Proposition 3. *Suppose that assumptions H2 and H6 hold, and there exist matrices* Γ_i, $i=1,\ldots,r-1$ *that solve the linear equations*

$$A_{11}\Gamma_1 + B_1\Gamma_2 + D_1 = \Gamma_1 S$$

$$\sum_{j=1}^{i} A_{ij}\Gamma_j + B_i\Gamma_{i+1} + D_i = \Gamma_i S, \qquad i=2,\ldots,r-1 \tag{2.47}$$

and

$$\sum_{k=1}^{r} M_k\Gamma_k - Q = 0 \tag{2.48}$$

Then the EFSMRP for a linear system in the BC- form is solvable.

Proof.   Note first that if conditions (2.47) are met then $R_i = 0$, $i=1,\ldots,r-1$, in (2.41), and we have therefore exactly the observer error system (2.46) which under assumption H6, can be stabilised by a proper choice of $\tilde{L}$.

A sliding manifold will be designed based on the system (2.45) considering the state $\hat{z}_{i+1}, i=1,\ldots,r-1$ as a fictitious control vector in the ith block of (2.45), and the term $L_i(e-\hat{e})$ as the perturbation. This procedure is outlined as follows.

We start by defining a new variable $\chi_1 = \hat{z}_1$. Taking the derivative of χ_1 along (2.45) yields

$$\dot{\chi}_1 = A_{11}\hat{z}_1 + B_1\hat{z}_2 + L_1(e-\hat{e}) \tag{2.49}$$

As mentioned above, $\hat{z}_2$ is considered as a quasi-control in (2.49), and must force the desired dynamics, $K_1\chi_1$ with design stable matrix K_1 for this block by the anticipation of its dynamics of the following form:

$$\dot{\chi}_1 = A_{11}\hat{z}_1 + B_1\hat{z}_2 + L_1(e-\hat{e}) = K_1\chi_1 \tag{2.50}$$

Now, $\hat{z}_2$ is calculated from (2.50) as a desired state $\hat{z}_2^d$ as follows:

$$\hat{z}_2^d = -B_1^+[A_{11}\hat{z}_1 + L_1(e-\hat{e}) - K_1\chi_1]$$

where $B_1^+ = B_1^T(B_1 B_1^T)^{-1}$ denotes the right pseudo-inverse matrix of B_1.

Proceeding in the same way, we define a second new variable χ_2 as $\chi_2 = \hat{z}_2 - \hat{z}_2^d$. Taking the derivative of χ_2 and anticipating its dynamics, we obtain the next block

$$\dot{\chi}_2 = A_{21}\hat{z}_1 + A_{22}\hat{z}_2 + B_2\hat{z}_3 + L_2(e-\hat{e}) - \dot{\hat{z}}_2^d = K_2\chi_2 \tag{2.51}$$

The desired state of $\hat{z}_3$ is calculated from (2.51) as follows:

$$\hat{z}_3^d = -B_2^+ \left(A_{21}\hat{z}_1 + A_{22}\hat{z}_2 + L_2(e-\hat{e}) - \dot{\hat{z}}_2^d - K_2\chi_2 \right)$$

where $B_2^+ = B_2^T(B_2 B_2^T)^{-1}$, and K_2 is a Hurwitz matrix.

This procedure may be performed iteratively defining the ith new state as $\chi_i = \hat{z}_i - \hat{z}_i^d$, and the ith block as follows:

$$\dot{\chi}_i = \sum_{j=1}^{i} A_{ij}\hat{z}_j + B_i\hat{z}_{i+1} + L_i(e - \hat{e}) - \dot{\hat{z}}_i^d = K_i\chi_i, \qquad i = 4, \ldots, r-1$$

and the desired state as

$$\hat{z}_{i+1}^d = -B_i^+ \left(\sum_{j=1}^{i} A_{ij}\hat{z}_j + L_i(e - \hat{e}) - \dot{\hat{z}}_i^d - K_i\chi_i \right)$$

where, again, $B_i^+ = B_i^T(B_i B_i^T)^{-1}$, and K_i is a Hurwitz matrix.

In the final step, $\hat{z}_r^d$ is known, and defining the last new variable $\chi_r = \hat{z}_r - \hat{z}_r^d$, the rth block is transformed as follows:

$$\dot{\chi}_r = \sum_{k=1}^{r} A_{rk}\hat{z}_k + B_r u + R_r\hat{w} + L_r(e - \hat{e}) - \dot{\hat{z}}_r^d$$

It should be noted that the new state $\chi = (\chi_1, \ldots, \chi_r)^T$ is derived by the nonsingular transformation

$$\chi_1 = \hat{z}_1, \quad \chi_i = \hat{z}_i - \hat{z}_i^d, \qquad i = 2, \ldots, r \tag{2.52}$$

This transformation simplifies system (2.45) to the following form:

$$\begin{aligned}
\dot{\chi}_1 &= K_1\chi_1 + B_1\chi_2 \\
\dot{\chi}_i &= K_i\chi_i + B_i\chi_{i+1}, \qquad i = 2, \ldots, r-1 \\
\dot{\chi}_r &= \sum_{k=1}^{r} A_{rk}\hat{z}_k + B_r u + R\hat{w} + L_r(e - \hat{e}) - \dot{\hat{z}}_r^d
\end{aligned} \tag{2.53}$$

A natural choice of the switching function for system (2.53) is $\sigma = \chi_r$. In order to generate a sliding mode in (2.53), we choose the control as $u = -k_r B_r^+ \,\text{sign}\,(\sigma)$. If $k_r > \|B_r u_{eq}\|$, the condition (SMS_{ef}) is guaranteed, where $B_r^+ = B_r^T(B_r B_r^T)^{-1}$, and u_{eq} is calculated from $\dot{\sigma} = 0$ as $u_{eq} = -B_r^+(\sum_{k=1}^{r} A_{rk}\hat{z}_k + R_r\hat{w} + L_r(e - \hat{e}) - \dot{\hat{z}}_r^d)$. The sliding mode motion on $\sigma = \chi_r = 0$ is described by the reduced order system

$$\begin{aligned}
\dot{\chi}_1 &= K_1\chi_1 + B_1\chi_2 \\
\dot{\chi}_i &= K_i\chi_i + B_i\chi_{i+1} \qquad i = 2, \ldots, r-2 \\
\dot{\chi}_{r-1} &= K_{r-1}\chi_{r-1}
\end{aligned} \tag{2.54}$$

$$\dot{w} = Sw \tag{2.55}$$

$$\dot{\epsilon} = (\tilde{A} - \tilde{L}\tilde{C})\epsilon \tag{2.56}$$

$$e = \sum_{k=1}^{r} M_k z_k + \left(\sum_{k=1}^{r} M_k \Gamma_k - Q\right) w \tag{2.57}$$

Since the diagonal matrices K_i, $i = 1,\ldots,r-1$ in (2.54) are Hurwitz , then the states of (2.54) tend asymptotically to zero, i.e., $\lim_{t\to\infty} \chi_i(t) = 0$, $i = 1,\ldots,r-1$. Hence, by transformation (2.52) $\lim_{t\to\infty} \hat{z}_i(t) = 0$, $i = 1,\ldots,r$. Now, by assumption H6 there is a matrix $\tilde{L}$ in (2.56) such that $\lim_{t\to\infty} \epsilon(t) = 0$, therefore $\lim_{t\to\infty} z_i(t) = 0$, $i = 1,\ldots,r$, satisfying condition (S_{ef}). In consequence, thanks to condition (2.48) the output tracking error $e(t)$ (2.57) tends asymptotically to zero, satisfying condition (R_{ef}).

Remark 1.   *Note that the Regular form conditions (2.35) and (2.36) are represented for the BC-form as (2.47) and (2.48), respectively.*

2.4   Discontinuous dynamic regulator for nonlinear systems

In this section, the EFSMRP solvability conditions will be derived for the nonlinear perturbed system (2.1)–(2.4), and in the sequel, a discontinuous regulator will be developed for nonlinear systems presented first in Regular form and then in Nonlinear Block Controllable form (NBC-form).

2.4.1   Solvability conditions

To achieve local asymptotic stability and output regulation, let us first introduce a C^k $(k \geq 2)$ mapping $x = \pi(w)$, with $\pi(0) = 0$, defined in neighbourhood W of 0. Then we define the steady-state error $z = x - \pi(w)$, which can be taken as a change of variables that transforms (2.1)–(2.4) into

$$\dot{z} = f(z + \pi(w)) + g(z + \pi(w))u + d(z + \pi(w))w - \frac{\partial \pi(w)}{\partial w} s(w) \tag{2.58}$$

$$\dot{w} = s(w) \tag{2.59}$$

$$e = h(z + \pi(w)) - q(w) \tag{2.60}$$

Setting the sliding manifold as $\sigma(z) = 0$ with $G(0) = \Sigma$ a constant matrix of proper dimension, $G(z) = \partial\sigma(z)/\partial z$, and calculating from $\dot{\sigma}(z) = 0$ the equivalent control u_{eq} as

$$u_{eq}(z, w) = [G(z)g(z + \pi(w)]^{-1}\left[f(z + \pi(w)) + d(z + \pi(w))w - \frac{\partial \pi(w)}{\partial w} s(w)\right]$$

$$\tag{2.61}$$

the sliding mode dynamics on $\sigma(z) = 0$ yields

$$\dot{z} = p(z + \pi(w)) \left[f(z + \pi(w)) + d(z + \pi(w))w - \frac{\partial \pi(w)}{\partial w} s(w) \right] \qquad (2.62)$$

where the nonlinear projector operator $p(\cdot)$ is defined as $p(\cdot) = I_n - g(z + \pi(w)) \times [G(z)g(z + \pi(w))]^{-1}$.

Lemma 2.   *The following relation:*

$$p(\pi(w)) \left[f(\pi(w)) + d(\pi(w))w - \frac{\partial \pi(w)}{\partial w} s(w) \right] = 0$$

is true if and only if there are $\pi(w)$ and $\lambda(w)$, such that

$$f(\pi(w)) + d(\pi(w))w - \frac{\partial \pi(w)}{\partial w} s(w) = g(\pi(w))\lambda(w)$$

The proof of this Lemma is similar to the linear case.

On the other hand, using the linearisation matrices $A = [\partial f(x)/\partial x]_{(0)}$, $B = g(0)$, $D = d(0)$, $S = [\partial s(w)/\partial w]_{(0)}$, $C = [\partial h/\partial x]_{(0)}$, $Q = [\partial q(w)/\partial w]_{(0)}$ and $\Pi = [\partial \pi(w)/\partial w]_{w=0}$, the system (2.58)–(2.60) can be represented as

$$\dot{\zeta} = \bar{A}\zeta + \bar{B}u + \Phi(\zeta) \qquad (2.63)$$

$$e = \bar{C}\zeta + \phi_e(\zeta)$$

where

$$\zeta = (z, w)^T, \qquad \bar{A} = \begin{pmatrix} A & A\Pi - \Pi S + D \\ 0 & S \end{pmatrix}, \qquad \bar{B} = \begin{pmatrix} B \\ 0 \end{pmatrix}$$

$$\bar{C} = \begin{pmatrix} C & (C\Pi - Q) \end{pmatrix}, \qquad \Phi(\zeta) = \begin{pmatrix} \phi(z, w) \\ \phi_w(w) \end{pmatrix}$$

and the functions $\phi(z, w)$, $\phi_w(w)$, $\phi_e(z, w)$ and their first derivatives vanish at the origin. The sliding mode dynamics (2.62) can be thus represented as

$$\dot{z} = PAz + P(A\Pi - \Pi S + D)w + \phi_s(z, w) \qquad (2.64)$$

where $\phi_s(z, w)$ and its first derivative vanish at the origin. Now, if assumption H4 holds, one can propose an asymptotic observer for (2.58) and (2.59) or (2.63) of the following form:

$$\dot{\xi} = \left[\begin{matrix} f(\xi) + g(\xi)u + d(\xi)\hat{w} - \dfrac{\partial \pi(\hat{w})}{\partial \hat{w}} s(\hat{w}) \\[1.5em] s(\hat{w}) \end{matrix} \right] + L(e - \hat{e}), \qquad \hat{e} = \bar{C}\xi + \phi_e(\xi)$$

$$(2.65)$$

where $\xi = (\hat{z}, \hat{w})^T$ is the estimate of $\zeta = (z, w)^T$; $f(\xi) + g(\xi)u + d(\xi)\hat{w} - (\partial \pi(\hat{w})/\partial \hat{w})s(\hat{w}) = \bar{A}\xi + \bar{B}u + \Phi(\xi)$, and the matrix L is chosen to stabilise the observer error dynamics as in the linear case (2.21):

$$\dot{\epsilon} = (\bar{A} - L\bar{C})\epsilon + \Phi_\epsilon(\xi, \epsilon) \tag{2.66}$$

with $\Phi_\epsilon(\xi, \epsilon) = \Phi(\zeta) - \Phi(\xi)$, and $\Phi_\epsilon(\xi, 0) = 0$. Using the state, $\hat{z}$, of the observer (2.65), the sliding manifold in terms of estimated states is set as $\hat{\sigma}(\hat{z}) = 0$ with $G(\hat{z}) = \partial\hat{\sigma}(\hat{z})/\partial\hat{z}$, $G(0) = \Sigma$. Similar to the linear case we establish the following result.

Proposition 4. *Under assumptions H1, H2 and H4, if there exists a C^k ($k \geq 2$) mapping $x = \pi(w)$, with $\pi(0) = 0$, defined in a neighbourhood W of 0 and satisfying the following conditions:*

$$f(\pi(w)) + g(\pi(w))\lambda(\rho(w)) + d(\pi(w))w = \frac{\partial\pi(w)}{\partial w}s(w) \tag{2.67}$$

$$h(\pi(w)) - q(w) = 0 \tag{2.68}$$

at $(z, \epsilon) = (0, 0)$. Then, the nonlinear EFSMRP is solvable.

Proof. Selecting the control as $u = -k[G(\hat{z})g(\hat{z} + \pi(\hat{w}))]^{-1}\,\text{sign}\,(\hat{\sigma})$, $k > 0$, a Lyapunov function $V = \frac{1}{2}\hat{\sigma}^T\hat{\sigma}$, and taking its derivative along the trajectories of (2.65), we can see that, if

$$k > \|[G(\hat{z})g(\hat{z} + \pi(\hat{w}))]u_{eq}(\xi)\|$$

with $u_{eq}(\xi) = -[G(\hat{z})g(\xi)]^{-1}[f(\xi) + d(\xi)\hat{w} - (\partial\pi(\hat{w})/\partial\hat{w})s(\hat{w}) + L_1(e - \hat{e})]$, then the requirement ($SMS_e f$) is fulfilled. After the sliding mode occurs, substituting $u_{eq}(\xi)$ in (2.58), and using (2.64) and (2.66), the closed-loop system motion on $\hat{\sigma}(\hat{z}) = 0$ can be described by

$$\dot{z} = PAz + P(A\Pi - \Pi S + D)w + \hat{\phi}_s(z, w, \epsilon), \quad \sigma(z) + \phi_\sigma(\epsilon) = 0 \tag{2.69}$$

$$\dot{w} = Sw + \phi_w(w) \tag{2.70}$$

$$\dot{\epsilon} = (\bar{A} - L\bar{C})\epsilon + \Phi_\epsilon(z, w, \epsilon) \tag{2.71}$$

$$e = h(z + \pi(w)) - q(w) \tag{2.72}$$

Here, $\hat{\phi}_s(z, w, \epsilon)$ and $\phi_\sigma(\epsilon)$ and its first derivatives vanish at the origin, and $\hat{\phi}_s(z, w, 0) = \phi_s(z, w)$; $P = [\partial p(\zeta)/\partial]_{\zeta=0}$ is the same operator defined as in the linear case, and the matrix Σ can be chosen (by assumption H2) such that the $(n - m)$ eigenvalues of PA are in C^-. Additionally, the matrix L can be selected, by assumption H4, such that the matrix $(\bar{A} - L\bar{C})$ in (2.71) is Hurwitz. We can easily see that for all sufficiently small initial states $(x(0), w(0), \epsilon(0))$, the condition ($S_{ef}$) is satisfied.

Now, if the mapping $\pi(w)$ satisfies the partial differential equation (2.67), then by Lemma 2 it follows that

$$P(A\Pi - \Pi S + D)w + \tilde{\phi}_s(w)$$

$$= p(\pi(w))\left[f(\pi(w)) + d(\pi(w))w - \frac{\partial \pi(w)}{\partial w}s(w)\right]$$

$$= 0$$

Therefore, under assumption H1, the system (2.69) and (2.70) has a (sliding) centre manifold [14]

$$\sigma(z) = 0, \qquad z = 0 \tag{2.73}$$

or in the original variables the graph of mappings

$$\sigma(x - \pi(w)) = 0, \qquad x = \pi(w) \tag{2.74}$$

which is locally invariant and attractive under the flow of (2.62). The restriction of this flow to manifold (2.73) or (2.74) is a diffeomorphic copy of the flow of the exosystem (2.70). Thus, $\lim_{t\to\infty} z(t) = 0$, and if condition (2.68) holds, then by continuity of $h(z + \pi(w))$ (2.72), $e(t) \to 0$ as $t \to \infty$, i.e., that condition (R_{ef}) is satisfied.

Remark 2.   *The sliding centre manifold (2.74) is rendered locally invariant by the effect of a suitable equivalent control $u_{eq}(w) = [G(0)g(\pi(w)]^{-1}[f(\pi(w)) + d(\pi(w))w - (\partial\pi(w)/\partial w)s(w)]$ (2.61), and this manifold is annihilated by the error map $e = h(x) - q(w)$ in a similar way as takes place in the classical regulator formulation.*

2.4.2   Sliding regulator for nonlinear systems in Regular form

Now, consider transformation of the nonlinear system (2.1) by a diffeomorphism $x' = \varphi(x)$ to the Regular form [15]:

$$\dot{x}_1 = f_1(x_1, x_2) + d_1(x_1, x_2)w \tag{2.75}$$

$$\dot{x}_2 = f_2(x') + g_2(x')u + d_2(x')w \tag{2.76}$$

$$\dot{w} = s(w) \tag{2.77}$$

$$e = h(x_1, x_2) - q(w) \tag{2.78}$$

where $x' = (x_1, x_2)^T$, $x_1 \in X_1 \subset \Re^{n-m}$, $x_2 \in X_2 \subset \Re^m$ and $\mathrm{rank}[g_2(x')] = m \; \forall x' \in X \subset \Re^n$.

Let us now introduce the steady state for x_1 and x_2 as $\pi_1(w)$ and $\pi_2(w)$, respectively. Then, defining the steady-state error

$$z = x' - \pi(w) = \begin{pmatrix} z_1 \\ z_2 \end{pmatrix} = \begin{pmatrix} x_1 \\ x_2 \end{pmatrix} - \begin{pmatrix} \pi_1(w) \\ \pi_2(w) \end{pmatrix} \tag{2.79}$$

the dynamic equation for (2.79) with tracking error e can be obtained from (2.75)–(2.78) as

$$\dot{z}_1 = f_1(z_1 + \pi_1(w), z_2 + \pi_2(w)) + d_1(z_1 + \pi_1(w), z_2 + \pi_2(w))w$$
$$- \frac{\partial \pi_1(w)}{\partial w} s(w) \tag{2.80}$$

$$\dot{z}_2 = f_2(z + \pi(w)) + g_2(z + \pi(w))u + d_2'(z, w) \tag{2.81}$$

$$e = h(z_1 + \pi_1(w), z_2 + \pi_2(w)) - q(w) \tag{2.82}$$

where $d_2'(z, w) = d_2(z_1 + \pi_1(w), z_2 + \pi_2(w))w - (\partial \pi_2(w)/\partial w)s(w)$. The proposed sliding manifold is expressed as

$$\sigma = z_2 - \sigma_1(z_1) = 0, \quad \sigma_1(0) = 0, \quad \left[\frac{\partial \sigma_1}{\partial z_1}\right]_{(0)} = \Sigma_1 \tag{2.83}$$

and the $(n - m)$th order sliding mode equation describing the motion on (2.83), is given by

$$\dot{z}_1 = f_1(z_1 + \pi_1(w), \sigma_1(z_1) + \pi_2(w)) + d_1(z_1 + \pi_1(w), \sigma_1(z_1) + \pi_2(w))w$$
$$- \frac{\partial \pi_1(w)}{\partial w} s(w) \tag{2.84}$$

To estimate the states of system (2.80), (2.81) and (2.77), the proposed nonlinear observer is designed as

$$\dot{\xi} = \begin{bmatrix} f_1(\hat{z}_1 + \pi_1(\hat{w}), \hat{z}_2 + \pi_2(\hat{w})) + d_1(\hat{z}_1 + \pi_1(\hat{w}), \hat{z}_2 + \pi_2(\hat{w}))\hat{w} - \dfrac{\partial \pi_1(\hat{w})}{\partial \hat{w}} s(\hat{w}) \\ f_2(\hat{z} + \pi(\hat{w})) + g_2(\hat{z} + \pi(\hat{w}))u + d_2'(\hat{z}, \hat{w}) \\ s(\hat{w}) \end{bmatrix}$$
$$+ L'(e - \hat{e}) \tag{2.85}$$

with $\xi = (\hat{z}_1, \hat{z}_2, \hat{w})^T$ the estimate of $\zeta = (z_1, z_2, w)^T$, and $\hat{e} = h(\hat{z}_1 + \pi_1(\hat{w}), \hat{z}_2 + \pi_2(\hat{w})) - q(w)$. To analyse the stability of the sliding dynamics (2.84) and the observer (2.85), we consider only the linear part in (2.80)–(2.82) and (2.77)

$$\begin{pmatrix} \dot{z}_1 \\ \dot{z}_2 \end{pmatrix} = \begin{pmatrix} A_{11} & A_{12} \\ A_{21} & A_{22} \end{pmatrix} \begin{pmatrix} z_1 \\ z_2 \end{pmatrix} + \begin{pmatrix} 0 \\ B_2 \end{pmatrix} u + \begin{pmatrix} R_1 \\ R_2 \end{pmatrix} w + \begin{pmatrix} \phi_1(\zeta) \\ \phi_2(\zeta) \end{pmatrix} \tag{2.86}$$

$$\dot{w} = Sw + \phi_w(w) \tag{2.87}$$

$$e = C_1 z_1 + C_2 z_2 + (C_1 \Pi_1 + C_2 \Pi_2 - Q)w + \phi_e(\zeta)$$

and sliding mode equation (2.84)

$$\dot{z}_1 = (A_{11} - A_{12}\Sigma_1)z_1 + R_1 w + \phi_{1s}(z_1, w)$$

where $R_1 = A_{11}\Pi_1 + A_{12}\Pi_2 - \Pi_1 S + D_1$ and $R_2 = A_{21}\Pi_1 + A_{22}\Pi_2 - \Pi_2 S + D_2$, with $A_{ij} = [\partial f_i/\partial x_j]_{(0,0)}$, $B_2 = g_2(0)$, $C_i = [\partial h/\partial x_i]_{(0,0)}$, $D_i = d_i(0,0), \Pi_i = [\partial \pi_i/\partial w]_{(0)}$ and functions $\phi_i(z,w)$, $\phi_w(w)$, $\phi_e(z,w)$ and ϕ_{1s} vanish at the origin with their first derivatives; $\forall i,j = \{1,2\}$, and the constant matrices S and Q are already defined. Then using (2.85)–(2.87), the observer error dynamics becomes

$$\dot{\varepsilon} = (A' - L'C')\varepsilon + \Phi'(\zeta, \varepsilon) \qquad (2.88)$$

where

$$\varepsilon = \zeta - \xi = (\varepsilon_1, \varepsilon_2, \varepsilon_3)^T, \quad A' = \begin{pmatrix} A_{11} & A_{12} & R_1 \\ A_{21} & A_{22} & R_2 \\ 0 & 0 & S \end{pmatrix}, \quad B' = \begin{pmatrix} 0 \\ B_2 \\ 0 \end{pmatrix}$$

$$L' = \begin{pmatrix} L_1 \\ L_2 \\ L_3 \end{pmatrix}, \quad \Phi'(\zeta, \varepsilon) = \begin{pmatrix} \phi_1(\zeta) - \phi_1(\xi) + L_1(\phi_e(\zeta) - \phi_e(\xi)) \\ \phi_2(\zeta) - \phi_2(\xi) + L_2(\phi_e(\zeta) - \phi_e(\xi)) \\ \phi_w(w) - \phi_w(\hat{w}) + L_3(\phi_e(\zeta) - \phi_e(\xi)) \end{pmatrix}$$

and

$$C' = (C_1 \quad C_2 \quad (C_1\Pi_1 + C_2\Pi_2 - Q))$$

Note that the detectability requirement H5 is the same required to stabilise (2.88) in a similar way to the linear case (2.34).

Before defining the sliding manifold and discontinuous control, we will establish the conditions that will solve the EFSMRP for the nonlinear system in Regular form.

Proposition 5.   *Under assumptions H1, H2 and H5, if there exists C^k ($k \geq 2$) mappings $x_1 = \pi_1(w)$ and $x_2 = \pi_2(w)$, with $\pi_1(0) = 0$ and $\pi_2(0) = 0$, defined in a neighbourhood W of 0, that satisfy the following conditions:*

$$f_1(\pi_1(w), \pi_2(w)) + d_1(\pi_1(w), \pi_2(w))w = \frac{\partial \pi_1(w)}{\partial w}s(w) \qquad (2.89)$$

$$h(\pi_1(w), \pi_2(w)) - q(w) = 0 \qquad (2.90)$$

at $(x_1, x_2, w, \varepsilon) = (0, 0, 0, 0)$ then, the EFSMRP for nonlinear systems in Regular form is solvable.

Proof.   We define the estimated sliding manifold and control as

$$u = -kB_2^{-1}\,\text{sign}(\hat{\sigma}), \quad \hat{\sigma} = \hat{z}_2 + \hat{\sigma}_1(\hat{z}_1) = 0, \quad \left[\frac{\partial \hat{\sigma}_1}{\partial \hat{z}_1}\right]_{(0)} = \Sigma_1$$

If the control gain k is chosen such that $k > \| g_2(\hat{z}, \hat{w}) u_{eq}(\hat{z}, \hat{w}) \|$ where $u_{eq}(\hat{z}, \hat{w})$ is a solution of $\dot{\hat{\sigma}} = 0$, then the condition (SMS_{ef}) holds. After the sliding mode occurs, we have $\hat{z}_2 = \hat{\sigma}_1(\hat{z}_1)$ and $z_2 = \sigma_1(z_1 - \varepsilon_1) - \varepsilon_2$, and the motion of the closed-loop system will be governed by

$$\dot{z}_1 = (A_{11} - A_{12}\Sigma_1)z_1 + R_1 w + \hat{\phi}_{1s}(z, w, \varepsilon)$$

$$\dot{w} = Sw + \phi_w(w)$$

$$\dot{\varepsilon} = (A' - L'C')\varepsilon + \Phi'(\zeta, \varepsilon)$$

$$e = h(z_1 + \pi_1(w), \sigma_1(z_1 - \varepsilon_1) - \varepsilon_2 + \pi_2(w)) - q(w)$$

were $\hat{\phi}_{1s}(z, w, \epsilon)$ vanishes at the origin with it first derivative, and $\hat{\phi}_{1s}(z, w, 0) = \phi_{1s}(z, w)$. Recalling that for the linear system in Regular form case, the matrices $(A_{11} - A_{12}\Sigma_1)$ and $(A' - L'C')$ are Hurwitz by a proper choice of Σ_1 and L', respectively, and, if condition (2.89) holds, then $R_1 w + \hat{\phi}_{1s}(z, w, 0) = f_1(\pi_1(w), \pi_2(w)) + d_1(\pi_1(w), \pi_2(w))w - (\partial \pi_1(w)/\partial w)s(w) = 0$, under the property of centre manifolds, we have $z_1(t) \to 0 \Rightarrow x_1(t) \to \pi_1(w(t))$, and $z_2(t) \to 0 \Rightarrow x_2(t) \to \pi_2(w(t))$ with $t \to \infty$. Thus, the requirement (S_{ef}) is fulfilled. So, by continuity, if condition (2.90) holds, then the output tracking error (2.78) converges to zero and condition (R_{ef}) holds too.

2.4.3 Nonlinear Block Controllable form with disturbances

In this section, a discontinuous control strategy will be investigated for a class of nonlinear systems in the Nonlinear Block Controllable form (NBC-form). The essential feature of the proposed method is the decoupling of the system motion into motions of lower dimension in order to simplify the control design [16], therefore, system (2.1) is decomposed into r blocks:

$$\dot{x}_1 = f_1(x_1) + b_1(x_1)x_2 + d_1(x_1)w$$

$$\dot{x}_i = f_i(x_1, \ldots, x_i) + b_i(x_1, \ldots, x_i)x_{i+1} + d_i(x_1, \ldots, x_i)w, \qquad i = 3, \ldots, r-1$$

$$\dot{x}_r = f_r(x_1, \ldots, x_r) + b_r(x_1, \ldots, x_r)u + d_r(x_1, \ldots, x_r)w \qquad (2.91)$$

$$\dot{w} = s(w)$$

$$e = h(x_1, \ldots, x_r) - q(w)$$

where the transformed state vector $\tilde{x}$ is decomposed as $\tilde{x} = (x_1, \ldots, x_r)^T$ and $x_i \in X_i \subset \Re^{n_i}$. In the ith block, the vector x_{i+1} is regarded as a fictitious control vector, where $\text{rank}[b_i(x_1, \ldots, x_i)] = n_i$ and $\|b_i(x_1, \ldots, x_i)\| \leq \beta_i > 0 \ \forall x \in X$. As in the linear case, the integers $(n_1, \ldots, n_r)$ define the plant structure by the condition $n_1 \leq n_2 \leq \cdots \leq n_r$ with $\sum_{i=1}^{r} n_i = n$. A convenient representation of system (2.91), where the linear part is explicitly expressed, is given by

$$\dot{x}_1 = A_{11}x_1 + B_1 x_2 + D_1 w + \psi_1(x_1, w)$$

$$\dot{x}_i = \sum_{j=1}^{i} A_{ij}x_j + B_i x_{i+1} + D_i w + \psi_i(x_1, \ldots, x_i, w), \qquad i = 2, \ldots, r-1$$

$$\dot{x}_r = \sum_{k=1}^{r} A_{rk}x_k + B_r u + D_r w + \psi_r(x_1, \ldots, x_r, w) \tag{2.92}$$

$$\dot{w} = Sw + \phi_w(w)$$

$$e = \sum_{k=1}^{r} M_k x_k + \phi_e(x_1, \ldots, x_r, w)$$

with $A_{jk} = (\partial f_j / \partial x_k)_{(0)}$, $S = (\partial s / \partial w)_{(0)}$, $M_k = (\partial h / \partial x_k)_{(0)}$, $B_j = b_j(0)$, $\mathrm{rank}(B_i) = n_i$, $D_j = d_j(0)\ \forall j, k \in \{1, \ldots, r\}$, and functions $(\psi_1, \ldots, \psi_r, \phi_w, \phi_e)$ that vanish at the origin with their first derivatives.

Now, we introduce the steady-state error z as

$$z = \tilde{x} - \pi(w) \tag{2.93}$$

where $\pi = (\pi_1, \ldots, \pi_r)^T$ is the steady state for $\tilde{x}$ with $[\partial \pi_i(w)/\partial w]_{(0)} = \Gamma_i, i = 1, \ldots, r$. The system (2.91) in the new coordinates (2.93) is of the following form:

$$\dot{z}_1 = f_1(z_1 + \pi_1(w)) + b_1(z_1 + \pi_1(w))z_2 + r_1(z_1 + \pi_1(w), \pi_2(w), w)$$

$$\dot{z}_i = f_i(z_1 + \pi_1(w), \ldots, z_i + \pi_i(w)) + b_i(z_1 + \pi_1(w), \ldots, z_i + \pi_i(w))z_{i+1}$$

$$\qquad + r_i(z_1 + \pi_1(w), \ldots, z_i + \pi_i(w), w), \qquad i = 2, \ldots, r-1 \tag{2.94}$$

$$\dot{z}_r = f_r(z_1 + \pi_1(w), \ldots, z_r + \pi_r(w)) + b_r(z_1 + \pi_1(w), \ldots, z_r + \pi_r(w))u$$

$$\qquad + r_r(z_1 + \pi_1(w), \ldots, z_r + \pi_r(w), w)$$

$$\dot{w} = Sw + \phi_w(w) \tag{2.95}$$

$$e = h(z_1 + \pi_1(w), \ldots, z_r + \pi_r(w)) - q(w)$$

where

$$r_1(\cdot) = b_1(z_1 + \pi_1(w))\pi_2(w) + d_1(z_1 + \pi_1(w))w - \frac{\partial \pi_1(w)}{\partial w}s(w)$$

$$r_i(\cdot) = b_i(z_1 + \pi_1(w), \ldots, z_i + \pi_i(w))\pi_{i+1}(w)$$

$$\qquad + d_i(z_1 + \pi_1(w), \ldots, z_i + \pi_i(w))w - \frac{\partial \pi_i(w)}{\partial w}s(w), \qquad i = 2, \ldots, r-1$$

$$r_r(\cdot) = d_r(z_1 + \pi_1(w), \ldots, z_r + \pi_r(w))w - \frac{\partial \pi_r(w)}{\partial w}s(w)$$

On the other hand, we propose an observer for system (2.94)–(2.95), of the form

$$\dot{\hat{z}}_1 = \hat{f}_1(\hat{z}_1 + \pi_1(\hat{w})) + \hat{b}_1(\hat{z}_1 + \pi_1(\hat{w}))\hat{z}_2 + \hat{r}_1(\hat{z}_1 + \pi_1(\hat{w}), \pi_2(\hat{w}), \hat{w})$$
$$+ L_1(e - \hat{e})$$

$$\dot{\hat{z}}_i = \hat{f}_i(\hat{z}_1 + \pi_1(\hat{w}), \ldots, \hat{z}_i + \pi_i(\hat{w})) + \hat{b}_i(\hat{z}_1 + \pi_1(\hat{w})), \ldots, \hat{z}_i + \pi_i(\hat{w}))\hat{z}_{i+1}$$
$$+ r_i(\hat{z}_1 + \pi_1(\hat{w})), \ldots, \hat{z}_i + \pi_i(\hat{w}), \hat{w}) + L_i(e - \hat{e})$$
$$i = 2, \ldots, r - 1 \tag{2.96}$$

$$\dot{\hat{z}}_r = \hat{f}_r(\hat{z}_1 + \pi_1(\hat{w}), \ldots, \hat{z}_r + \pi_r(\hat{w})) + \hat{b}_r(\hat{z}_1 + \pi_1(\hat{w}), \ldots, \hat{z}_r + \pi_r(\hat{w}))u$$
$$+ r_r(\hat{z}_1 + \pi_1(\hat{w})), \ldots, \hat{z}_r + \pi_r(\hat{w}), \hat{w}) + L_r(e - \hat{e})$$

$$\dot{\hat{w}} = \hat{s}(\hat{w}) + L_{r+1}(e - \hat{e}) \tag{2.97}$$

$$\hat{e} = \hat{h}(\hat{z}_1 + \pi_1(\hat{w})), \ldots, \hat{z}_r + \pi_r(\hat{w})) - q(\hat{w})$$

where $\xi = (\hat{z}_1, \ldots, \hat{z}_r, \hat{w})^T$ is the estimate of $\zeta = (z_1, \ldots, z_r, w)^T$ and matrix $\tilde{L} = (L_1, \ldots, L_{r+1})^T$ is the observer gain. Using the linearisation matrices defined in (2.92) and (2.93), the observer error ($\epsilon = \xi - \zeta$) system is derived from (2.94)–(2.95) and (2.96)–(2.97), as

$$\dot{\epsilon} = (\tilde{A} - \tilde{L}\tilde{C})\epsilon + \tilde{\Psi}(\zeta, \epsilon)$$

where the matrices $\tilde{A}$ and $\tilde{C}$ are defined in (2.46), and

$$\tilde{\Psi}(\zeta, \epsilon) = \begin{pmatrix} \psi_1(\zeta) - \psi_1(\xi) + L_1[\phi_e(\zeta) - \phi_e(\xi)] \\ \cdots \\ \psi_r(\zeta) - \psi_r(\xi) + L_r[\phi_e(\zeta) - \phi_e(\xi)] \\ \phi_w(w) - \phi_w(\hat{w}) + L_{r+1}[\phi_e(\zeta) - \phi_e(\xi)] \end{pmatrix}$$

From the previous discussion, we can derive the conditions that allow the EFSMRP to be solved for systems in NBC-form.

Proposition 6. *Under assumptions H1, H2 and H5, if there exists C^k ($k \geq 2$) mappings $x_i = \pi_i(w)$, $i = 1, \ldots, r$ with $\pi_i(0) = 0$ defined in a neighbourhood W of 0, that satisfy the following conditions:*

$$\frac{\partial \pi_1(w)}{\partial w} s(w) = f_1(\pi_1(w)) + b_1(\pi_1(w))\pi_2(w) + d_1(\pi_1(w))w \tag{2.98}$$

$$\frac{\partial \pi_2(w)}{\partial w} s(w) = f_2(\pi_1(w), \pi_2(w)) + b_2(\pi_1(w), \pi_2(w))\pi_3(w)$$
$$+ d_2(\pi_1(w), \pi_2(w))w \tag{2.99}$$

$$\frac{\partial \pi_i(w)}{\partial w} s(w) = f_i(\pi_1(w), \ldots, \pi_i(w)) + b_i(\pi_1(w), \ldots, \pi_i(w))\pi_{i+1}(w)$$

$$+ d_i(\pi_1(w), \ldots, \pi_i(w))w, \qquad i = 3, \ldots, r-1 \tag{2.100}$$

$$0 = h(\pi_1(w), \ldots, \pi_r(w)) - q(w) \tag{2.101}$$

at $(\tilde{x}, w, \epsilon) = (0, 0, 0)$, then the EFSMRP for NBC-form systems is solvable.

At this point, a procedure for the designing of the sliding manifold and the discontinuous control based on the block control technique is possible. Referring to system (2.96)–(2.97), we consider the state $\hat{z}_{i+1}, i = 1, \ldots, r-1$ as a fictitious control vector in the ith block. This yields the following iterative procedure.

Let us define a new variable $\chi_1 = \hat{z}_1$. Then considering $\hat{z}_2$ as a control input in the first block of (2.96), we anticipate the dynamics of this block as follows:

$$\dot{\chi}_1 = \hat{f}_1(\hat{z}_1 + \pi_1(\hat{w})) + \hat{b}_1(\hat{z}_1 + \pi_1(\hat{w}))\hat{z}_2 + \hat{r}_1(\hat{z}_1 + \pi_1(\hat{w}), \pi_2(\hat{w}), \hat{w})$$

$$+ L_1(e - \hat{e})$$

$$= K_1 \chi_1 \tag{2.102}$$

where K_1 is a Hurwitz matrix. The state $\hat{z}_2$ is calculated from $\dot{\chi}_1 = 0$ (2.102) as a desired state $\hat{z}_2^d$ of the following form:

$$\hat{z}_2^d = -\hat{b}_1^+(\cdot)[\hat{f}_1(\hat{z}_1 + \pi_1(\hat{w})) + \hat{r}_1(\hat{z}_1 + \pi_1(\hat{w}), \pi_2(\hat{w}), \hat{w}) + L_1(e - \hat{e}) - K_1\chi_1]$$

$$\tag{2.103}$$

where $\hat{b}_1^+ = \hat{b}_1^T(\hat{b}_1\hat{b}_1^T)^{-1}$ denotes the right pseudo-inverse matrix of $\hat{b}_1$. Proceeding in the same way, we define a second new variable χ_2 as follows:

$$\chi_2 = \hat{z}_2 - \hat{z}_2^d \tag{2.104}$$

then, taking its derivative, and anticipating its dynamics, we have that the next block is:

$$\dot{\chi}_2 = f_2(\hat{z}_1 + \pi_1(\hat{w}), \hat{z}_2 + \pi_2(\hat{w})) + \hat{b}_2(\hat{z}_1 + \pi_1(\hat{w})), \hat{z}_2 + \pi_2(\hat{w}))\hat{z}_3$$

$$+ \hat{r}_2(\hat{z}_1 + \pi_1(\hat{w})), \hat{z}_2 + \pi_2(\hat{w}), \hat{w}) + L_2(e - \hat{e}) - \dot{\hat{z}}_2^d$$

$$= K_2 \chi_2 \tag{2.105}$$

where K_2 is a Hurwitz matrix, and the desired state of $\hat{z}_3$ is calculated from (2.105) as follows:

$$\hat{z}_3^d = -\hat{b}_2^+(\cdot)[f_2(\cdot)) + \hat{r}_2(\cdot) + L_2(e - \hat{e}) - \dot{\hat{z}}_2^d - K_2\chi_2] \tag{2.106}$$

with $\hat{b}_2^+ = \hat{b}_2^T(\hat{b}_2\hat{b}_2^T)^{-1}$, and a third new variable is defined as

$$\chi_3 = \hat{z}_3 - \hat{z}_3^d \tag{2.107}$$

This procedure is performed iteratively, i.e., an ith new state is defined as $\chi_i = \hat{z}_i - \hat{z}_i^d$, and the ith block formed as follows:

$$\dot{\chi}_i = \hat{f}_i(\hat{z}_1 + \pi_1(\hat{w}), \ldots, \hat{z}_i + \pi_i(\hat{w})) + \hat{b}_i(\hat{z}_1 + \pi_1(\hat{w}), \ldots, \hat{z}_i + \pi_i(\hat{w}))\hat{z}_{i+1}$$
$$+ \hat{r}_i(\hat{z}_1 + \pi_1(\hat{w}), \ldots, \hat{z}_i + \pi_i(\hat{w}), \hat{w}) + L_i(e - \hat{e}) - \dot{\hat{z}}_i^d$$
$$= K_i \chi_i, \qquad i = 3, \ldots, r - 1$$

and the desired state as

$$\hat{z}_{i+1}^d = -\hat{b}_i^+(\cdot)[\hat{f}_i(\cdot) + \hat{r}_i(\cdot) + L_i(e - \hat{e}) - \dot{\hat{z}}_i^d - K_i \chi_i]$$

where $\hat{b}_i^+ = \hat{b}_i^T(\hat{b}_i \hat{b}_i^T)^{-1}$, and K_i is a Hurwitz matrix.

In the final step, the vector $\hat{z}_r^d$ is calculated. Therefore the last new variable to be defined is $\chi_r = \hat{z}_r - \hat{z}_r^d$. As in the linear case, the new state $\chi = (\chi_1, \ldots, \chi_r)^T$ is derived by the nonsingular transformation

$$\chi_1 = \hat{z}_1$$

$$\chi_i = \hat{z}_i - \hat{z}_i^d, \qquad i = 2, \ldots, r$$

which simplifies the system (2.96) to the following form:

$$\dot{\chi}_1 = K_1 \chi_1 + b_1 \chi_2$$
$$\dot{\chi}_i = K_i \chi_i + b_i \chi_{i+1}, \qquad i = 3, \ldots, r - 1 \qquad (2.108)$$
$$\dot{\chi}_r = \hat{f}_r(\hat{z}_1 + \pi_1(\hat{w}), \ldots, \hat{z}_r + \pi_r(\hat{w})) + \hat{b}_r(\hat{z}_1 + \pi_1(\hat{w}), \ldots, \hat{z}_r + \pi_r(\hat{w}))u$$
$$+ r_r(\hat{z}_1 + \pi_1(\hat{w}), \ldots, \hat{z}_r + \pi_r(\hat{w}), \hat{w}) + L_r(e - \hat{e}) - \dot{\hat{z}}_r^d$$

Taking advantage of the structure of system (2.108), the sliding manifold is appropriately selected as $\sigma = \chi_r = 0$, and, in order to generate a sliding mode in (2.108), we choose the control as $u = -k_r \hat{b}_r^+(\cdot)\,\text{sign}\,(\sigma)$. If $k_r > \|b_r(\cdot)u_{eq}\|$ the condition (SMS_{ef}) is guaranteed, where $\hat{b}_r^+ = \hat{b}_r^T(\hat{b}_r \hat{b}_r^T)^{-1}$, and u_{eq} is calculated from $\dot{\sigma} = \dot{\chi}_r = 0$. Then, the motion along the manifold $\sigma = 0$ is described by

$$\dot{\chi}_1 = K_1 \chi_1 + b_1 \chi_2$$
$$\dot{\chi}_i = K_i \chi_i + b_i \chi_{i+1}, \qquad i = 2, \ldots, r - 2 \qquad (2.109)$$
$$\dot{\chi}_{r-1} = K_{r-1} \chi_{r-1}$$

$$\dot{w} = s(w)$$

$$\dot{\epsilon} = (\tilde{A} - \tilde{L}\tilde{C})\epsilon + \tilde{\Psi}(\zeta, \epsilon) \qquad (2.110)$$

$$e = h(z_1 + \pi_1(w), \ldots, z_r + \pi_r(w)) - q(w) \qquad (2.111)$$

Due to the block triangular form, the fact that the diagonal matrices K_i are Hurwitz and that the $b_i(\cdot)$ matrices are bounded, there exists a neighbourhood of the origin $\chi = 0$, and the states of (2.109) tend asymptotically to zero, i.e., $\lim_{t \to \infty} \chi_i(t) = 0$, $i = 1, \ldots, r - 1$. Under assumption H6 similar to the linear case, there is a matrix $\tilde{L}$ such that $(\tilde{A} - \tilde{L}\tilde{C})$ in (2.110) is Hurwitz, therefore for sufficiently small $\epsilon(0)$ we have $\lim_{t \to \infty} \epsilon(t) = 0$.

Now, the following statements are derived step by step under the assumption that the estimated state $\xi = (\hat{z}_1, \ldots, \hat{z}_r, \hat{w})^T$ has converged to the real state $\zeta = (z_1, \ldots, z_r, w)^T$.

Since $z_1(t) = \hat{z}_1(t) = \chi_1(t) \to 0$ with $t \to \infty$, and if condition (2.98) holds, one can easily see in (2.103) that $\hat{z}_2^d(t) \to 0$ with $t \to \infty$. Therefore, from (2.104) and $\chi_2(t) \to 0$, it follows that $z_2(t) = \hat{z}_2(t) \to 0$ with $t \to \infty$.

For the second step, based on results obtained during the first step and using the condition (2.99), we can see from (2.106) that $\hat{z}_3^d(t) \to 0$ with $t \to \infty$. Hence, from (2.107) it follows $z_3(t) = \hat{z}_3(t) \to 0$ with $t \to \infty$.

Performing iteratively this procedure for the subsequent steps, one can verify that under condition (2.100), then $\hat{z}_i^d(t) \to 0$ with $t \to \infty$ and as a consequence the ith state $z_i(t) = \hat{z}_i(t) \to 0$ with $t \to \infty$, $i = 4, \ldots, r$, and thus this implies that (S_{ef}) is satisfied.

Finally, thanks to condition (2.101), the output tracking error (2.111) tends asymptotically to zero, satisfying condition (R_{ef}).

2.5 Conclusions

The Error Feedback Sliding Mode Regulation Problem has been introduced. Solution conditions are derived for linear systems and different classes of nonlinear systems including systems presented in the Regular and NBC-forms. In particular, the combination of VSS and block control techniques allows straightforward solutions to be obtained, specially when compared to the classical solutions of the error feedback regulator problem. Additionally the sliding mode based controller achieves robustness with respect to the uncertainty.

2.6 Acknowledgement

This work was supported by CONACYT (Mexico) under grants 36960A and 37687A.

2.7 References

1   FRANCIS, B. A.: 'The linear multivariable regulator problem', *SIAM J. Control Optimiz.*, 1977, **15**, pp. 486–505

2   ISIDORI, A. and BYRNES, C. I.: 'Output regulation of nonlinear systems', *IEEE Trans. Aut. Control*, 1990, **35**(2), pp. 131–140

3   UTKIN, V. I.: 'Sliding modes in control and optimization' (Springer-Verlag, London, 1992)

4   ELMALI, H. and OLGAC, N.: 'Robust output tracking control of nonlinear MIMO systems via sliding mode technique', *Automatica*, 1992, **28**(1), pp. 145–151

5   ELMALI, H. and OLGAC, N.: 'Tracking nonlinear nonminimum phase systems using sliding control', *International Journal of Control*, 1993, **57**(5), pp. 1141–1158

6   CASTILLO-TOLEDO, B. and CASTRO-LINARES, R.: 'On robust regulation via sliding mode for nonlinear systems', *Systems and Control Letters*, 1995, **24**, pp. 361–371

7   EDWARDS, C. and SPURGEON, S. K.: 'Robust output tracking using a sliding-mode controller/observer scheme', *International Journal of Control*, 1996, **64**(5), pp. 967–983

8   SIRA-RAMIREZ, H.: 'A dynamical variable structure control strategy in asymptotic output tracking problem', *IEEE Trans. Aut. Control*, 1993, **38**, pp. 615–620

9   BONIVENTO, C., MARCONI, L., and ZANASI, R.: 'Output regulation of nonlinear systems by sliding mode', *Automatica*, 2001, **37**, pp. 535–542

10 El-CHESAWI, O. M. E., ZINOBER, A. S. I., and BILLINGS, S. A.: 'Analysis and design of variable structure systems using a geometric approach', *International Journal of Control* 1983, **38**, pp. 657–671

11 DRAGENOVIC, B.: 'The invariance conditions in variable structure systems', *Automatica*, 1969, **5**(3), pp. 287–295

12 UTKIN, V. I. and YOUNG K.-K. D.: 'Methods for constracting discontinuity planes in multidimensional variable-structure systems', *Automation and Remote Control*, 1978, **39**(10) pp. 1466–1470

13 DRAKUNOV, C. V., IZOSIMOV, D. B., LOUKIANOV, A. G., UTKIN, V. A., and UTKIN, V. I.: 'Block control principle, I and II', *Automation and Remote Control*, 1990, **51**(5), pp. 601–609; 1990, **51**(6), pp. 737–746

14 CARR, J.: 'Applications of centre manifold theory' (Springer-Verlag, New York, 1981)

15 LUK'YANOV, A. G. and UTKIN, V. I.: 'Methods for reducing dynamic systems to regular form', *Automation and Remote Control*, 1981, **42**(4), pp. 413–420

16 LOUKIANOV, A. G.: 'Nonlinear block control with sliding mode', *Automation and Remote Control*, 1998, **59**(7), pp. 916–933

Chapter 3

Deterministic output noise effects in sliding mode observation

Alex S. Poznyak

3.1 Preliminaries

The *state observation problem* arises during *Identification* or *Feedback Control* when the current system states cannot be directly measured and the only available information at each time instant is the output of the system. This is a function of the current state that may be corrupted by 'output noise' of a deterministic or even stochastic nature. Usually the dimension of the output signal is less than that of the corresponding state space vector. The following questions turn out to be extremely important for the control-designer:

- *Observability problem*: whether or not the output signal contains enough information to provide successful state-estimation.
- *Observer structure*: if the system is observable, then the problem is how to construct an estimating process.
- *Sliding mode observers (SMO)*: the specific advantages and disadvantages of such observers.
- *Output noise effects*: how the *SMO* work in the presence of noise disturbances in the output signal.
- *Stochastic specifics*: is there a difference between output observation effects that occur due to deterministic noise (usually bounded) or stochastic noise (practically, never bounded)?

All of these issues will be discussed later. The main principles of the design of special devices, namely *SMO*, generating signals ('state estimates') close to the current state vector of the process, will be emphasised.

This chapter consists of two parts: the first surveys the problem under consideration, and the second deals with *Deterministic Output Noise*. All necessary background information is given in the Appendix concluding the presented material.

3.2   State-estimation as a component of identification theory: a short survey

Modern *Identification Theory* [1–3] basically deals with the problem of the efficient extraction of signal and systems dynamic properties based on available data measurements.

Nonlinear system identification is traditionally concerned with two issues:

- estimation of parameters based on direct and complete state space measurements;
- state space estimation (filtering) of completely known nonlinear dynamics.

3.2.1   Parameter estimation

Parameter identification for different classes of nonlinear systems has been extensively studied during the last three decades. Basically, the class of linear and nonlinear systems whose dynamics depends linearly on the unknown parameters was considered, and external noise was assumed to be of a stochastic nature (see, for example [1, 4]). In the paper by Poznyak [5] the convergence properties of a least mean square (LMS) discrete time procedure in the presence of stochastic noise were studied. The relationship between a nonlinear function and a maximum value of the noise density function was established to guarantee convergence. In Sheikholeslam [6] a family of observer-based parameter identifiers that exploited the a priori known parameter dependencies was introduced to improve the identification performance. The Lyapunov-like approach was suggested to construct a stable adaptive algorithm for parameter estimation for the case when no external perturbations were present [7]. The gradient type procedure together with an additional state space estimator of Luenberger structure was applied to realise asymptotic parameter convergence.

Remark 1.   *A general feature of these publications described above is that exact state space vector measurements are assumed to be available.*

3.2.2   State-estimation

Contributions to the observer construction problem for nonlinear systems in the presence of complete information about the nonlinear dynamics have been reported [8–11]. Most of these results deal with the situation where it is possible to obtain a set of rather restrictive conditions when the dynamics of the observation errors is linear and there is no observation noise. In Reference 12, a class of observers for nonlinear systems subjected to bounded nonlinearities or uncertainties was suggested. A canonical form and a necessary and sufficient observability condition for a class of nonlinear systems that are linear with respect to the inputs was established by Gauthier and Bornard [13]. The extended Luenberger observer for a class of SISO nonlinear systems was designed by Zeitz [14]. These results were extended in Birk and Zeitz [15] for a class of MIMO nonlinear systems. An exponentially convergent observer was derived in Gauthier et al. [16] for nonlinear systems that are observable for any input signal. More advanced results were obtained in Ciccarella et al. [17] where, based on simple assumptions of regularity, global asymptotic convergence of the estimated states to the true states was shown.

Remark 2. *All of these papers consider the case where the given dynamic description does not contain any unknown parameters.*

3.2.3 Simultaneous state and parameter estimation

A much more difficult situation arises in the case where it is required to construct *state and parameter estimates simultaneously* in the presence of both internal (unknown parameters and unmodelled dynamics) and external (observation noise) uncertainties. The traditional approach for dealing with such problems is called *Adaptive Filtering* (see [3, 18, 19] where the state observer uses current estimates of parameters) or *Adaptive Identification* (see [20–26] when the identifier is constructed based on current state estimates). To solve this difficult problem, high-gain type observers were suggested by Tornambè [19]. Ljung [3] studied the asymptotic behaviour of the extended Kalman filter when applied to the identification of linear stochastic discrete time systems but, unfortunately, the conditions for convergence in a mean square sense turn out to be very complex for verification. In Haykin [18], a variety of recursive estimation algorithms that converge to the optimum Wiener solution (in some statistical sense) were considered and a number of engineering applications of adaptive filters were discussed. The identifiability concept is constructively discussed in Grewal and Glover [21], Siferd and Maybeck [22] and Tunali and Tarn [23] where necessary and sufficient conditions for identifiability were investigated in the light of the relationship between nonlinear observability, functional expansion and the uniqueness theorem on nonlinear realisation theory. The approach based on H^∞-theory results and applied to the parameter identification problem in the presence of non-parametric dynamic uncertainty was suggested in Krause and Khargonekar [25]. The augmented system is introduced in Bortoff and Spong [26] and an identifier based on an extended Luenberger observer is constructed as well as the sufficient conditions for global convergence of its parameter estimates. The most advanced techniques for determining the observability and identifiability properties are based on differential algebra; this approach has been extensively developed by Diop and Fliess [27] and, particularly, the concept of identifiability employing the notion of characteristic sets, was suggested by Ljung and Glad [28]. A comprehensive survey concerning the continuous-time approaches to system identification, studied before 1990, can be found in Unbehauen and Rao [24]. In the recent papers of Poznyak and Correa [29, 30], based on the work of Ciccarella et al. [17] and Bortoff and Spong [26], a switched structure robust state and parameter estimator for a class of MIMO nonlinear systems was designed and an upper bound was derived for the corresponding estimate error functional which turned out to be a linear combination of the external and internal uncertainty levels. In the absence of any uncertainties and noise perturbations, the global asymptotic stability of the error directly follows from the main theorem ('sharpness' property).

Remark 3. *It was shown (and this is an important note for this chapter) that this simultaneous state and parameter estimation problem can be converted (by a special procedure) to the problem of state estimation only, but such transformation implies a singularity effect within a certain domain. This work asserts that the switched structure observers may be applied to avoid this singularity effect.*

3.2.4   Observations under uncertainties

A further line of investigation relates to the observation problem subjected to bounded nonlinearities or *uncertainties* (see [12, 31]). In the situation when the plant model is incomplete or uncertain, the implementation of high-gain observers seems to be convenient [19, 32–34]. In Yaz and Azemi [35] a novel robust/adaptive observer is presented for state reconstruction of nonlinear systems with uncertainty having unknown bounds. The observer uses a nonlinear gain that is continuously adapted to guarantee a uniformly bounded and convergent observation error. A robust adaptive observer for a class of nonlinear systems is proposed in Ruijun et al. [36] based on generalised dynamic recurrent neural networks. This does not require an off-line training phase. A method for fault detection of a nonlinear system by means of a nonlinear observer is proposed by Preston et al. [37]. The observer is designed such that the error dynamics are independent of the state, input, output and unknown disturbances. The conditions necessary for the observer to exist and to be robust with respect to the unknown input are given. A robust nonlinear observer is considered in Shields [38] for a class of singular nonlinear descriptor systems subject to unknown inputs. This class is partly characterised by globally Lipschitz nonlinearities. A suboptimal robust filtering of states for finite dimensional linear systems with time-varying parameters under nonrandom disturbances was considered in Poznyak and Osorio-Cordero [39].

3.2.5   Sliding mode observation

Many of the theoretical developments in the area of sliding mode control assume that the system state vector is available for use by the control scheme. In order to exploit such strategies, a suitable estimate of the states should be constructed for use in the original control law. Despite fruitful research and development activity in the area of variable structure control theory, few authors have considered the application of the main principles of sliding mode control to the problem of observer design. For deterministic systems the earliest work in this field appeared originally in Utkin [40, 41]. The approach described in the book of Edwards and Spurgeon [42] is conceptually similar to that proposed by Slotine [43]. The papers of Walcott and Zak [12, 44] seek global error convergence for a class of uncertain systems using some algebraic manipulations to effectively solve an associated constrained Lyapunov problem for systems of reasonable order. This approach is discussed in detail in Zak and Walcott [45].

This collection also describes a hyperstability approach to observer design by Balestino and Innocenti [46], based on the concept of positive realness.

3.3 Estimation problem statement: formalism

3.3.1 The consistent class of nonlinear systems

Consider the class Σ of non-stationary nonlinear systems (NLS) with multi inputs and multi outputs (MIMO) containing mixed uncertainties

$$\Sigma : \begin{cases} \dot{z}_t = f(t, z_t, u_t, c) + \zeta_1(t, z_t, u_t, c), & z_{t=0} = z_0 \\ y_t = h(t, z_t, u_t, c) + \zeta_2(t, z_t, u_t, c) \end{cases} \tag{3.1}$$

where $t \in \Re^+ := \{t : t \geq 0\}$ corresponds to a time variable, $z_t \in \Re^n$ is a state vector at time t, $c \in \mathcal{C} \subseteq \Re^q$ is the constant vector of unknown parameters defined within a connected set $\mathcal{C}$, $y_t \in \Re^p$ is an output vector at time t, $u_t \in U \subseteq \Re^m$ is a vector of control actions at time t.

The functions $\zeta_1(\cdot) \in D_1 \subseteq \Re^n$, $\zeta_2(\cdot) \in D_2 \subseteq \Re^m$ characterise *mixed uncertainties* that may include both unmodelled dynamics and deterministic or stochastic noises.

The class Σ of NLS is assumed to be *consistent*, that is, for any fixed pair $(c, z_0) \in \mathcal{C} \times \Re^n$ and for any input sequence $\{u_t\}$ there exists a strong solution $\{z(t, z_0, u_t, c)\}$ of the corresponding Cauchy problem (3.1).

Definition 1. *A control strategy $\{u_t\}$ is said to be admissible if it is smooth enough and provides the consistency condition for Σ.*

3.3.2 The extended system and problem formulation

For $N := n + q$, define the extended state vector

$$x_t := \begin{bmatrix} z_t \\ c \end{bmatrix} \in \Re^N \tag{3.2}$$

and rewrite Σ in *the extended form* as the uncertain system

$$\begin{cases} \dot{x}_t = F(t, x_t, u_t) + \xi_x(t, x_t, u_t), & x_{t=0} = x_0 \\ y_t = H(t, x_t, u_t) + \xi_y(t, x_t, u_t) \end{cases} \tag{3.3}$$

where

$$F(\cdot) := \begin{bmatrix} f(\cdot) \\ 0 \end{bmatrix}, \quad \xi_x(\cdot) := \begin{bmatrix} \zeta_x(\cdot) \\ 0 \end{bmatrix}, \quad x_0 := \begin{bmatrix} z_0 \\ c \end{bmatrix} \tag{3.4}$$

$$\begin{cases} f(t, x_t, u_t) := f(t, z_t, u_t, c)|_x, & H(t, x_t, u_t) := h(t, z_t, u_t, c)|_x \\ \xi_x(t, x_t, u_t) := \zeta_1(t, z_t, u_t, c)|_x, & \xi_y(t, x_t, u_t) := \zeta_2(t, z_t, u_t, c)|_x \end{cases} \tag{3.5}$$

Definition 2. *For the consistent class Σ of nonlinear uncertain systems, given by (3.3), define the function $\hat{x}_t$ ($t \in [0, \infty)$), named below the 'ε-state estimate',* *which satisfies the following conditions:*

1. *$\hat{x}_t$ has the same dimension as x_t and may only be a function of time t as well as the past input-output information u_t and y_t, available up to this time, that is,*

$$\hat{x}_t \in \Re^N$$

and

$$\hat{x}_t = \hat{x}_t(u_\tau, y_\tau \mid \tau \in [0, t])$$

2. *The distance $\|\hat{x}_t - x_t\|^2$ remains bounded 'on average' over all possible trajectories x_t uniformly with respect to the initial conditions $x_0 \in \mathcal{C} \times \Re^n$ and the given input $\zeta_1(\cdot) \in D_1 \subseteq \Re^n$ and output $\zeta_2(\cdot) \in D_2 \subseteq \Re^m$ uncertainties, that is,*

$$\sup_{\substack{x_0 \in \mathcal{C} \times \Re^n \\ \zeta_1 \in D_1, \zeta_2 \in D_2}} \limsup_{T \to \infty} \frac{1}{T} \int_{t=0}^{T} \|\hat{x}_t - x_t\|^2 \, dt \leq \varepsilon < \infty \tag{3.6}$$

If for two different ε-state estimates $\hat{x}_t'$ and $\hat{x}_t''$ the corresponding values ε' and ε'' of the tolerance levels (3.6) are arranged in such a way that $\varepsilon' < \varepsilon''$, we say that *the estimate $\hat{x}_t'$ is better than $\hat{x}_t''$.*

Remark 4.   *Note that if an estimate $\hat{x}_t$ is ε-uniformly bounded, that is,*

$$\sup_{\substack{x_0 \in \mathcal{C} \times \Re^n \\ \zeta_1 \in D_1, \zeta_2 \in D_2}} \limsup_{t \to \infty} \|\hat{x}_t - x_t\|^2 \leq \varepsilon < \infty \tag{3.7}$$

then, it is ε-uniformly bounded 'on average', satisfying (3.6).

Now we are ready to formulate the problem of '*Simultaneous State and Parameter Estimation*'.

The Problem.   *For the consistent class Σ of nonlinear uncertain systems, given by (3.3), construct an ε-state estimate $\hat{x}_t$ $(t \in [0, \infty))$ such that the corresponding tolerance level ε (3.6) is minimised.*

3.4   The nominal (nondisturbed) system and observability property

3.4.1   *Nondisturbed system*

Based on (3.3), define the nominal (nondisturbed) extended system related to the consistent class Σ as follows:

$$\begin{cases} \dot{\bar{x}}_t = F(t, \bar{x}_t, u_t) \\ \bar{y}_t = H(t, \bar{x}_t, u_t) \end{cases} \tag{3.8}$$

Here $\bar{x}_t$ and $\bar{y}_t$ have the dimensions $\Re^N$ and $\Re^P$ which are the same as for x_t and y_t, respectively. This system will play a key role in highlighting the observability notion and its relation with the state-estimation problem.

Let us consider the most interesting situation when $p < N$, that is, when the number of measurable outputs is less than the number of extended states. In view of this, the vector $\bar{x}_t$ cannot be found from the vectors $\bar{y}_t$ and u_t since in this case for any u_t and t

$$\det \frac{\partial}{\partial \bar{x}_t} H(t, \bar{x}_t, u_t) = 0$$

and the basic theorem from real analysis '*On the existence of the inverse function*' cannot be applied and hence $\bar{x}_t$ cannot be found from $\bar{y}_t$ and u_t. Thus, more measurable outputs are needed to realise this invertion operation.

3.4.2   *Output differentiation as a generator of new outputs*

The natural way to obtain this information is given below. It is based on the concepts introduced in References 13, 15, 23 and 26. They will be fundamental throughout this section.

Suppose below that the input vector u_t is at least k-times differentiable where the integer k satisfies the inequality

$$km \geq N - m, k \geq \max \left\{ \text{int} \left(\frac{N}{m} - 1 \right); 0 \right\}$$

Let us then calculate the corresponding derivatives of the output signal $\bar{y}_t$ along the trajectories of the nominal system (3.8):

$$\dot{\bar{y}}_t = \frac{\partial}{\partial t} H(t, \bar{x}_t, u_t) + \sum_{i=1}^{N} \frac{\partial}{\partial \bar{x}_{i,t}} H(t, \bar{x}_t, u_t) F_i(t, \bar{x}_t, u_t)$$

$$+ \sum_{j=1}^{m} \frac{\partial}{\partial u_{j,t}} H(t, \bar{x}_t, u_t) \dot{u}_{j,t} := L_F H(t, \bar{x}_t, u_t) \tag{3.9}$$

and, thus, by induction we obtain

$$\frac{d^s}{dt^s} \bar{y}_t = (L_F)^s H(t, \bar{x}_t, u_t), \qquad s = 1, \ldots, k \tag{3.10}$$

Remark 5.   *Here, the operator L_F on the right-hand side of (3.9) is called the 'Lie derivative operator' applied to the matrix function $H(t, \bar{x}_t, u_t)$ in the direction of the vector-field $F(t, \bar{x}_t, u_t)$.*

Combining all the relations obtained for $s = 0, \ldots, k$ leads to the following system of differential equations

$$Y_t := \begin{pmatrix} \bar{y}_t \\ \dot{\bar{y}}_t \\ \vdots \\ \dfrac{d^k}{dt^k} \bar{y}_t \end{pmatrix} = \begin{pmatrix} \bar{y}_t \\ H(t, \bar{x}_t, u_t) \\ \vdots \\ (L_F)^k H(t, \bar{x}_t, u_t) \end{pmatrix} := \Phi(t, \bar{x}_t, \mathcal{U}_t^k) \tag{3.11}$$

where the '*extended input vector*' $\mathcal{U}_t^k$ is defined as

$$\mathcal{U}_t^k := \begin{bmatrix} u_{1,t} & \dot{u}_{1,t} & \cdots & u_{1,t}^{(k)} & \cdots & u_{m,t} & \dot{u}_{m,t} & \cdots & u_{m,t}^{(k)} \end{bmatrix}^T \tag{3.12}$$

We will refer to the vector Y_t as '*the extended output vector*'. Only one strict condition on the dimension of Y_t should be fulfilled: the size of the additional vector components should be exactly equal to $(N - p)$, i.e.,

$$\left(\dot{\bar{y}}_t^{\mathsf{T}} \quad \cdots \quad \frac{d^k}{dt^k} \bar{y}_t^{\mathsf{T}} \right)^{\mathsf{T}} \in \Re^{N-p}$$

The '*extended output vector*' $Y_t \in \Re^N$ can be associated with the information available (measurable) at time t.

3.4.3   Observability matrix

Definition 3.   *The consistent class Σ of nonlinear systems (3.3) is said to be completely uniformly (with respect to the inputs $\mathcal{U}_t^k$) locally observable in a neighbourhood of the point $\bar{x}_t$ at time t, if the vector-field (3.11) defining the corresponding nominal system (3.8)*

$$\Phi(t, \bar{x}_t, \mathcal{U}_t^k) : \Re^+ \times \Re^N \times \Re^{km} \to \Re^N$$

is a diffeomorphism (one-to-one relation) between a neighbourhood of the point $\bar{x}_t$ and $\Re^N$.

Remark 6.   *This means that the point $\bar{x}_t$, or in another words, the extended state vector $\bar{x}_t \in \Re^N$ at the given time t can be uniquely defined based on the available information on the extended output vector $Y_t \in \Re^N$ for any possible input vector $\mathcal{U}_t^k \in \Re^{km}$.*

It is well known from the theorem on the existence of the inverse function, that $\Phi(t, \bar{x}_t, \mathcal{U}_t^k)$ is a diffeomorphism at a fixed $\bar{x}_t \in \Re^N$ and $t \geq 0$ if and only if the so-called '*observability matrix*' defined as

$$Q(t, \bar{x}_t, \mathcal{U}_t^k) := \frac{\partial}{\partial \bar{x}_t} \Phi(t, \bar{x}_t, \mathcal{U}_t^k) \tag{3.13}$$

is *nonsingular* for any $\mathcal{U}_t^k \in \Re^{km}$, that is, when

$$\det Q(t, \bar{x}_t, \mathcal{U}_t^k) \neq 0 \tag{3.14}$$

Definition 4.   *We say that the consistent class Σ of nonlinear systems (3.3) is completely uniformly (with respect to the inputs $\mathcal{U}_t^k$) globally observable in $\Re^N$ if it is completely uniformly locally observable in each point $\bar{x}_t \in \Re^N$ at any time $t \geq 0$.*

The main test to check the complete uniform observability property for the given class Σ of extended nonlinear systems (3.3) consists of verifying the inequality (3.14) everywhere in $\Re \times \Re^N \times \Re^{km}$. This may also serve as a tool to determine the set of *'singular (bad) times and inputs'* which make the extended system unobservable.

3.5 Examples of observability analysis

Consider several examples (from [29]) illustrating the notions given before.

3.5.1 Simple pendulum

Consider the simple pendulum (with non atomic mass, friction and without input) given in Fig. 3.1. Suppose g and m are known. The aim is to estimate $\{\theta_t, \dot{\theta}_t\}$ and l. The dynamical model is as follows:

$$\ddot{\theta}_t + \frac{mgl}{ml^2 + I} \sin(\theta_t) + \frac{kl^2}{ml^2 + I} \dot{\theta}_t = n_{1,t} \qquad \theta_{t=0} = \theta_0, \quad \dot{\theta}_{t=0} = \dot{\theta}_0$$

where $n_{1,t}$ is an unknown external noise. The previous expression can be rewritten in another form as

$$\ddot{\theta}_t + \frac{g}{l} \sin(\theta_t) - \zeta_1(t, \theta_t, \dot{\theta}_t, l) = 0 \qquad \theta_{t=0} = \theta_0, \quad \dot{\theta}_{t=0} = \dot{\theta}_0$$

where

$$\zeta_1(t, \theta_t, \dot{\theta}_t, l) := n_{1,t} - \frac{1}{ml^2 + I} \left[\frac{gI \sin(\theta_t)}{l} + kl^2 \dot{\theta}_t \right]$$

describes the unmodelled dynamics. Furthermore, suppose that the available measurements are only the angular position θ_t disturbed by the output observation noise, that is,

$$y_t = \theta_t + \zeta_{2,t}$$

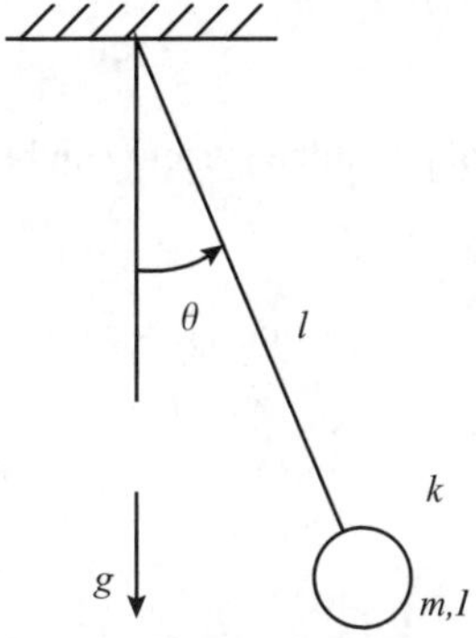

Figure 3.1   Simple pendulum

where $\zeta_{2,t}$ is the noise in the observation. Define the state vector

$$z_t = [z_{1,t} \quad z_{2,t}]^T = [\theta_t \quad \dot{\theta}_t]^T$$

Then it follows

$$\begin{cases} \dot{z}_t = \begin{bmatrix} z_{2,t} \\ a\sin(z_{1,t}) \end{bmatrix} + \begin{bmatrix} 0 \\ \zeta_1 \end{bmatrix}, & z_{t=0} = z_0 \\ y_t = z_{1,t} + \zeta_{2,t} \end{cases} \tag{3.15}$$

where $a := -gl^{-1}$ is a system parameter. Thus, the nominal (unperturbed) nonextended system is given by

$$\begin{cases} \dot{\bar{z}}_t = \begin{bmatrix} \bar{z}_{2,t} \\ a\sin(\bar{z}_{1,t}) \end{bmatrix}, & \bar{z}_{t=0} = \bar{z}_0 \\ \bar{y}_t = \bar{z}_{1,t} \end{cases}$$

Evidently, when the parameter a is known, this system is completely uniformly observable and the corresponding observability matrix is the identity matrix of order 2. Indeed,

$$Y_t := \begin{pmatrix} \bar{y}_t \\ \dot{\bar{y}}_t \end{pmatrix} = \begin{pmatrix} \bar{z}_{1,t} \\ \bar{z}_{2,t} \end{pmatrix} := \Phi(\bar{z}_t), \quad Q(\bar{z}_t) := \frac{\partial}{\partial \bar{z}_t}\Phi(\bar{z}_t) = I$$

If we need to estimate the parameter a (indeed, for estimation of l a similar analysis can be performed) define the extended state vector

$$x := [z_1 \quad z_2 \quad a]^T$$

Then (3.15) takes the form

$$\begin{cases} \dot{x}_t = \begin{bmatrix} x_{2,t} \\ x_3\sin(w_{1,t}) \\ 0 \end{bmatrix} + \begin{bmatrix} 0 \\ \zeta_1 \\ 0 \end{bmatrix}, & x_{t=0} = x_0 := \begin{bmatrix} z_0 \\ a \end{bmatrix} \\ y_t = x_{1,t} + \zeta_{2,t} \end{cases}$$

and the corresponding nominal extended system can be expressed as

$$\dot{\bar{x}}_t = \begin{bmatrix} \bar{x}_{2,t} \\ \bar{x}_3\sin(\bar{w}_{1,t}) \\ 0 \end{bmatrix}, \quad \bar{x}_{t=0} = \bar{x}_0, \quad \bar{y}_t = \bar{x}_{1,t}$$

Then

$$Y_t := \begin{pmatrix} \bar{y}_t \\ \dot{\bar{y}}_t \\ \ddot{\bar{y}}_t \end{pmatrix} = \begin{pmatrix} \bar{x}_{1,t} \\ \bar{x}_{2,t} \\ \bar{x}_3\sin(\bar{x}_{1,t}) \end{pmatrix} := \Phi(\bar{x}_t)$$

and the observability matrix and its determinant are given by

$$Q(\bar{x}_t) := \frac{\partial}{\partial \bar{x}_t} \Phi(\bar{x}_t) = \begin{bmatrix} 1 & 0 & 0 \\ 0 & 1 & 0 \\ \bar{x}_3 \cos(\bar{x}_1) & 0 & \sin(\bar{x}_1) \end{bmatrix}, \quad \det Q(\bar{x}_t) = \sin(\bar{x}_1)$$

The nonobservable (singular) manifold is given by $\bar{x}_1 = n\pi$, n is an integer. Thus, at any point $\theta_t = n\pi$ the system loses observability in the extended space which also includes the unknown parameter to be estimated.

3.5.2   Duffing equation

Consider the mechanical spring-mass system depicted in Fig. 3.2.

The spring is considered to be of the hard type, where the restoration force is given by $F_s = k(1 + b^2 s_t^2)s_t$, where s_t is the horizontal coordinate of the centre of mass. The friction force is assumed to satisfy $F_f = c\dot{s}_t$. The mass M is assumed to be known and the output is given by $y_t = s_t + \zeta_{2,t}$ with $\zeta_{2,t}$ as an unknown output noise. The objective is to estimate $\{s_t, \dot{s}_t\}$ and k.

The corresponding dynamical model including the unknown noise $n_{1,t}$ is as follows:

$$\ddot{s}_t + \frac{k}{M} s_t - \zeta_1(t, s_t, \dot{s}_t, k) = \frac{u_t}{M}, \qquad s_{t=0} = s_0, \quad \dot{s}_{t=0} = \dot{s}_0$$

where the unmodelled dynamics ζ_1 is given by

$$\zeta_1(t, s_t, \dot{s}_t, k) := \frac{1}{M}[kb^2 s_t^3 + c\dot{s}_t] - n_{1,t}$$

Define the state vector as $z_t = [z_{1,t} \quad z_{2,t}]^T = [s_t \quad \dot{s}_t]^T$. Then the representation of the system in terms of z-variables is as follows:

$$\begin{cases} \dot{z}_t = \begin{bmatrix} z_{2,t} \\ az_{1,t} + \dfrac{u_t}{M} \end{bmatrix} + \begin{bmatrix} 0 \\ \zeta_1(t, z_t, a) \end{bmatrix}, & z_{t=0} = z_0 \\ y_t = z_{1,t} + \zeta_{2,t} \end{cases} \tag{3.16}$$

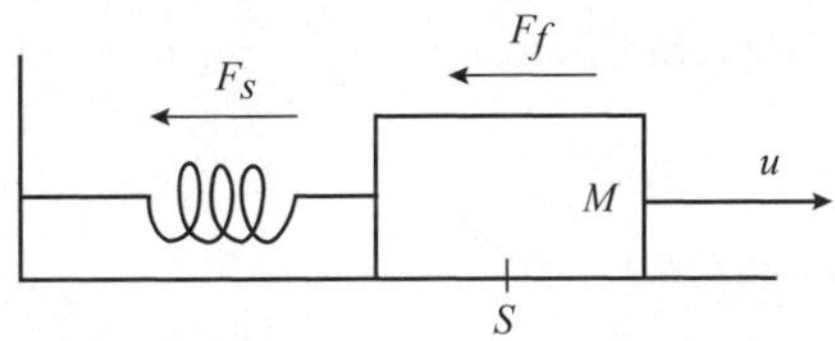

Figure 3.2   Spring-mass system

with $a := -kM^{-1}$ and $\zeta_{2,t}$ as an output noise term. Then, the nominal nonextended system is given by

$$\begin{cases} \dot{\bar{z}}_t = \begin{bmatrix} \bar{z}_{2,t} \\ a\bar{z}_{1,t} + \dfrac{u_t}{M} \end{bmatrix}, & \bar{z}_{t=0} = z_0 \\ \bar{y}_t = \bar{z}_{1,t} \end{cases} \tag{3.17}$$

for which the observability matrix $Q(\bar{z}_t)$ is equal to the identity matrix of order 2. Indeed,

$$Y_t := \begin{pmatrix} \bar{y}_t \\ \dot{\bar{y}}_t \end{pmatrix} = \begin{pmatrix} \bar{z}_{1,t} \\ \bar{z}_{2,t} \end{pmatrix} := \Phi(\bar{z}_t), \quad Q(\bar{z}_t) := \frac{\partial}{\partial \bar{z}_t} \Phi(\bar{z}_t) = I$$

Hence, the system (3.17) is completely uniformly observable.

Considering the parameter a as the additional state (assuming that it is a priori unknown), the extended state vector x can be expressed as $x := [z_1 \quad z_2 \quad a]^T$. Then, (3.16) can be rewritten as

$$\dot{x}_t = \begin{bmatrix} x_{2,t} \\ x_3 x_{1,t} + \dfrac{u_t}{M} \\ 0 \end{bmatrix} + \begin{bmatrix} 0 \\ \zeta_1(t, w_t) \\ 0 \end{bmatrix}, \quad x_{t=0} = x_0 := \begin{bmatrix} z_0 \\ a \end{bmatrix}$$

$$y_t = x_{1,t} + \zeta_{2,t}$$

where the nominal extended system is given by

$$\dot{\bar{x}}_t = \begin{bmatrix} \bar{x}_{2,t} \\ \bar{x}_3 \bar{x}_{1,t} + \dfrac{u_t}{M} \\ 0 \end{bmatrix}, \quad \bar{y}_t = \bar{x}_{1,t}, \quad \bar{x}_{t=0} = \bar{x}_0$$

It follows that

$$Y_t := \begin{pmatrix} \bar{y}_t \\ \dot{\bar{y}}_t \\ \ddot{\bar{y}}_t \end{pmatrix} = \begin{pmatrix} \bar{x}_{1,t} \\ \bar{x}_{2,t} \\ \bar{x}_3 \bar{x}_{1,t} + \dfrac{u_t}{M} \end{pmatrix} := \Phi(\bar{x}_t)$$

Hence, the observability matrix $Q(\bar{x}_t)$ is equal to

$$Q(\bar{x}_t) := \frac{\partial}{\partial \bar{x}_t} \Phi(\bar{x}_t) = \begin{bmatrix} 1 & 0 & 0 \\ 0 & 1 & 0 \\ \bar{x}_3 & 0 & \bar{x}_{1,t} \end{bmatrix}$$

The given system is observable almost everywhere. It is not observable at $\bar{x}_{1,t} = 0$ since $\det Q(\bar{x}_t) = \bar{x}_{1,t}$.

3.5.3   Van der Pol oscillator

The dynamic model of this system is given by

$$\ddot{s}_t - a[(1 - s_t^2)\dot{s}_t - s_t] - \zeta_1(t, s_t, \dot{s}_t) = u_t, \quad s_{t=0} = s_0, \quad \dot{s}_{t=0} = \dot{s}_0$$

where $\zeta_1(t, s_t, \dot{s}_t, u_t)$ represents the presence of noise and unmodelled dynamics. It is assumed that the observable variable is s_t, but that this is contaminated by noise $\zeta_{2,t}$, dependent on time and on the current state; that is, $y_t = s_{1,t} + \zeta_{2,t}$. Define the state vector $z_t = [z_{1,t} \quad z_{2,t}]^T = [s_t \quad \dot{s}_t]^T$. Then, in terms of the variable z, the system has the following representation:

$$\begin{cases} \dot{z}_t = \begin{bmatrix} z_{2,t} \\ a[(1 - z_{1,t}^2)z_{2,t} - z_{1,t}] + u_t \end{bmatrix} + \begin{bmatrix} 0 \\ \zeta_1(t, z_t) \end{bmatrix}, \quad z_{t=0} = z_0 \\ y_t = z_{1,t} + \zeta_2(t, z_t) \end{cases} \tag{3.18}$$

where the constant a will be considered as the significant system parameter. The corresponding nominal nonextended system for (3.18) is

$$\begin{cases} \dot{\bar{z}}_t = \begin{bmatrix} \bar{z}_{2,t} \\ a[(1 - \bar{z}_{1,t}^2)\bar{z}_{2,t} - \bar{z}_{1,t}] + u_t \end{bmatrix}, \quad \bar{z}_{t=0} = z_0, \quad \bar{y}_t = \bar{z}_{1,t} \end{cases}$$

Analogously to the previous examples, this system turns out to be completely uniformly observable and its observability matrix is equal to the identity matrix of order 2. Considering a as an unknown parameter, define the extended state vector $x := [z_1 \quad z_2 \quad a]^T$. Then, (3.18) can be rewritten as

$$\begin{cases} \dot{x}_t = \begin{bmatrix} x_{2,t} \\ x_{3,t}[(1 - x_{1,t}^2)x_{2,t} - x_{1,t}] + u_t \\ 0 \end{bmatrix} + \begin{bmatrix} 0 \\ \zeta_1(t, w_t) \\ 0 \end{bmatrix} \\ w_{t=0} = w_0 := [z_0^T \quad a]^T, \quad y_t = x_{1,t} + \zeta_2(t, w_t) \end{cases}$$

The corresponding nominal extended system is given by

$$\dot{\bar{x}}_t = \begin{bmatrix} \bar{x}_{2,t} \\ \bar{x}_{3,t}[(1 - \bar{x}_{1,t}^2)\bar{x}_{2,t} - \bar{x}_{1,t}] + u_t \\ 0 \end{bmatrix}, \quad \bar{y}_t = \bar{x}_{1,t}, \quad \bar{x}_{t=0} = \bar{x}_0$$

The observability matrix $Q(\bar{x}_t)$ and its determinant can be shown to be equal to

$$Q(\bar{x}_t) := \begin{bmatrix} 1 & 0 & 0 \\ 0 & 1 & 0 \\ -2\bar{x}_{1,t}\bar{x}_{2,t}\bar{x}_{3,t} & \bar{x}_{3,t}(1 - \bar{x}_{1,t}^2) & (1 - \bar{x}_{1,t}^2)\bar{x}_{2,t} - \bar{x}_{1,t} \end{bmatrix}$$

$$\det Q(\bar{x}_t) = (1 - \bar{x}_{1,t}^2)\bar{x}_{2,t} - \bar{x}_{1,t}$$

Thus, the nominal extended system is not observable over the manifold

$$(1 - \bar{x}_{1,t}^2)\bar{x}_{2,t} - \bar{x}_{1,t} = 0$$

Summary 1.   *The main conclusions of the sections above may be formulated as follows: if the given class of nonlinear systems is completely observable within some set of the state variables or in the whole space, then it is desirable to discuss questions relating to observer design.*

Relevant basic issues to consider are:

- possible observer structures;
- the admissible choice of design parameters selection within a given structure;
- the selection of optimal (or, sub-optimal) parameters.

The next section will tackle the problem of choosing an appropriate structure for the design of state-space observers for systems in the presence of external noise disturbances.

3.6   Observer structure

3.6.1   *Asymptotic nonlinear observers*

Let us suppose below that the given consistent class Σ of nonlinear systems (3.3), containing noise uncertainty, is completely uniformly (with respect to the inputs U_t^k) globally observable in $\Re^N$.

Definition 5.   *We say that the 'ε-state estimate' $\hat{x}_t \in \Re^N$ is generated by the global asymptotic nonlinear observer G if it satisfies the following conditions:*

1. *(ODE property): the function $\hat{x}_t$ is the solution of the following ordinary differential equation*

$$\frac{d}{dt}\hat{x}_t = G(t, \hat{x}_t, u_t, y_t), \qquad \hat{x}_0 \text{ is fixed} \tag{3.19}$$

2. *(The exact mapping property): the dynamics of the given system (3.3) and (3.19) coincide for all $t \geq 0$, that is,*

$$x_t = \hat{x}_t, \quad \dot{x}_t = \dot{\hat{x}}_t \tag{3.20}$$

   if the initial states of the original model (3.3) and the estimating model (3.19) coincide ($x_0 = \hat{x}_0$) and there are no disturbances at all, that is,

$$\xi_x(t, x_t, u_t) = \xi_y(t, x_t, u_t) = 0 \qquad \forall t \geq 0$$

3. *(The asymptotic consistency property): if the initial states of the original model and the estimating model do not coincide, that is,*

$$x_0 \neq \hat{x}_0$$

   but still there are no disturbances, then the estimates $\hat{x}_t$ should satisfy

$$\|x_t - \hat{x}_t\| \underset{t \to \infty}{\to} 0 \tag{3.21}$$

The '*ODE*' and '*exact mapping*' properties imply that for any $x_t \in \Re^N$, $u_t \in \Re^m$ and any $t \in \Re^+$ the following identity holds:

$$\frac{d}{dt}\hat{x}_t = G(t, \hat{x}_t, u_t, y_t) = G(t, \hat{x}_t, u_t, H(x_t, u_t)) = F(t, x_t, u_t) = \frac{d}{dt}x_t \tag{3.22}$$

The '*asymptotic consistency*' property ensures that for the case when $\hat{x}_t \neq x_t$, but their difference is small ($\|\hat{x}_t - x_t\|$ is small for large enough t), it follows that

$$G(t, \hat{x}_t, u_t, y_t) = G(t, \hat{x}_t, u_t, H(\hat{x}_t, u_t) + [y_t - H(\hat{x}_t, u_t)])$$

$$\cong G(t, \hat{x}_t, u_t, H(\hat{x}_t, u_t)) + \frac{\partial}{\partial \hat{x}_{1,t}} G(t, \hat{x}_t, u_t, H(\hat{x}_t, u_t))$$

$$\times [y_t - H(\hat{x}_t, u_t)] + O(\|H(x_t, u_t) - H(\hat{x}_t, u_t)\|^2) \tag{3.23}$$

By (3.22) we have

$$G(t, \hat{x}_t, u_t, H(\hat{x}_t, u_t)) = F(t, \hat{x}_t, u_t) \tag{3.24}$$

For the function $H(\hat{x}_t, u_t)$, satisfying the *Lipschitz condition* with respect to the first argument, uniformly on u_t, it follows that

$$O(\|H(x_t, u_t) - H(\hat{x}_t, u_t)\|^2) \leq O(L_H \|x_t - \hat{x}_t\|^2)$$

$$= O(\|x_t - \hat{x}_t\|^2) \qquad \text{(small enough)} \tag{3.25}$$

In view of (3.19), (3.22), (3.23), (3.24) and (3.25) we obtain

$$\frac{d}{dt}\hat{x}_t = F(t, \hat{x}_t, u_t) + \frac{\partial}{\partial \hat{x}_t} F(t, \hat{x}_t, u_t)[y_t - H(\hat{x}_t, u_t)] \tag{3.26}$$

The last ODE defines the so-called '*local structure*' of the state-space observer. To fulfil the '*asymptotic consistency*' property for all possible initial states $\hat{x}_0, x_0 \in \mathfrak{R}^N$, we generalise the local structure (3.26) changing the gain-matrix $(\partial / \partial \hat{x}_t) F(t, \hat{x}_t, u_t)$ to a general one, namely, to $K(t, \hat{x}_t, u_t) \in \mathfrak{R}^{N \times P}$, which leads to the following '*global nonlinear observer structure*':

$$\frac{d}{dt}\hat{x}_t = F(t, \hat{x}_t, u_t) + K(t, \hat{x}_t, u_t)[y_t - H(\hat{x}_t, u_t)] \tag{3.27}$$

The gain-matrix $K(t, \hat{x}_t, u_t)$ should be selected to provide the property (3.6) with a lower possible tolerance level.

This structure from (3.27) is called a '*Luenberger-structure observer*'. The class of this type of observers, corresponding to the class of linear systems, is called the '*Kalman-type structure*' and the observer itself is named a '*Kalman Filter*'.

3.6.2   Output noise

There exist two situations involving output-noise properties.

1. The output-noise $\xi_{2,t}$ is assumed to be *bounded*, but, may not be differentiable, that is,

$$\limsup_{t \to \infty} \|\xi_{2,t}\| \leq \varepsilon_2 < \infty \tag{3.28}$$

2. The output-noise $\xi_{2,t}$ is assumed to be the *output of a stable filter* whose input is supplied by a bounded disturbance $w_{2,t}$, that is,

$$\dot{\xi}_{2,t} = A_f \xi_{2,t} + w_{2,t} \tag{3.29}$$

where $A_f \in R^{p \times p}$ is a constant Hurwitz (stable) matrix and $w_{2,t} \in R^p$ is an input bounded vector-disturbance such that

$$\limsup_{t \to \infty} \|w_{2,t}\| \leq \bar{w}_2 < \infty \tag{3.30}$$

In this case the effective output-noise $\xi_{2,t}$ turns out to be also bounded but, in addition, is differentiable. To realise the corresponding analysis assume two distinct cases:

- the matrix parameter A_f of the (forming) filter is a priori known;
- the parameter A_f is assumed to be a priori unknown.

3.7   Standard high-gain observer

3.7.1   A specific class of dynamic models

The nonlinear system has state vector

$$x_t := (x_{1,t}, x_{2,t})^\mathsf{T} \in R^{2n}, \qquad x_{1,t}, x_{2,t} \in R^n \tag{3.31}$$

satisfying the following nonlinear dynamics

$$\begin{cases} \dot{x}_{1,t} = x_{2,t} \\ \dot{x}_{2,t} = f_2(x_{1,t}, x_{2,t}, t) + \zeta_{1,t} \\ y_t = x_{1,t} + \xi_{2,t}, \qquad x_{1,0}, x_{2,0} \text{ are given} \end{cases} \tag{3.32}$$

Here, $\xi_{1,t}, \xi_{2,t} \in R^n$ are, as before, the noise terms disturbing the dynamics of the system itself ($\xi_{2,t}$) and the output signal ($\xi_{1,t}$); $f_2 : R^n \times R^n \times R^1 \to R^n$ is a given mapping (function) providing the existence and uniqueness of the solution of the system of ODE given by (3.32).

Below, we discuss the problem of the estimation of the $x_{2,t}$-states using observers of the Luenberger structure (3.27) with a specific selection of the gain-matrix $K(t, \hat{x}_t, u_t)$.

3.7.2   Mechanical example

Nonlinear models given in the form (3.32) are often encountered in practice especially as models of mechanical systems where the position vector is available at each time and the corresponding velocities are not measurable. The following example illustrates this fact.

Example 1.   *The dynamic model of the simplest mechanical system, dealing with the movement of a solid bar over a surface in the presence of friction, is as follows*

$$m\ddot{x}_t + k\dot{x}_t = F_t + \zeta_{1,t} \tag{3.33}$$

where $m > 0$ is the mass of a bar, x_t is the horizontal position of a rigid bar, k is a friction coefficient, F is a given external force (control) and $\zeta_{1,t}$ is an uncontrollable *input (dynamic noise). The output y_t is assumed to be given as*

$$y_t = x_t + \xi_{2,t} \tag{3.34}$$

where $\xi_{2,t}$ is an observation (output) noise. Introduce two new variables:

$$x_{1,t} := x_t, \qquad x_{2,t} := \dot{x}_t \tag{3.35}$$

This directly leads to the presentation of (3.33) in the form (3.32), that is,

$$\begin{cases} \begin{pmatrix} \dot{x}_{1,t} \\ \dot{x}_{2,t} \end{pmatrix} = \begin{pmatrix} f_1(x_{1,t}, x_{2,t}) \\ f_2(x_{1,t}, x_{2,t}) \end{pmatrix} + \begin{pmatrix} \xi_{1,t}^{(1)} \\ \xi_{1,t}^{(2)} \end{pmatrix} \\ \qquad := \begin{pmatrix} x_{2,t} \\ -km^{-1}x_{2,t} + m^{-1}F_t + m^{-1}\zeta_{1,t} \end{pmatrix} + \begin{pmatrix} 0 \\ m^{-1}\zeta_{1,t} \end{pmatrix} \\ y_t = x_{1,t} + \xi_{2,t} \end{cases} \tag{3.36}$$

3.7.3 High-gain observer structure

Definition 6. *A global asymptotic nonlinear observer with the Luenberger structure (3.27), when the gain-matrix is selected as the constant matrix,*

$$K(t, \hat{x}_t, u_t) := \begin{bmatrix} K \in R^{n \times n} \\ KL \in R^{n \times n} \end{bmatrix}$$

is named a high-gain observer (HGO). It has the following structure

$$\begin{cases} \dfrac{d}{dt}\hat{x}_{1,t} = \hat{x}_{2,t} + K(y_t - \hat{x}_{1,t}) \\ \dfrac{d}{dt}\hat{x}_{2,t} = f_2(\hat{x}_{1,t}, \hat{x}_{2,t}) + KL(y_t - \hat{x}_{1,t}) \\ \hat{y}_t = \hat{x}_{1,t} \end{cases} \tag{3.37}$$

The problem now is to select the constant matrices K and L to guarantee the 'asymptotic consistency property' (3.21) if no noise is present in the system and to have a finite upper bound for the corresponding average quadratic error if there is noise in the given dynamics.

3.7.4 Upper bound for estimation error and asymptotic consistency property

3.7.4.1 Main theorem

Define the state-estimation error Δ_t as

$$\Delta_t := \hat{x}_t - x_t \tag{3.38}$$

The next theorem states the conditions that the HGO (3.37) must satisfy to fulfil the asymptotic consistency property (3.21).

Theorem 1.   *If*

1. *the nonlinear system (3.36) satisfies the global Lipschitz condition, that is,*

$$\| f_2(x_1 + \Delta_1, x_2 + \Delta_2) - f_2(x_1, x_2) \| \le L_f \left\| \begin{array}{c} \Delta_1 \\ \Delta_2 \end{array} \right\|$$

   for any $x_1, x_2, \Delta_1, \Delta_2 \in R^n$;
2. *the noise disturbances are bounded, that is,*

$$\| \xi_{1,t} \|^2 \le \varepsilon_1, \quad \| \xi_{2,t} \|^2 \le \varepsilon_2 \qquad \forall t \ge 0$$

3. *the gain matrices K and KL provide the existence of a positive solution*

$$0 < P = P^{\mathsf{T}} \in R^{2n \times 2n}$$

   to the following matrix Riccati equation

$$P\mathcal{K} + \mathcal{K}^{\mathsf{T}} P + PRP + Q = 0$$

$$\mathcal{K} := \begin{bmatrix} -K & I \\ -KL & 0 \end{bmatrix} \text{ is a stable (Hurwitz) matrix}$$

$$R := \Lambda, \qquad Q := \| \Lambda^{-1} \| L_f^2 I + Q_0$$

   for some positive definite matrices Q_0 and Λ, that is,

$$0 < Q_0 = Q_0^{\mathsf{T}} \in R^{2n \times 2n}, \qquad 0 < \Lambda = \Lambda^{\mathsf{T}} \in R^{2n \times 2n}$$

   then the HGO (3.37) provides error convergence to the μ-zone:

$$\left[\sqrt{\Delta_t^{\mathsf{T}} P \Delta_t} - \mu \right]_+ \underset{t \to \infty}{\longrightarrow} 0 \tag{3.39}$$

where

$$\mu = \sqrt{\frac{\beta}{\alpha}}, \qquad \alpha := \lambda_{\min}(P^{-1/2} Q_0 P^{-1/2}) \tag{3.40}$$

$$\beta := 3[(\| K^{\mathsf{T}} \Lambda_1^{-1} K \| + \| L^{\mathsf{T}} K^{\mathsf{T}} \Lambda_2^{-1} KL \|)\varepsilon_2^2 + \| \Lambda_2^{-1} \| \varepsilon_1^2]$$

and the function $[\cdot]_+$ is defined as

$$[z]_+ := \begin{cases} z & \text{if } z \ge 0 \\ 0 & \text{if } z < 0 \end{cases}$$

Proof. Define the Lyapunov function $L(\Delta_t)$ as $V(\Delta_t) = \|\Delta_t\|_P^2 := \Delta_t^{\mathsf{T}} P \Delta_t$. Then, by (3.32) and (3.37), it follows that

$$
\dot{\Delta}_t = \begin{pmatrix} \hat{x}_{2,t} - x_{2,t} + K(y_t - \hat{x}_{1,t}) \\ f_2(\hat{x}_{1,t}, \hat{x}_{2,t}) - f_2(x_{1,t}, x_{2,t}) + KL(y_t - \hat{x}_{1,t}) - \xi_{1,t}^{(2)} \end{pmatrix}
$$

$$
= \begin{bmatrix} -K & I \\ -KL & 0 \end{bmatrix} \Delta_t
$$

$$
+ \begin{pmatrix} K\xi_{2,t} \\ f_2(x_{1,t} + \Delta_{1,t}, x_{2,t} + \Delta_{2,t}) - f_2(x_{1,t}, x_{2,t}) - \xi_{1,t}^{(2)} + KL\xi_{2,t} \end{pmatrix}
$$

and hence,

$$
\frac{d}{dt} V(\Delta_t) = 2\Delta_t^{\mathsf{T}} P \dot{\Delta}_t = 2\Delta_t^{\mathsf{T}} P
$$

$$
\times \left(\mathcal{K}\Delta_t + \begin{bmatrix} K\xi_{2,t} \\ f(x_{1,t} + \Delta_{1,t}, x_{2,t} + \Delta_{2,t}) - f(x_{1,t}, x_{2,t}) - \xi_{1,t} + KL\xi_{2,t} \end{bmatrix} \right)
$$

In view of the matrix inequality

$$
X^{\mathsf{T}} Y + Y^{\mathsf{T}} X \leq X^{\mathsf{T}} \Lambda X + Y^{\mathsf{T}} \Lambda^{-1} Y \tag{3.41}
$$

valid for any $X, Y \in R^{k \times m}$ and any $0 < \Lambda = \Lambda^{\mathsf{T}} \in R^{k \times k}$, and by the assumptions 1, 2 and 3 of this theorem, we obtain

$$
\frac{d}{dt} V(\Delta_t) \leq \Delta_t^{\mathsf{T}} (P\mathcal{K} + \mathcal{K}^{\mathsf{T}} P) \Delta_t + \Delta_t^{\mathsf{T}} P \Lambda P \Delta_t
$$

$$
+ \left\| \begin{bmatrix} K\xi_{2,t} \\ f(x_{1,t} + \Delta_{1,t}, x_{2,t} + \Delta_{2,t}) - f(x_{1,t}, x_{2,t}) - \xi_{1,t}^{(2)} + KL\xi_{2,t} \end{bmatrix} \right\|_{\Lambda^{-1}}^2
$$

$$
\leq \Delta_t^{\mathsf{T}} (P\mathcal{K} + \mathcal{K}^{\mathsf{T}} P + P\Lambda P) \Delta_t + \|K^{\mathsf{T}} \Lambda_1^{-1} K\| \|\xi_{2,t}\|^2
$$

$$
+ 3\|\Lambda_2^{-1}\| L_f^2 \|\Delta_t\|^2 + 3\|\Lambda_2^{-1}\| \|\xi_{1,t}^{(2)}\|^2 + 3\|L^{\mathsf{T}} K^{\mathsf{T}} \Lambda_2^{-1} KL\| \|\xi_{2,t}\|^2
$$

$$
\leq \Delta_t^{\mathsf{T}} (P\mathcal{K} + \mathcal{K}^{\mathsf{T}} P + P\Lambda P + \|\Lambda_2^{-1}\| L_f^2 I + Q_0) \Delta_t - \Delta_t^{\mathsf{T}} Q_0 \Delta_t
$$

$$
+ 3\|K^{\mathsf{T}} \Lambda_1^{-1} K\| \varepsilon_2^2 + 3\|\Lambda_2^{-1}\| \varepsilon_1^2 + 3\|L^{\mathsf{T}} K^{\mathsf{T}} \Lambda_2^{-1} KL\| \varepsilon_2^2
$$

$$
= -\Delta_t^{\mathsf{T}} P^{1/2} (P^{-1/2} Q_0 P^{-1/2}) P^{1/2} \Delta_t + \beta
$$

$$
\leq -\lambda_{\min}(P^{-1/2} Q_0 P^{-1/2}) \Delta_t^{\mathsf{T}} P \Delta_t + \beta
$$

$$
= -\lambda_{\min}(P^{-1/2} Q_0 P^{-1/2}) V(\Delta_t) + \beta
$$

where β is as in (3.40), which implies (see Lemma 1 in Appendix of this chapter) that $[\sqrt{V(\Delta_t)} - \mu]_+ \underset{t \to \infty}{\to} 0$. The theorem is thus proven.

Remark 7.   *If there is no noise in the dynamics and contaminating the output of the system, it follows that $\beta = \mu = 0$ and, hence, the property (3.39) of the asymptotic consistency is verified; that is, $V(\Delta_t) \underset{t\to\infty}{\to} 0$.*

3.7.5   Analysis of the matrix Riccati equation

Consider now Assumption 3 and its key role in the study of the HGO. It is known [47] (see also Appendix A in Reference 48) that if the matrix $\mathcal{K}$ is stable, the pair $(\mathcal{K}, R^{1/2})$ is controllable, the pair $(Q^{1/2}, \mathcal{K})$ is observable, and the special *local frequency condition* (Yakubovitch-Kalman condition) holds, then the matrix Riccati equation

$$\mathcal{K}^T P + P\mathcal{K} + PRP + Q = 0 \tag{3.42}$$

has a single positive symmetric solution P giving stability to the matrix $[\mathcal{K} - R^{1/2}P]$. To fulfil this local frequency condition it is sufficient (see Appendix A [48]) that the following matrix inequality holds:

$$A^T R^{-1} A - Q > \tfrac{1}{4}[A^T R^{-1} - R^{-1}A]R[A^T R^{-1} - R^{-1}A]^T \tag{3.43}$$

Consider verifying the conditions for the existence of a solution for the special case when the gain matrices are diagonal with equal nonzero elements, that is,

$$K = kI, \quad L = lI \tag{3.44}$$

In view of (3.44) and applying Shur's formula

$$\det \begin{bmatrix} A & B \\ C & D \end{bmatrix} = \det A \, \det(D - CA^{-1}B) = \det D \, \det(A - BD^{-1}C)$$

it follows that

$$\mathcal{K} := \begin{bmatrix} -kI & I \\ -klI & 0 \end{bmatrix}, \quad \det[\mathcal{K} - \lambda I] = \det \begin{bmatrix} -(k+\lambda)I & I \\ -klI & -\lambda I \end{bmatrix}$$

$$= -(k+\lambda)\det(-\lambda I - kl(k+\lambda)^{-1}I)$$

$$= [(k+\lambda)\lambda + kl]^n$$

$$= (\lambda^2 + k\lambda + kl)^n = 0$$

and, hence, the matrix $\mathcal{K}$ is stable (i.e., Re $\lambda_{1,2}(\mathcal{K}) < 0$) for any $k > 0$ and $0 < l \le k/4$, since $\det[\mathcal{K} - \lambda] = 0$ for $\lambda_{1,2} = \tfrac{1}{2}(-k \pm \sqrt{k^2 - 4kl})$ with Re$\lambda_{1,2} < 0$ $\forall k > 0$ and $l \le k/4$. Obviously, the pair $(\mathcal{K}, R^{1/2})$ is controllable and the pair $(Q^{1/2}, \mathcal{K}^T)$ is observable. The last step is to check when the matrix inequality (3.43) holds. For this case if, in addition, the matrices R and Q_0 are diagonal, that is,

$$R := \Lambda = \mu I, \quad Q_0 = q_0 I, \quad Q := qI, \quad q = \lambda^{-1}L_f^2 + q_0 \qquad (\mu, q_0, \lambda > 0)$$

it can be written as

$$\mu^{-1}\mathcal{K}^T \mathcal{K} - (\mu^{-1}L_f^2 + q_0)\begin{bmatrix} I & 0 \\ 0 & I \end{bmatrix} > \frac{1}{4}\mu^{-1}[\mathcal{K}^T - \mathcal{K}][\mathcal{K} - \mathcal{K}^T] \tag{3.45}$$

Since

$$\mathcal{K}^T\mathcal{K} = \begin{bmatrix} -kI & -klI \\ I & 0 \end{bmatrix}\begin{bmatrix} -kI & I \\ -klI & 0 \end{bmatrix} = \begin{bmatrix} k^2(1+l)I - kI & \\ -kI & I \end{bmatrix}$$

$$[\mathcal{K}^T - \mathcal{K}][\mathcal{K} - \mathcal{K}^T] = \begin{bmatrix} 0 & -(1+kl)I \\ (1+kl)I & 0 \end{bmatrix}\begin{bmatrix} 0 & (1+kl)I \\ -(1+kl)I & 0 \end{bmatrix}$$

$$= \begin{bmatrix} (1+kl)^2 I & 0 \\ 0 & (1+kl)^2 I \end{bmatrix} = (1+kl)^2 \begin{bmatrix} I & 0 \\ 0 & I \end{bmatrix}$$

and the matrix inequality (3.45) is converted to

$$\mu^{-1}\begin{bmatrix} k^2(1+l)I & -kI \\ -kI & I \end{bmatrix} - (\mu^{-1}L_f^2 + q_0)\begin{bmatrix} I & 0 \\ 0 & I \end{bmatrix} > \frac{1}{4}\mu^{-1}(1+kl)^2\begin{bmatrix} I & 0 \\ 0 & I \end{bmatrix}$$

or, in the equivalent form, to

$$\begin{bmatrix} \left[k^2(1+l)I - (L_f^2 + \mu q_0) \\ -\frac{1}{4}(1+kl)^2 \right]I & -kI \\ -kI & \left[1 - (L_f^2 + \mu q_0) \\ -\frac{1}{4}(1+kl)^2 \right]I \end{bmatrix} > 0$$

The symmetric block-matrix $\begin{bmatrix} M_{11} & M_{12} \\ M_{12}^T & M_{22} \end{bmatrix}$ given above is positive if and only if the matrix inequalities, given below, hold:

$$M_{11} > 0, \quad M_{22} > 0, \quad M_{11} - M_{12}M_{22}^{-1}M_{12}^T > 0, \quad M_{22} - M_{12}^T M_{11}^{-1}M_{12} > 0$$

which implies the conditions

$$\min\{k^2(1+l); 1\} - (L_f^2 + \mu q_0) - \frac{1}{4}(1+kl)^2 > 0;$$

$$\frac{k^2(1+l) - (L_f^2 + \mu q_0) - (1/4)(1+kl)^2 - k^2}{1 - (L_f^2 + \mu q_0) - (1/4)(1+kl)^2} > 0$$

Take for simplicity $l = k \geq 1$. In this case the last inequalities will be fulfilled if

$$|1 - (L_f^2 + \mu q_0) - \frac{1}{4}(1+k^2)^2| - k > 0$$

which for big enough k implies (see Fig. 3.3)

$$F(k) := k^4 + 2k^2 - 4k - 3 > \rho := 4(L_f^2 + \mu q_0) \tag{3.46}$$

For the parameters k, satisfying the last inequality (3.46) (belonging to the so-called *high-gain parameter zone* $\{k : F(k) > \rho\}$), the asymptotic consistency property is guaranteed.

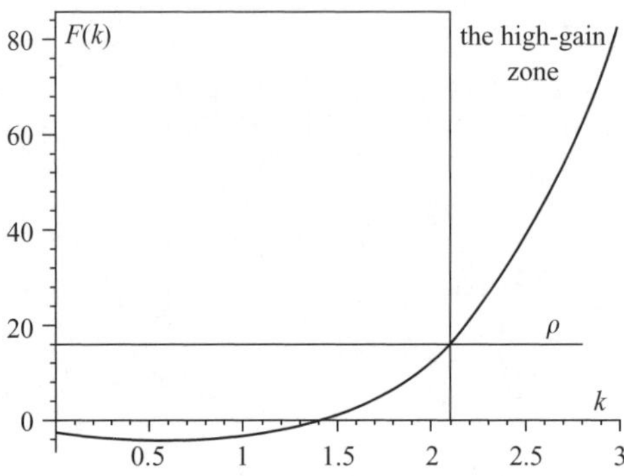

Figure 3.3   The function $F(k)$

3.7.6   Noise generated by stable filters

Now the output noise $\xi_{2,t}$ is generated by (3.29) and (3.30). In this case, the following dimensional identity holds: $p = n$.

Case 2: the filter generating the noise is unknown.   For this situation, any additional constructions are not needed since the HGO (3.37) can be directly applied. The only single difference consists in the output noise effect: instead of $\|\xi_{2,t}\|^2 \le \varepsilon_2$ $\forall t \ge 0$ we have

$$\xi_{2,t} = e^{A_f t}\xi_{2,0} + \int_{\tau=0}^{t} e^{A_f(t-\tau)} w_{2,\tau}\, d\tau, \qquad \limsup_{t \to \infty} \|\xi_{2,t}\|^2 \le \bar{w}_2^2 \|A_f^{-1}\|$$

Thus, in the β-definition (3.40) the value ε_2 should be changed to $\bar{\varepsilon}_2 := \bar{w}_2^2 \|A_f^{-1}\|$.

3.8   Sliding mode observers

3.8.1   Structure of sliding mode observers

Consider the solution of the state-estimation problem for the nonlinear system given by

$$\begin{cases} \dot{x}_t = f(x_t, u_t) + \xi_{1,t} \\ y_t = Cx_t + \xi_{2,t}, \qquad x_0 \text{ is fixed} \end{cases} \tag{3.47}$$

($x_t \in R^n$ is the state of the system, $u_t \in U \subseteq R^m$ is its controlled input, $y_t \in R^p$ is the output) using the *sliding mode type observer* (SLMO) having the following structure:

$$\frac{d}{dt}\hat{x}_t = f(\hat{x}_t, u_t) + K\,\text{sign}(y_t - C\hat{x}_t) \tag{3.48}$$

where the vector function

$$\text{sign}(z) := (\text{sign}(z_1), \text{sign}(z_2), \ldots, \text{sign}(z_n))^{\mathsf{T}} \tag{3.49}$$

is defined by its components

$$\operatorname{sign}(z_i) := \begin{cases} 1 & \text{if } z_i > 0 \\ -1 & \text{if } z_i < 0 \\ \text{undefined} & \text{if } z_i = 0 \end{cases} \tag{3.50}$$

Here $K \in R^{n \times m}$ is a gain-matrix to be selected.

3.8.2 *Fundamental properties*

Consider the following simple two dimensional dynamic system and a corresponding observer of the sliding mode type (see [41]):

$$\begin{cases} \text{Dynamic system} \\ \dot{x}_1 = x_2 + \xi_1 \\ \dot{x}_2 = -bx_2 + \xi_2, \qquad b > 0 \\ y = x_1 + \eta \end{cases} \qquad \begin{cases} \text{Sliding mode observer} \\ \dfrac{d}{dt}\hat{x}_1 = \hat{x}_2 - v \\ \dfrac{d}{dt}\hat{x}_2 = -b\hat{x}_2 + Lv \\ v := M\operatorname{sign}(\sigma), \qquad \sigma := \hat{y} - y = e_1 \\ \hat{y} = \hat{x}_1 \end{cases}$$

$$\tag{3.51}$$

Here ξ_1, ξ_2 and η denote state and output noise, respectively. Below we will briefly analyse three important situations that provide direct motivation for the work to be developed.

3.8.2.1   Case 1: no noise present, i.e., $\xi_1 = \xi_2 = \eta = 0$

The error $(e_i := \hat{x}_i - x_i, i = 1, 2)$ and 'sliding function' $(\sigma := \hat{y} - y = e_1)$ dynamics are as follows:

$$\begin{cases} \text{Error dynamics} \\ \dot{e}_1 = e_2 - v \\ \dot{e}_2 = -be_2 + Lv \\ v = M\operatorname{sign}(e_1) \end{cases} \qquad \begin{cases} \text{'Sliding function' dynamics} \\ \dfrac{d}{dt}\left(\dfrac{\sigma^2}{2}\right) = \sigma\dot{\sigma} \\ \qquad = \sigma(e_2 - v) \leq |\sigma||e_2| - M\sigma\operatorname{sign}(\sigma) \\ \qquad = |\sigma||e_2| - M|\sigma| = -|\sigma|(M - |e_2|) \end{cases}$$

Taking M significantly large (fulfilling $M > |e_2(0)| + \rho, \; \rho > 0$), we may guarantee the finite time convergence of σ to 0, that is, $\sigma(t) = 0$ for any $t \geq t_f = |\sigma(0)|/\rho$. Using the concept of '*the equivalent control*' (see [41]), define $\operatorname{sign}(0)$ in such a way that $\dot{\sigma} = 0$ when $\sigma = 0$, which leads to the following:

$$\dot{\sigma} = e_2 - v = 0, \qquad v = v_{eq} = e_2$$

$$\dot{e}_2 = -be_2 + Lv_{eq} = -(b - L)e_2, \qquad e_2 \underset{t \to \infty}{\to} 0 \text{ if take } L < b$$

Summary

- the convergence to the 'sliding surface' $\sigma = 0$ in *a finite time* takes place if the sliding gain parameter M is selected to be sufficiently large;

- the sliding mode approach is *successfully workable* ($e_1 \underset{t \geq t_f}{=} 0, e_2 \underset{t \to \infty}{\to} 0$) in the no noise situation.

3.8.2.2   Case 2: no output noise present, i.e., $\xi_1 \neq 0$, $\xi_2 \neq 0$ and $\eta = 0$

In this case the dynamics are as follows

$$\begin{cases} \text{Error dynamics} \\ \dot{e}_1 = e_2 - v - \xi_1 \\ \dot{e}_2 = -be_2 + Lv - \xi_2 \\ v = M \operatorname{sign}(e_1) \end{cases} \qquad \begin{cases} \text{'Sliding function' dynamics} \\ \dfrac{d}{dt}\left(\dfrac{\sigma^2}{2}\right) = \sigma\dot{\sigma} = \sigma(e_2 - v - \xi_1) \\ \qquad \leq |\sigma||e_2 - \xi_1| - M\sigma \operatorname{sign}(\sigma) \\ \qquad = |\sigma|(|e_2| + |\xi_1|) - M|\sigma| \\ \qquad = -|\sigma|(M - |e_2| - |\xi_1|) \end{cases}$$

If the noise ξ_1 is bounded, i.e., $|\xi_1| \leq \varepsilon_1$, and if M is sufficiently large ($M > |e_2(0)| + \varepsilon_1 + \rho$, $\rho > 0$), we may guarantee finite time convergence of σ to 0; that is, $\sigma(t) = 0$ for any $t \geq t_f = |\sigma(0)|/\rho$. The concept of '*the equivalent control*' leads directly to the following:

$$\dot{\sigma} = e_2 - v - \xi_1 = 0, \qquad v = v_{eq} = e_2 - \xi_1$$

$$\dot{e}_2 = -be_2 + Lv_{eq} - \xi_2 = -(b - L)e_2 - L\xi_1 - \xi_2$$

$$e_2 \text{ converges to the dead-zone } e_2^+ = \frac{L\varepsilon_1 + \varepsilon_2}{b + |L|} \qquad \text{if } |\xi_2| \leq \varepsilon_2 \text{ with } L < 0$$

Summary

- the convergence to the 'sliding surface' $\sigma = 0$ in *a finite time* takes place if the sliding gain parameter M is sufficiently large;
- the sliding mode approach is *still workable* ($e_1 \underset{t \geq t_f}{=} 0$, e_2 converges to the dead-zone e_2^+ which can be made small enough by selecting a large enough $|L|$) if no output noise is present and the state noise is bounded.

3.8.2.3   Case 3: output noise present, i.e., $\xi_1 \neq 0$, $\xi_2 \neq 0$ and $\eta \neq 0$

In this case the dynamics are as follows

$$\begin{cases} \text{Error dynamics} \\ \dot{e}_1 = e_2 - v - \xi_1 \\ \dot{e}_2 = -be_2 + Lv - \xi_2 \\ v = M \operatorname{sign}(e_1 - \eta) \end{cases} \qquad \begin{cases} \text{'Sliding function' dynamics} \\ \dfrac{d}{dt}\left(\dfrac{\sigma^2}{2}\right) = \sigma\dot{\sigma} = \sigma(e_2 - v - \xi_1) \\ \qquad \leq |\sigma||e_2 - \xi_1| - M\sigma \operatorname{sign}(\sigma) \\ \qquad = |\sigma|(|e_2| + |\xi_1|) - M|\sigma| \\ \qquad = -|\sigma|(M - |e_2| - |\xi_1|) \end{cases}$$

If the noise ξ_1 is bounded, i.e., $|\xi_1| \leq \varepsilon_1$, and if M is sufficiently large ($M > |e_2(0)| + \varepsilon_1 + \rho$, $\rho > 0$), we may guarantee finite time convergence of σ to 0, that is, $\sigma(t) = 0$ for any $t \geq t_f = |\sigma(0)|/\rho$. In this case the concept of '*the equivalent control*' leads directly to the following:

(a) if η is not differentiable, then

$$\sigma = e_1 - \eta \quad \text{and} \quad \dot{\sigma} \text{ does not exist}$$

(b) if η is differentiable and $|\dot{\eta}| \leq \varepsilon_\eta$, then

$$\sigma = e_1 - \eta, \quad \dot{\sigma} = e_2 - v - \xi_1 - \dot{\eta} = 0, \quad v = v_{eq} = e_2 - \xi_1 - \dot{\eta}$$

$$\dot{e}_2 = -be_2 + Lv_{eq} - \xi_2 = -(b-L)e_2 - L(\xi_1 + \dot{\eta}) - \xi_2$$

$$e_2 \text{ converges to the dead-zone } e_2^+ = \frac{L(\varepsilon_1 + \varepsilon_\eta) + \varepsilon_2}{b + |L|} \text{ if } |\xi_2| \leq \varepsilon_2 \text{ and } L < 0$$

Summary

- the convergence to the 'sliding surface' $\sigma = 0$ in *a finite time* takes place by appropriate selection of the sliding gain parameter M (this should be large enough), and $e_1 = \eta$ but there is *no filtering of the noise*;
- the existing sliding mode approach is *not desirable* ($e_1 = \eta$, $e_2 \underset{t \geq t_f}{\text{ converges to}}$

  the dead-zone e_2^+ which can be made small by selecting a large enough $|L|$, if the output noise is smooth enough, and the method is completely unapplicable if the noise is non-smooth).

The following questions of interest arise:

- If we deal with output noise, how can we modify the concept of 'sliding mode observation' to obtain more acceptable results? Is it possible to do this in principle?
- If the sliding mode method can be modified, how can the gain coefficients L and M be selected?

These questions motivate the following study.

By (3.47) and (3.48), it follows that the state-estimation error $\Delta_t := \hat{x}_t - x_t$ satisfies

$$\dot{\Delta}_t = f(\hat{x}_t, u_t) - f(x_t, u_t) + K \,\text{sign}(y_t - C\hat{x}_t) - \xi_{1,t}$$

$$= A\Delta_t - K \,\text{sign}(C\Delta - \xi_{2,t}) + \Delta f_t - \xi_{1,t} \tag{3.52}$$

where the term Δf_t is defined as

$$\Delta f_t := f(\hat{x}_t, u_t) - f(x_t, u_t) - A\Delta_t \tag{3.53}$$

Here $A \in R^{n \times n}$ is any stable (Hurwitz) matrix.

At this stage the main question is: '*how can one analyse this differential that contains in the right-hand side the principal term* $\text{sign}(C\Delta_t - \xi_{2,t})$'?

The following assumptions concerning the class of nonlinear systems and the noise properties are needed.

Assumption 1. There exist nonnegative constants L_{0f}, L_{1f} such that for any $\hat{x}$, $x \in R^n$ and any $u \in U \subseteq R^m$ the following *generalised Lipschitz inequality* holds

$$\| f(\hat{x}, u) - f(x, u) \| \leq L_{0f} + L_{1f} \|\hat{x} - x\| \tag{3.54}$$

This inequality implies that

$$\|\Delta f_t\| \le L_{0f} + (L_{1f} + \|A\|)\|\Delta_t\| \tag{3.55}$$

3.8.3 Bounded output-noise

In this subsection we will consider the case of *bounded output-noise*, that is,

Assumption 2.

$$\|\xi_{i,t}\|^2_{\Lambda_{\xi_i}} := (\xi_{i,t})^\mathsf{T} \Lambda_{\xi_i} \xi_{i,t} \le \varepsilon_i < \infty, \quad 0 \le \Lambda_{\xi_i} = \Lambda_{\xi_i}^\mathsf{T} \ (i = 1, 2) \tag{3.56}$$

Below we represent some technical assumptions used for gain-matrix construction.

Assumption 3.   There exists a positive constant k and a positive definite matrix $Q = Q^\mathsf{T} > 0$ such that the following *Matrix Riccati Equation* has a positive solution $P = P^\mathsf{T} > 0$:

$$PA + A^\mathsf{T} P + PRP + Q = 0$$

$$R := \Lambda_f^{-1} + \Lambda_{\xi_1}^{-1} + 2\|\Lambda_f\|(L_{1f} + \|A\|)^2 I$$

Assumption 4.   The gain-matrix K is selected as

$$K = kP^{-1}C^\mathsf{T} \tag{3.57}$$

where k is a positive constant.

Define the Lyapunov function $L(\Delta_t)$ as before

$$V(\Delta_t) = \|\Delta_t\|_P^2 := \Delta_t^\mathsf{T} P \Delta_t \tag{3.58}$$

with a positive weighting matrix $P \in R^{n \times n}$. In view of Assumptions 1, 2 and 4, (3.52) and (3.41) it follows that:

$$\dot{V}(\Delta_t) = 2\Delta_t^\mathsf{T} P \dot{\Delta}_t$$

$$= 2\Delta_t^\mathsf{T} P[A\Delta_t - K\,\mathrm{sign}(C\Delta - \xi_{2,t}) + \Delta f_t - \xi_{1,t}]$$

$$= 2\Delta_t^\mathsf{T} PA\Delta_t - 2k\Delta_t^\mathsf{T} C^\mathsf{T}\,\mathrm{sign}(C\Delta - \xi_{2,t}) + 2\Delta_t^\mathsf{T} P\Delta f_t - 2\Delta_t^\mathsf{T} P\xi_{1,t}$$

$$\le \Delta_t^\mathsf{T}(PA + A^\mathsf{T} P)\Delta_t - 2k\Delta_t^\mathsf{T} C^\mathsf{T}\,\mathrm{sign}(C\Delta - \xi_{2,t})$$

$$\quad + \Delta_t^\mathsf{T} P\Lambda_f^{-1} P\Delta_t + (\Delta f_t)^\mathsf{T} \Lambda_f \Delta f_t + \Delta_t^\mathsf{T} P\Lambda_{\xi_1}^{-1} P\Delta_t + (\xi_{1,t})^\mathsf{T} \Lambda_{\xi_1} \xi_{1,t}$$

$$\le \Delta_t^\mathsf{T}(PA + A^\mathsf{T} P + P[\Lambda_f^{-1} + \Lambda_{\xi_1}^{-1}]P + Q)\Delta_t$$

$$\quad + \varepsilon_1 - \Delta_t^\mathsf{T} Q\Delta_t + 2\|\Lambda_f\|(L_{0f}^2 + (L_{1f} + \|A\|)^2\|\Delta_t\|^2)$$

$$\quad - 2k\Delta_t^\mathsf{T} C^\mathsf{T}\,\mathrm{sign}(C\Delta - \xi_{2,t}) = \Delta_t^\mathsf{T}(PA + A^\mathsf{T} P + PRP + Q)\Delta_t$$

$$\quad + \varepsilon_1 - \Delta_t^\mathsf{T} Q\Delta_t + 2\|\Lambda_f\|L_{0f}^2 - 2k(C\Delta_t)^\mathsf{T}\,\mathrm{sign}(C\Delta - \xi_{2,t}) \tag{3.59}$$

The following inequalities will now be employed:

$$(x, \operatorname{sign}(x+z)) = ((x+z), \operatorname{sign}(x+z)) - (z, \operatorname{sign}(x+z))$$

$$\geq \sum_{i=1}^{n} |(x+z)_i| - \sum_{i=1}^{n} |z_i| \geq \sum_{i=1}^{n} |x_i| - 2\sum_{i=1}^{n} |z_i|$$

$$\geq \sum_{i=1}^{n} |x_i| - 2\sqrt{n}\|z\| \tag{3.60}$$

Here, we have used the fact that $|(x+z)_i| \geq |x_i| - |z_i|$ and $\sum_{i=1}^{n} |z_i| \leq \sqrt{n}\|z\|$. The last condition results from the Cauchy-Bounyakowski inequality $\sum_{i=1}^{n} a_i b_i \leq \sqrt{\sum_{i=1}^{n} a_i^2}\sqrt{\sum_{i=1}^{n} b_i^2}$ for $a_i := n^{-1}$, $b_i := |z_i|$. Applying (3.60) to (3.59) and in view of Assumption 3, it follows that

$$\dot{V}(\Delta_t) \leq \Delta_t^{\mathsf{T}}(PA + A^{\mathsf{T}}P + P[\Lambda_f^{-1} + \Lambda_{\xi_1}^{-1}]P + Q)\Delta_t$$

$$+ \varepsilon_1 - \Delta_t^{\mathsf{T}}Q\Delta_t + 2\|\Lambda_f\|L_{0f}^2 - 2k(C\Delta_t)^{\mathsf{T}}\operatorname{sign}(C\Delta - \xi_{2,t})$$

$$\leq \Delta_t^{\mathsf{T}}(PA + A^{\mathsf{T}}P + P[\Lambda_f^{-1} + \Lambda_{\xi_1}^{-1}]P + Q)\Delta_t$$

$$+ \varepsilon_1 - \Delta_t^{\mathsf{T}}Q\Delta_t + 2\|\Lambda_f\|L_{0f}^2 - 2k\left(\sum_{i=1}^{n} |(C\Delta_t)_i| - 2\sqrt{n}\|\xi_{2,t}\|\right)$$

$$\leq \Delta_t^{\mathsf{T}}(PA + A^{\mathsf{T}}P + P[\Lambda_f^{-1} + \Lambda_{\xi_1}^{-1}]P + Q)\Delta_t$$

$$- \Delta_t^{\mathsf{T}}Q\Delta_t - 2k\sum_{i=1}^{n} |(C\Delta_t)_i| + \tilde{\rho}(k) \tag{3.61}$$

where

$$\tilde{\rho}(k) := \varepsilon_1 + 2\|\Lambda_f\|L_{0f}^2 + 4k\sqrt{n}\|\Lambda_{\xi_2}^{-1}\|\sqrt{\varepsilon_2} \tag{3.62}$$

Since

$$\left(\sum_{i=1}^{n} |(C\Delta_t)_i|\right)^2 \geq \sum_{i=1}^{n} |(C\Delta_t)_i|^2 = \|C\Delta_t\|^2 = \|CP^{-1/2}P^{1/2}\Delta_t\|^2$$

$$= \Delta_t^{\mathsf{T}}P^{1/2}P^{-1/2}[C^{\mathsf{T}}C + \delta I]P^{-1/2}P^{1/2}\Delta_t - \delta\Delta_t^{\mathsf{T}}\Delta_t$$

$$\geq \alpha_P \Delta_t^{\mathsf{T}}P\Delta_t - \delta\Delta_t^{\mathsf{T}}\Delta_t$$

with

$$\alpha_P := \lambda_{\min}(P^{-1/2}[C^{\mathsf{T}}C + \delta I]P^{-1/2}) \geq \delta\lambda_{\min}(P^{-1}) > 0 \qquad \text{if } \delta > 0 \tag{3.63}$$

and in view of the relations

$$\sqrt{a^2 - b^2} \geq |a| - |b| \qquad \text{(here } |a| \geq |b|)$$

$$a^2 - b^2 \geq a^2 - 2|a| \cdot |b| + b^2$$

$$0 \geq 2b^2 - 2|a| \cdot |b| = 2|b|(|b| - |a|)$$

from (3.61), we finally derive that for any scalar $\Lambda_1 > 0$

$$\dot{V}(\Delta_t) = \frac{d}{dt}(\|\Delta_t\|_P^2)$$

$$\leq \Delta_t^\mathsf{T}(PA + A^\mathsf{T}P + P[\Lambda_f^{-1} + \Lambda_{\xi_1}^{-1}]P + Q)\Delta_t$$

$$- \|\Delta_t\|_Q^2 - 2k(\sqrt{\alpha_P}\|\Delta_t\|_P - \sqrt{\delta\Delta_t^\mathsf{T}\Delta_t}) + \rho(k)$$

$$\leq \Delta_t^\mathsf{T}(PA + A^\mathsf{T}P + P[\Lambda_f^{-1} + \Lambda_{\xi_1}^{-1}]P + Q + k\delta\Lambda_1 I)\Delta_t$$

$$- \alpha_Q\|\Delta_t\|_P^2 - 2k\sqrt{\alpha_P}\|\Delta_t\|_P + \rho(k) \tag{3.64}$$

where

$$\alpha_Q := \lambda_{\min}(P^{-1/2}QP^{-1/2}) > 0, \qquad \rho(k) := k\Lambda_1 + \tilde{\rho}(k) \tag{3.65}$$

At this point we are ready to formulate the main result.

Theorem 2 (on the sliding mode observer). *If the Assumptions 1–4 are fulfilled, then*

$$\left[1 - \frac{\tilde{\mu}}{V_t}\right]_+ \to 0$$

with

$$\tilde{\mu} = \tilde{\mu}(k) := \left(\frac{\rho(k)}{\sqrt{k^2\alpha_P + \rho(k)\alpha_Q} + k\sqrt{\alpha_P}}\right)^2 \tag{3.66}$$

Proof. This follows directly from (3.64) in view of Lemma 2 in the Appendix.

Corollary 1. *If the matrix $C^\mathsf{T}C$ is a rank deficient, that is, $\lambda_{\min}(C^\mathsf{T}C) = 0$, then $\alpha_P = 0$ and, hence, $\tilde{\mu}(k) = \rho(k)/\alpha_Q$ which provides a smaller upper bound for the state-estimation error for smaller values of the gain parameter k.*

Corollary 2. *If the matrix $C^\mathsf{T}C$ has full rank, that is, $\lambda_{\min}(C^\mathsf{T}C) > 0$, then $\alpha_P > 0$ and the parameter $\sqrt{\tilde{\mu}(k)}$ as a function of k (for $\alpha_P = \alpha_Q = 1$) has the following dependence (see Fig. 3.4).*

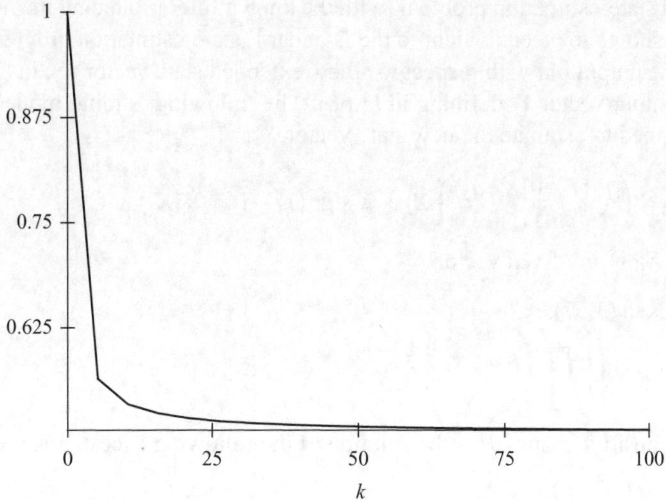

Figure 3.4   The dependence $\sqrt{\tilde{\mu}(k)}$

In general, for any $k > 0$

$$\tilde{\mu}(k) \geq \tilde{\mu}(\infty) = \left(\frac{4\sqrt{n}\|\Lambda_{\xi_2}^{-1}\|\sqrt{\varepsilon_2}}{\sqrt{\alpha_P + 4\sqrt{n}\|\Lambda_{\xi_2}^{-1}\|\sqrt{\varepsilon_2}\alpha_Q} + \sqrt{\alpha_P}} \right)^2$$

Thus, greater k provides a better guaranteed upper bound $\tilde{\mu}(k)$ of the state-estimation error. The best possible upper bound is $\tilde{\mu}(\infty)$.

3.8.4   Output noise formed by a stable filter

Consider now the case where the output noise $\xi_{2,t}$ is an output of a stable filter (3.29) with input supplied by a bounded disturbance $w_{2,t}$ (3.30).

Case 1: the forming filter is known.   Assume, first, that the matrix parameters A_f of this filter are known. Then, introducing the extended state vector X_t defined by

$$X_t := (x_t, \quad \xi_{2,t})^\mathsf{T} \tag{3.67}$$

and defining the new output Y_t as

$$Y_t = [C \quad I]X_t \tag{3.68}$$

we can rewrite the given dynamic system (3.47) as follows

$$\begin{cases} \dot{X}_t = \begin{pmatrix} f([I \quad 0]X_t, u_t) \\ A_f \end{pmatrix} X_t + \begin{pmatrix} \xi_{1,t} \\ w_{2,t} \end{pmatrix} \\ Y_t = [C \quad I]X_t \end{cases} \tag{3.69}$$

The initial state-estimation problem (with the known filter producing the noise at the output) is shown to be equivalent to the 'standard' state-estimation problem without noise in the output but with respect to a new extended state vector X_t, in (3.67), and the new output vector Y_t defining in (3.68). The following 'sliding mode' observer can be applied to estimate the new state vector X_t:

$$\frac{d}{dt}\hat{X}_t = \begin{pmatrix} f([I \quad 0]\hat{X}_t, u_t) \\ A_f \end{pmatrix} \hat{X}_t + K\,\mathrm{sign}(Y_t - [C \quad I]\hat{X}_t)$$

(3.70)

$\qquad \hat{X}_0$ is any fixed vector

Select K as in (3.57):

$$K = kP^{-1}\begin{bmatrix} C^{\mathsf{T}} \\ I \end{bmatrix}$$

where the positive matrix P is the solution of the following Riccati equation

$$P\bar{A} + \bar{A}^{\mathsf{T}}P + PRP + Q = 0$$

$$\bar{A} := \begin{bmatrix} A \\ A_f \end{bmatrix}, \quad R := \Lambda_f^{-1} + \Lambda_{\xi_1}^{-1} + 2\|\Lambda_f\|(L_{1f} + \|\bar{A}\|)^2 I$$

In view of the previous theorem, the state-estimation error $\Delta_t := \hat{X}_t - X_t$ converges to the zone $\tilde{\mu}(k)$ equal to

$$\tilde{\mu}(k) = \left(\frac{\varepsilon_1 + 2\|\Lambda_f\|L_{0f}^2}{\sqrt{k^2\alpha_P + (\varepsilon_1 + 2\|\Lambda_f\|L_{0f}^2)\alpha_Q} + k\sqrt{\alpha_P}} \right)^2$$

with

$$\alpha_P := \lambda_{\min}(P^{-1/2}[C \quad I]^{\mathsf{T}}[C \quad I]P^{-1/2})$$

$$= \lambda_{\min}\left(P^{-1/2}\begin{bmatrix} C^{\mathsf{T}}C & C^{\mathsf{T}} \\ C & I \end{bmatrix}P^{-1/2} \right) = 0$$

Thus, in this case, for any $k > 0$ it follows that

$$\tilde{\mu}(k) = \tilde{\mu} = \frac{\varepsilon_1 + 2\|\Lambda_f\|L_{0f}^2}{\alpha_Q}$$

Varying Q within the constraint (3.43), it is possible to obtain the maximal $\alpha_Q = \lambda_{\min}(P^{-1/2}QP^{-1/2})$, which corresponds with the minimal upper bound $\tilde{\mu}$.

Case 2: the forming filter is unknown.   Here, analogously to the high-gain case, any additional constructions are not needed since the SLMO (3.48) can be directly applied. The unique difference involves the output noise effect: in (3.62) $\bar{\varepsilon}_2 := \bar{w}_2^2\|A_f^{-1}\|$ should be used instead of ε_2.

3.9 Conclusion

In this chapter it has been shown that

- The modified (with the linear correction term and δ-regularisation) concept of 'sliding mode observation' *does really work*, in principle, and provides acceptable quality of the state-estimation process for output noise of a deterministic nature: the averaged state-estimation error norm is shown to be bounded asymptotically.
- The correct selection of the gain-matrix K in the SMO is related to the corresponding algebraic Riccati equation.
- The convergence zone is dependent on the process and observer properties and can be minimised by appropriate selection of the gain matrices.

3.10 Appendix

Lemma 1.   *Let a nonnegative function V_t satisfy the following differential inequality*

$$\frac{d}{dt}V_t \le -\alpha V_t + \beta$$

where $\alpha > 0$ and $\beta \ge 0$. Then $[1 - \mu/\sqrt{V_t}]_+ \to 0$ with $\mu = \sqrt{\beta/\alpha}$ and the function $[\cdot]_+$ defined as

$$[z]_+ := \begin{cases} z & \text{if } z \ge 0 \\ 0 & \text{if } z < 0 \end{cases}$$

Proof.   Introduce the function $G_t := [\sqrt{V_t} - \mu]_+^2 = V_t[1 - \mu/\sqrt{V_t}]_+^2$ where $[z]_+$ is a '*cutting function*' or a '*dead zone*'. For the derivative of this function we obtain

$$\begin{aligned}
\dot{G}_t &:= \frac{[\sqrt{V_t} - \mu]_+}{\sqrt{V_t}}\dot{V}_t = \left[1 - \frac{\mu}{\sqrt{V_t}}\right]_+ \dot{V}_t \\
&\le \left[1 - \frac{\mu}{\sqrt{V_t}}\right]_+ (-\alpha V_t + \beta) \\
&= -\alpha V_t \left[1 - \frac{\mu}{\sqrt{V_t}}\right]_+ \left(1 - \frac{\beta}{(\alpha V_t)}\right) \\
&\le -\alpha V_t \left[1 - \frac{\mu}{\sqrt{V_t}}\right]_+ \left(1 - \frac{\mu^2}{V_t}\right) \le 0
\end{aligned} \tag{3.71}$$

if $\mu := \sqrt{\beta/\alpha}$. By the Weiershtass theorem, the last inequality implies that G_t converges (since it is non-increasing and bounded from below), that is, $G_t \to G^* < \infty$. The integration of (3.71) from 0 to T yields

$$G_T - G_0 \le -\alpha \int_0^T V_t^2 \left[1 - \frac{\mu}{\sqrt{V_t}}\right]_+ \left(1 - \frac{\mu^2}{V_t}\right) dt$$

which leads to the following inequality

$$\alpha \int_0^T V_t \left[1 - \frac{\mu}{\sqrt{V_t}}\right]_+ \left(1 - \frac{\mu^2}{V_t}\right) dt \leq G_0 - G_T \leq G_0 \qquad (3.72)$$

Dividing by T and taking the upper limits of both sides, we finally obtain:

$$\overline{\lim_{T \to \infty}} \frac{1}{T} \int_0^T V_t \left[1 - \frac{\mu}{\sqrt{V_t}}\right]_+ \left(1 - \frac{\mu^2}{V_t}\right) dt \leq 0$$

and, hence, there exists a subsequence t_k such that

$$0 \leq V_{t_k} \left[1 - \frac{\mu}{\sqrt{V_{t_k}}}\right]_+ \left(1 - \frac{\mu^2}{V_{t_k}}\right) = V_{\phi_k} G_{t_k} \left[1 + \frac{\mu}{\sqrt{V_{t_k}}}\right]_+ \to 0$$

or, $G_{\phi_k} \xrightarrow[k \to \infty]{} 0$. It follows that $G^* = 0$, which is equivalent to the fact that $[1 - \mu/\sqrt{V_t}]_+ \to 0$. The theorem is hence proven.

Lemma 2.   *Let a nonnegative function V_t satisfy the following differential inequality*

$$\frac{d}{dt} V_t \leq -\alpha V_t - \delta \sqrt{V_t} + \beta$$

where $\alpha > 0$ and δ, $\beta \geq 0$. Then $[1 - \tilde{\mu}/V_t]_+ \to 0$ with

$$\tilde{\mu} = \frac{(\beta/\alpha)^2}{(\sqrt{(\delta/2\alpha)^2 + \beta/\alpha} + \delta/2\alpha)^2}$$

Proof.   Consider the nonnegative function $\tilde{V}_t$ function satisfying the equality

$$\frac{d}{dt} \tilde{V}_t = -\alpha \tilde{V}_t - \delta \sqrt{\tilde{V}_t} + \beta$$

The equilibrium point $\tilde{V}^*$ of this equation, satisfying

$$-\alpha \tilde{V}^* - \delta \sqrt{\tilde{V}^*} + \beta = 0,$$

is as follows

$$\tilde{V}^* = \left(\sqrt{\left(\frac{\delta}{2\alpha}\right)^2 + \frac{\beta}{\alpha}} - \frac{\delta}{2\alpha}\right)^2 = \frac{(\beta/\alpha)^2}{(\sqrt{(\delta/2\alpha)^2 + \beta/\alpha} + \delta/2\alpha)^2}$$

Defining $\Delta_t := (V_t - \tilde{V}^*)^2$, we derive

$$\dot{\Delta}_t = 2(V_t - \tilde{V}^*)\dot{V}_t \leq 2(V_t - \tilde{V}^*)[-\alpha V_t - \delta\sqrt{V_t} + \beta]$$

$$= 2(V_t - \tilde{V}^*)[-\alpha V_t - \delta\sqrt{V_t} + \beta + (\alpha\tilde{V}^* + \delta\sqrt{\tilde{V}^*} - \beta)]$$

$$\times 2(V_t - \tilde{V}^*)[-\alpha(V_t - \tilde{V}^*) - \delta(\sqrt{V_t} - \sqrt{\tilde{V}^*})]$$

$$= -2(\sqrt{V_t} - \sqrt{\tilde{V}^*})^2(\sqrt{V_t} + \sqrt{\tilde{V}^*})[\alpha(\sqrt{V_t} + \sqrt{\tilde{V}^*}) + \delta] < 0$$

for any $V_t \neq \tilde{V}^*$, which implies: $V_t \underset{t\to\infty}{\to} \tilde{V}^*$. As in Lemma 1, for

$$\tilde{G}_t := [V_t - \tilde{\mu}]_+^2 = V_t\left[1 - \frac{\tilde{\mu}}{V_t}\right]_+^2$$

we obtain

$$\frac{d}{dt}\tilde{G}_t := 2[V_t - \tilde{\mu}]_+\dot{V}_t = 2V_t\left[1 - \frac{\tilde{\mu}}{V_t}\right]_+\dot{V}_t$$

$$\leq 2V_t\left[1 - \frac{\tilde{\mu}}{V_t}\right]_+(-\alpha V_t - \delta\sqrt{V_t} + \beta)$$

$$= -2V_t\left[1 - \frac{\tilde{\mu}}{V_t}\right]_+\left[\alpha(V_t - \tilde{V}^*) + \delta(\sqrt{V_t} - \sqrt{\tilde{V}^*})\right] \leq 0$$

The last inequality implies that $\tilde{G}_t$ converges, that is, $\tilde{G}_t \to \tilde{G}^* < \infty$. The integration of the last inequality from 0 to T yields

$$\tilde{G}_T - \tilde{G}_0 \leq -2\int_0^T V_t\left[1 - \frac{\tilde{\mu}}{V_t}\right]_+\left[\alpha(V_t - \tilde{V}^*) + \delta(\sqrt{V_t} - \sqrt{\tilde{V}^*})\right]dt$$

which leads to the following inequality

$$2\int_0^T V_t\left[1 - \frac{\tilde{\mu}}{V_t}\right]_+\left[\alpha(V_t - \tilde{V}^*) + \delta(\sqrt{V_t} - \sqrt{\tilde{V}^*})\right]dt \leq \tilde{G}_0 - \tilde{G}_T \leq \tilde{G}_0$$

Dividing by T and taking the upper limits of both sides, we obtain:

$$\overline{\lim_{T\to\infty}}\frac{1}{T}\int_0^T V_t\left[1 - \frac{\tilde{\mu}}{V_t}\right]_+\left[\alpha(V_t - \tilde{V}^*) + \delta(\sqrt{V_t} - \sqrt{\tilde{V}^*})\right]dt \leq 0$$

and, hence, there exists a subsequence t_k such that

$$V_{t_k}\left[1 - \frac{\tilde{\mu}}{V_{t_k}}\right]_+\left[\alpha(V_{t_k} - \tilde{V}^*) + \delta\left(\sqrt{V_{t_k}} - \sqrt{\tilde{V}^*}\right)\right] \to 0$$

or, $\tilde{G}_{t_k} \underset{k\to\infty}{\to} 0$. So, it follows that $\tilde{G}^* = 0$, which is equivalent to the fact that

$$\left[1 - \frac{\tilde{\mu}}{V_t}\right]_+ \to 0$$

The theorem is then proven.

3.11   References

1 LJUNG, L. and GUNNARSSON, S.: 'Adaptation and tracking in system identification – a survey', *Automatica*, 1990, **26**, pp. 7–21

2 EYKHOFF, P. and PARKS, P. C.: 'Identification and system parameter estimation; where do we stand now?' (Editorial), *Automatica*, 1990, **26**, pp. 3–5

3 LJUNG, L.: 'Asymptotic behavior of the extended Kalman filter as a parameter estimator for linear systems', *IEEE Trans. Aut. Control*, 1979, **AC-24**, pp. 37–50

4 LEI, G. and CHEN, H. F.: 'Identification and stochastic adaptive control' (Birkhauser, Boston, 1991)

5 POZNYAK, A. S.: 'Estimating the parameters of autoregression process by the method of least squares', *Int. J. Sci.*, 1980, **11**, pp. 577–588

6 SHEIKHOLESLAM, S.: 'Observer based parameter identifier for nonlinear systems with parameter dependencies', *IEEE Trans. Aut. Control*, 1995, **AC-40**, pp. 382–387

7 SONG, Y. D.: 'Adaptive parameter estimators for a class of nonlinear systems', *Int. J. Adapt. Control Signal Process.*, 1997, **11**, pp. 641–648

8 WILLIAMSON, D.: 'Observation of bilinear systems with application to biological control', *Automatica*, 1977, **13**, pp. 243–254

9 KRENER, A. J. and ISIDORI, A.: 'Linearization by output injection and nonlinear observers', *Syst. Control Lett.*, 1983, **3**, pp. 47–52

10 KRENER, A. J. and RESPONDEK, W.: 'Nonlinear observers with linearizable error dynamics', *SIAM J. Control and Optim.*, 1985, **23**, pp. 197–216

11 XIA, X.-U. and GAO, W.-B.: 'Nonlinear observer design by observer error linearization', *SIAM J. Control Optim.*, 1989, **27**, pp. 199–216

12 WALCOTT, B. L. and ZAK, H.: 'State observation of nonlinear uncertain dynamical systems', *IEEE Transactions on Automatic Control*, 1987, **AC-32**, pp. 166–170

13 GAUTHIER, J. P. and BORNARD, G.: 'Observability for any $u(t)$ of a class of nonlinear systems', *IEEE Trans. Aut. Control*, 1981, **AC-26**, pp. 922–926

14 ZEITZ, M.: 'The extended Luenberger observer for nonlinear systems', *Syst. Control Lett.*, 1987, **9**, pp. 149–156

15 BIRK, J. and ZEITZ, M.: 'Extended Luenberger observer for nonlinear multivariable systems', *Int. J. Control*, 1988, **47**, pp. 1823–1836

16 GAUTHIER, J. P., HAMMOURI, H., and OTHMAN, S.: 'A simple observer for nonlinear systems applications to bioreactors', *IEEE Trans. Aut. Control*, 1992, **37**, pp. 875–880

17 CICCARELLA G., DALLA MORA, M., and GERMANI, A.: 'A Luenberger-like observer for nonlinear systems', *Int. J. Control*, 1993, **57**, pp. 537–556

18 HAYKIN, S.: 'Adaptive filter theory' (Prentice-Hall, 1991)

19 TORNAMBÈ, A.: 'Use of asymptotic observers having high-gains in the state and parameter estimation', Proc. 28th Conf. *Dec. Control*, Tampa, Florida, December 1989, pp. 1791–1794

20 BERNTSEN, H. E. and BALCHEN, J. G.: 'Identifiability of linear dynamical systems', Proc. 3rd IFAC Symp. *Identif. and Syst. Param. Estim.*, Hague, Netherlands, 1983, pp. 871–874

21 GREWAL, M. S. and GLOVER, K.: 'Identifiability of linear and nonlinear dynamical systems', *IEEE Trans. Aut. Control*, 1976, **AC-21**, pp. 833–837

22 SIFERD, R. E. and MAYBECK, P. S.: 'Identifiability of nonlinear dynamical systems', *Proc. Conf. Dec. Control*, 1982, pp. 1167–1171

23 TUNALI, E. T. and TARN, T.-J.: 'New results for identifiability of nonlinear systems', *IEEE Trans. Aut. Control*, 1987, **AC-32**, pp. 146–154

24 UNBEHAUEN, H. and RAO, G.: 'Continuous-time approaches to system identification', *Automatica*, 1990, **26**, pp. 23–35

25 KRAUSE, J. M. and KHARGONEKAR, P. P.: 'Parameter identification in the presence of non-parametric dynamic uncertainty', *Automatica*, 1990, **26**, pp. 113–23

26 BORTOFF, S. A. and SPONG, M. W.: 'Parameter identification for nonlinear systems', Proc. 29th Conf. *Dec. Control*, Honolulu, Hawaii, December 1990, pp. 772–777

27 DIOP, S. and FLIESS, M.: 'Nonlinear observability, identifiability and persistent trajectories', Proc. 30th Conf. *Dec. Control*, 1991, pp. 714–719

28 LJUNG, L. and GLAD, T.: 'On global identifiability for arbitrary model parametrizations', *Automatica*, 1994, **30**, pp. 265–276

29 POZNYAK, A. and CORREA, J.: 'Variable structure robust state and parameter estimator', *Adaptive Control and Signal Processing*, special issue dedicated to the memory of Ya. Z. Tzypkin, 2001, **15**(4), pp. 179–208

30 POZNYAK, A. and CORREA, J.: 'Switching structure robust state and parameter estimator for MIMO nonlinear systems', *Int. Journal of Control*, 2001, **74**(2), pp. 175–189

31 WALCOTT, B. L., CORLESS, M. J., and ZAK, S.: 'Comparative study of non-linear state-observation techniques', *Int. J. Control*, 1987, **45**, pp. 2109–2132

32 DABROOM, A. and KHALIL, H.: 'Numerical differentiation using high-gain observers', Proc. IEEE Conf. on *Dec. and Contr. (CDC97)*, San Diego, CA, 1997, pp. 4790–4794

33 NICOSIA, S., TOMEI, P., and TORNAMBE, A.: 'A nonlinear observer for elastic robot', *IEEE Journal of Robotics and Automation*, 1988, **4**, pp. 45–52

34 BULLINGER, E. and ALLGOWER, F.: 'An adaptive high-gain observer for nonlinear systems', Proc. IEEE Conf. *Dec. and Contr.*, San Diego CA, USA, 1997, pp. 4348–4353

35   YAZ, E. and AZEMI, A.: 'Robust/adaptive observers for systems having uncertain functions with unknown bounds', Proceedings of *Amer. Contr. Conf.*, NY, USA, 1994, **1**, pp. 73–74

36   RUIJUN, Z., TIANYOU, C., and CHENG, S.: 'Robust nonlinear adaptive observer design using dynamic recurrent neural networks', *Proceedings of the 1997 American Control Conference*, IL, USA, 1997, **2**, pp. 1096–1100

37   PRESTON, G. J., SHIELDS, D. N., and DALEY, S.: 'Application of a robust nonlinear fault detection observer to a hydraulic system', International Conference on *Control '96,* UK, 1996, **2**, pp. 1484–1489

38   SHIELDS, D. N.: 'Observer design for nonlinear descriptor systems for control and fault detection', *Systems Science*, 1996, **22**(1), pp. 5–20

39   POZNYAK, A. S. and OSORIO-CORDERO, A.: 'Suboptimal robust filtering of states of finite dimensional linear systems with time-varying parameters under nonrandom disturbances', Proc. IEEE Conf. on *Dec. and Contr.*, 1994, Lake Buena Vista, FL, pp. 3931–3936

40   UTKIN, V. I.: 'Principles of identification using sliding regimes', *Soviet Physics Doclady*, 1981, **26**, pp. 271–272

41   UTKIN, V. I.: 'Sliding modes in control optimization' (Springer-Verlag, Berlin, 1992)

42   EDWARDS, E. and SPURGEON, S.: 'Sliding mode control: theory and application' (Taylor & Francis, London, 1998)

43   SLOTINE, J. J. E.: 'Sliding controller design for nonlinear systems', *Int. J. of Control*, 1984, **40**, pp. 421–434

44   WALCOTT, B. L. and ZAK, H.: 'Combined observer-controller synthesis for uncertain dynamical systems with applications', *IEEE Transactions on Systems, Man and Cybernetics*, 1988, **18**, pp. 88–104

45   ZAK, H. and WALCOTT, B. L.: 'State observation of nonlinear control systems via the method of Lyapunov', in ZINOBER, A. S. I. (Ed.): 'Deterministic control of uncertain systems' (Peter Peregrinus, Stevenage, UK, 1990) pp. 333–350

46   BALLESTRINO, A. and INNOCENTI, M.: 'The hyperstability approach to VSCS design', in ZINOBER, A. S. I. (Ed.): 'Deterministic control of uncertain systems' (Peter Peregrinus, Stevenage, UK, 1990) pp. 170–199

47   WILLEMS, J. C.: 'Least squares optimal control and algebraic Riccati equations', *IEEE Transactions on Automatic Control*, 1971, **16**, pp. 621–634

48   POZNYAK, A. S., SANCHEZ, E. N., and YU, W.: 'Dynamic neural networks for nonlinear control: identification, state estimation and trajectory tracking' (World Scientific, London, 2001)

Chapter 4

Stochastic output noise effects in sliding mode observation

Alex S. Poznyak

4.1 Introduction

An area of intensive study is that of *state estimation for stochastic processes*, or, in the mathematical slang, *filtering*. Deng and Krstic [1] perform a Lyapunov-like analysis to prove the stability of the estimation error of a nonlinear system whose dynamics is perturbed by stochastic noise and whose measured output has no noise. The suggested output-feedback controller is robust with respect to disturbances (of the standard Wiener process type), but the effect of any model uncertainties on the output as well as the effect of unmodelled dynamics is not considered. The state estimation problem with observations which may or may not contain a signal at any sample time is considered by Nanacara and Yaz [2] from a covariance assignment viewpoint. A receding horizon Kalman finite-impulse response (FIR) filter is suggested in Wook et al. [3] for continuous-time systems, combining the Kalman filter with the receding horizon strategy. The suggested filter turns out to be a remarkable deadbeat observer. An observer design methodology that is applicable to a general class of nonlinear stochastic systems and measurement models is given in References 4–6. In the last paper, the authors presented a design methodology for state estimation of nonlinear stochastic systems and measurement models with coloured noise processes. The method is based on an extension of deterministic variable-structure observer schemes. A robust filtering of the states for closed-loop systems was also studied in Reference 7.

For the case of sliding observation design for *stochastic systems*, only two (but very significant) papers are available: Drakunov [8] and Yaz and Azemi [5]. In Reference 8 the estimation of the states of a linear dynamic system is considered using a sliding mode observer. Based on 'the averaging approach', it is shown that if the gain parameter of this sliding mode observer increases, then the corresponding estimates tend to the optimal values generated by a Kalman filter. Reference 5 seems to be more advanced and closer to the approach applied here. The authors show the mean-square convergence of the estimation error with a zone depending on the considered process properties. The observer structure contains both linear and signum-type correction terms. The noise disturbing the measured output is assumed to be generated by a known forming filter and enables the problem to be reformulated without noise in the output by state-space extension. This publication is the background for the approach applied in this chapter.

The *stochastic specifics* of the considered models are related to the application of stochastic calculus to analyse (from the mathematical point of view) the behaviour of the state observers which, in turn, are pure deterministic models. This calculus includes *the differential Itô rule* (which, when no stochastic noise is present, coincides with the usual Lie-derivative), the *conditional* (with respect to a fixed prehistory given by a corresponding σ-algebra) *mathematical expectation operator* and the, so-called, *Wiener processes* or *Brownian motion* corresponding to the stochastic noise acting on the input as well as the output of the considered system. Within the engineering profession this is commonly called *white noise* which is interpreted as the 'time-derivative' of a Wiener process. This is incorrect from the mathematical point of view since the random trajectories generated by a Brownian motion are not differentiable practically everywhere. This is why the direct use of the standard Lyapunov technique, related to the differentiation of a Lyapunov ('energy') function over the trajectories of a given dynamic model, is impossible. Another mathematical tool is needed. The stochastic calculus is created to meet this need. Another specific property of stochastic models is that noise of a stochastic nature is practically unbounded with probability one. This means that all upper estimates including the upper bounds of the deterministic noise do not serve in this case. However, engineers use observers of the same structure for the case of stochastic noise as for the deterministic case. The central question discussed in this chapter is as follows: is there a difference between output observation effects when they occur due to deterministic noise (usually bounded) or stochastic noise (practically, never bounded with probability one and non-differentiable)? It will be shown that in some 'averaged' sense deterministic and stochastic noise effects are similar.

This chapter is structured as follows: first, we introduce the class of stochastic models to be considered with a very detailed discussion of the input and output stochastic noise properties. Then we formulate the state estimation problem as the determination of an upper bound for the 'averaged' estimation error if a mixed high-gain, sliding mode (HG-SM) observer is applied. After that, the main results for two possible observation schemes are presented and the corresponding upper bound analysis concludes this study.

4.2   Problem setting

4.2.1   Stochastic continuous-time system

Let $(\Omega, \mathcal{F}, \{\mathcal{F}_t\}_{t\geq 0}, \mathbf{P})$ be a given filtered probability space, that is,

- the probability space $(\Omega, \mathcal{F}, \mathbf{P})$ is complete;
- the sigma-algebra $\mathcal{F}_0$ contains all the $\mathbf{P}$-null sets in $\mathcal{F}$;

- the flow $\{\mathcal{F}_t\}_{t\geq 0}$ of sigma-algebras $\mathcal{F}_t$ (or, filtration) is right continuous:

$$\mathcal{F}_{t+} := \bigcap_{s>t} \mathcal{F}_s = \mathcal{F}_t$$

On this probability space an m-dimensional standard *Brownian motion* is defined, i.e., $(W(t), t \geq 0)$ (with $W(0) = 0$) is an $\{\mathcal{F}_t\}_{t\geq 0}$-adapted $\mathbb{R}^m$-valued process such that

$$E\{W(t) - W(s) \mid \mathcal{F}_s\} = 0 \qquad \mathbf{P} - a.s.$$

$$E\{[W(t) - W(s)][W(t) - W(s)]^{\mathsf{T}} \mid \mathcal{F}_s\} = (t-s)I \qquad \mathbf{P} - a.s.$$

$$\mathbf{P}\{\omega \in \Omega : W(0) = 0\} = 1$$

Here the operator $E\{\varphi(t) \mid \mathcal{F}_s\}$ means the conditional mathematical expectation applied to the random variable $\varphi(t)$ under the fixed 'prehistory' $\mathcal{F}_s$ $(s < t)$.

Consider the stochastic nonlinear continuous-time system with the state $x(t)$ and output $y(t)$ dynamics given by

$$\begin{cases} x(t) = x(0) + \int\limits_{s=0}^{t} f(x(s),s)\,dt + \int\limits_{s=0}^{t} \sigma_x(x(s),s)\,dW(s) \\[2em] y(t) = Cx(t) + \int\limits_{s=0}^{t} \sigma_y(x(s),s)\,dW(s) \end{cases} \tag{4.1}$$

or, in the abstract (symbolic) form,

$$\begin{cases} dx(t) = f(x(t),t)\,dt + \sigma_x(x(t),t)\,dW(t) \\ dy(t) = Cdx(t) + \sigma_y(x(t),t)\,dW(t) \\ x(0) = x_0, \quad y(0) = Cx_0, \qquad t \in [0,T]\,(T>0) \end{cases} \tag{4.2}$$

The first integral in (4.1) is a stochastic ordinary integral and the second one is an Itô integral [9]. In the above $f : [0,T] \times \mathbb{R}^n \to \mathbb{R}^n$, $\sigma_x : [0,T] \times \mathbb{R}^n \to \mathbb{R}^{n \times m}$ and $\sigma_y : [0,T] \times \mathbb{R}^n \to \mathbb{R}^{k \times m}$. Hereafter, the time interval T is supposed to be infinitely large $(T \to \infty)$.

Remark 1.   *The state-output mapping (4.2) considered here is a non-classical one in the sense that the classical problem statement [10], starting from the pioneering works of Kalman, deals with the following state-output transformation*

$$dy(t) = Cx(t)\,dt + \sigma_y(x(t),t)\,dW(t)$$

which corresponds to the situation when the output process $y(t)$ contains the integral of the past information, that is,

$$y(t) = C \int\limits_{s=0}^{t} x(s)\,ds + \int\limits_{s=0}^{t} \sigma_y(x(s),s)\,dW(s)$$

Remark 2.   *If, within the initial setting, the random processes $W_x(t)$ and $W_y(t)$, perturbing the state $x(t)$ and output $y(t)$ dynamics, are different, we may introduce the joint Wiener process $W(t) := [W_x^{\mathsf{T}}(t) \vdots W_y^{\mathsf{T}}(t)]^{\mathsf{T}}$ and the corresponding extending matrices $\tilde{\sigma}_x(x,t)$ and $\tilde{\sigma}_y(x,t)$ defined as*

$$\tilde{\sigma}_x(x,t) := \big[\tilde{\sigma}_x(x,t) \vdots 0\big], \qquad \tilde{\sigma}_y(x,t) := \big[0 \vdots \sigma_x(x,t)\big]$$

in such a way that the identities

$$\sigma_x(x,t)\,dW_x(t) = \tilde{\sigma}_x(x,t)\,dW(t)$$
$$\sigma_y(x,t)\,dW_y(t) = \tilde{\sigma}_y(x,t)\,dW(t)$$

are satisfied. Without loss of generality we may consider the unique random disturbance $W(t)$ in (4.1).

It is assumed that

A1. $\{\mathcal{F}_t\}_{t \geq 0}$ is the natural filtration generated by $(W(t), t \geq 0)$ and augmented by the **P**-null sets from $\mathcal{F}$.

The following definition is used.

Definition 1.   *The function $g : [0,T] \times \mathbb{R}^n \to \mathbb{R}^{n \times m}$ is said to be an $L_{A,\Lambda,\phi}(C^2)$-mapping if*

1.   *it is Borel measurable;*
2.   *it is C^2 in x for any $t \in [0,T]$;*
3.   *there exists a constant L, a symmetric positive matrix Λ and a matrix $A \in \mathbb{R}^{n \times m}$ such that for any $t \in [0,T]$ and for any $x, \hat{x} \in \mathbb{R}^n \times \mathbb{R}^n$ the following inequalities hold:*

$$\|g(x,t) - g(\hat{x},t) - A(x-\hat{x})\|_\Lambda \leq L\|x-\hat{x}\|_\Lambda, \qquad \|f(0,t)\|_\Lambda \leq L$$
$$\|g_x(x,t) - g_x(\hat{x},t)\|_\Lambda \leq L\|x-\hat{x}\|_\Lambda$$
$$\|g_{xx}(x,t) - g_{xx}(\hat{x},t)\|_\Lambda \leq \phi(\|x-\hat{x}\|_\Lambda)$$

(here $g_x(\cdot,x,\cdot)$ and $g_{xx}(\cdot,x,\cdot)$ are partial derivatives of first and second order and $\phi(\cdot)$ is a module of continuity).

We will refer to the condition 3 above as the *quasi-Lipschitz condition*. In view of this definition, it is also assumed that

A2.   The vector function $f(x,t)$ is $L_{A,\Lambda,\phi}(C^2)$-mapping and the matrices $\sigma_x(x,t)$, $\sigma_y(x,t)$ are $L_{0,I,\phi}(C^2)$-mappings.

The assumptions **A1** and **A2** ensure the existence to the solution of the stochastic differential equation (4.2) (see, for example [9]).

Remark 3. *The only source of uncertainty in this system description is the system random noise $W(t)$.*

4.2.2 Noise properties

Assume that the noise acting on the process (4.2) satisfies the following assumptions:

A3. the state noise-effect matrix $\sigma_x(x,t)$ is uniformly bounded; that is, for any $x \in \mathbb{R}^n$ and any $t \geq 0$

$$\sigma_x(x,t)\sigma_x^\mathsf{T}(x,t) \leq \bar{\sigma}_x < \infty \tag{4.3}$$

and the output noise-effect matrix $\sigma_y(x,t)$ satisfies the condition

$$\limsup_{T \to \infty} T^{-1} \int_{t=0}^{T} E\{W_t^y W_t^{y\mathsf{T}}\}\, dt = \Xi < \infty$$

$$W_t^y := \int_{s=0}^{t} \sigma_y(x(s),s)\, dW(s) = (W_{1,s}^y, \ldots, W_{m,s}^y)^\mathsf{T} \tag{4.4}$$

Remark 4. *1) Taking into account the properties of a Wiener process [9], for any $t \geq 0$ it follows that $E\{W_t^y\} = 0$ and*

$$E\{W_t^y W_t^{y\mathsf{T}}\} = E\left\{ \int_{s=0}^{t}\int_{\tau=0}^{t} \sigma_y(x(s),s)\, dW(s)\, dW^\mathsf{T}(\tau)\sigma_y^\mathsf{T}(x(\tau),\tau) \right\}$$

$$= \int_{s=0}^{t} E\{\sigma_y(x(s),s)\sigma_y^\mathsf{T}(x(s),s)\}\, ds$$

In view of (4.4), boundedness of the last integral implies

$$\operatorname{tr}\{\sigma_y(x(s),s)\} \underset{t \to \infty}{\to} 0$$

2) Several classes of processes W_t^y are usually considered:

- *moving average type:*

$$\sigma_y(x(s),s) := \begin{cases} \check{\sigma}_y(x(s),s) & \text{if } s \in [t, t-\tau_0] \\ 0 & \text{if } s \in [0, t-\tau_0) \end{cases}$$

(here τ_0 is a fixed averaging interval);

- *exponential decreasing correlation:*

$$\sigma_y(x(s),s) := \sigma_y^0 \exp(-Ns)$$

- *'inverse-root' decreasing:*

$$\sigma_y(x(s),s) := \frac{\sigma_y^0}{\sqrt{t+a}}, \qquad a > 0$$

- *a standard Wiener process (which does not satisfy A3):*

$$\sigma_y(x(s),s)\sigma_y^{\mathsf{T}}(x(s),s) := I_{k\times k} \qquad \text{for all } s \geq 0$$

4.2.3   Observer structures

As has been shown in Chapter 3, a pure sliding mode observer does not work well in the presence of output noise even if the noise is deterministic and bounded. Following [5, 11–13], we will study two so-called *'linear high gain-sliding mode'* (LHG-SM) observer structures.

4.2.3.1   1st LHG-SM structure

This contains a linear correction term proportional to the output estimation error and is given by

$$\frac{d}{dt}\hat{x}(t) = f(\hat{x}(t),t) + K_0(y(t) - C\hat{x}(t)) + K\,\mathrm{sign}(y(t) - C\hat{x}(t))$$

$$\hat{x}(0) = \hat{x}_0, \qquad \hat{x}(t) \in \mathbb{R}^n$$

(4.5)

Here K and K_0 are fixed matrices of appropriate dimension. The function $\mathrm{sign}(z)$ is defined as

$$\mathrm{sign}(z) := (\mathrm{sign}(z_1),\ldots,\mathrm{sign}(z_m))^{\mathsf{T}}$$

$$\mathrm{sign}(z_i) := \begin{cases} 1 & \text{if } z_i > 0 \\ -1 & \text{if } z_i < 0 \\ \text{not defined} & \text{if } z_i = 0 \end{cases}$$

(4.6)

When $K = 0$, this structure corresponds to a *linear high gain observer*. If $K_0 = 0$ a *sliding mode observer* results. If both matrices are non-zero, we obtain a *joint observer structure*. Below we show that a sliding mode term added to the linear high-gain term essentially helps to decrease the convergence zone of the state estimation error.

4.2.3.2   2nd LHG-SM structure

This contains a linear correction term proportional to the *derivative* of the output estimation error and is given by

$$d\hat{x}(t) = f(\hat{x}(t),t)\,dt + K_0(dy(t) - C\,d\hat{x}(t)) + K\,\mathrm{sign}(y(t) - C\hat{x}(t))\,dt$$

$$\hat{x}(0) = \hat{x}_0, \qquad \hat{x}(t) \in \mathbb{R}^n$$

(4.7)

or

$$d\hat{x}(t) = [I + K_0 C]^{-1} f(\hat{x}(t), t)\, dt + [I + K_0 C]^{-1} K_0\, dy(t)$$
$$+ [I + K_0 C]^{-1} K \operatorname{sign}(y(t) - C\hat{x}(t))\, dt \qquad (4.8)$$
$$\hat{x}(0) = \hat{x}_0, \qquad \hat{x}(t) \in \mathbb{R}^n, \quad \det[I + K_0 C] \neq 0$$

4.2.4   Problem formulation

Our aim is to define the class of gain matrices K and K_0 that guarantee (in some probabilistic sense) the existence of a finite upper bound to *the time-averaged state estimation error* defined as

$$\tilde{\Delta}(t) := \frac{1}{t} \int_{\tau=0}^{t} \|\Delta(\tau)\|_P\, d\tau$$

where $\Delta(t) := \hat{x}(t) - x(t)$ and P is a weighting matrix. This defines a class $\mathcal{K}$ of matrices K and K_0 where there exists a function $\mu = \mu(K, K_0)$ such that

$$\limsup_{t \to \infty} E\{\tilde{\Delta}(t)\} = \limsup_{t \to \infty} \frac{1}{t} \int_{\tau=0}^{t} E\{\|\Delta(\tau)\|_P\}\, d\tau \leq \mu = \mu(K, K_0) \qquad (4.9)$$

Here μ is a guaranteed upper bound on the zone defining where the averaged state estimation error $\tilde{\Delta}(t)$ converges. The function $\mu = \mu(K, K_0)$ will be analysed for both types of observers.

4.3   Main result

Below we will present theorems on the convergence of the state estimation error for both observer schemes. They consider the mean-square convergence of the time-averaged estimation error to a fixed zone depending on the characteristics of both the process and the observer: it is proven that the mean-square error trajectories, even if they leave the given zone for a 'short time', have the property that their 'time-averaged' weighted norm remains bounded asymptotically within this zone.

4.3.1   Convergence analysis for the first observer scheme

Theorem 1 (the time-averaged convergence of the first scheme).   *If under the assumptions A1–A3,*

1.   *there exist positive definite matrices* $\Lambda, \Lambda_1, Q_0 > 0$ *and constants* $\lambda > 0$, $k, \delta \geq 0$
     such that the following algebraic matrix Riccati equation

$$P[A - K_0 C] + [A - K_0 C]^\mathsf{T} P + P[\Lambda^{-1} + K_0 \Lambda_1^{-1} K_0^\mathsf{T}]P$$
$$+ [L\Lambda + k\delta\lambda I + Q_0] = 0 \qquad (4.10)$$

has a positive solution $P = P^{\mathsf{T}} > 0$ *(this demands the stability of the matrix* $[A - K_0 C]$ *which, in turn, requires that the pair* (A, C) *is observable)*;

2. *the mixed linear-sliding observer (4.5) has the sliding gain matrix*

$$K = kP^{-1}C^{\mathsf{T}} \tag{4.11}$$

then the mixed linear-sliding observer (4.5) provides convergence on average of the normalised state estimation error $\|\Delta(t)\|_P$ *to the* μ*-zone:*

$$\limsup_{t \to \infty} T^{-1} \int_{t=0}^{T} E\{\|\Delta(t)\|_P\}\, dt \overset{a.s.}{\leq} \mu := \frac{\tilde{\beta}/\tilde{\alpha}}{\sqrt{(\tilde{\delta}/2\tilde{\alpha})^2 + \tilde{\beta}/\tilde{\alpha} + \tilde{\delta}/2\tilde{\alpha}}} \tag{4.12}$$

with

$$\tilde{\alpha} := \lambda_{\min}(P^{-1/2} Q_0 P^{-1/2}), \quad \tilde{\delta} := 2k\sqrt{\alpha_{P,\delta}}$$

$$\alpha_{P,\delta} := \lambda_{\min}(P^{-1/2}[C^{\mathsf{T}}C + \delta I]P^{-1/2})$$

$$\tilde{\beta} := k(\lambda^{-1} + 4\sqrt{n}\sqrt{\operatorname{tr}\Xi}) + \lambda \operatorname{tr}\Xi + \operatorname{tr}\{\bar{\sigma}_x P\}$$

$$\|\Delta(t)\|_P := \sqrt{\Delta^{\mathsf{T}}(t)P\Delta(t)}$$

4.3.1.1 Proof of Theorem 1

Consider the following Lyapunov function

$$V(z) := \|P^{1/2}z\|^2 = \|z\|_P^2 = z^{\mathsf{T}} Pz$$

where P is a positive definite matrix. Considering (4.2) and (4.5), derive the differential equation for the state estimation error $\Delta(t)$:

$$dV(\Delta) = [f(\hat{x}, t) - f(x, t) + K_0(-C\Delta + W_t^y) + K\operatorname{sign}(-C\Delta + W_t^y)]\, dt$$
$$- \sigma_x(x, t)\, dW(t) \tag{4.13}$$

The Itô formula [9]

$$dV(\Delta) \overset{a.s.}{=} \nabla^{\mathsf{T}} V(\Delta)\, d\Delta + \tfrac{1}{2}\operatorname{tr}\{\sigma_x^{\mathsf{T}}(x, t)\nabla^2 V(\Delta)\sigma_x(x, t)\}\, dt \tag{4.14}$$

is applied to $V(z)$. Given (4.13), this leads to the following equality (in symbolic form):

$$dV(\Delta) \overset{a.s.}{=} 2\Delta^{\mathsf{T}} P[f(\hat{x}, t) - f(x, t) + K_0(-C\Delta + W_t^y)$$
$$+ K\operatorname{sign}(-C\Delta + W_t^y)]\, dt - 2\Delta^{\mathsf{T}} P\sigma_x(x, t)\, dW(t)$$
$$+ \operatorname{tr}\{\sigma_x^{\mathsf{T}}(x, t)P\sigma_x(x, t)\}\, dt \tag{4.15}$$

1) To estimate the term containing the signum-function, use the simple inequality

$$(x, \operatorname{sign}(x + z)) = ((x + z), \operatorname{sign}(x + z)) - (z, \operatorname{sign}(x + z))$$

$$\geq \sum_{i=1}^{n} |(x + z)_i| - \sum_{i=1}^{n} |z_i| \tag{4.16}$$

Since $|(x+z)_i| \geq |x_i| - |z_i|$, we have

$$(x, \text{sign}(x+z)) \geq \sum_{i=1}^{n} |x_i| - 2 \sum_{i=1}^{n} |z_i| \tag{4.17}$$

The Cauchy-Bounyakowski inequality

$$\sum_{i=1}^{n} a_i b_i \leq \sqrt{\sum_{i=1}^{n} a_i^2} \sqrt{\sum_{i=1}^{n} b_i^2}$$

for $a_i := n^{-1}$, $b_i := |z_i|$ is converted to $\sum_{i=1}^{n} |z_i| \leq \sqrt{n} \|z\|$, which together with (4.17) implies

$$(x, \text{sign}(x+z)) \geq \sum_{i=1}^{n} |x_i| - 2 \sum_{i=1}^{n} |z_i| \geq \sum_{i=1}^{n} |x_i| - 2\sqrt{n} \|z\| \tag{4.18}$$

By Assumption 2 (4.11), taking $K = kP^{-1}C^\mathsf{T}$, the inequality (4.18) implies

$$2\Delta^\mathsf{T} PK \, \text{sign}(-C\Delta + W_t^y) = -2k\Delta^\mathsf{T} C^\mathsf{T} \text{sign}(C\Delta - W_t^y)$$

$$\leq -2k \sum_{i=1}^{n} |C\Delta|_i + 4\sqrt{n} k \|W_t^y\| \tag{4.19}$$

2) To estimate the remaining term in (4.15), use the matrix inequality

$$X^\mathsf{T} Y + Y^\mathsf{T} X \leq X^\mathsf{T} \Lambda^{-1} X + Y^\mathsf{T} \Lambda Y \tag{4.20}$$

(valid for any $X, Y \in \mathbb{R}^{m \times n}$ and any $0 < \Lambda = \Lambda^\mathsf{T} \in \mathbb{R}^{m \times m}$). In view of this and applying A2, we derive

$$\begin{aligned}
2\Delta^\mathsf{T} P[f(\hat{x},t) - f(x,t)] &= 2\Delta^\mathsf{T} P[f(\hat{x},t) - f(x,t) - A\Delta] + 2\Delta^\mathsf{T} PA\Delta \\
&\leq \Delta^\mathsf{T} P\Lambda^{-1} P\Delta + [f(\hat{x},t) - f(x,t) - A\Delta]^\mathsf{T} \\
&\quad \times \Lambda[f(\hat{x},t) - f(x,t) - A\Delta] + 2\Delta^\mathsf{T} PA\Delta \\
&\leq \Delta^\mathsf{T} P\Lambda^{-1} P\Delta + L\Delta^\mathsf{T} \Lambda\Delta + 2\Delta^\mathsf{T} PA\Delta \\
&= \Delta^\mathsf{T} (PA + A^\mathsf{T} P + P\Lambda^{-1} P + L\Lambda)\Delta
\end{aligned} \tag{4.21}$$

and for some $\Lambda_1 > 0$

$$2\Delta^\mathsf{T} PK_0 W_t^y \leq \Delta^\mathsf{T} PK_0 \Lambda_1^{-1} K_0^\mathsf{T} P\Delta + W_t^{y\mathsf{T}} \Lambda_1 W_t^y \tag{4.22}$$

3) Applying the obtained estimates (4.19), (4.21) and (4.22) to (4.15), we obtain

$$dV(\Delta) \overset{a.s.}{\leq} \Delta^{\mathsf{T}}(P[A - K_0 C] + [A - K_0 C]^{\mathsf{T}} P$$
$$+ P[\Lambda^{-1} + K_0 \Lambda_1^{-1} K_0^{\mathsf{T}}]P + L\Lambda)\Delta\, dt$$
$$\times \left(-2k \sum_{i=1}^{n} |C\Delta|_i + 4\sqrt{n}k\|W_t^y\| + \|W_t^y\|^2 \cdot \|\Lambda_1\|\right) dt$$
$$+ \operatorname{tr}\{\sigma_x^{\mathsf{T}}(x,t) P \sigma_x(x,t)\}\, dt - 2\Delta^{\mathsf{T}} P \sigma_x(x,t)\, dW(t) \tag{4.23}$$

Since

$$\left(\sum_{i=1}^{n} |(C\Delta_t)_i|\right)^2 \geq \sum_{i=1}^{n} |(C\Delta_t)_i|^2 = \|C\Delta_t\|^2 = \Delta_t^{\mathsf{T}} C^{\mathsf{T}} C \Delta_t$$
$$= \Delta_t^{\mathsf{T}} (C^{\mathsf{T}} C + \delta I) \Delta_t - \delta \|\Delta_t\|^2$$
$$\geq [\lambda_{\min}(P^{-1/2}[C^{\mathsf{T}} C + \delta I]P^{-1/2})]\| P^{1/2}\Delta_t\|^2 - \delta\|\Delta_t\|^2$$

with

$$\alpha_{P,\delta} := \lambda_{\min}(P^{-1/2}[C^{\mathsf{T}} C + \delta I]P^{-1/2}) > 0, \qquad \delta > 0 \tag{4.24}$$

and, in view of the inequality $\sqrt{a^2 - b^2} \geq \sqrt{|a|} - \sqrt{|b|}$ valid for any $|a| \geq |b|$, by (4.20) it follows that

$$\sum_{i=1}^{n} |(C\Delta_t)_i| \geq \sqrt{\alpha_{P,\delta}\| P^{1/2}\Delta_t\|^2 - \delta\|\Delta_t\|^2} \geq \sqrt{\alpha_{P,\delta}}\sqrt{V(\Delta)} - \sqrt{\delta}\|\Delta_t\|$$
$$\geq \sqrt{\alpha_{P,\delta}}\sqrt{V(\Delta)} - \tfrac{1}{2}\lambda^{-1} - \tfrac{1}{2}\Delta_t^{\mathsf{T}}(\delta\lambda)\Delta_t$$

for any scalar $\lambda > 0$. Then, the inequality (4.23) is transformed to

$$dV(\Delta) \overset{a.s.}{\leq} \Delta^{\mathsf{T}}(P[A - K_0 C] + [A - K_0 C]^{\mathsf{T}} P$$
$$+ P[\Lambda^{-1} + K_0 \Lambda_1^{-1} K_0^{\mathsf{T}}]P + L\Lambda + k\delta\lambda I)\Delta\, dt$$
$$\times \left(-2k\sqrt{\alpha_{P,\delta}}\sqrt{V(\Delta)} + k\lambda^{-1} + 4\sqrt{n}k\|W_t^y\| + \|W_t^y\|^2 \cdot \|\Lambda_1\|\right) dt$$
$$+ \operatorname{tr}\{\sigma_x^{\mathsf{T}}(x,t) P \sigma_x(x,t)\}\, dt - 2\Delta^{\mathsf{T}} P \sigma_x(x,t)\, dW(t)$$

or, by Assumption 1 of Theorem (see (4.10)), to the following inequality

$$dV(\Delta) \overset{a.s.}{\leq} -\Delta^{\mathsf{T}} Q_0 \Delta\, dt$$
$$\times \left(-2k\sqrt{\alpha_{P,\delta}}\sqrt{V(\Delta)} + k\lambda^{-1} + 4\sqrt{n}k\|W_t^y\| + \|W_t^y\|^2 \cdot \|\Lambda_1\|\right) dt$$
$$+ \operatorname{tr}\{\sigma_x^{\mathsf{T}}(x,t) P \sigma_x(x,t)\}\, dt - 2\Delta^{\mathsf{T}} P \sigma_x(x,t)\, dW(t)$$

which yields the final expression

$$dV(\Delta) \overset{a.s.}{\leq} -\big(\tilde{\alpha} V(\Delta) + \tilde{\delta}\sqrt{V(\Delta)} - \tilde{\beta}(t)\big)\, dt - 2\Delta^{\mathsf{T}} P \sigma_x(x,t)\, dW(t) \tag{4.25}$$

with

$$\tilde{\alpha} := \lambda_{\min}(P^{-1/2}Q_0 P^{-1/2}), \quad \tilde{\delta} := 2k\sqrt{\alpha_{P,\delta}}$$

$$\alpha_{P,\delta} := \lambda_{\min}(P^{-1/2}[C^{\mathsf{T}}C + \delta I]P^{-1/2})$$

$$\tilde{\beta}(t) := k\lambda_1^{-1} + 4\sqrt{n}k\|W_t^y\| + \|W_t^y\|^2 \cdot \|\Lambda_1\| + \mathrm{tr}\{\sigma_x^{\mathsf{T}}(x,t)P\sigma_x(x,t)\}$$

The symbolic stochastic inequality (4.25) in its direct form is as follows:

$$V(\Delta(T)) - V(\Delta(0)) \overset{a.s.}{\leq} - \int_{t=0}^{T} (\tilde{\alpha}V(\Delta(t)) + \tilde{\delta}\sqrt{V(\Delta)(t)})\,dt$$

$$+ \int_{t=0}^{T} \tilde{\beta}(t)\,dt - 2\int_{t=0}^{T} \Delta^{\mathsf{T}}(t)P\sigma_x(x,t)\,dW(t) \tag{4.26}$$

where the third integral in the right-hand side is the Itô integral that satisfies (see [9])

$$E\left\{ \int_{t=0}^{T} \Delta^{\mathsf{T}}(t)P\sigma_x(x,t)\,dW(t) \right\} = 0 \tag{4.27}$$

Applying the mathematical expectation operator $E\{\cdot\}$ to both sides of (4.26), and in view of (4.27), we derive

$$E\{V(\Delta(T))\} - E\{V(\Delta(0))\}$$

$$\leq - \int_{t=0}^{T} (\tilde{\alpha}E\{V(\Delta(t))\} + \tilde{\delta}E\{\sqrt{V(\Delta)(t)}\} - E\{\tilde{\beta}(t)\})\,dt \tag{4.28}$$

Here the change in the order of integration and the mathematical expectation is correctly used since the integral $\int_{t=0}^{T} E\{V(\Delta(t))\}\,dt$ exists for any finite T as can be directly seen from the previous inequality. By Jensen's inequality

$$E\{V(\Delta)\} = E\{(\sqrt{V(\Delta)})^2\} \geq (E\{\sqrt{V(\Delta)}\})^2$$

from (4.28) it follows that

$$E\{V(\Delta(T))\} - E\{V(\Delta(0))\}$$

$$\leq - \int_{t=0}^{T} (\tilde{\alpha}(E\{\sqrt{V(\Delta)}\})^2 + \tilde{\delta}E\{\sqrt{V(\Delta)(t)}\} - E\{\tilde{\beta}(t)\})\,dt \tag{4.29}$$

The term $E\{\tilde{\beta}(t)\}$ satisfies

$$\limsup_{t\to\infty} T^{-1} \int_{t=0}^{T} E\{\tilde{\beta}(t)\}\, dt \le k\lambda^{-1} + 4\sqrt{n}k \limsup_{t\to\infty} T^{-1} \int_{t=0}^{T} E\{\|W_t^y\|\}\, dt$$

$$+ \|\Lambda_1\| \limsup_{t\to\infty} T^{-1} \int_{t=0}^{T} E\{\|W_t^y\|^2\}\, dt$$

$$+ \operatorname{tr}\{\sigma_x^{\mathsf{T}}(x,t)\, P\sigma_x(x,t)\} \le \tilde{\beta} \tag{4.30}$$

4) Calculating the upper limits from the right- and then left-hand sides of the inequality (4.28) and in view of (4.30), we obtain

$$0 \le \limsup_{t\to\infty} T^{-1} \int_{t=0}^{T} \left(\tilde{\alpha}\big(E\{\sqrt{V(\Delta)}\}\big)^2 + \tilde{\delta}E\{\sqrt{V(\Delta)}\}\right) dt \le \tilde{\beta} \tag{4.31}$$

which implies

$$\tilde{\beta} \ge \limsup_{t\to\infty} T^{-1} \int_{t=0}^{T} \left(\tilde{\alpha}\big(E\{\sqrt{V(\Delta)}\}\big)^2 + \tilde{\delta}E\{\sqrt{V(\Delta)}\}\right) dt \ge \tilde{\alpha}\gamma^2 + \tilde{\delta}\gamma$$

$$\tag{4.32}$$

$$\gamma := \limsup_{t\to\infty} T^{-1} \int_{t=0}^{T} E\{\sqrt{V(\Delta)}\}\, dt = \limsup_{t\to\infty} E\{\tilde{\Delta}(t)\}$$

Solving the quadratic inequality (4.32) with respect to γ, we obtain the final result (4.12). The theorem is then proven.

4.3.2   Convergence analysis for the second observer scheme

Theorem 2 (the time-averaged convergence of the second scheme).   *If, under the assumptions A1–A3,*

1. *there exist positive definite matrices $\Lambda, Q > 0$ and nonnegative constants $\lambda > 0$, $k, \delta \ge 0$ such that the following algebraic matrix Riccati equation*

$$P[\bar{A} - K_0 C] + [\bar{A} - K_0 C]^{\mathsf{T}} P + P[\Lambda^{-1} + K_0 \Lambda_1^{-1} K_0^{\mathsf{T}}]P$$

$$+ [L([I + K_0 C]^{-1})^{\mathsf{T}} \Lambda [I + K_0 C]^{-1} + k\delta\lambda + Q] = 0 \tag{4.33}$$

$$\bar{A} := [I + K_0 C]^{-1} A, \quad \det[I + K_0 C] \ne 0$$

has a positive solution $P = P^{\mathsf{T}} > 0$ (this ensures the stability of the matrix $[\bar{A} - K_0 C]$ if the pair $(\bar{A}, C)$ is observable);

2. *the mixed linear-sliding observer (4.5) has the following sliding gain matrix*

$$K = k[I + K_0 C]P^{-1}C^\mathsf{T} \tag{4.34}$$

then the mixed linear-sliding observer (4.5) provides mean-square convergence in a time-averaged sense of the normalised state estimation error $\|\Delta(t)\|_P$ to the μ-zone:

$$\limsup_{t \to \infty} T^{-1} \int_{t=0}^{T} E\{\|\Delta(t)\|_P\}\, dt \overset{a.s.}{\leq} \mu \tag{4.35}$$

$$\mu := \frac{\tilde{\beta}^{(2)}/\tilde{\alpha}^{(2)}}{\sqrt{(\tilde{\delta}^{(2)}/2\tilde{\alpha}^{(2)})^2 + \tilde{\beta}^{(2)}/\tilde{\alpha}^{(2)} + \tilde{\delta}^{(2)}/2\tilde{\alpha}^{(2)}}}$$

with

$$\tilde{\alpha}^{(2)} := \lambda_{\min}(P^{-1/2}QP^{-1/2}), \quad \tilde{\delta}^{(2)} := 2k\sqrt{\alpha_{P,\delta}^{(2)}}$$

$$\alpha_{P,\delta}^{(2)} := \lambda_{\min}(P^{-1/2}[C^\mathsf{T}C + \delta I]P^{-1/2})$$

$$\tilde{\beta}^{(2)} := k\lambda^{-1} + 4\sqrt{n}k\sqrt{\operatorname{tr}\Xi} + \|\Lambda_1\|\operatorname{tr}\Xi \tag{4.36}$$

$$+ \operatorname{tr}\{[K_0\sigma_y(x,t) - \sigma_x(x,t)]^\mathsf{T}([I + K_0C]^{-1})^\mathsf{T}$$

$$\times P[I + K_0C]^{-1}[K_0\sigma_y(x,t) - \sigma_x(x,t)]\}$$

4.3.2.1   Proof of Theorem 2

This almost repeats the proof of the previous theorem if it is considered that for (4.7) we have

$$d\Delta = [f(\hat{x},t) - f(x,t) + K_0(-Cd\Delta + \sigma_y(x,t)\,dW(t))$$

$$+ K\operatorname{sign}(-C\Delta + W_t^y)]\,dt - \sigma_x(x,t)\,dW(t)$$

which implies ($\det[I + K_0C] \neq 0$)

$$d\Delta = [I + K_0C]^{-1}[f(\hat{x},t) - f(x,t) + [I + K_0C]^{-1}K\operatorname{sign}(-C\Delta + W_t^y)]\,dt$$

$$+ [I + K_0C]^{-1}[K_0\sigma_y(x,t) - \sigma_x(x,t)]\,dW(t)$$

and

$$dV(\Delta) \overset{a.s.}{\leq} 2\Delta^\mathsf{T}P[[I + K_0C]^{-1}(f(\hat{x},t) - f(x,t))$$

$$+ [I + K_0C]^{-1}K\operatorname{sign}(-C\Delta + W_t^y)]\,dt$$

$$+ 2\Delta^\mathsf{T}P[I + K_0C]^{-1}[K_0\sigma_y(x,t) - \sigma_x(x,t)]\,dW(t)$$

$$+ \operatorname{tr}\{[K_0\sigma_y(x,t) - \sigma_x(x,t)]^\mathsf{T}([I + K_0C]^{-1})^\mathsf{T}\,dt$$

$$\times P[I + K_0C]^{-1}[K_0\sigma_y(x,t) - \sigma_x(x,t)]\}$$

Then, using the same technique as in Theorem 1, we derive

$$dV(\Delta) \overset{a.s.}{\leq} -(\tilde{\alpha}^{(2)}V(\Delta) + \tilde{\delta}^{(2)}\sqrt{V(\Delta)})\,dt + dI_{1,t} + dI_{2,t}$$

where, in view of (4.36),

$$dI_{1,t} := 2\Delta^{\mathsf{T}} P[I + K_0 C]^{-1}[K_0 \sigma_y(x,t) - \sigma_x(x,t)]\,dW(t), \quad dI_{2,t} := \tilde{\beta}^{(2)}(t)\,dt$$

$$\tilde{\beta}^{(2)}(t) := \operatorname{tr}\{[K_0\sigma_y(x,t) - \sigma_x(x,t)]^{\mathsf{T}}([I + K_0 C]^{-1})^{\mathsf{T}} \cdot P[I + K_0 C]^{-1}$$

$$\times [K_0\sigma_y(x,t) - \sigma_x(x,t)]\} + k\lambda^{-1} + 4\sqrt{nk}\|W_t^y\| + \|W_t^y\|^2 \cdot \|\Lambda_1\|$$

Integrating the last inequality and applying the mathematical expectation operator, we obtain

$$E\{V(\Delta(T))\} - E\{V(\Delta(0))\}$$

$$\leq - \int_{t=0}^{T} (\tilde{\alpha}^{(2)} E\{V(\Delta(t))\} + \tilde{\delta}^{(2)} E\{\sqrt{V(\Delta)(t)}\})\,dt + \int_{t=0}^{T} E\tilde{\beta}^{(2)}(T)\,dt$$

$$\leq - \int_{t=0}^{T} (\tilde{\alpha}^{(2)}(E\{\sqrt{V(\Delta(t))}\})^2 + \tilde{\delta}^{(2)} E\{\sqrt{V(\Delta)(t)}\})\,dt + \int_{t=0}^{T} E\tilde{\beta}^{(2)}(T)\,dt$$

which yields (see (4.29)) the result of the theorem.

4.4   Convergence zone analysis

Consider here the first observer scheme.

1. If the matrix $C^{\mathsf{T}}C$ has incomplete rank, so that, $\lambda_{\min}(C^{\mathsf{T}}C) = 0$, then $\alpha_{P,\delta} := \lambda_{\min}(P^{-1/2}[C^{\mathsf{T}}C + \delta I]P^{-1/2}) = \delta\lambda_{\min}(P^{-1}) > 0$ if and only if the regularising parameter δ is strictly positive, i.e., $\delta > 0$. If $\delta = 0$, then $\tilde{\delta} = 0$ and, hence, there is no sliding mode effect at all. In this case ($\delta = 0$), the zone of convergence is defined by

$$\mu = \mu_{\delta=0} = \sqrt{\frac{\tilde{\beta}}{\tilde{\alpha}}} \tag{4.37}$$

The case of $\delta > 0$ and $\lambda_{\min}(C^{\mathsf{T}}C) = 0$ provides a smaller convergence zone with

$$\mu = \mu_{\delta>0} = \frac{\sqrt{\tilde{\beta}/\tilde{\alpha}}}{\sqrt{1 + (\tilde{\delta}/2\sqrt{\tilde{\alpha}\tilde{\beta}})^2} + \tilde{\delta}/2\sqrt{\tilde{\alpha}\tilde{\beta}}} < \mu_{\delta=0} \tag{4.38}$$

2. If the matrix $C^{\mathsf{T}}C$ has complete rank, so that, $\lambda_{\min}(C^{\mathsf{T}}C) > 0$, then $\alpha_{P,\delta} > 0$ for any $\delta \geq 0$ and we have a sliding mode effect ($\tilde{\delta} > 0$ always).

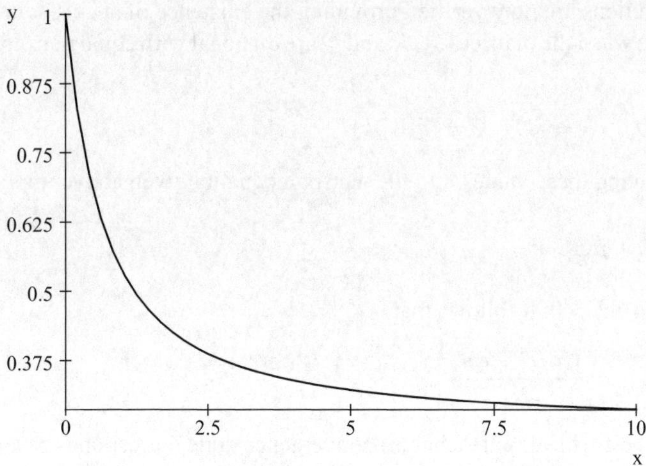

Figure 4.1   The function $\mu = \mu(k)$

3. Considering the size μ of the convergence zone (4.38) as a function of k (the sliding mode parameter), we have

$$\mu = \mu(k) = \frac{ak + b}{\sqrt{c^2 k^2 + ak + b} + ck}$$

$$a := \frac{\left(\Lambda_1 + 4\sqrt{n}\sqrt{\mathrm{tr}\,\Xi}\right)}{\tilde{\alpha}} \tag{4.39}$$

$$b := \|\Lambda_1\|\mathrm{tr}\,\Xi + \mathrm{tr}\{\bar{\sigma}_x P\}/\tilde{\alpha}, \quad c := \frac{\sqrt{\alpha_{P,\delta}}}{\tilde{\alpha}}$$

(P as well as P^{-1} has a low sensitivity to k for small enough δ)

The dependence $\mu = \mu(k)$ in (4.39) is shown (for $a = b = c = 1$) in Fig. 4.1. One can see that better estimation corresponds to higher values of k.

But for large k the positivity of the solution P of (4.10) may be lost. This property is guaranteed for an algebraic Riccati equation

$$\mathcal{A}^T P + P\mathcal{A} + P\mathcal{R}P + \mathcal{Q} = 0$$

if the matrix $\mathcal{A}$ is stable, the pair $(\mathcal{A}, \mathcal{R}^{1/2})$ is controllable, the pair $(\mathcal{Q}^{1/2}, \mathcal{A}^\mathsf{T})$ is observable and a particular local frequency condition holds (see [14]). To satisfy this condition, it is sufficient (see Appendix A [7]) that the following matrix inequality holds:

$$\mathcal{A}^T \mathcal{R}^{-1}\mathcal{A} - \mathcal{Q} > \tfrac{1}{4}[\mathcal{A}^T \mathcal{R}^{-1} - \mathcal{R}^{-1}\mathcal{A}]\mathcal{R}[\mathcal{A}^T \mathcal{R}^{-1} - \mathcal{R}^{-1}\mathcal{A}]^T$$

In our case,

$$\mathcal{A} = \bar{A} - K_0 C, \quad \mathcal{R} = [\Lambda^{-1} + K_0 \Lambda_1^{-1} K_0^\mathsf{T}]$$

$$\mathcal{Q} = L([I + K_0 C]^{-1})^\mathsf{T} \Lambda[I + K_0 C]^{-1} + k\delta\lambda + Q$$

These conditions are now verified providing the existence of such a solution for the special case when the matrices $\mathcal{A}$, $\mathcal{R}$ and $\mathcal{Q}$ are diagonal with equal nonzero elements, that is,

$$\mathcal{A} = \tilde{a}I, \quad \mathcal{R} = \tilde{r}I, \quad \mathcal{Q} = (\tilde{q} + \tilde{p}k)I$$

Substituting these values into the matrix inequality given above implies

$$\frac{\tilde{a}}{\tilde{r}} - (\tilde{q} + \tilde{p}k) > 0$$

or, for $(\tilde{a}/\tilde{r}) - \tilde{q} > 0$, it follows that

$$k < k_{\max} := \frac{((\tilde{a}/\tilde{r}) - \tilde{q})}{\tilde{p}}$$

Thus, the 'best' (ideal, unreachable) convergence zone corresponds to $k = k_{\max}$ and is equal to

$$\mu = \mu(k_{\max}) = \left(\frac{ak_{\max} + b}{\sqrt{c^2 k_{\max}^2 + ak_{\max} + b} + ck} \right)^2$$

4.5   Conclusion

It has been shown that:

- The modified (with a linear correction term and δ-regularisation) concept of 'sliding mode observation' does work and provides good quality state estimates for the case of stochastic output noise: the average of the state estimation error is shown to be bounded asymptotically.
- Correct selection of the gain matrices K_0 and K of the mixed observer is related to a corresponding algebraic Riccati equation.
- The convergence zone is dependent on the process and observer properties and can be minimised by appropriate selection of the gain matrices.

4.6   References

1   DENG, H. and KRSTIC, M.: 'Output-feedback stochastic nonlinear stabilization', *IEEE Trans. on Automat. Contr.*, 1999, **44**, pp. 328–333

2   NANACARA, W. and YAZ, E.: 'Linear and nonlinear estimation with uncertain observations', Proceedings of *American Control Conference*, NY, USA, 1994, **2**, pp. 1429–1433

3   WOOK, K. H., PYUNG K. S., and GYEON, P. P.: 'A receding horizon Kalman FIR filter for linear continuous-time systems', *IEEE Transactions on Automatic Control*, 1999, **44**(11), pp. 2115–2120

4   YAZ, E. and AZEMI, A.: 'A numerical procedure for discrete-time nonlinear stochastic observer design', Proceedings of the 32nd IEEE Conference on *Decision and Control*, NY, USA, 1993, **2**, pp. 1237–1238

5   YAZ, E. and AZEMI, A.: 'Observer design for discrete and continuous non-linear stochastic systems', *International Journal of Systems Science*, 1993, **24**(12), pp. 2289–2302

6   YAZ, E. and AZEMI, A.: 'Extensions of deterministic and stochastic variable structure observers with applications to disturbance minimization', Proceedings of the *1992 American Control Conference*, IL, USA, 1992, **1**, pp. 881–885

7   POZNYAK, A., MARTINEZ-GUERRA, R., and OSORIO-CORDERO, A.: 'Robust high-gain observer for nonlinear closed-loop stochastic systems', *Mathematical Methods in Engineering Practice*, 2000, **6**, pp. 31–60

8   DRAKUNOV, S. V.: 'An adaptive quasi-optimal filter with discontinuous parameters', *Automation and Remote Control*, 1983, **44**, pp. 1167–1175

9   GARD, T.: 'Introduction to stochastic differential equations' (Marcel Dekker, NY and Basel, 1988)

10 BENSOUSSAN A.: 'Stochastic control of partially observable systems' (Cambridge University Press, Cambridge, UK, 1992). 'Nonlinear Systems', *Int. J. Control*, 1993, **57**, pp. 537–556

11 WALCOTT, B. L. and ZAK, H.: 'State observation of nonlinear uncertain dynamical systems', *IEEE Transactions on Automatic Control*, 1987, **AC-32**, pp. 166–170

12 WALCOTT, B. L. and ZAK, H.: 'Combined observer-controller synthesis for uncertain dynamical systems with applications', *IEEE Transactions on Systems, Man and Cybernetics*, 1988, **18**, pp. 88–104

13 ZAK, H. and WALCOTT, B. L.: 'State observation of nonlinear control systems via the method of Lyapunov', in ZINOBER, A. S. I. (Ed.); 'Deterministic control of uncertain systems' (Peter Peregrinus, Stevenage, UK, 1990, pp. 333–350)

14 WILLEMS, J. C.: 'Least squares optimal control and algebraic Riccati equations', *IEEE Transactions on Automatic Control*, 1971, **16**, pp. 621–634

Chapter 5

Discrete-time VSS

Čedomir Milosavljević

This chapter reviews some basic results obtained in the study of discrete-time (DT) variable structure control systems (DVSCS) theory during its twenty-five year history. For this purpose, the chapter is organised as follows: in Section 5.1 basic definitions, assumptions and remarks are introduced that are necessary for the connection with continuous-time (CT) variable structure control systems (CVSCS) and form an introduction of terminologies for DVSCS. Section 5.2 is a brief overview of the more notable works in DVSCS. Section 5.3 gives the definition of a quasi-sliding mode (QSM) and a DT sliding mode (DSM). In Section 5.4, the Lyapunov stability concept is used to define invariant sets in DVSCS. Section 5.5 gives DSM existence conditions as a new motion phenomenon that is not possible in CVSCS. In Section 5.6, a basic concept of DVSCS, which is founded on the DT equivalent control method and a boundary layer concept for the system with nominal parameters, is presented, while Section 5.7 introduces some disturbance estimation methods. Section 5.8 describes two methods of DVSCS with sliding sectors. In Section 5.9 basic properties of DVSCS are given. Design methods to establish sliding surfaces are summarised in Section 5.10. Section 5.11 gives numerical examples that illustrate the properties of some DVSCS algorithms. Some issues in the practical realisation of DVSCS are given in Section 5.12 and Section 5.13 contains a list of published papers and other work that has been used for the preparation of this chapter.

5.1 Discrete-time variable structure control (DVSC)

Consider the following CT plant

$$\dot{\mathbf{x}}(t) = (\mathbf{A} + \Delta\mathbf{A}(t))\mathbf{x}(t) + (\mathbf{B} + \Delta\mathbf{B}(t))\mathbf{u}(t) + \mathbf{D}\mathbf{v}(t), \tag{5.1}$$

where $\mathbf{x}(t) \in \Re^n$, $\mathbf{u}(t) \in \Re^m$ and $\mathbf{v}(t) \in \Re^l$ are, respectively, vectors of the state, control and external disturbance; the pair $(\mathbf{A}, \mathbf{B})$ are controllable, the matrices $\mathbf{A}, \mathbf{B}, \mathbf{D}$ are of appropriate dimensions and rank $\mathbf{B} = m$. The objective is to design a DT controller that will govern the given plant using VSC techniques with a DSM or more properly with a quasi-sliding mode (QSM).

Assumption 1.   Elements of $\mathbf{A}, \mathbf{B}, \mathbf{D}$ and upper and lower bounds of $\mathbf{v}(t)$ are known, the parameter variations $(\Delta\mathbf{A}(t), \Delta\mathbf{B}(t))$ and external disturbances are matched [1], i.e. rank $[\mathbf{B}|\Delta\mathbf{A}|\Delta\mathbf{B}|\mathbf{D}] = $ rank $\mathbf{B}$ is satisfied, and the plant (5.1) is minimum phase.

As it is well known, and established in detail in the previous chapters, the sliding mode control (SMC) technique, in the class of CVSC design, consists of the following two basic steps:

Step 1.   Choose a set of m hypersurfaces: $\mathbf{S}(\mathbf{x}) = [S_1(\mathbf{x}), S_2(\mathbf{x}), \ldots, S_m(\mathbf{x})]$, $S_i = \{\mathbf{x}|s_i(\mathbf{x}) = 0\}$, $i = 1, 2, \ldots, m$, each of which crosses the origin of the state space, $\mathbf{x} = \mathbf{0}$, in such a way that intersections of all the given hypersurfaces, denoted by $S_1(\mathbf{x}) \cap S_2(\mathbf{x}) \cap \cdots \cap S_m(\mathbf{x})$ produce an $n - m$ dimensional subspace

$$S_E = \{\mathbf{x}|S_E(\mathbf{x}) = 0\}, \tag{5.2}$$

which represents a desired stable dynamics of the plant (5.1).

Step 2.   Choose elements of the switching control vector $\mathbf{u}(t) = [u_1(t), u_2(t), \ldots, u_m(t)]^{\mathrm{T}}$:

$$u_i(t) = \begin{cases} u_i^+(t) & \text{if } s_i(t) > 0, \\ u_i^-(t) & \text{if } s_i(t) < 0, \end{cases} \qquad i = 1, 2, \ldots, m; \tag{5.3}$$

where $u_i^+(t) \neq u_i^-(t)$ are continuous functions of time, which provides:

a)   reaching of the subspace S_E in a finite time from any initial state $\mathbf{x}(0)$, and
b)   existence of the stable SM on the hypersurface S_E.

Digital realisation of the described design algorithm may be affected by partial or full introduction of DT signal processing:

D1)   the discretisation process is introduced only for the determination of the hypersurfaces, i.e. $s_i(\mathbf{x}(kT))$, $k \in N_+$, T-sampling time, but elements of the control (5.3) remain CT, $u_i = u_i(t)$;
D2)   the discretisation process is performed only on the control vector elements: $u_i = u_i(kT)$, $s_i = s_i(t)$;
D3)   complete discretisation is used: a) with dual-rate sampling time: $u_i = u_i(kT)$, $s_i = s_i(kT_1)$; b) with uniform sampling time: $u_i = u_i(kT)$, $s_i = s_i(kT)$, i.e.

$$u_i(kT) = \begin{cases} u_i^+(kT) & \text{if } s_i(kT) > 0, \\ u_i^-(kT) & \text{if } s_i(kT) < 0, \end{cases} \qquad i = 1, 2, \ldots, m. \tag{5.4}$$

Remark 1.   It is assumed that DT signal processing is realised by microprocessors, with A/D and D/A converters as zero-order-hold (ZOH) elements. Then the controls, $u_i^+(kT)$, $u_i^-(kT)$, in cases D2 and D3, remain constant over the sampling interval $kT \leq t < (k+1)T$, with first order breaks occurring at the sampling time instants $t = kT$. They may be treated as continuous functions (see Definition 2) and, therefore, satisfy the necessary SM existence conditions.

Because digital signal processing introduces a transport delay, the system with D1 is equal to a CVSCS with a QSM. Such systems have been analysed in many papers (for example [2, 3]), and will not be further explored in this chapter.

A QSM or ideal SM may be established on S_E in the systems of type D2 [4, 5], but these systems are not of wide practical interest and along with the systems of type D3a, whose characteristics are similar to those of D1 type systems, shall not be subject to further analysis. Our further interest shall be directed to the systems of type D3b. From this point of view, it is necessary to introduce a DT mathematical model of the plant (5.1). This model is

$$\mathbf{x}[(k+1)T] = \mathbf{A}_d \mathbf{x}(kT) + \mathbf{B}_d \mathbf{u}(kT) + \mathbf{d}_d(kT), \tag{5.5a}$$

$$\mathbf{A}_d = e^{\mathbf{A}T_s}; \quad \mathbf{B}_d = \int_0^T e^{\mathbf{A}\tau} d\tau \mathbf{B}; \quad \mathbf{d}_d = \mathbf{d}_v + \mathbf{d}_A + \mathbf{d}_B;$$

$$\mathbf{d}_v = \int_0^T e^{\mathbf{A}\tau} \mathbf{D}\mathbf{v}((k+1)T - \tau) d\tau;$$

$$\mathbf{d}_A = \int_0^T e^{\mathbf{A}\tau} \Delta\mathbf{A}((k+1)T - \tau)\mathbf{x}((k+1)T - \tau) d\tau; \tag{5.5b}$$

$$\mathbf{d}_B = \int_0^T e^{\mathbf{A}\tau} \Delta\mathbf{B}((k+1)T) d\tau$$

Remark 2.   If $(\mathbf{A}, \mathbf{B})$ is controllable, the pair $(\mathbf{A}_d, \mathbf{B}_d)$ is controllable for almost all choices of T.

Remark 3.   In general, the fact that the matching conditions hold for CVSCS does not necessarily mean that the same conditions also hold in a DVSCS, because the ZOH does not take place in the disturbance channels. However, the corresponding error introduced is $O(T^2)$ if a disturbance estimator is used [6]. From this point of view, it is reasonable to choose the sampling time T as small as possible.

Definition 1 [7].   The sampling time T shall be small, where the sampling time is considered small if any function that is expanded in powers of T can be approximated to some degree of accuracy by keeping only terms up to and including those of order T^2.

Definition 2 [6].   The DT control $u(kT)$ is said to be equivalent to the discontinuous one if $\nabla u(kT) = O(1)$, continuous if $\nabla u(kT) = O(T)$, smooth if $\nabla^2 u(kT) = O(T^2)$, where ∇ denotes the backward difference operator.

Remark 4.   By introducing discretisation, the DT plant model may become nonminimum-phase for some sampling time T. In the subsequent analysis, it is assumed that the sampling frequency is chosen such that the DT plant model remains minimum phase.

In the above definition of the control task, it was assumed that the plant (5.1) was subject to stabilisation. In practical applications of control systems, we also have requirements for regulation and tracking. The tracking problem may be defined in this way. A reference vector is defined as

$$\mathbf{x}_r(kT) = [x_{r1}, x_{r2}, \ldots, x_{rn}]^{\mathrm{T}} \tag{5.6}$$

and this should be tracked by the plant (5.1) using a DVSCS. For the given task, we introduce the error vector

$$\mathbf{e}(kT) = \mathbf{x}(kT) - \mathbf{x}_r(kT). \tag{5.7}$$

From (5.7), $\mathbf{x}(kT) = \mathbf{e}(kT) + \mathbf{x}_r(kT)$, and (5.5a) becomes

$$\mathbf{e}[(k+1)T] = \mathbf{A}_d e(kT) + \mathbf{B}_d \mathbf{u}(kT) + \mathbf{d}_d(kT) - \mathbf{x}_r((k+1)T) + \mathbf{A}_d \mathbf{x}_r(kT). \tag{5.8}$$

For the regulation problem, $\mathbf{x}_r(kT) = \text{const}$, $(\mathbf{x}((k+1)T) = 0)$. In both cases, additional terms in (5.8), which are the consequence of the reference vector, may be interpreted as an additional disturbance. Then (5.8) becomes

$$e[(k+1)T] = \mathbf{A}_d e(kT) + \mathbf{B}_d \mathbf{u}(kT) + \mathbf{d}(kT), \tag{5.9}$$

$$\mathbf{d}(k) = \mathbf{d}_d(k) + \mathbf{A}_d \mathbf{x}_r(k) - \mathbf{x}_r(k+1), \tag{5.10}$$

which is of the form in (5.5a), therefore, in the subsequent explanation (5.5a) will be used.

Remark 5.   For notational convenience, in the following sections, $\bullet(k)$ stands for $\bullet(kT)$.

5.2   Control for discrete-time systems (review of early works)

In this section, we give a brief review of the published papers and other work that is the basic source for development of DVSC as a subclass of VSS.

The first work in the area of DVSCS was published in Russia [8]. In this paper, the type D1 discretisation has been considered. The main intention was to optimise the sliding hyperplane to obtain the minimum difference between CSM and DSM. Stability problems were also considered. Viktorova [9] analysed a hardware realisation of CVSC with digital equipment. Up to 1980, no other paper had been published in this area. Papers published in the 1980s had been oriented to the compilation of CVSCS algorithms, to establish the real SM existence conditions and to determine a SM sector width [3, 4, 10–14]. For occurrence of a real SM in DVSC systems of D1–D3 type, the term zigzag motion was introduced [14], but the term QSM [15] is further commonly used in the literature. Similar research was carried out [16–18] which introduced the term pseudo-sliding mode (PSM) for QSM.

Necessary and sufficient conditions for the existence of the bounded QSM sector have also been studied [14, 19–22].

The DT equivalent control, Lyapunov stability method and DSM were first introduced by Bučevac [23] and Salihbegović [24], and further established by numerous researchers [25–28, 35].

In the 1990s a number of publications in the area of DVSCS were quickly established. By using a discrete Lyapunov function, Furuta [28] introduced a two-term control. The first term is the equivalent control, and the second is a three level relay action with a dead zone. Finally, the motion of Furuta's system occurred in a pre-defined sector. Furuta's approach was further exploited in a great number of papers [30–34]. This design concept will be detailed in Section 5.8.

Gao et al. [35] proposed a new approach to DVSCS synthesis by controlling the system dynamics not only in a QSM but in the reaching phase, too. The method is based on the so-called reaching law method for CVSCS [29, 36]. In this paper, the problem of load rejection was studied, and a QSM sector was determined. This concept includes, as a partial case, some previous published approaches. In this manner, Bartoszewicz [37] proposed a DVSCS with a non-stationary sliding surface and additional integral action. His work provides a control signal with lower maximum values and a reduced QSM band width compared to Gao's method.

Bartolini et al. [38] have shown how SMC with an adaptive control enables generation of motion close to the ideal DSM for system operation under uncertain conditions. DVSCS with adaptation have also been investigated [39].

The algorithm proposed by Golo and Milosavljević [40, 41] is based on Gao's reaching law method, but uses a difference equation with a δ-operator. This algorithm is robust and chatter-free even if significant non-modelled inertial dynamics in the control object occur. A model with a δ-operator was also used [39]. Approaches using a δ-operator are very useful for DVSC with a high sampling frequency and, therefore, this modelling approach will become more interesting in the future due to increases in microprocessor speed. One of the interesting approaches, which is called the CVSC chatter-free approach [42], may be used with DT signal processing without any adaptation and remains chatter-free for the nominal plant.

Because DVSCS do not provide invariance to load disturbances, for improving their robustness, it is necessary to introduce a disturbance observer. One effective method is that of the delayed disturbance estimator [6, 43, 44]. This method will be explained in Section 5.7. For disturbance rejection, other methods were proposed by Gao *et al.* [35] and Bartoszewicz [37]. Tang and Misawa [45] studied the multivariable system with unmatched additive uncertainties using one sliding hyperplane.

The main difference between the design of DVSCS and CVSCS is in the determination of the switching hyperplane parameters. Some design suggestions may be found [34, 41, 46]. Tracking DVSCS was analysed [30, 39].

The previously mentioned contributions assume that all the state coordinates are available for direct measurement. Unfortunately, in most real systems, this is not possible. Then it is necessary to use state observers. Problems of observer design for DVSCS were studied [47–49]. One of the problems in DVSCS with or without state observers is computational time delay [50, 51]. DVSC using measured outputs is also analysed [31, 45, 52].

In the last few years, attention has been directed to DTVSC with a second-order sliding mode [53, 54].

There are plenty of other published papers and contributions in the area of DVSCS. In the above brief review, it was not possible to mention them all.

5.3   Definition of sliding mode and quasi-sliding modes in discrete-time

In this section, QSM and DSM in the systems described by the model (5.5) and (5.4) shall be defined. Generally speaking, because of sampling and the fact that the matching conditions are not fully satisfied, it is not possible to obtain a SM in the sense of a CSM, but a QSM [15] or PSM [18] will result. In DVSCS, motions that are not possible in CVSCS may occur. This motion is called ideal DSM (which will be abbreviated to DSM in what follows); this only occurs in the nominal system. The terms QSM and DSM need further definition.

Definition 3 [35].   The QSM is the motion that satisfies the following conditions:

a)   once the trajectory of the system first crosses the switching hypersurface S_E, it will cross again at every successive sampling time, resulting in a zigzag motion around S_E;

b)   the size of each successive zigzagging step is not increasing and hence the trajectory stays within a specified band.

This definition is restrictive. Motion in a predefined sector may have occurred without the sliding manifold being crossed at every successive sampling time. Those systems are of important practical interest [15, 18, 28, 32, 37].

Definition 4 [37].   The QSM is the motion in a predefined ε-vicinity of the sliding hypersurface $S_E = 0$ such that the system trajectory, after entering this band, never abandons it, i.e. $|S_E(k)| \leq \varepsilon$, where the positive constant ε is called the quasi-sliding mode band width.

Definition 5 [27, 38].   In the DVSC: $\mathbf{x}(k+1) = \mathbf{f}(\mathbf{x}(k))$, $\mathbf{x} \in \mathfrak{R}^n$ a DSM takes place on the subset M of the manifold $S_E = \{\mathbf{x} | S_E(\mathbf{x}) = 0\}$ if there exists an open neighbourhood U of this subset such that for $\mathbf{x} \in U$ it follows that $S_E(\mathbf{f}(\mathbf{x})) \in M$.

It is clear that a DSM may be defined concisely as:

Definition 6.   DSM is such a motion that for $\mathbf{x}(k = k^*) \in S_E$ implies $\mathbf{x}(k^* + j) \in S_E$, $\forall j \in N_0$.

In DVSC, regardless of the occurrence of a DSM, trajectories of motion of the system, in inter-sample time intervals, are in an open neighbourhood of the sliding manifold S_E. From this point of view, Definition 4 incorporates all previous definitions of QSM as well as the definition of DSM.

5.4   Lyapunov stability and invariant sets in discrete-time systems

For the stability of the DSMCS (5.5), (5.4) and to have a desired QSM motion, described by $S_E = 0$, it is necessary to bring the system state to the manifold (5.2) or to its neighbourhood, from any initial condition $\mathbf{x}(0)$, and to steer the state in the prescribed ε-vicinity of (5.2), regardless of the action of any bounded disturbance. In order to fulfil these requirements, as in the case of CVSCS, there are different approaches, called switching schemes (SS) [35]:

SS1.   Fixed-order switching scheme: $\mathbf{x}(0) \to S_1 \to S_1 \cap S_2 \to S_1 \cap S_2 \cap S_3 \to \cdots \to S_E$.

SS2.   Free-order switching scheme: $\mathbf{x}(0) \to S_i \to S_p \cap S_q \to \cdots \to S_E$, where S_i denotes any of the m hypersurfaces and is used to label that which is reached first.

SS3.   Eventual sliding mode switching scheme: $\mathbf{x}(0) \to S_E$, without prescription of a QSM arising on other switching hypersurfaces.

SS4.   Decentralised switching scheme. The system is treated as m single-input subsystems, each having a scalar switching function and an associated sliding mode. The systems are coupled in general. However, any interaction is treated as a disturbance or a precompensator may be used to obtain a non-interactive or diagonally dominant plant.

For further explanation, for simplicity, assume an SS4 switching scheme, i.e. scalar type control and organisation of a QSM on any of the m switching hypersurfaces $S_i = S$.

In general, we may decompose DSMCS motion into three phases: reaching phase, QSM phase and steady-state phase [35]. Another way [55] is to divide motion of the DSMCS into three phases: reaching phase, switching phase and chattering phase.

Definition 7.   The DVSCS is said to be in the reaching phase if

$$\operatorname{sgn}(s(k+1)) = \operatorname{sgn}(s(k)), \qquad k \in (0, K) \quad \text{and} \quad |s(k+1)| < |s(k)|. \qquad (5.11)$$

Definition 8.   The DVSCS is said to be in the chattering phase if

$$\operatorname{sgn}(s(k+1)) = -\operatorname{sgn}(s(k)), \qquad \forall k. \qquad (5.12)$$

Definition 9. Steady-state motion of a DSMC system is bounded motion in the ε-vicinity of the system error equilibrium.

Remark 6. For further explanation, conventionally, a chattering mode will denote any motion given by Definition 3; PSM – any motion given by Definition 4, and a QSM – any motion in the ε-vicinity of switching hypersurface S_i including DSM.

For the stability of the DSMCS, it is necessary to satisfy conditions given by the following definitions [56]:

Definition 10. S is stable, relative to the system (5.5), (5.4) iff $\forall \varepsilon \in R_+$, $\exists \delta = \delta(\varepsilon) \in R_+$ so that the distance $d[\mathbf{x}(0), S] < \delta$ implies that $\mathbf{X}(k, \mathbf{x}(0), u(\cdot))$ exists $\forall k \geq 0$ and $d_{N_o}[\mathbf{X}(k, \mathbf{x}(0), u(\cdot)), S] < \varepsilon$.

Definition 11. S is attractive (globally) relative to the system (5.5), (5.4), iff $\forall \Delta \in (0, \infty)$ such that $d[\mathbf{x}(0), S] < \Delta$ implies $\lim_{k \to \infty}\{d[\mathbf{X}(k, \mathbf{x}(0), u_i(\cdot)), S]\} = \lim_{k \to \infty}\{d[(\mathbf{x}(k)), S]\} = 0$.

Definition 12. S is (globally) stable relative to the system (5.5), (5.4), iff S is stable and (globally) attractive at the same time.

5.5 'Sliding conditions' in discrete-time

For the system (5.1), (5.3), SM existence conditions on the hypersurface S are given by the relation

$$s(\mathbf{x})\dot{s}(t) < 0, \tag{5.13}$$

proposed in 1955 by Dolgolenko [see 57] which may be obtained by using the second Lyapunov stability method with the Lyapunov function $V(\mathbf{x}) = \frac{1}{2}s^2$. In the earlier works of CVSCS, instead of (5.13), the following local sliding mode conditions

$$\lim_{s \to 0^+} \dot{s}(\mathbf{x}) < 0; \quad \lim_{s \to 0^-} \dot{s}(\mathbf{x}) > 0, \tag{5.14}$$

have been very often used. The conditions (5.13) and (5.14) were translated into the DT domain as

$$s(k)\Delta s(k) < 0 \quad \text{and} \quad \lim_{s \to 0^+} \Delta s(\mathbf{x}) < 0; \quad \lim_{s \to 0^-} \Delta s(\mathbf{x}) > 0;$$

$$\Delta s = s(k+1) - s(k). \tag{5.15}$$

These conditions are necessary but not sufficient [15]. They do not guarantee a stable (convergent) QSM. The necessity of (5.15) was effectively proved by Sira-Ramirez [21]. Using Lyapunov's second stability method for DT systems and taking the following positive definite Lyapunov function candidate $V(\mathbf{x}(k)) = |s(k)|$, necessary and sufficient conditions for the existence of a QSM have been derived in the following form [24]

$$|s(k+1)| < |s(k)|, \tag{5.16}$$

which was decomposed [19] into the two inequalities:

$$[s(\mathbf{x}(k+1)) - s(\mathbf{x}(k))]\mathrm{sgn}\{s(\mathbf{x}(k))\} < 0,$$
$$[s(\mathbf{x}(k+1)) + s(\mathbf{x}(k))]\mathrm{sgn}\{s(\mathbf{x}(k))\} \geq 0. \tag{5.17}$$

The first inequality in (5.17) is only another form of (5.15) and therefore denotes the necessary sliding mode existence conditions. The second inequality gives sufficient conditions for the convergence, or stability, of the QSM. This relation indicates that if a stable QSM exists, then a phase trajectory hypersurface crossing will occur in every successive sampling interval and the distance of the phase point from the sliding surface at the $(k+1)$th sampling interval is not greater than that at the previous sampling time. The conditions (5.17) actually impose upper and lower bounds on the control, which depend on the distance of the system state from the sliding surface [20]. The same conditions may be derived by using the Lyapunov function candidate $V(\mathbf{x}(k)) = s^2(k)$, which yields

$$\Delta V(\mathbf{x}(k)) < 0 \Rightarrow s^2(k+1) - s^2(k) \Rightarrow [s(k+1) - s(k)][s(k+1) + s(k)] < 0. \tag{5.18}$$

Multiplying (5.18) by $\mathrm{sgn}^2(s)$, taking into account the necessary conditions, we may obtain (5.17). Furuta [28] derives QSM existence conditions in the form

$$\Delta V(\mathbf{x}(k)) = s^2(\mathbf{x}(k+1)) - s^2(\mathbf{x}(k)) = 2s(\mathbf{x}(k))\Delta s(\mathbf{x}(k)) + \Delta s^2(\mathbf{x}(k)) < 0. \tag{5.19}$$

Finally, a convergent QSM regime exists on S iff [21]

$$|s(k+1)s(k)| < s^2(k).$$

Taking into account Definitions 5 and 6, the DSM may be defined by

$$s(k+1) = s(k) = 0, \quad \forall k \geq k^* \in N_o \subset \Re^+. \tag{5.20}$$

5.6   DVSC with attractive boundary layer

In this section, we define the equivalent control for DVSCS and some methods for reaching phase organisation.

Assumption 2.   It is assumed that in the system (5.5), (5.4) the disturbances are measurable and therefore may be fully compensated. For the simplicity of explanation, we also assume a linear switching function

$$s(\mathbf{x}) = \mathbf{c}^{\mathrm{T}}\mathbf{x}(k); \quad \mathbf{c}^{\mathrm{T}} \in \Re^{1 \times n}. \tag{5.21}$$

Our intention is to reach the hyperplane $s(\mathbf{x}) = 0$ from any arbitrary state $\mathbf{x}(0)$ in a finite number of sampling-time periods. For the given system, this task may be realised by using the so-called one-step control. For one-step reaching we have

$$s(k+1) = \mathbf{c}^T\mathbf{x}(k+1) = 0 \Rightarrow \mathbf{c}^T\mathbf{A}_d\mathbf{x}(k) + \mathbf{c}^T\mathbf{b}_d u(k) = 0. \tag{5.22}$$

Solving for u, assuming that $\det(\mathbf{c}^T\mathbf{b}_d) \neq 0$, we obtain

$$u = u^{out} = -(\mathbf{c}^T\mathbf{b}_d)^{-1}\mathbf{c}^T\mathbf{A}_d\mathbf{x}(k). \tag{5.23}$$

This control is here termed the outside control. We further want to keep the state on the sliding surface $s = 0$, i.e. to satisfy condition (5.20):

$$\mathbf{c}^T\mathbf{x}(k+1) = \mathbf{c}^T\mathbf{x}(k) = 0 \Rightarrow \mathbf{c}^T\mathbf{A}_d\mathbf{x}(k) + \mathbf{c}^T\mathbf{b}_d u(k) = \mathbf{c}^T\mathbf{x}(k). \tag{5.24}$$

Solving for the control u, one can obtain

$$u = u^{in} = u_{eq} = u^{sl} = -(\mathbf{c}^T\mathbf{b}_d)^{-1}[\mathbf{c}^T\mathbf{A}_d\mathbf{x}(k) - \mathbf{c}^T\mathbf{x}(k)], \tag{5.25}$$

which is the sliding control, equivalent control or inside control. This control steers the system state onto the switching hyperplane $s = 0$ at the sampling-time moments $t = kT$.

Remark 7.   The u^{out} control and the u_{eq} control for the given linear DVSCS is the same control. Indeed, the term $\mathbf{c}^T\mathbf{x}(k)$ in (5.25) is equal to zero, because the system state is on the switching manifold. Consequently (5.25) is equal to (5.23), i.e.

$$u^{out}(k) = u_{eq}(k) = -(\mathbf{c}^T\mathbf{b}_d)^{-1}\mathbf{c}^T\mathbf{A}_d\mathbf{x}(k). \tag{5.26}$$

In this way, for a linear DVSCS, the reaching phase control and the SM phase control are unique and linear. This is the important difference between DVSCS and CVSCS.

Applying (5.26) to the system (5.5), we obtain a difference equation that describes the motion of the given system not only in the SM. Unfortunately, control (5.26), as a reaching phase control, may not be used in general, because its value is inversely proportional to the sampling time period [44] and may be very high if the sampling time is small and there is a big distance between the initial state and the hyperplane. To overcome this drawback, it is necessary to formulate a control dependent on distance. Let us introduce a boundary layer around the hyperplane S defined as

$$S_\sigma = \{\mathbf{x} \,|\, \|s(\mathbf{x})\| \leq \sigma\} \tag{5.27}$$

and a nonlinear control

$$u(k) = \begin{cases} u^{out}(k) & \text{if } \mathbf{x}(k) \notin S_\sigma, \\ u_{eq}(k) & \text{if } \mathbf{x}(k) \in S_\sigma. \end{cases} \tag{5.28}$$

This type of nonlinear control has different realisations. For example, Bučevac [23] used

$$u^{out}(k) = -(\mathbf{c}^T\mathbf{b}_d)^{-1}[\mathbf{c}^T\mathbf{A}_d\mathbf{x}(k) - \alpha s(k)], \qquad \alpha > 0, \tag{5.29}$$

for a system without disturbance; Su et al. [6] suggested

$$u^{out}(k) = -(\mathbf{c}^T\mathbf{b}_d)^{-1}[\mathbf{c}^T\mathbf{A}_d\mathbf{x}(k) - s(k) + K\operatorname{sgn}(s(k))]. \tag{5.30}$$

By choosing adequate α in (5.29) or K in (5.30), it is possible to determine the step size for the state to approach the boundary layer S_σ. But, because limitations are present in any real control system, it is more convenient to use a nonlinear control in the form

$$u(k) = \begin{cases} u_{eq}(k) & \text{if } |u_{eq}(k)| \le u_0, \\[2mm] u_0\dfrac{u_{eq}(k)}{|u_{eq}(k)|} & \text{if } |u_{eq}(k)| > u_0, \end{cases} \tag{5.31}$$

where u_0 is the maximum control allowed for the given system. It is proved by Bartolini et al. [38] that the control (5.31) ensures that the sliding hyperplane is attractive. In this way, the control system will be as fast as possible in the reaching phase. After the control enters the linear zone, a one step control is used and the systems state reaches the sliding surface in a finite time, remains on it and asymptotically moves to the equilibrium state. The control (5.31) is the so-called boundary layer control. This type of control is often recommended for CVSCS for chattering avoidance.

Another approach with the boundary layer concept is proposed by Golo and Milosavljević [41]. This method is based on a reaching law concept, introduced by Gao et al. [35]. The basic intention of Gao's method is to prescribe the dynamics of the system motion in the reaching phase. The reaching law is given in the form

$$s(k+1) - s(k) = -qTs(k) - \varepsilon T\operatorname{sgn}(s(k)), \qquad \varepsilon,\ q,\ (1-qT) > 0 \tag{5.32}$$

and always satisfies the reaching condition (Definition 7). A desirable reaching mode response can be achieved by judicious choice of parameters k and q and the width of the QSM band by choice of parameters ε, q and T.

From the given DT model (5.5), (5.4), (5.21) for the nominal plant, we first obtain

$$s(k+1) - s(k) = \mathbf{c}^T\mathbf{A}_d\mathbf{x}(k) + \mathbf{c}^T\mathbf{b}_d u(k) - \mathbf{c}^T\mathbf{x}(k). \tag{5.33}$$

By equalising the right sides of (5.32) and (5.33) and solving for $u(k)$ we obtain the control law

$$u(k) = -(\mathbf{c}^T\mathbf{b}_d)^{-1}[\mathbf{c}^T\mathbf{A}_d\mathbf{x}(k) - \mathbf{c}^T\mathbf{x}(k) + qT\mathbf{c}^T\mathbf{x}(k) + \varepsilon T\operatorname{sgn}(\mathbf{c}^T\mathbf{x}(k))]. \tag{5.34}$$

Substituting (5.34) into (5.5) gives the response of the DVSCS

$$\mathbf{x}(k+1) = [\mathbf{A}_d - \mathbf{b}_d(\mathbf{c}^T\mathbf{b}_d)^{-1}\mathbf{c}^T(\mathbf{A}_d - \mathbf{I} + qT\mathbf{I})]\mathbf{x}(k)$$
$$- \mathbf{b}_d(\mathbf{c}^T\mathbf{b}_d)^{-1}[\varepsilon T\operatorname{sgn}(\mathbf{c}^T\mathbf{x}(k))]. \tag{5.35}$$

The QSM band and the steady-state band are given, respectively, by

$$\left\{\mathbf{x}\,\middle|\,|s(\mathbf{x})| < \frac{\varepsilon T}{1-qT}\right\}, \qquad \{\mathbf{x}\,|\,|s(\mathbf{x})| < \varepsilon T\}. \tag{5.36}$$

Reference 41 starts from a mathematical model in the form

$$\delta \mathbf{x}(k) \overset{\Delta}{=} \frac{\mathbf{x}(k+1) - \mathbf{x}(k)}{T} = \mathbf{A}_\delta \mathbf{x}(k) + \mathbf{b}_\delta u(k); \qquad \mathbf{A}_\delta = \frac{\mathbf{A}_d - \mathbf{I}_n}{T}, \quad \mathbf{b}_\delta = \frac{\mathbf{b}_d}{T},$$

(5.37)

with sliding hyperplane

$$s = \mathbf{c}_\delta^{\mathrm{T}}(T)\mathbf{x}; \qquad \mathbf{c}_\delta^{\mathrm{T}} \mathbf{b}_\delta = 1$$

(5.38)

The reaching law is defined as:

$$\delta s(k) \overset{\Delta}{=} \frac{s(k+1) - s(k)}{T} = \mathbf{c}_\delta^{\mathrm{T}} \delta \mathbf{x}(k) = -\Phi(s(k), \mathbf{X}(k)),$$

$$\mathbf{X}(k) = \begin{bmatrix} \mathbf{x}(k) \\ \hat{\mathbf{x}}(k) \end{bmatrix} = \begin{bmatrix} \mathbf{x}(k) \\ \mathbf{x}(k-1) \end{bmatrix}; \quad \hat{\mathbf{x}}(0) \overset{\Delta}{=} \mathbf{x}(0).$$

(5.39)

From (5.39), taking into account (5.37), it is obtained that

$$\mathbf{c}_\delta^{\mathrm{T}} \delta \mathbf{x}(k) = \mathbf{c}_\delta^{\mathrm{T}} \mathbf{A}_\delta \mathbf{x}(k) + u(k) = -\Phi(s(k), \mathbf{X}(k))$$

$$\Rightarrow u(k) = -\mathbf{c}_\delta^{\mathrm{T}} \mathbf{A}_\delta \mathbf{x}(k) - \Phi(s(k), \mathbf{X}(k)).$$

(5.40)

Defining the boundary layer

$$S(T) = \{ \mathbf{X} \in \mathfrak{R}^{2n} \,|\, |\mathrm{s}| < T\varepsilon + Tf(\mathbf{x}(k), \mathbf{x}(k-1)) \}, \qquad \varepsilon > 0$$

(5.41)

it is proved that by using

$$\Phi(s(k), \mathbf{X}(k)) = \min\left(\frac{|s|}{T}, \sigma + q|s| \right) \mathrm{sgn}(s)$$

(5.42)

as the nonlinear part of the control (5.40) and the linear part $(-\mathbf{c}_\delta^{\mathrm{T}} \mathbf{A}_\delta \mathbf{x}(k))$, the subspace (5.38) is attractive from any initial conditions $\mathbf{x}(0)$. After reaching the given subspace, the hyperplane (5.38) is reached in one sampling-time period.

It also proved that the given system (5.37) with the control (5.40) is robust to bounded parameter variations in the state matrix $\mathbf{A}_\delta$ and any exogenous bounded disturbance $\mathbf{d}_d(t)$ if the matching conditions are satisfied. Moreover, the system may be designed with a given degree of exponential stability, the control signal is smooth and the system is chatter-free which is not a feature of Gao's approach. The disturbance rejection capability of this algorithm [41] may be improved by introducing a proportional-integral action or compensation of disturbance effects using a one step delayed disturbance estimator.

5.7 DVSC with disturbance estimation

The approaches given in the previous section have one drawback – the impossibility of measuring the disturbance, $\mathbf{d}_d(k)$, in almost all real systems. To solve this problem, different ways have been recommended in the literature: control law for robust control

[35, 37], adaptation mechanism using a model-following control system [38], or one step delayed disturbance estimator [6, 43].

It is clear that if the control (5.31) is used, in the reaching phase, excluding the last sampling-time period, it is not possible to introduce any action to totally compensate for the disturbances. But, (5.5) may be used to estimate $\mathbf{d}_d(k)$ in any motion phase. From (5.5) one can obtain

$$\mathbf{d}_d(k) = \mathbf{x}(k+1) - \mathbf{b}_d u(k) - \mathbf{A}_d \mathbf{x}(k). \tag{5.43}$$

This simple relation may not be used because the state $\mathbf{x}(k+1)$ cannot be predicted. If the disturbance is bounded and smooth (with bounded $\dot{\mathbf{d}}(t)$), it may be predicted as

$$\mathbf{d}_d(k) \approx \mathbf{d}_d(k-1) = \mathbf{x}(k) - \mathbf{b}_d u(k-1) - \mathbf{A}_d \mathbf{x}(k-1). \tag{5.44}$$

The error is [6]

$$\mathbf{d}(k) - \mathbf{d}(k-1) = \int_0^T e^{\mathbf{A}\tau} \int_{kT=\lambda}^{(k+1)T-\lambda} \dot{\mathbf{d}}(\lambda) d\tau d\lambda = O(T^2). \tag{5.45}$$

Now, the control in the linear zone will be

$$\mathbf{u}^l(k) = -(\mathbf{c}^{\mathrm{T}} \mathbf{b}_d)^{-1} [\mathbf{c}^{\mathrm{T}} \mathbf{A}_d \mathbf{x}(k) + \mathbf{c}_\delta^{\mathrm{T}} \mathbf{d}_d(k-1)], \tag{5.46}$$

which replaces u_{eq} in (5.28).

For Golo's algorithm [41], instead of (5.40) the control will be

$$u(k) = -\mathbf{c}_\delta^{\mathrm{T}} \mathbf{A}_\delta \mathbf{x}(k) - \Phi(s(k), \mathbf{X}(k)) - \mathbf{c}^{\mathrm{T}} \mathbf{d}(k-1). \tag{5.47}$$

The control algorithm proposed by Bartoszewicz [37], which is also based on a reaching law approach, gives similar characteristics to Golo's algorithm [41]. The main difference between this algorithm and those of Gao and Golo is the introduction of a non-stationary sliding hyperplane taking into account bounded state matrix uncertainty and exogenous disturbances. It is assumed that the lower d_l and the upper d_u disturbance bounds are known constants and

$$d_l \leq d(k) = \mathbf{c}^{\mathrm{T}} \Delta \mathbf{A}_d \mathbf{x}(k) + \mathbf{c}^{\mathrm{T}} \mathbf{d}_d(k) \leq d_u. \tag{5.48}$$

Then auxiliary values are introduced

$$d_o = \frac{d_l + d_u}{2}; \quad \delta_d = \frac{d_u - d_l}{2}. \tag{5.49}$$

The proposed reaching law strategy is

$$s(k+1) = d(k) - d_o + s_d(k+1), \tag{5.50}$$

where $d(k)$ is unknown and given by (5.48) and $s_d(k)$ is an a priori known function that satisfies some conditions specified in the original paper. Bartoszewicz [37] recommends a desired non-stationary hyperplane in the form

$$s_d(k) = \left(1 - \frac{k}{k^*}\right) s(0), \qquad k^* < \frac{s(0)}{2\delta_d}, \quad k = 0, 1, \ldots, k^*, \tag{5.51}$$

where k^* defines the number of sampling-time steps necessary to reach the given $s(\mathbf{x}(k)) = 0$.

In order to determine the control from (5.50) we have

$$\mathbf{c}^T \mathbf{A}_d \mathbf{x}(k) + \mathbf{c}^T \mathbf{b}_d u(k) + d(k) = d(k) - d_o + s_d(k+1)$$

$$\Rightarrow u(k) = -(\mathbf{c}^T \mathbf{b}_d)^{-1} [\mathbf{c}^T \mathbf{A}_d \mathbf{x}(k) - d_o + s_d(k+1)]. \tag{5.52}$$

By substituting the obtained control from (5.52), we obtain

$$s(k) = \mathbf{c}^T \mathbf{A}_d \mathbf{x}(k-1) - \mathbf{c}^T \mathbf{b}_d (\mathbf{c}^T \mathbf{b}_d)^{-1} [\mathbf{c}^T \mathbf{A}_d \mathbf{x}(k-1) - d_o + s_d(k)] + d(k-1)$$

$$\Rightarrow s(k) = d(k-1) - d_o + s_d(k) \stackrel{s_d(k)=0}{=} d(k-1) - d_o. \tag{5.53}$$

It is clear that for $k \geq k^*$ the system state satisfies the following inequality

$$|s(k)| = |d(k-1) - d_o| \leq \delta_d. \tag{5.54}$$

The obtained QSM band width is smaller than the QSM band width of Gao's approach [35] which is for the system with uncertainties and disturbance

$$|s(k)| \leq 2\delta_d + \varepsilon T. \tag{5.55}$$

For improving characteristics of the closed-loop control system, Bartoszewicz proposes [37] a modified strategy, which yields a control signal of the form

$$u(k) = -(\mathbf{c}^T \mathbf{b}_d)^{-1} \left[\mathbf{c}^T \mathbf{A}_d \mathbf{x}(k) - d_o + s_d(k+1) + \sum_{i=0}^{k} (s(i) - s_d(i)) \right] \tag{5.56}$$

By using (5.56) the QSM band is

$$|s(k)| < \delta_d. \tag{5.57}$$

5.8 DVSC with sliding sectors

In this section, we give some algorithms in which, in general, a QSM occurs in the non-stationary sliding sector around the hyperplane. There are two methods with different motion control inside sliding sector: M1) motion is generated by the equivalent control; M2) the system is in free motion.

The basic case M1 [28] considers applying a control signal with two components. The first component is the outside control (u^{out}) which provides reaching conditions, and the second is the inside control (u^{in}), the equivalent control (5.25), which leads the system inside the sector $\tilde{\delta}$. Wang and Wu [33] have suggested a simplification of Furuta's method, starting from a sliding mode model, obtained by using the control (5.25)

$$\mathbf{x}(k+1) = \hat{\mathbf{A}}\mathbf{x}(k); \qquad \hat{\mathbf{A}} = \mathbf{A}_d - \mathbf{b}_d(\mathbf{c}^T \mathbf{b}_d)^{-1} \mathbf{c}^T (\mathbf{A}_d - \mathbf{I}),$$

$$s(k) = \mathbf{c}^T \mathbf{x}(k) = 0, \tag{5.58}$$

in which elements of the switching hyperplane vector $\mathbf{c}^{\mathrm{T}}$ should be chosen such that (5.58) is stable and the matrix $\hat{\mathbf{A}}$ has distinct eigenvalues. Then there exists a transformation matrix $\mathbf{N}$ so that $\tilde{\mathbf{x}}(k) = \mathbf{N}^{-1}\mathbf{x}(k)$, and

$$\tilde{\mathbf{A}} = \mathbf{N}^{-1}\hat{\mathbf{A}}\mathbf{N} = \mathrm{diag}\,\{\lambda_i\}; \qquad |\lambda_i| < 1, \quad i = 1, 2, \ldots, n-1. \tag{5.59}$$

The transformed system (5.58) becomes

$$\tilde{\mathbf{x}}(k+1) = \tilde{\mathbf{A}}\tilde{\mathbf{x}}(k)$$

$$\tilde{\mathbf{c}}^{\mathrm{T}}\tilde{\mathbf{x}}(k) = 0; \qquad \tilde{\mathbf{c}}^{\mathrm{T}} = \mathbf{c}^{\mathrm{T}}\mathbf{N} = [\tilde{c}_1, \tilde{c}_2, \ldots, \tilde{c}_n]. \tag{5.60}$$

The outside control is

$$\mathbf{u}^{out} = K_d\tilde{\mathbf{x}}(k); \qquad K_d = f_0\mathbf{e}^{\mathrm{T}}; \qquad \mathbf{e}^{\mathrm{T}} = [1, 1, \ldots, 1] \in R^{1 \times n} \tag{5.61}$$

and the switching state-dependent gain f_0 is determined by

$$f_0 = \begin{cases} -\tilde{\delta} < f_0 < 0 & \text{for } \omega > \tilde{\delta}, \\ 0 & \text{for } -\tilde{\delta} \leq \omega \leq \tilde{\delta}, \\ 0 > f_0 < \tilde{\delta} & \text{for } \omega < \tilde{\delta}, \end{cases} \tag{5.62}$$

where

$$\omega = \frac{2s(k)}{\mathbf{c}^{\mathrm{T}}\mathbf{b}_d \sum_{i=1}^{n} x_i(k)}; \quad 0 < \tilde{\delta} < \left| \frac{2\tilde{c}_n}{\mathbf{c}^{\mathrm{T}}\mathbf{b}_d \|\mathbf{N}\|_1} \right|; \qquad \|\mathbf{N}\|_1 = \max_j \sum_{i=1}^{n} n_{ij}. \tag{5.63}$$

The algorithm for controller design is as follows:

Step 1.   Determine $\mathbf{c}^{\mathrm{T}}$ so that the system (5.58) is stable with distinct eigenvalues $|\lambda_i| < 1$, $i = 1, 2, \ldots, n-1$, $\lambda_n = 1$.

Step 2.   Find matrix $\mathbf{N}$ so that $\mathbf{N}^{-1}\hat{\mathbf{A}}\mathbf{N} = \tilde{\mathbf{A}} = \mathrm{diag}\{\lambda_i, 1\}$ and calculate $\tilde{\mathbf{c}}^{\mathrm{T}} = \mathbf{c}^{\mathrm{T}}\mathbf{N}$.

Step 3.   Choose $\tilde{\delta}$ satisfying (5.63).

Step 4.   Combine (5.25), (5.61), (5.62) to get the desired controller.

The M2 sliding sector control design method, proposed by Furuta and Pan [32, 58], is based on introducing a so-called P_dR_d-sliding sector. Inside the P_dR_d-sliding sector, the system is without control but is quadratically stable and chatter free. The P_dR_d-sliding sector is defined by

$$\mathsf{L}_d = \{\mathbf{x} | \mathbf{x}^{\mathrm{T}}(k)[\mathbf{A}_d^{\mathrm{T}}\mathbf{P}_d\mathbf{A}_d - \mathbf{P}_d]\mathbf{x}(k) \leq -\mathbf{x}^{\mathrm{T}}(k)\mathbf{R}_d\mathbf{x}(k), \ \mathbf{x}(k) \in \Re^n\}, \tag{5.64}$$

where $\mathbf{P}_d$ is an $n \times n$ positive-definite symmetric matrix, $\mathbf{R}_d$ is an $n \times n$ positive-semi-definite symmetric matrix, $\mathbf{R}_d = \mathbf{G}_d^{\mathrm{T}}\mathbf{G}_d$, $\mathbf{G}_d \in \Re^{l \times n}$ and $(\mathbf{G}_d\mathbf{A}_d)$ is an observable pair. Inside the given sector, the forward difference of the Lyapunov function $V(\mathbf{x}(k)) = \mathbf{x}^{\mathrm{T}}\mathbf{P}_d\mathbf{x}(k) > 0$ is

$$\Delta V(k) = V(k+1) - V(k) = \mathbf{x}^{\mathrm{T}}(k)[\mathbf{A}_d^{\mathrm{T}}\mathbf{P}_d\mathbf{A}_d - \mathbf{P}_d)\mathbf{x}(k)$$

$$\leq -\mathbf{x}^{\mathrm{T}}(k)\mathbf{R}_d\mathbf{x}(k), \qquad \forall \mathbf{x}(k) \in \mathsf{L}_d. \tag{5.65}$$

Given that a $P_d R_d$-sliding sector exists for any $\mathbf{P}_d$ and $\mathbf{R}_d$ defined before, it may be rewritten as

$$\mathsf{L}_d = \{\mathbf{x}|s^2(k) \le \delta^2(k),\ \mathbf{x}(k) \in \Re^n\}, \qquad s^2(k) = \mathbf{x}^{\mathrm{T}}(k)\mathbf{P}_{d1}\mathbf{x}(k) \ge 0;$$

$$\delta^2(k) = \mathbf{x}^{\mathrm{T}}(k)\mathbf{P}_{d2}\mathbf{x}(k) \ge 0, \tag{5.66}$$

where $\mathbf{P}_{d1}$ and $\mathbf{P}_{d2}$ are positive-semi-definite $n \times n$ matrices.

If a positive-definite-symmetric matrix $\mathbf{P}_d$ is used, obtained from the discrete Riccati equation

$$\mathbf{P}_d = \mathbf{Q} + \mathbf{A}_d^{\mathrm{T}}\mathbf{P}_d\mathbf{A}_d - \mathbf{A}_d^{\mathrm{T}}\mathbf{P}_d\mathbf{b}_d[1 + \mathbf{b}_d^{\mathrm{T}}\mathbf{P}_d\mathbf{b}_d]^{-1}\mathbf{b}_d^{\mathrm{T}}\mathbf{P}_d\mathbf{A}_d, \tag{5.67}$$

where $\mathbf{Q} \in \Re^{n \times n}$ is a positive-definite-symmetric matrix, then the $P_d R_d$-sliding sector may be defined as

$$\mathsf{L}_d = \{\mathbf{x}||s(k)| \le \delta(k),\ \mathbf{x}(k) \in \Re^n\}, \tag{5.68}$$

$$s(k) = \mathbf{c}_d^{\mathrm{T}}\mathbf{x}(k), \qquad \mathbf{c}_d^{\mathrm{T}} = \frac{\mathbf{b}_d^{\mathrm{T}}\mathbf{P}_d\mathbf{A}_d}{(1 - \mathbf{b}_d^{\mathrm{T}}\mathbf{P}_d\mathbf{b}_d)^{1/2}}, \tag{5.69}$$

$$\delta_k = [\mathbf{x}^{\mathrm{T}}(k)(\mathbf{Q} - \mathbf{R}_d)\mathbf{x}(k)]^{1/2}. \tag{5.70}$$

The control law

$$u(k) = \begin{cases} 0, & \mathbf{x}(k) \in \mathsf{L}_d, \\ -(\mathbf{c}_d^{\mathrm{T}}\mathbf{b}_d)^{-1}[\mathbf{c}_d^{\mathrm{T}}\mathbf{A}_d + K_d\,\mathrm{sgn}(\mathbf{c}_d^{\mathrm{T}}\mathbf{b}_d s(k)]\delta(k), & \mathbf{x}(k) \notin \mathsf{L}_d, \end{cases} \tag{5.71}$$

enables the system to be quadratically stable if $\mathbf{c}_d^{\mathrm{T}}\mathbf{b}_d$ is invertible, and

$$0 < K_d \le \min\left\{1, |\mathbf{c}_d^{\mathrm{T}}\mathbf{b}_d|\frac{\sqrt{1 + \mathbf{b}_d^{\mathrm{T}}\mathbf{P}_d\mathbf{b}_d}}{\mathbf{b}_d^{\mathrm{T}}\mathbf{P}_d\mathbf{b}_d}\right\}; \qquad K_d^2\mathbf{R}_d > (\mathbf{c}_d^{\mathrm{T}}\mathbf{A}_d)^{\mathrm{T}}(\mathbf{c}_d^{\mathrm{T}}\mathbf{A}_d).$$

$$\tag{5.72}$$

In the methods developed from the work of Furuta, robust stability with respect to exogenous disturbances has not been considered. Naturally, the disturbance may be compensated for by using the one step delayed disturbance estimator given above. The main drawback of Furuta's M2 approach is the non-smooth control and the inability to select the sliding hyperplane in advance.

5.9 Properties of DVSC

As mentioned above, the motion in a DVSCS consists of three different phases. The basic question is how the system's motion in these phases exists, what are its main characteristics and how much does it differ from those of a CVSCS?

In the reaching phase, characterised by relations (5.11), the sign of the control signal does not change. Depending on the control algorithm, the control signal during the dominant period of the reaching-time may be unchanged, if the control of type (5.31) is applied. For this type of control, the controlled variable in this period of motion has the same form as in the equivalent CVSCS with boundary layer or relay control. In the other case, when the control decreases, the motion is like that of a CVSCS with linearly dependent control. In any case, reaching-phase trajectories are smooth and differ little from the case of CVSCS. At the end of the reaching phase, depending on the control algorithm, the motion may be different. For a nominal system without uncertainties, the so-called soft descent to the sliding hyperplane may occur, if control algorithms of type (5.31), (5.40) and (5.52) is applied.

The QSM phase, generally speaking, occurs in a sector, whose width is dependent on the control algorithm used. Even if control algorithms with soft descent are applied, and the system has nominal conditions, motion in this phase, for continuous plant, takes place in a sector. Some of the above given algorithms, for example Gao's, were designed to chatter about the sliding hyperplane, according to relation (5.12). The width of the QSM sector is a design parameter in the nominal case. The algorithms of Bartolini et al. [38] (5.31), with adaptation or with disturbance estimation (5.44), give motion with smooth control and a QSM sector width of $O(T^2)$. The algorithm proposed by Golo (5.40) has similar characteristics without disturbance estimation; Bartoszewicz's algorithm (5.52) incorporates disturbance estimation. In any case, with or without disturbance estimation, the given algorithms, except that of Gao, prefer a higher sampling-time frequency. In DVSC systems with equivalent control or free motion in sliding sector, defined by relations (5.62) and (5.71), chattering does not exist.

The steady-state motion of DVSCS is not often investigated. Only a few papers have been published covering this problem [55, 59, 60]. This motion depends on the applied control algorithm as well as on the control system type: stabilisation, regulation or tracking type. Some interesting research was carried out [61]. The steady-state accuracy estimation of the VSCS was given and shown to depend on the system as well as on the switching function parameters.

The influence of the quantisation effects of A/D converters have not yet been analysed. The increase of DVSCS capabilities is limited by quantisation errors and by stochastic noise [6].

Taking into account that in real CVSCS, non-idealities such as a dead zone, hysteresis, time-delay, etc. may exist, real SM characteristics of these systems are not much better that those of QSM in DVSCS. Taking into account the flexibility of DT signal processing yields significant capabilities in signal processing for observation and estimation, it is likely that DVSCS will achieve a very notable place in control engineering practice.

5.10   Approaches to design the 'sliding surface' in discrete-time

There are different approaches to the design of the switching manifold. If a CT hyperplane is known, a primitive approach to DT hyperplane design uses a differential mapping method. In any case, the eigenvalues of the obtained DT hyperplane must be stable. A standard approach may be obtained by transformation of the original nominal system (5.5) into normal form. Using a transformation $\mathbf{x} = \mathbf{P}_1 \bar{\mathbf{x}}$, with [57]

$$
\mathbf{P}_1 = \mathbf{M}_c \begin{bmatrix} a_2 & a_3 & \cdots & a_n & 1 \\ a_3 & a_4 & \cdots & 1 & \\ \cdots & \cdots & & & \\ a_n & 1 & & \mathbf{0} & \\ 1 & & & & \end{bmatrix},
\tag{5.73}
$$

where $\mathbf{M}_c$ is the controllability matrix and a_i, $i = 1, 2, 3, \ldots, n$, are the characteristic polynomial coefficients of the system (5.5)

$$
D(z) = \det[z\mathbf{I} - \mathbf{A}_d] = z^n + a_n z^{n-1} + \cdots + a_2 z + a_1.
\tag{5.74}
$$

The original system (5.5) is transformed to

$$
\begin{bmatrix} \bar{\mathbf{x}}_1(k+1) \\ \bar{x}_2(k+1) \end{bmatrix} = \begin{bmatrix} \mathbf{A}_{11} & \mathbf{A}_{12} \\ \mathbf{A}_{21} & \mathbf{A}_{22} \end{bmatrix} \begin{bmatrix} \bar{\mathbf{x}}_1(k) \\ \bar{x}_2(k) \end{bmatrix} + \begin{bmatrix} 0 \\ 1 \end{bmatrix} u(k); \qquad \begin{bmatrix} \mathbf{A}_{11} & \mathbf{A}_{12} \\ \mathbf{A}_{21} & \mathbf{A}_{22} \end{bmatrix} = \mathbf{P}_1^{-1} \mathbf{A}_d \mathbf{P},
\tag{5.75a}
$$

$$
\bar{\mathbf{c}}_0^{\mathsf{T}} \bar{\mathbf{x}}_1(k) + \bar{x}_2 = 0; \qquad \bar{\mathbf{c}}^{\mathsf{T}} = \mathbf{c}^{\mathsf{T}} \mathbf{P}_1.
\tag{5.75b}
$$

In the equation (5.75a), x_2 is a scalar and plays the role of a control. Replacing $\bar{x}_2(k)$ from (5.75b) in (5.75a) and using the equivalent control method, we get

$$
\bar{\mathbf{x}}_1(k+1) = [\mathbf{A}_{11} - \mathbf{A}_{12} \bar{\mathbf{c}}_0^{\mathsf{T}}] \bar{\mathbf{x}}_1(k),
$$
$$
\bar{x}_2(k+1) = -\bar{\mathbf{c}}_0^{\mathsf{T}} \bar{\mathbf{x}}_1(k+1),
\tag{5.76}
$$

which is the equation of a DSM. If the pair $(\mathbf{A}_d, \mathbf{b}_d)$ is controllable, then $(\mathbf{A}_{11}, \mathbf{A}_{12})$ is also controllable. Under this condition, by choosing a vector $\mathbf{c}$, the eigenvalues of the system matrix $\tilde{\mathbf{A}} = [\mathbf{A}_{11} - \mathbf{A}_{12} \bar{\mathbf{c}}_0^{\mathsf{T}}]$ can be arbitrarily assigned. As a consequence, the stability of the DSM can be guaranteed.

An asymptotically stable system, ensuring $\bar{\mathbf{x}} \to 0$ as $t \to \infty$, is guaranteed by choosing the eigenvalues of the matrix $\tilde{\mathbf{A}}$ to lie within the unit circle. Obviously, the eigenvalues of $\tilde{\mathbf{A}}$ are only determined by $\bar{\mathbf{c}}_0^{T}$ since the characteristic polynomial is given by $z^{n-1} + \sum_{i=1}^{n-1} c_i z^{i-1}$. Let the eigenvalues be distinct and given by $z_i = e^{-\alpha_i T}$, $\alpha_i > 0$, then the elements of $\bar{\mathbf{c}}_0^{T}$ are determined by

$$
\bar{c}_i = \frac{1}{(i-1)!} \frac{d^{i-1} \prod_{j=1}^{n-1} (z - z_j)}{dz^{i-1}} \Big|_{z=0}.
\tag{5.77}
$$

Now, the vector $\mathbf{c}^{\mathsf{T}}$ defining the sliding hyperplane for the non-transformed system is given by

$$
\mathbf{c}^{\mathsf{T}} = [\bar{\mathbf{c}}^{\mathsf{T}} | 1] \mathbf{P}_1^{-1}.
\tag{5.78}
$$

Remark 8.   For the system given by model (5.37), the distinct eigenvalues are given by

$$
z_i = \frac{e^{-\alpha_i T} - 1}{T}, \qquad \alpha_i > 0, \quad \alpha_i \neq \alpha_j \qquad \text{for } i \neq j.
\tag{5.79}
$$

This traditional hyperplane design philosophy is focused on the asymptotic stability of the SM in the nominal system. It is known that the disturbance rejection capability depends on switching function design too. Spurgeon [46] proposed a method of hyperplane design taking into account the disturbance rejection capability. Starting from a Lyapunov function candidate to analyse the stability of the uncertain system (5.5), $V(\mathbf{x}) = \mathbf{x}^T\mathbf{P}\mathbf{x}$ with positive-definite-symmetric $n \times n$ matrix $\mathbf{P}$, and defining

$$\mathbf{c}^T = \mathbf{b}_d^T\mathbf{P}, \tag{5.80}$$

leads to the sliding mode equivalent control equation $\mathbf{x}(k+1) = \mathbf{A}_{eq}\mathbf{x}(k)$ with

$$\mathbf{A}_{eq} = [\mathbf{I} - \mathbf{b}_d(\mathbf{b}_d^T\mathbf{P}\mathbf{b}_d)^{-1}\mathbf{b}_d^T\mathbf{P}]\mathbf{A}_d. \tag{5.81}$$

It is proved that if $\|\mathbf{A}_{eq}\|_p = \sqrt{eig_{\max}(\mathbf{P}^{-1}\mathbf{A}_{eq}^T\mathbf{P}\mathbf{A}_{eq})} < 1$ is satisfied, then the given system is globally uniformly asymptotic stable. Moreover, if the disturbance is bounded by the relation

$$\|\mathbf{d}_d\|_p \leq \rho_o + \rho_1\|\mathbf{x}\|_p; \qquad \rho_o, \rho_1 > 0, \tag{5.82}$$

then the perturbed system is globally uniformly asymptotically stable about the ball centred at $\mathbf{0}$ and with radius, r, given by

$$r = \frac{\rho_o}{\sqrt{1 - \|\mathbf{A}_{eq}\|_p^2} - \rho_1}. \tag{5.83}$$

5.11   Numerical examples

Example 1.   Assume the continuous-time plant is a DC-motor whose position will be controlled. Its mathematical model (5.1) is given by: $\mathbf{A} = [0, 1; 0, -16]$; $\mathbf{b} = [0; -680]$; $\mathbf{d}(t) = [0 : d(t)]$ with neglected electrical time constant. The goal is to design a control system of regulator type with a QSM motion. We will design two different types of controllers. The first one, proposed in Reference 41, represents a reaching law boundary-layer concept, and the second, proposed in Reference 32, represents the sliding sector concept. These controllers will be termed as Golo's and Furuta's controllers, respectively, in the subsequent discussion. The advantages and main differences between these two approaches will be demonstrated.

Golo's controller.   For this controller, as was indicated in the original paper, it is better to have the sampling time as small as possible. Let the sampling time be $T = 0.0004$ s. The corresponding parameters of the discrete-time plant model according to (5.5) are: $\mathbf{A}_d = [1, 0.000\,3987; 0, 0.993\,62]$, $\mathbf{b}_d = [0.000\,05; 0.271\,13]$, and $\mathbf{A}_\delta = [0, 0.996\,81; 0, -15.948\,91]$, $\mathbf{b}_\delta = [0.135\,71; 677.8286]$, according to (5.37); $\mathbf{c}_\delta^T = [-0.0221, -0.001\,47]$ for $\alpha = 15$; $\mathbf{c}_\delta^T\mathbf{A}_\delta = [0, -0.001\,466\,18]$. Let $q = 10$, $\sigma = 20$. The control, according to (5.40) and (5.42), is

$$u(k) = -0.001\,466\,18x_2(k) - \min\{2500|s(k)|, 20 + 10|s(k)|\}.$$

Furuta's M2 controller.   In the original paper [32], no recommendations on the choice of sampling-time were given. Let the sampling-time be the same as for the previous controller: $T = 0.0004$ s. For $\mathbf{Q} = \mathbf{I}_2$, $\mathbf{R}_d = r\mathbf{Q}$, $r = 0.05$ the parameters of Furuta's controller, according to (5.67), (5.69) and (5.72) are: $\mathbf{P}_d = [2504.9, 3.7; 3.7, 4.13]$, $\mathbf{c}_d^{\mathrm{T}} = [-1.0000 - 0.9751]$, $\mathbf{c}_d^{\mathrm{T}}\mathbf{b}_d = 0.2644$; $(\mathbf{c}_d^{\mathrm{T}}\mathbf{b})^{-1} = 3.7817$; $\mathbf{c}_d^{\mathrm{T}}\mathbf{A}_d = [-1, -0.9693]$; $K_d \le 0.994$. Using (5.71) with the given parameters, the control is obtained as

$$u(k) = \begin{cases} 0, & \mathbf{x}(k) \in \mathsf{L}_d, \\ -3.782[x_1(k) + 0.969x_2(k) + K_d\,\mathrm{sgn}(0.2644s(k))\delta(k), & \mathbf{x}(k) \notin \mathsf{L}_d, \end{cases}$$

$$s(k) = \mathbf{c}_d^{\mathrm{T}}\mathbf{x}(k) = x_1(k) + 0.975x_2(k); \quad 0 < K_d \le 0.994;$$

$$\mathbf{x}(k) \in \mathsf{L}_d : |s(k)| \le \delta(k) = \sqrt{r}\sqrt{x_1^2 + x_2^2}.$$

In Figs 5.1, 5.2 and 5.3 phase plane plots, error signal, switching function dynamics and control signals for both Golo's and Furuta's control systems are given, respectively. It is evident, from Figs 5.1 and 5.2, that Furuta's approach gives a more sluggish response than Golo's approach for the given sampling time. The responses of Furuta's system for different sampling-times are given in Fig 5.2. From this figure it is evident that for rise-time improvement in the system with Furuta's controller, the sampling-time should be increased. It may be seen from Fig. 5.2 that Furuta's system with a sampling-time of 0.4 s gives a rise-time close to Golo's system and remains quadratically stable. This is an excellent feature of Furuta's approach. However, the control signal and switching function in Furuta's system are not smooth (Fig. 5.3b). Further, Furuta's fundamental algorithm is not robust to load disturbances. It may be concluded that Furuta's system is very useful for small sampling frequency and, therefore, for controlling low speed plants. Because the given plant – the DC-motor – is a relatively high-speed plant, in the further discussion only the simulation results of Golo's

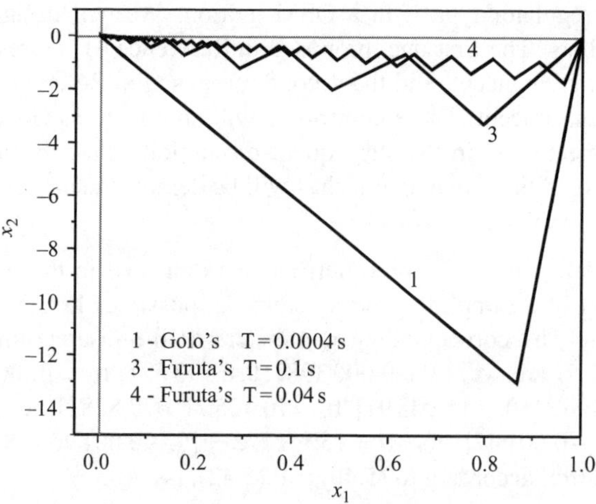

Figure 5.1   Phase plane

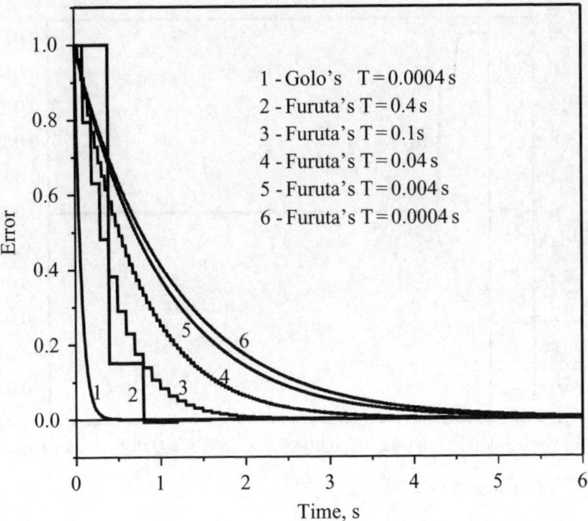

Figure 5.2   Error dynamics

approach will be presented. The goal is to show that the DSMCS has satisfactory robustness to parameter variations, external disturbances and unmodelled dynamics.

For robustness assessment of this control algorithm, variations in plant parameters were realised in the wide range (200–1000 for gain and 30–120 ms for time-constant) without changing the regulation process character. From Fig. 5.3a, it is evident that for the nominal plant model, the sliding line is reached in finite time and with soft descent. Figure 5.4 shows the disturbance rejection capabilities of the given nominal control system (curve NCS), and by using disturbance compensation: with one step delayed estimator (curve DE), or with proportional-integral compensator, $u_{PI}(k) = u(k) + 16T(u(k) - u(k-1))$, on the control object input (curve PI).

If the controller is applied on the real object with electrical time constant of 4 ms, the following results are obtained: Fig. 5.5 shows the phase plot from the real plant with the designed controller, and Fig. 5.6 displays the switching function dynamics and control signal. It is evident that chattering does not exist. The controlled variable is close to that given in Fig. 5.2 (marked 'Golo's').

Example 2.   Let the above designed Golo control system be used as a system for tracking a desired angular trajectory given by the relation

$$\mathbf{x}_d(t) = \begin{cases} \dfrac{At^2}{2t_1} & \text{for } 0 < t \le t_1, \\[2ex] A\left(\dfrac{t-t_1}{2}\right) & \text{for } t_1 < t \le t_2, \\[2ex] -\dfrac{A[t-(t_1+t_2)]^2}{2t_1} & \text{for } t_2 < t \le t_3. \end{cases}$$

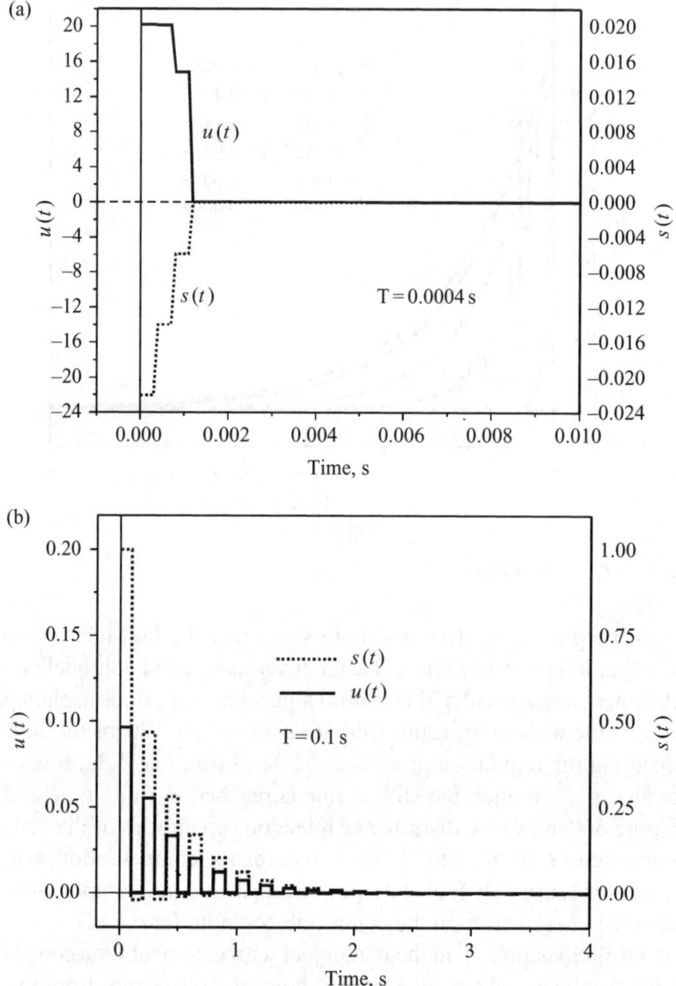

Figure 5.3   Switching function dynamics s(t) and control signal u(t) for the nominal plant model: a) Golo's system; b) Furuta's system

For this requirement we introduce new state coordinates: error: $x_1 - x_d$, and its differential: $x_2 - \dot{x}_d$.

Let $A = 2$, $t_1 = 1$, $t_2 = 3$, $t_3 = 4$. Figures 5.7a and 5.7b display the desired trajectory (x_d), the output trajectory (x_1) and the error signal of the system with the designed controller in the presence of the above given exogenous disturbance without (Fig. 5.7a) and with a disturbance estimator or PI compensator (Fig. 5.7b).

It may be concluded that the illustrated DVSCS is robust and has characteristics similar to an appropriately designed CVSCS.

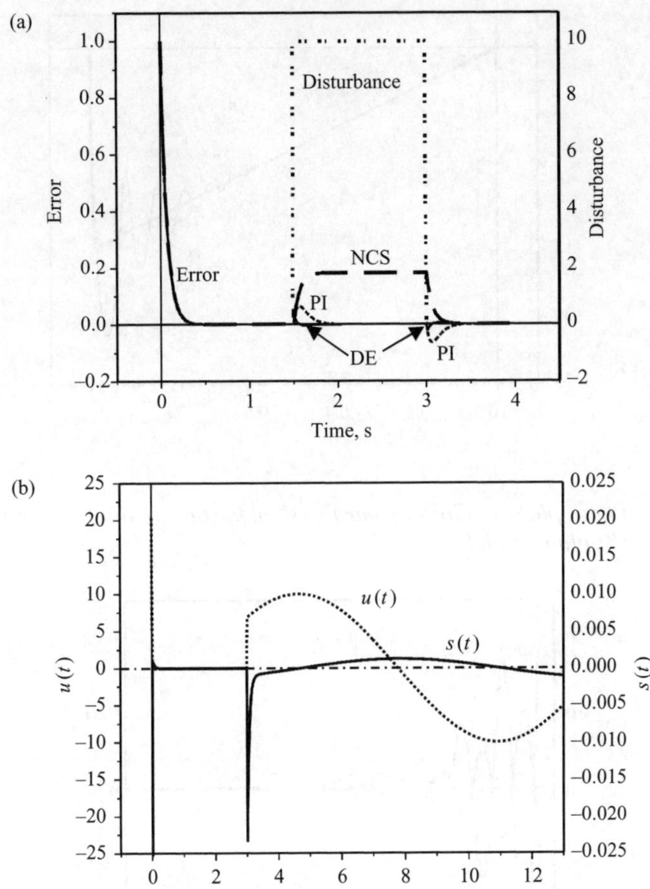

Figure 5.4   *Disturbance rejection capabilities of Golo's nominal system: a) error dynamics: NCS = non-compensated system, DE = compensated system with disturbance estimator, PI compensated system with PI compensator; b) switching function dynamics and control signal with periodic disturbance and PI compensator. a) $d(t) = 10[h(t-1.5) - h(t-3)]$; b) $d(t) = 10\sin(0.5t)h(t-3)$*

5.12   Issues in the realisation of DT SMC

From the above discussion and the given numerical examples, it may be seen that a DVSCS can be realised based on different approaches:

1. By using CVSC algorithms for control design and elements of DT signal processing for practical realisation. Such an approach may lead to instability of the control system if the sampling-time period is not sufficiently small. Conversely, a high sampling frequency may cause excitation of unmodelled dynamics of

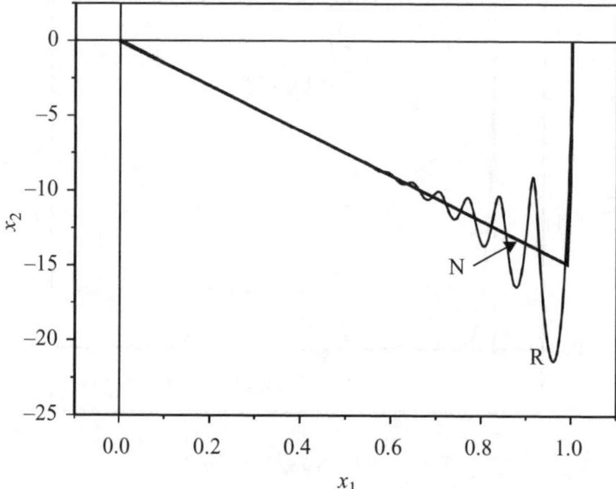

Figure 5.5   Phase plane of Golo's control system for the nominal (N) and of the real (R) plant model

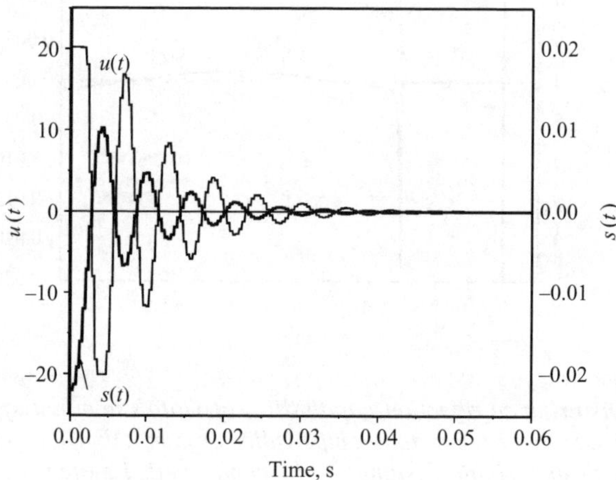

Figure 5.6   Switching function dynamics s(t) and control u(t) for the system with real plant

the plant. This type of DVSCS should be used preferably for control only of non-electromechanical systems.

2.   By using DVSC algorithms for the control synthesis and a corresponding fully digital realisation of the controller. This realisation approach has two basic concepts:

   a)   A boundary layer approach with soft descent, obtained by using the discrete-time equivalent control method for the nominal plant model and applying disturbance observers or other adaptive control.

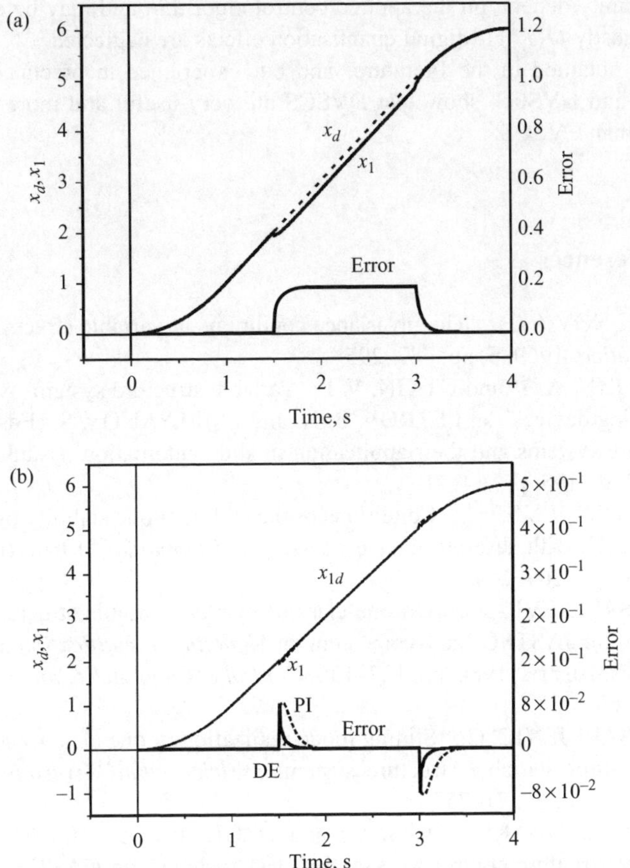

Figure 5.7   Tracking trajectories and error signals: a) without disturbance compensation; b) with disturbance compensation. DE = by disturbance estimator; PI = by PI compensator

b)   A sector control approach where the main feature is the existence of a QSM in a predetermined sector around the switching hyperplane. There are two basic sector methods: (i) Gao's method, which provides only stable motion in the sense of Lyapunov, and (ii) Furuta's method, which provides a quadratically asymptotically stable system. There are two approaches by Furuta: using in the sliding sector (a) only the equivalent control or (b) free motion control.

Furuta's sliding sector methods are suitable for systems with low sampling frequency whereas Gao's sector method demands a higher sampling frequency. However, a too small sampling-time period may cause problems with chattering.

Without considering the control algorithm, a DVSCS with a continuous plant theoretically may not be asymptotically stable but only stable, because the steady-state motion is always in a bounded ε-vicinity of the equilibrium state. The dimensions of the ε-vicinity depend on the applied control algorithm and may be estimated as $O(T^2)$ or mostly $O(T^3)$ if digital quantisation effects are neglected.

Results obtained in the literature, and our experience in practical realisation of CVSCS and DVSCS, show that DVSCS are very useful and more suitable for realisation than CVSCS.

5.13   References

1 DRAŽENOVIĆ, B.: 'The invariance conditions in variable structure systems', *Automatica*, 1969, **5**, pp. 287–295

2 BAKAKIN, A. V. and UTKIN, V. I.: 'Variable structure systems with delay in switching devices', in PETROV, B. N. and EMELYANOV, S. (Eds): 'Variable structure systems and their application in slide automation' (Nauka, Moscow, 1968), vol. 198, pp. 64–71

3 MILOSAVLJEVIĆ, Č.: 'Conditions of the sliding mode stability of the second order VSS with discrete data processing', *Automatika*, 1980, (in Serbian), **21**(5–6), pp. 269–274

4 MILOSAVLJEVIĆ, Č.: 'On one class of discrete variable structure systems', Proc. of the IASTED Int. Symposium on *Modeling, Identification and Control*, Insbruck, Austria, 1984, pp. 127–130; *Control & Computer (Can.)*, 1988, **16**(3), pp. 56–60

5 MILOSAVLJEVIĆ, Č.: 'Sliding mode realization in one class of second order discrete-time variable structure systems', *Elektrotehnički vestnik*, 1986, (in Slovenian), **2**, pp. 71–75

6 SU, W.-C., DRAKUNOV, S. V., and ÖZGÜNER, Ü.: 'Implementation of variable structure control for sampled data systems', in GAROFALO, F. and GLIELMO, L. (Eds): 'Robust control via variable structure and Lyapunov techniques', 1996 (Springer Verlag, London, 1996), pp. 87–106

7 CORRADINI M. L. and ORLANDO, G.: 'Variable structure control of discretized continuous-time systems', *IEEE Trans.*, 1998, **IE-43**(9), pp. 1329–1334

8 BAKAKIN, A. V. and TARAN, V. A.: 'Digital equipments used in control systems of variable structure', Proc. of *Automatic Control and Components of Computers*, 1967, (in Russian), pp. 30–39

9 VIKTOROVA, V. S.: 'Digital variable structure controllers', in PETROV, B. N. and EMELYANOV, S. V. (Eds): 'Variable structure systems and their use in flight control' (Nauka, Moscow, 1968 (in Russian)), pp. 198–207

10 DOTE, Y. and HOFT, R. G.: 'Microprocessor base sliding mode controller for DC motor drives', *IEEE IAS Conference Record*, Cincinnati, U.S.A., 1980, pp. 641–645

11 DOTE, Y., MUNABE, T., and MURAKAMI, S.: 'Microprocessor-based force control for manipulator using variable structure with sliding mode', Proc. IFAC Symposium on *Control in Power Electronics and Electrical Drives*, Lausanne, Switzerland, 1983, pp. 145–149

12   ESPAÑA, M. D. and ORTEGA, R. S.: 'Un nuevo metodo de sintesis de sistemas de estramatura variable con reduccion del castaneo', 3rd Congresso Brasileiro de Automatica, Rio de Janeiro, 1980, pp. 93–99

13   LIN, S.-C. and TSAI, S.-J.: 'A microprocessor-based incremental servo system with variable structure', *IEEE Trans.*, 1984, **IE-31**(4), pp. 313–316

14   MILOSAVLJEVIĆ, Č.: 'Some problems of the discrete variable structure systems control law realization', Ph.D. Thesis, University of Sarajevo (YU), 1982 (in Serbian)

15   MILOSAVLJEVIĆ, Č.: 'General conditions for the existence of a quasi-sliding mode on the switching hyperplane in discrete variable structure systems', *Automatic Remote Control*, 1985, **46**, pp. 307–314

16   POTTS, R. B. and YU, X.: 'Discrete variable structure system with pseudo-sliding mode', *J. Austral. Math. Soc.* Ser. B, 1991, pp. 365–376

17   POTTS, R. B. and YU, X.: 'Difference equation modeling of a variable structure systems', *Computer Math. Applic.*, 1994, **28**(1–3), pp. 281–289

18   YU, X. and POTTS, R. B.: 'Computer-controlled variable-structure systems', *Austral. Math. Soc. Ser. B*, 1992, pp. 1–17

19   SARPTURK, S. Z., ISTEFANOPULOS, Y., and KAYNAK, O.: 'On the stability of discrete-time sliding mode control systems', *IEEE Trans.*, 1987, **AC-32**(10), pp. 930–932

20   KOTTA, U.: 'Comments on "On the stability of discrete-time sliding mode control systems" ', *IEEE Trans.*, 1989, **AC-34**(9), pp. 1021–1022

21   SIRA-RAMIREZ, H.: 'Non-linear discrete variable structure systems in quasi-sliding mode', *Int. J. Control*, 1991, **54**(5), pp. 1171–1187

22   WESTPHAL, L. C.: 'Lessons from an example in "On the stability of discrete-time sliding mode control systems" ', *IEEE Trans.*, 1982, **AC-44**(7), pp. 1444–1445

23   BUČEVAC, Z.: 'Design of digital discrete control systems with sliding mode', Ph.D. Dissertation, Mech. Eng. Faculty University of Belgrade, 1985, (in Serbian)

24   SALIHBEGOVIĆ, A.: 'Contribution to analysis and synthesis of discrete realized systems with switched control', Ph.D. Dissertation, University of Sarajevo (YU), 1985 (in Bosnian)

25   MAGAÑA, M. E. and ZAK, S. H.: 'The control of discrete-time uncertain dynamical systems', Research Report TR-EE 87-32. School of Electrical Engn. Purdue University, West Lafayette, Indiana 47907, 1987

26   DRAKUNOV, S. V. and UTKIN, V. I.: 'On discrete-time sliding mode', Proc. IFAC Symposium on *Nonlinear Control System Design*, Capry (Italy), 1989, pp. 484–489

27   DRAKUNOV, S. V. and UTKIN, V. I.: 'Sliding mode control in dynamic systems', *Int. J. Control*, 1992, **55**(4), pp. 1029–1037

28   FURUTA, K.: 'Sliding mode control of a discrete system', *System & Control Letters*, 1990, **14**, pp. 145–142

29   GAO, W., and HUNG, J. Y.: 'Variable structure control of nonlinear systems: a new approach', *IEEE Trans.*, 1993, **IE-40**(1), pp. 45–55

30   CHAN, C. Y.: 'Robust discrete quasi-sliding mode tracking controller', *Automatica*, 1995, **31**(10), pp. 1509–1511

31   FURUTA, K.: 'VSS type self-tuning control', *IEEE Trans.*, 1993, **IE-40**, pp. 37–74

32   FURUTA, K. and PAN, Y.: 'Variable structure control with sliding sector', *Automatica*, 2000, **36**, pp. 211–228

33   WANG, W.-J. and WU, G.-H.: 'Variable structure control design on discrete-time systems from another viewpoint', *Control Theory and Advanced Technology*, 1992, **8**(1), pp. 1–16

34   WANG, W.-J., WU, G.-H., and YANG, D.-C.: 'Variable structure control design for uncertain discrete-time systems', *IEEE Trans.*, 1994, **AC-39**(1), pp. 99–102

35   GAO, W., WANG, Y., and HOMAIFA, A.: 'Discrete-time variable structure control systems', *IEEE Trans.*, 1995, **IE-42**, pp. 117–122

36   HUNG, J. Y., GAO, W., and HUNG, J. C.: 'Variable structure control: a survey', *IEEE Trans.*, 1993, **IE-42**(2), pp. 2–22

37   BARTOSZEWICZ, A.: 'Discrete-time quasi-sliding-mode control strategies', *IEEE Trans.*, 1998, **IE-45**(4), pp. 633–637

38   BARTOLINI, G., FERRARA, A., and UTKIN, V. I.: 'Adaptive sliding mode control in discrete-time systems', *Automatica*, 1995, **31**(5), pp. 769–773

39   CHAN, C. Y.: 'Discrete adaptive sliding-mode tracking controller', *Automatica*, 1998, **33**(5), pp. 999–1002

40   GOLO, G. and MILOSAVLJEVIĆ, Č.: 'Two-phase triangular wave oscillator based on discrete-time sliding mode control', *Electronic Letters*, 1997, **33**(22), pp. 1838–1839

41   GOLO, G. and MILOSAVLJEVIĆ, Č.: 'Robust discrete-time chattering-free sliding mode control', *Systems & Control Letters*, 2000, **41**, pp. 19–28

42   ŠABANOVIĆ, A., JEZERNIK, K., and WADA, K.: 'Chattering-free sliding modes in robotic manipulators control', *Robotica*, 1996, **14**, pp. 17–29

43   SU, W.-C., DRAKUNOV, S. V., and ÖZGÜNER, Ü.: 'An $O(T^2)$ boundary layer in sliding mode for sampled-data systems', *IEEE Trans.*, 2000, **AC-45**(3), pp. 482–485

44   YOUNG, K. D., UTKIN, V. I., and ÖZGÜNER, Ü.: 'A control engineer's guide to sliding mode control', *IEEE Trans.*, 1999, **CST-7**(3), pp. 328–342

45   TANG, Y. C. and MISAWA, E.: 'Discrete variable structure control for linear multivariable systems: the state feedback case', Oklahoma State University, School of Mechanical & Aerospace Engineering, Advanced control laboratory, Report ACL-98-007, 1998. Also, 'Discrete variable structure control for linear multivariable systems: the output feedback case', Oklahoma State University, School of Mechanical & Aerospace Engineering, Advanced control laboratory, Report ACL-98-008, 1998

46   SPURGEON, S. K.: 'Hyperplane design techniques for DT variable structure control systems', *Int. J. Control*, 1992, **55**(2), pp. 445–456

47   SIRA-RAMIREZ, H., SPURGEON, S., and Zinober, A. S. I.: 'Robust observer-controller design for linear systems', in 'Variable structure and lyapunov control' (Springer-Verlag, London, 1994) pp. 161–180

48 KORONDI, P., HASHIMOTO, H., and UTKIN, V. I.: 'Direct torsion control of flexible shaft in an observer-based discrete-time sliding mode', *IEEE Trans.*, 1998, **IE-45**(2), pp. 291–296

49 MISAWA, E. A.: 'Boundary layer eigenvalues in observer-based discrete-time sliding mode control', Proc. of the American Control Conference, Anchorage, AK, May 2002, pp. 2935–2936

50 MISAWA, E. A.: 'Observer-based discrete-time sliding mode control with computational time delay: the linear case', Proc. of the American Control Conference, Seattle, Washington, June 1995, pp. 1323–1327

51 WU, S.-T.: 'On digital high-gain and sliding-mode control', *Int. J. Control*, 1997, **66**(1), pp. 65–83

52 MITIĆ, D. and MILOSAVLJEVIĆ, Č.: 'Sliding mode based generalized minimum variance control with $O(T^3)$ accuracy', in ŠABANOVIĆ, A. (Ed.): 'Advances in variable structure systems – theory and application', Proc. of the 7th Int. Workshop on *VSS*, University of Sarajevo, 17–19 July 2002 (Bosnia and Herzegovina), pp. 69–76

53 BARTOLINI, G., PISANO, A., and USAI, E.: 'Digital sliding mode control with $O(T^3)$ accuracy', in YU, X. and XU, J.-X. (Eds): 'Advances in variable structure systems – analysis, integration and application', Proc. of the 6th IEEE Int. Workshop on *VSS*, Gold Coast, Queensland, Australia, Dec. 7–9, 2000, pp. 103–112

54 BARTOLINI, G., PISANO, A., and USAI, E.: 'Digital second-order sliding mode control for uncertain nonlinear systems', *Automatica*, 2001, **37**, pp. 1371–1377

55 XU, J.-X., ZHENG, F., and LEE, T.: 'On sampled data variable structure control systems', in YOUNG, K. D. and ÖZGÜNER, Ü. (Eds): 'Lecture notes in control and information sciences', (Springer, London, 1999), pp. 69–92

56 BUČEVAC, Z.: 'A stabilizing discrete digital variable structure control algorithm applied to linear plant', Proc. of the Int. Conf. of Technical Informatics, Timisoara, Romania, 1996, vol. 2, pp. 105–112

57 UTKIN, V. I.: 'Sliding modes in control and optimization' (Springer, Berlin, 1982)

58 FURUTA, K. and PAN, Y.: 'Discrete-time VSS control for continuous-time systems', Proc. of the First Asian Control Conference, Tokyo, 1994, pp. 377–380

59 MILOSAVLJEVIĆ, Č., MIHAJLOVIĆ, N., and GOLO, G.: 'Static accuracy of the variable structure system', Proc. of VI Int. SAUM Conf. on *Systems Automatic Control and Measurements*, Niš, YU, Sept. 28–30, 1988, pp. 464–469

60 YU, X. and CHEN, G.: 'Discretization analysis of a class of second order SMC systems', in ŠABANOVIĆ, A. (Ed.): 'Advances in variable structure systems – theory and application', Proc. of the 7th Int. Workshop on VSS, University of Sarajevo, 17–19 July 2002 (Bosnia and Herzegovina)

61 YU, X. and CHEN, G.: 'Discretization behaviors of equivalent control based sliding mode control systems', *IEEE Trans.*, **AC-48**(9), 2003, pp. 1641–1646

Part II
New trends in sliding mode control

Chapter 6

Robustness issues of 2-sliding mode control

Arie Levant and Leonid Fridman

6.1 Introduction

The sliding mode control approach [1, 2] is based on keeping exactly a properly chosen constraint by means of high frequency switching of the control. The approach exploits the main features of the sliding mode: its insensitivity to external and internal disturbances, ultimate accuracy and finite-time transient. However, the use of standard sliding modes has some restrictions. If the task is to keep an output variable σ at zero, the standard sliding mode can be implemented only when the relative degree of σ is 1. In other words, the control has to appear explicitly in the first total derivative $\dot{\sigma}$. Also, high frequency control switching leads to the so-called chattering effect which is exhibited by high frequency vibration of the controlled plant and can be dangerous in some applications.

A number of methods were proposed to overcome these difficulties. In particular, high gain control with saturation approximates the sign-function and diminishes the chattering; while on-line estimation of the so-called equivalent control [1] is used to reduce the discontinuous-control component [3], the sliding-sector method [4] is suitable to control disturbed linear time-invariant systems. Yet, the most comprehensive approach seems to be the sliding mode order approach [5–7], which allows all the above restrictions to be removed, while preserving the main sliding mode features and improving its accuracy in the presence of switching imperfections. Independently developed dynamical [8] and terminal [9] sliding modes are closely related to this approach.

In particular, 2-sliding modes are used to remove the chattering or to keep constraints of the second relative degree and have already been successfully implemented for the solution of various problems [7, 10–17] (see also Part III of this book). The current chapter deals with the robustness aspects of 2-sliding mode control.

In particular, the influence of measurement noise and of unaccounted-for fast actuator dynamics are considered.

Most 2-sliding controllers explicitly use $\dot{\sigma}$ or its sign. It is shown [18] that 2-sliding controllers are very robust with respect to the sampling noise of σ and $\dot{\sigma}$. Unfortunately, $\dot{\sigma}$ is often unavailable. The first difference of σ is usually used instead of $\dot{\sigma}$ in order to overcome the difficulty [6, 19], but the resulting performance critically depends on the sampling step that has to be chosen with respect to the often unknown measurement-noise magnitude. Thus, the robustness of the controller is partially lost. This chapter considers two main methods to solve the problem. The first one is to use a variable sampling step dependent on the real-time output measurements [20]. The other one is new and suggests a recently developed robust exact differentiator [18, 21] to be used as a natural part of the standard 2-sliding controllers. The resulting controllers preserve the ultimate accuracy and finite-time convergence of the original controllers and do not require any information on the noise. Corresponding theorems and simulation results are presented.

Introduction of the actuator dynamics causes the relative degree to exceed 2. It is shown in the chapter that the arising higher-order sliding mode is never stable, but the instability is local and not crucial if the actuator is fast and stable. The case of a linear autonomous control system is considered. It is shown by the method of description functions [22] that fast stable actuators cause oscillations in a small vicinity of the 2-sliding manifold. Correspondent simulation results are presented.

6.2 Main notions and the problem statement

6.2.1 Definitions

Let us first recall that according to the definition by Filippov [23], any discontinuous differential equation $\dot{x} = v(x)$, where $x \in \mathbf{R}^n$ and v is a locally bounded measurable vector function, is replaced by an equivalent differential inclusion $\dot{x} \in V(x)$ (see Chapter 1). In the simplest case, when v is continuous almost everywhere, $V(x)$ is the convex closure of the set of all possible limits of $v(y)$ as $y \to x$, while $\{y\}$ are continuity points of v. Solutions of the equation are defined as absolutely continuous functions $x(t)$, satisfying the differential inclusion almost everywhere. In the following, the equation $\dot{x} = v(x)$ can be considered as a result of closing a smooth dynamic system by some possibly-dynamical discontinuous feedback.

Let σ be a smooth output function. Then, provided that

- successive total time derivatives $\sigma, \dot{\sigma}, \ldots, \sigma^{(r-1)}$ are continuous functions of the closed-loop system state space variables, and
- the set $\sigma = \dot{\sigma} = \cdots = \sigma^{(r-1)} = 0$ is non-empty and consists locally of Filippov trajectories [24] (Fig. 6.1),

the motion on the set $\sigma = \dot{\sigma} = \cdots = \sigma^{(r-1)} = 0$ is called an r-sliding mode (rth order sliding mode).

The additional condition of the Filippov velocity set containing more than 1 vector may be imposed in order to exclude some trivial cases. It is natural to call the sliding

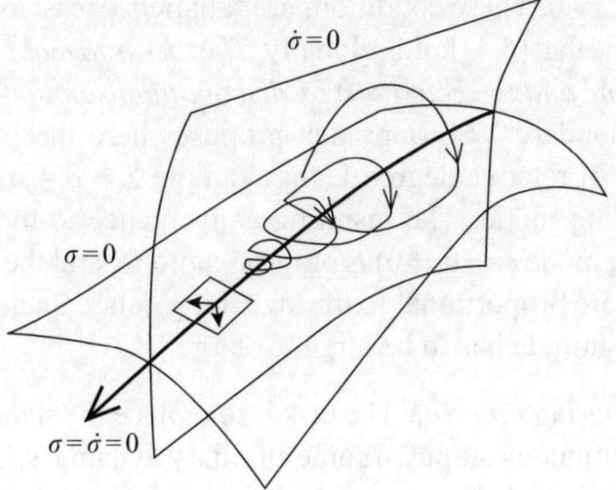

Figure 6.1   *2-sliding mode*

order r *strict* if $\sigma^{(r)}$ is discontinuous or does not exist in a vicinity of the r-sliding point set, but sliding mode orders are mostly considered strict by default.

6.2.2   2-sliding control problem

There are two main statements of the control problem leading to the 2-sliding mode solution.

Sliding mode with relative degree 2. In order to standardise the problem, let an uncertain dynamic system have the form

$$\dot{x} = a(t, x) + b(t, x)u, \quad \sigma = \sigma(t, x), \tag{6.1}$$

where $x \in \mathbf{R}^n$, $u \in \mathbf{R}$ is control; smooth functions a, b, σ and the dimension n are unknown. The relative degree of the system is assumed to be 2, which means that $(\partial/\partial u)\ddot\sigma(t, x, u) \neq 0$. The task is to nullify the measured output σ. The output σ is called the sliding variable and can be understood to be the tracking deviation of the system output from some desired smooth signal given in real time. The system trajectories are supposed to be infinitely extendible in time for any bounded Lebesgue-measurable input u.

Calculating the second total time derivative $\ddot\sigma$ along the trajectories of (6.1) shows that under these conditions

$$\ddot\sigma = h(t, x) + g(t, x)u, \qquad h = \ddot\sigma|_{u=0}, \quad g = \frac{\partial}{\partial u}\ddot\sigma \neq 0, \tag{6.2}$$

where the functions g, h are some unknown smooth functions that can be expressed by means of Lie derivatives [25]. Suppose that

$$0 < K_{\mathrm{m}} \leq \frac{\partial}{\partial u}\ddot\sigma \leq K_{\mathrm{M}}, \qquad |\ddot\sigma|_{u=0}| \leq C \tag{6.3}$$

for some K_{m}, K_{M}, $C > 0$. These conditions are satisfied at least locally for any smooth system (6.1). Assume that (6.3) holds globally. *The task is to make the measured output σ vanish in finite time and to keep $\sigma \equiv 0$ by discontinuous bounded feedback control.*

Note that the standard VSS approach proposes here the introduction of some auxiliary constraint of relative degree 1, for example $\Sigma = \sigma + \dot{\sigma}$, which is to be kept in the standard sliding mode. The resulting convergence to the desired state $\sigma \equiv 0$ (i.e., to the 2-sliding mode $\sigma = \dot{\sigma} = 0$) is only asymptotic, and the resulting accuracies $\sup |\sigma|$ and $\sup |\dot{\sigma}|$ are proportional to the switching delay. Such a control is also not bounded, for its magnitude has to be large for large $|\dot{\sigma}|$.

1-sliding mode chattering removal. The task is to replace the standard relay algorithm $u = -\mathrm{sign}\,\sigma$ by a continuous output of some auxiliary dynamic subsystem. To simplify and detail the constraint fulfillment problem, consider the dynamic system given by the equation

$$\dot{x} = f(t, x, u), \tag{6.4}$$

where $x \in \mathbf{R}^n$ is a state variable, t is time, $u \in \mathbf{R}$ is control, f is a C^1-function. Let $\sigma(t, x) \in \mathbf{R}$ be a C^2-function. *The goal is to force the sliding variable σ to vanish in finite time by means of control continuously dependent on time and not exceeding 1 in absolute value.*

Let K_{m}, K_{M}, C_0 be positive constants, $K_{\mathrm{m}} < K_{\mathrm{M}}$, and assume the following:

1. $|u| \leq \kappa$, $\kappa = \mathrm{const} > 1$. Any solution of (6.4) is well defined for all t, provided $u(t)$ is continuous and $|u(t)| \leq \kappa$ for each t.
2. There exists $u_0 \in (0, 1)$ such that for any continuous function $u(t)$ with $|u(t)| \geq u_0$, there is t_1, such that $\sigma(t)u(t) > 0$ for each $t > t_1$. Hence, the control $u(t) = -\mathrm{sign}\,\sigma(t_0)$, where t_0 is the initial time, ensures the manifold $\sigma = 0$ is reached in finite time.

$$\text{Denote } \Lambda_u(\cdot) = \frac{\partial}{\partial t}(\cdot) + \frac{\partial}{\partial x}(\cdot) f(t, x, u), \quad \dot{\sigma}(t, x, u) = \Lambda_u \sigma(t, x).$$

3. There are positive constants σ_0, K_{m}, K_{M} such that if $|\sigma(t, x)| < \sigma_0$, then

$$0 < K_{\mathrm{m}} < \frac{\partial}{\partial u}\dot{\sigma}(t, x, u) < K_{\mathrm{M}}$$

    for all u, $|u| \leq \kappa$, and the inequality $|u| > u_0$ entails $\dot{\sigma}u > 0$.
4. Within the region $|\sigma| < \sigma_0$, for all t, x, and u, the inequality $|\Lambda_u \Lambda_u \sigma(t, x)| < C_0$ holds. This means that the second time derivative of the constraint function σ, which is calculated with fixed values of the control u, is uniformly bounded.

It follows from the implicit function theorem that there is a function $u_{\mathrm{eq}}(t, x)$ (equivalent control [1]) satisfying the equation $\dot{\sigma} = 0$. Once $\sigma = 0$ is achieved, the control $u = u_{\mathrm{eq}}(t, x)$ would provide for exact constraint fulfillment. Conditions 3 and 4 mean that $|\sigma| < \sigma_0$ implies $|u_{\mathrm{eq}}| < u_0 < 1$, and that the rate of change of u_{eq} is bounded. This provides the possibility to approximate u_{eq} by a Lipschitz control. Note also that linear dependence on the control u is not required.

Consider u as an additional state variable and define the new control $v = \dot{u}$. The resulting dynamic system linearly depends on v and is actually of the form (6.1) with relative degree 2. Nevertheless, the problem of chattering removal cannot be completely reduced to the output control problem with relative degree 2. Indeed, the system satisfies the conditions of the relative-degree-2 problem only within the specified region $|\sigma| < \sigma_0$, and it is required that $|\sigma| < \sigma_0$ be maintained during the transient period and that the real control $u = \int v(t)\,dt$ does not exceed 1 in absolute value.

The variable structure system theory deals usually with systems (6.1) with the linear growth of the right-hand side. Under conventional assumptions [6, 26], the task of keeping the constraint $\sigma(t,x) = 0$ of relative degree 1 is reduced to the described problem of chattering removal. A new control v and a new constraint function φ are defined in this case by the substitution

$$u = k\Phi(x)v, \quad \varphi = \frac{\sigma(t,x)}{\Phi(x)},$$

where $\Phi(x) = \sqrt{xDx^t + h}$, k, $h > 0$, are constants, D is a non-negative definite matrix.

In the simple case when $\dot{x} = A(t)x + b(t)u$, $\sigma = c(t)x + \xi(t)$, all conditions are reduced to the boundedness of c, $\dot{c}$, $\ddot{c}$, $\dot{\xi}$, $\ddot{\xi}$, A, $\dot{A}$, b, $\dot{b}$ and to the inequality $cb > \text{const} > 0$ [6]. The corresponding constants determine the controlled class.

6.3   Standard 2-sliding controllers in systems with relative degree 2

Homogeneity approach to sliding mode control. Consider the first problem statement dealing with finite-time stabilisation of an output σ of relative degree 2 to 0. Note that under the given assumptions, solutions of (6.1) exist for any bounded Lebesgue-measurable control. As follows from (6.2) and (6.3) all such solutions satisfy the differential inclusion

$$\ddot{\sigma} \in [-C, C] + [K_{\mathrm{m}}, K_{\mathrm{M}}]u. \tag{6.5}$$

Most 2-sliding controllers may be considered as controllers for (6.5) steering σ, $\dot{\sigma}$ to 0 in finite time. The inclusion (6.5) does not 'remember' the original system (6.1). Thus, such controllers are obviously robust with respect to any perturbations preserving (6.3).

Hence, the problem is to find a feedback

$$u = \varphi(\sigma, \dot{\sigma}), \tag{6.6}$$

such that all the trajectories of (6.5) and (6.6) converge in finite time to the origin $\sigma = \dot{\sigma} = 0$ of the phase plane σ, $\dot{\sigma}$. Here φ is a locally bounded Borel-measurable function (all functions used in sliding mode control satisfy this restriction). The differential inclusion (6.5) and (6.6) is understood in the Filippov sense [23], which means that the right-hand vector set is enlarged in a special way in order to satisfy certain convexity and semicontinuity conditions (see Chapter 1). As a result new unfeasible solutions can appear, but all of them also have to converge to the origin. We now introduce a few simple auxiliary notions to be used further.

The inclusion (6.5) and (6.6) and the controller (6.6) itself are called 2-*sliding homogeneous* if for any $\kappa > 0$ the combined time-coordinate transformation

$$G_\kappa: (t, \sigma, \dot\sigma) \mapsto (\kappa t, \kappa^2 \sigma, \kappa \dot\sigma) \tag{6.7}$$

transfers its solutions into the solutions of the transformed inclusion. Indeed let $\sigma_1 = \kappa^2 \sigma$, $\sigma_2 = \kappa \dot\sigma$, $t_1 = \kappa t$, then

$$\frac{d\sigma_1}{dt_1} = \frac{d\kappa^2 \sigma}{d\kappa t} = \kappa \dot\sigma = \sigma_2,$$

$$\frac{d\sigma_2}{dt_1} = \frac{d\kappa \dot\sigma}{d\kappa t} = \ddot\sigma \in [-C, C] + [K_m, K_M]\varphi(\sigma, \dot\sigma).$$

Thus (6.6) is 2-sliding homogeneous iff almost everywhere

$$\varphi(\kappa^2 \sigma, \kappa \dot\sigma) \equiv \varphi(\sigma, \dot\sigma). \tag{6.8}$$

Recall that in this case the closed differential inclusion (6.5) and (6.6) is homogeneous of degree -1 with the dilation $d_\kappa: (\sigma, \dot\sigma) \mapsto (\kappa^2 \sigma, \kappa \dot\sigma)$ [22]. It is easy to check that this is the only possible homogeneity with $C \neq 0$. Similarly the r-sliding homogeneity is defined for any natural r. The homogeneity features can greatly facilitate the 2-sliding controller design and the convergence proofs [27].

Standard 2-sliding controllers. Following are a few of the most well known controllers. All of them are 2-sliding homogeneous.

The twisting controller [6, 26] and the convergence condition are given by

$$u = -r_1 \operatorname{sign}\sigma - r_2 \operatorname{sign}\dot\sigma, \qquad r_1 > r_2 > 0, \tag{6.9}$$

$$(r_1 + r_2)K_m - C > (r_1 - r_2)K_M + C, \qquad (r_1 - r_2)K_m > C. \tag{6.10}$$

The corresponding trajectories of the inclusion (6.5) and (6.6) are shown in Fig. 6.2a. A particular case of the controller with prescribed convergence law [6, 26] (Fig. 6.2b) is given by

$$u = -\alpha \operatorname{sign}(\dot\sigma + \lambda|\sigma|^{1/2} \operatorname{sign}\sigma), \qquad \alpha, \lambda > 0, \quad \alpha K_m - C > \frac{\lambda^2}{2}. \tag{6.11}$$

Controller (6.11) is close to a terminal sliding mode controller [9]. The so-called sub-optimal controller [11, 19] is given by

$$u = -r_1 \operatorname{sign}\left(\sigma - \frac{\sigma^*}{2}\right) + r_2 \operatorname{sign}\sigma^*, \qquad r_1 > r_2 > 0, \tag{6.12}$$

$$2[(r_1 + r_2)K_m - C] > (r_1 - r_2)K_M + C, \qquad (r_1 - r_2)K_m > C, \tag{6.13}$$

where σ^* is the value of σ detected at the closest time when $\dot\sigma$ was 0. The initial value of σ^* is 0. Any computer implementation of this controller requires successive measurements of $\dot\sigma$ or σ with some time step. Usually, the detection of the moments when

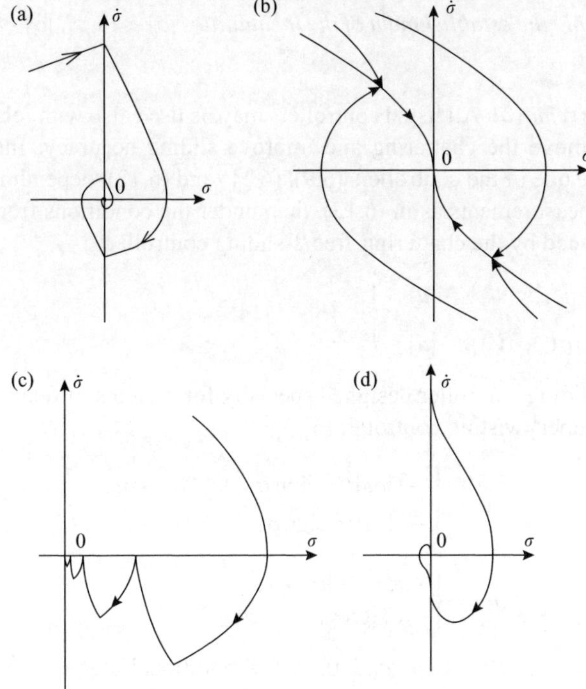

Figure 6.2   Phase trajectories of the standard 2-sliding controllers

$\dot{\sigma}$ changes its sign is performed, calculating the finite difference $\Delta\sigma_i$ at each sampling step t_i. Note that the slightly stronger condition (6.10) excludes the overregulation possibility (Fig. 6.2c).

Strictly speaking, the sub-optimal controller does not have the form (6.6). Indeed, the control value u depends on the whole history of $\dot{\sigma}$ and σ measurements, i.e., on $\dot{\sigma}(\cdot)$ and $\sigma(\cdot)$, and has the form $u = \varphi(\sigma(\cdot), \dot{\sigma}(\cdot))$. Nevertheless, it is naturally considered as a 2-sliding homogeneity controller, for it is invariant with respect to the transformation (6.7).

Theorem 1 [6, 19].   *2-sliding controllers (6.9), (6.11) and (6.12) provide for the finite-time convergence of any trajectory of (6.1), (6.3) into the 2-sliding mode $\sigma \equiv 0$. The convergence time is a locally bounded function of the initial conditions.*

Let the measurements be carried out at times t_i with constant step $\tau > 0$, $\sigma_i = \sigma(t_i, x(t_i))$, $\Delta\sigma_i = \sigma_i - \sigma_{i-1}, t \in [t_i, t_{i+1})$. Substituting σ_i for σ, sign $\Delta\sigma_i$ for sign $\dot{\sigma}$, and sign $(\Delta\sigma_i - \lambda\tau|\sigma_i|^{1/2} \operatorname{sign}\sigma_i)$ for sign $(\dot{\sigma} - \lambda|\sigma|^{1/2} \operatorname{sign}\sigma)$ discrete-measurement versions of the controllers are obtained. For example, the discrete-sampling version of the twisting controller is

$$u = -r_1 \operatorname{sign}\sigma - r_2 \operatorname{sign}\Delta\sigma_i, \qquad r_1 > r_2 > 0, \tag{6.14}$$

Theorem 2 [6, 19].   *The discrete-sampling versions of the controllers (6.9), (6.11) and (6.12) enable the establishment of the inequalities* $|\sigma| < \mu_0 \tau^2$, $|\dot{\sigma}| < \mu_1 \tau$ *for some positive* μ_0, μ_1.

Chattering removal. All listed controllers may be used also with relative degree 1 in order to remove the chattering and improve sliding accuracy. Indeed, let $u = \varphi(\sigma(\cdot), \dot{\sigma}(\cdot))$ be one of the controllers (6.9), (6.11) and (6.12), depending possibly on the previous measurements as in (6.12), then under the conditions from Section 6.2 it may be replaced by the chattering-free 2-sliding controller

$$\dot{u} = \begin{cases} -u, & |u| > 1, \\ \varphi(\sigma(\cdot), \dot{\sigma}(\cdot)), & |u| \le 1. \end{cases} \tag{6.15}$$

A new 2-sliding controller designed specially for systems of relative degree 1 is the so-called super-twisting controller [6]:

$$u = u_1 + u_2, \qquad \begin{aligned} u_1 &= \begin{cases} -\lambda |\sigma_0|^{1/2} \operatorname{sign} \sigma, & |\sigma| > \sigma_0, \\ -\lambda |\sigma|^{1/2} \operatorname{sign} \sigma, & |\sigma| \le \sigma_0, \end{cases} \\ \dot{u}_2 &= \begin{cases} -u, & |u| > 1, \\ -\alpha \operatorname{sign} \sigma, & |u| \le 1, \end{cases} \end{aligned} \tag{6.16}$$

where $\alpha_M > \alpha_m > C/K_m$, λ, α, $\sigma_0 > 0$. A few additional algebraic restrictions [6] involving $\alpha_M, \alpha_m, \rho, \lambda, \alpha, C, K_m, K_M$ can be easily fulfilled with sufficiently large λ, α, α_m, α_M/α_m, and are omitted here. Note that (6.16) does not require measurements of $\dot{\sigma}$, thus, this controller is obviously robust with respect to measurement noise. The main drawback of this controller is the lack of the Lipschitz property of the produced control, which may interfere with its implementation in complicated MIMO control systems. Its characteristic trajectory in the coordinates σ, $\dot{\sigma}$ is shown in Fig. 6.2c.

Theorems 1 and 2 are valid also for the analogous controllers of the form (14) with relative degree 1 and, after obvious reformulation, for the super-twisting controller (6.16) [6, 11].

Plan of the proof of Theorem 1.   It is obvious that with the *controller* (6.11) $\operatorname{sign} \ddot{\sigma} = \operatorname{sign} u$. Thus, the trajectory hits the curve $\dot{\sigma} + \lambda |\sigma|^{1/2} \operatorname{sign} \sigma = 0$ in finite time (Fig. 6.2b). Afterwards the point keeps moving in a 1-sliding mode along that curve to the origin, which proves the theorem.

Controller (6.9). Consider the successive intersections σ_i of a trajectory with the axis $\dot{\sigma} = 0$ (Fig. 6.2a). It is easily seen that

$$\left| \frac{\sigma_{i+1}}{\sigma_i} \right| \le \frac{[(r_1 - r_2)K_M + C]}{[(r_1 + r_2)K_m - C]},$$

thus (6.10) provides for the convergence of the trajectory to the origin. Due to the negative homogeneity degree of the inclusion (6.5) and (6.9) the convergence time is finite [28].

Controller (6.13). It is easily seen (Fig. 6.2c) that

$$\left|\frac{(\sigma_{i+1}-0.5\sigma_i)}{(\sigma_i-0.5\sigma_i)}\right|=2\left|\frac{(\sigma_{i+1}-0.5\sigma_i)}{\sigma_i}\right|\leq\left|\frac{[(r_1-r_2)K_M+C]}{[(r_1+r_2)K_m-C]}\right|,$$

thus (6.13) provides for the convergence of the trajectory to the origin. The convergence time is estimated by a geometric series with a finite sum (the same can be done for controller (6.9)).

Proof of Theorem 2.   It follows from the Lagrange Theorem that $\Delta\sigma_i=\dot{\sigma}(t)\tau+\varepsilon(t)$, where $|\varepsilon(t)|\leq2\sup|\ddot{\sigma}|\tau^2$, $t\in[t_i,t_{i+1})$. Thus, in the absence of input noise, the discrete-measurement versions of controllers (6.9) and (6.11) can be considered as controllers (6.9) and (6.11) with noisy measurements of $\dot{\sigma}$. The measurement error magnitude is uniformly bounded by the constant $N(\tau)=2\sup|\ddot{\sigma}|\tau$ which tends to zero with $\tau\to0$. Note that the constant $\sup|\ddot{\sigma}|$ depends on the controller choice, its parameters and the actual values of K_M, K_m, C_0 (Section 6.2).

According to Theorem 1, trajectories of (6.5) and (6.6) (i.e., with 'exact' measurements) starting from any closed disk D_0 centred at the origin O terminate at the origin in a finite time T and stay there forever. As follows from the continuous dependence of the Filippov inclusion solutions on *the graph* of the right-hand side [16], if τ_1 is sufficiently small, the trajectories of the controller with measurement noise magnitude $N(\tau_1)$ will terminate in some small closed disk $D_1'\subset D_0$ for any $\tau\leq\tau_1$. In their turn, the trajectories of (6.5) and (6.6) starting from D_1' terminate at O in time T. With sufficiently small $\tau_2\leq\tau_1$, the trajectories with measurement noise magnitude $N(\tau_1)$ terminate in some other small disk D_1'' in time T, $D_1''\subset D_1'$. Let D_1 be some disk containing all the trajectories' segments with the measurement noise magnitude $N(\tau_2)$ starting from D_1' with t varying in the range $[0,\ T]$, $D_1\subset D_0$ with sufficiently small τ_1. Obviously, with the measurement noise magnitude $N(\tau_2)$ any trajectory that starts from D_0 enters D_1 in time T to stay there forever. In particular, it is true with respect to the trajectories of the discrete-sampling controller versions with $\tau\leq\tau_2$.

The Filippov theory cannot be applied to the controller (6.12), nevertheless it can be directly shown that with the measurement noise magnitude $N(\tau_2)$ any trajectory that starts from D_0 enters D_1 in time T to stay there forever for some $\tau_2>0$, $D_1\subset D_0$ (actually it is obvious from Fig. 6.2c). Now, note that the time-coordinate transformation (6.7) transfers the trajectories of the discrete-sampling versions of controllers (6.9), (6.11) and (6.12) into the trajectories of the same controllers but with the sampling step changed to $\kappa\tau$. Thus with any $\kappa>0$ and the $\dot{\sigma}-$ measurement noise magnitude $N(\kappa\tau_2)$, all trajectories that start in $G_\kappa D_0$ enter $G_\kappa D_1$ in time κT to stay there forever. Let $\kappa>1$, then since it is true with the noise magnitude $N(\kappa\tau_2)$, it is also true with the less noise magnitude $N(\tau_2)$. Now choose $\kappa>1$ such that $G_\kappa D_1\subset D_0$. We produce the sequence of the embedded compact regions $D_1\subset D_0\subset G_\kappa D_0\subset G_\kappa^2 D_0\subset\cdots$ covering the whole plane σ, $\dot{\sigma}$. Thus, with some sampling step τ_2, any trajectory enters D_1 in finite time to stay there forever.

Let $D_1 \subset \{\sigma, \dot{\sigma} \mid |\sigma| \le c_1, \; |\dot{\sigma}| \le c_2\}$, and let τ be some arbitrary sampling step. Applying G_κ with $\kappa = \tau/\tau_2$ shows that with the sampling step τ any trajectory enters the region $\{\sigma, \dot{\sigma} \mid |\sigma| \le (c_1/\tau_2^2)\tau^2, |\dot{\sigma}| \le (c_2/\tau_2)\tau\}$ to stay there forever.

The described controllers depend on a few constant parameters. These parameters must be tuned in order to control the whole class of processes and constraint functions defined by the actual values of K_M, K_m, C_0 (and σ_0 with the second problem statement). Increasing the constants K_M, C_0 and reducing K_m, σ_0 at the same time, we enlarge the controlled class too. Such algorithms are obviously insensitive to any model perturbations and external disturbances that do not move the dynamic system from the given class.

6.4 Sampling noise and variable sampling step

Let $\delta > 0$ be the maximum of the possible error in the measurements of σ. It is obvious from the proof of Theorem 2 that with the sampling step τ fixed and δ sufficiently small, the measurement errors do not interfere with the performance of controllers (6.9), (6.11) and (6.12). But the sliding accuracy deteriorates when τ decreases and takes on values $\tau \le 0.5\delta/\sup|\dot{\sigma}|$, where $\sup|\dot{\sigma}|$ is some practical bound of $|\dot{\sigma}|$. Indeed $\Delta\sigma_i = \dot{\sigma}(\xi_i)\tau + \eta(t_i) - \eta(t_{i-1})$, where $\xi_i \in (t_{i-1}, t_i), \eta(t_i), \eta(t_{i-1})$ are the measurement errors not exceeding δ in absolute value, and the measurement error is certain to exceed the increment of σ. Note that in the case of chattering removal $|\dot{\sigma}| \le K_M|u - u_{eq}| \le 2K_M$ holds (Section 6.2). The problem is aggravated when δ cannot be estimated. A typical dependence of the sliding error on δ is shown qualitatively in Fig. 6.3a in the case of the twisting controller. With the other two controllers, the graphs are similar with large τ or small δ, but the stability loss is total.

To overcome the problem, introduce the following measurement step feedback [20]:

$$\tau = t_{i+1} - t_i = \begin{cases} \lambda|\sigma(t_i)|^{1/2}, & \lambda|\sigma(t_i)|^{1/2} > \tau_m, \\ \tau_m, & \lambda|\sigma(t_i)|^{1/2} \le \tau_m, \end{cases} \tag{6.17}$$

where $0 < \tau_m < \tau_M, \lambda > 0$.

Theorem 3. *With sufficiently small λ after a finite-time transient process, the trajectories of the system (6.1), (6.14) and (6.17) satisfy the inequalities*

$$|\sigma| \le \max(a_1 \tau_m^2, b_1 \delta), \quad |\dot{\sigma}| \le \max(a_2 \tau_m, b_2 \delta^{1/2}),$$

where a_1, a_2, b_1, b_2 are some positive constants depending on λ and the problem-statement parameters.

In the case of the problem of removing chattering, τ is to be made bounded from above by some $\tau_M > 0$, τ_M being sufficiently small [20], otherwise the invariance of the linearity region cannot be assured. While this theorem differs from the theorem proved [20], the proof is very similar and, being involved, is omitted here, for the problem is solved further in a better way. Theorem 3 means that

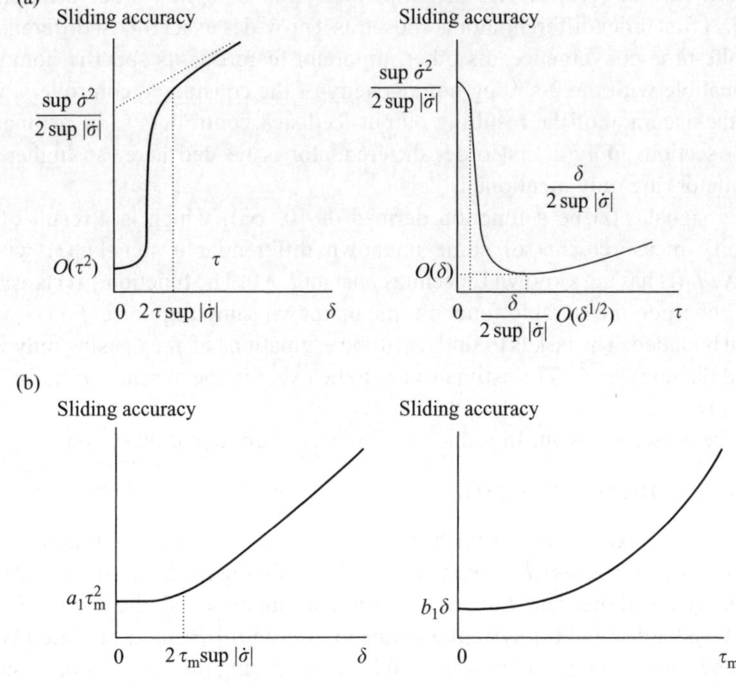

Figure 6.3   *Twisting controller (6.14): dependence of sliding accuracy on measurement error and sampling step: a) standard performance; b) performance with variable sampling step*

algorithm (6.14), (6.17) is a second order real sliding algorithm [6] that is robust with respect to measurements errors. The new typical dependence of the sliding error on δ is shown qualitatively in Fig. 6.3b. Note that this algorithm does not need any evaluation of the measurement errors.

Naturally, the algorithm may be simplified when δ is given a priori. In that case $\tau = \lambda_0 \delta^{1/2}$, $\lambda_0 > 0$, guarantees inequalities of the form $|\sigma| < a_1 \delta$, $|\dot{\sigma}| < a_2 \delta^{1/2}$ are established for some positive constants a_1, a_2 [20].

Unfortunately, the variable measurement step feedback is not always easy to implement in practice. Moreover it does not completely solve the problem of chattering removal, for in that case τ must be bounded from above by some possibly rather small number. The ideal solution proposed in this chapter is to estimate the derivative of σ in real time by means of a robust differentiator that is exact in the absence of noise.

6.5   Robust exact differentiation

Controllers (6.9), (6.11) and (6.12) require real-time exact calculation or direct measurement of $\dot{\sigma}$. The inclusion (6.5) causes boundedness of $\ddot{\sigma}$, which allows the implementation of robust differentiators [21, 29–33]. The boundedness of the right-hand side of (6.5) allows the implementation of a first-order differentiator [18, 21]. A first order differentiator is chosen as it provides exact robust differentiation with finite-time convergence. Its other important feature is its specific homogeneity compatible with the 2-sliding homogeneity of the considered controllers, which allows the accuracy of the resulting output-feedback controllers to be estimated in the next section. Only a first order differentiator is needed here, so higher order differentiators are only mentioned.

Let a signal $f(t)$ be a function defined on $[0, \infty)$, which is a result of real-time noisy measurements of some unknown differentiable signal $f_0(t)$ with the derivative $\dot{f}_0(t)$ having a known Lipschitz constant $L > 0$. The function $f(t)$ is assumed to be a Lebesgue-measurable function, the unknown sampling noise $f(t) - f_0(t)$ is assumed bounded. The task is to find real-time estimations of f_0, $\dot{f}_0$ using only values of f and the number L. The estimates are to be exact in the absence of noise, when $f(t) = f_0(t)$.

Let the noise be absent. Introduce a simple auxiliary dynamic system

$$\dot{z}_0 = u, \quad \sigma(t, z_0) = z_0 - f(t).$$

The task is to make σ and $\dot{\sigma}$ vanish in finite time by means of continuous control using only measurements of σ, i.e., to establish a 2-sliding mode. That is the standard chattering removal problem, but a few restrictions are removed: the control does not need to be bounded and the system is affine in the control. Thus, a modified version of the super-twisting controller is applied here, producing the closed-loop system

$$\dot{z}_0 = -\lambda_0 |z_0 - f|^{1/2} \operatorname{sign}(z_0 - f) + z_1,$$

$$\dot{z}_1 = -\lambda_1 \operatorname{sign}(z_0 - f).$$

Here $\lambda_1 > L$, and λ_0 is taken sufficiently large with respect to λ_1 [21]. The 2-sliding mode $\sigma = z_0 - f(t) = 0$, $\dot{\sigma} = -\lambda_0 |\sigma|^{1/2} \operatorname{sign} \sigma + z_1 - \dot{f} = z_1 - \dot{f} = 0$ is established in finite time. Thus, in the presence of noise, z_0 and z_1 are considered as estimates of f_0 and $\dot{f}_0$ respectively.

There is a simple, though rather conservative, algebraic criterion for the choice of λ_0 and λ_1 and a practically-exact simply-verified integral criterion is also available [21]. It is proved that the parameters can be taken in the form $\lambda_0 = \lambda_{00} L^{1/2}$, $\lambda_1 = \lambda_{01} L$ once λ_{0i} were adjusted for $L = 1$. It is proved [21] that with $|f(t) - f_0(t)| \le \varepsilon$ the inequalities

$$|z_0 - f_0(t)| \le \mu_1 \varepsilon, \quad |z_1 - \dot{f}_0(t)| \le \mu_2 L^{1/2} \varepsilon^{1/2},$$

are ensured in finite time, where the constants μ_1, $\mu_2 > 1$ depend entirely on the choice of λ_{00} and λ_{01}. That asymptotics have been proved to be the best possible when the Lipshitz constant L is known. One of the good choices is to take $\lambda_0 = 1.5 L^{1/2}$, $\lambda_1 = 1.1L$.

With discrete sampling, the inequalities of the form

$$|z_0 - f_0(t)| \le \nu_1 \tau^2, \quad |z_1 - \dot{f}_0(t)| \le \nu_2 \tau, \qquad \nu_1, \nu_2 > 0$$

are provided in the absence of noise.

It is proved [18] that the nth-order differentiator is realised in the form

$$\dot{z}_0 = v_0, \qquad v_0 = -\lambda_0 |z_0 - f(t)|^{n/(n+1)} \, \text{sign} \, (z_0 - f(t)) + z_1,$$

$$\dot{z}_1 = v_1, \qquad v_1 = -\lambda_1 |z_1 - v_0|^{(n-1)/n} \, \text{sign} \, (z_1 - v_0) + z_2,$$

$$\dots$$

$$\dot{z}_{n-1} = v_{n-1}, \qquad v_{n-1} = -\lambda_{n-1} |z_{n-1} - v_{n-2}|^{1/2} \, \text{sign} \, (z_{n-1} - v_{n-2}) + z_n,$$

$$\dot{z}_n = -\lambda_n \, \text{sign} \, (z_n - v_{n-1}),$$

where the $\lambda_i > 0$ are chosen sufficiently large in the reverse order. Note that the above contains all the lower-order differentiators and increasing the differentiation order by one requires tuning one parameter only. With $n = 1$, the above first-order differentiator is obtained.

6.6   Robust output-feedback control: differentiator in the feedback

The described 2-sliding controllers require real-time calculation or direct measurement of $\dot{\sigma}$ which is not always possible. The most natural way to solve this problem is to calculate $\dot{\sigma}$ in real time by means of the described robust exact differentiator [18, 21], whose application is possible due to the boundedness of $\ddot{\sigma}$ following from (6.5) and (6.6) and the control boundedness. Substitute the differentiator outputs z_0, z_1 for σ and $\dot{\sigma}$ respectively in the controllers (6.9), (6.11) and (6.12). The resulting controller has the form

$$u = \varphi(z_0(\cdot), z_1(\cdot)),$$

$$\dot{z}_0 = -\lambda_0 L^{1/2} |z_0 - \sigma|^{1/2} \, \text{sign} \, (z_0 - \sigma) + z_1, \qquad (6.18)$$

$$\dot{z}_1 = -\lambda_1 L \, \text{sign} \, (z_0 - \sigma),$$

where λ_0, $\lambda_1 > 0$, $L > C + K_M U_M$ are the parameters, U_M being the corresponding maximal absolute value of the control. Adjustment of λ_0, λ_1 is described in detail [21]; as has been mentioned, one of the reasonable choices is to take $\lambda_0 = 1.5$, $\lambda_1 = 1.1$. The outputs z_0, z_1 converge in finite time to σ and $\dot{\sigma}$ respectively in the absence of measurement noise [21].

Theorem 4. *Provided the parameters of the controller (6.18) are chosen as described above, each of the controllers (6.9), (6.11) and (6.12) provides in the absence of measurement noise for finite-time convergence of all trajectories to the 2-sliding mode $\sigma = \dot{\sigma} = 0$, otherwise convergence to a set defined by the inequalities $|\sigma| < \mu_0 \delta$, $|\dot{\sigma}| < \mu_1 \delta^{1/2}$ is assured for some positive constants μ_0, μ_1.*

Theorem 5. *Under the conditions of Theorem 4, the discrete-measurement versions of the modified controllers (6.9), (6.11) and (6.12) provide, in the absence of measurement noise, for the establishment of the inequalities $|\sigma| < \mu_0 \tau^2$, $|\dot{\sigma}| < \mu_1 \tau$ with some μ_0, $\mu_1 > 0$.*

The theorems are true also with respect to the chattering-elimination versions of the controllers and to any 2-sliding controller, satisfying the 2-sliding homogeneity property as shown below. In order to shorten the transient, the initial value of z_0 is taken equal to the first measured value of σ, while the initial value of z_1 is taken to be 0. In order to avoid crude and sometimes problematic estimates of the constants C, K_m, K_M, the controller parameters are usually adjusted during computer simulation.

Proof.   Let $\xi_i = z_i - \sigma^{(i)}$, $i = 0$, 1, then

$$u = \varphi(\sigma(\cdot) + \xi_0(\cdot), \dot{\sigma}(\cdot) + \xi_1(\cdot)), \tag{6.19}$$

$$\dot{\xi}_0 \in -\lambda_0 |\xi_0 + [-\delta, \delta]|^{1/2} \operatorname{sign}(\xi_0 + [-\delta, \delta]) + \xi_1, \tag{6.20}$$

$$\dot{\xi}_1 \in -\lambda_1 \operatorname{sign}(\xi_0 + [-\delta, \delta]), \tag{6.21}$$

where $u = \varphi(\sigma(\cdot), \dot{\sigma}(\cdot))$ is one of controllers (6.9), (6.11) and (6.12). Consider now differential inclusion (6.5) and (6.19)–(6.21) instead of (6.1), (6.12)–(6.14). That inclusion is understood in the sense of Filippov [23], which means that the right-hand vector set is enlarged in a special way. With $\delta = 0$ variables ξ_0, ξ_1 vanish in finite time [21]. Thus the first part of Theorem 4 is a trivial consequence of Theorem 1.

Now let the noise magnitude be $\delta > 0$. It is easy to see that the transformation

$$G_\kappa : \quad (t, \sigma, \dot{\sigma}, \xi_0, \xi_1, \delta) \mapsto (\kappa t, \kappa^2 \sigma, \kappa \dot{\sigma}, \kappa^2 \xi_0, \kappa \xi_1, \kappa^2 \delta)$$

transfers the trajectories of (6.5) and (6.19)–(6.21) into the trajectories of (6.5), (19)–(21) but with the changed noise magnitude $\kappa^2 \delta$. It is shown exactly as in the proof of Theorem 2 that for some small δ_0, the trajectories gather in a small set centred at the origin $|\sigma| \le \gamma$, $|\dot{\sigma}| \le \gamma$, $|\xi_1| \le \gamma$, $|\xi_2| \le \gamma$. Apply now the transformation G_κ with $\kappa = (\delta/\delta_0)^{1/2}$, and obtain for any δ the required asymptotics of the attracting set with $\mu_0 = \gamma/\delta_0$, $\mu_1 = \gamma/\delta_0^{1/2}$. Theorem 5 is similarly proved.

6.7   Output feedback: simulation results

Consider a variable-length pendulum control problem. All motions are restricted to some vertical plane. A load of some known mass m is moving along the pendulum rod (Fig. 6.4). Its distance from O equals $R(t)$ and is not measured. There is no friction. An engine transmits a torque w that is considered as the control. The task is to track some function x_c given in real time by the angular coordinate x of the rod. The system is described by the equation

$$\ddot{x} = -2\frac{\dot{R}}{R}\dot{x} - g\frac{1}{R}\sin x + \frac{1}{mR^2}w, \tag{6.22}$$

where $g = 9.81$ is the gravitational constant and $m = 1$. Let $0 < R_m \le R \le R_M$, $\dot{R}$, $\ddot{R}$, $\dot{x}_c$ and $\ddot{x}_c$ be bounded and assume $\sigma = x - x_c$ is available. The initial conditions

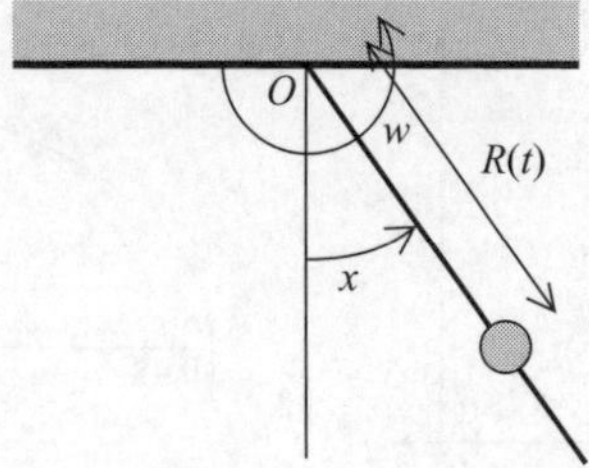

Figure 6.4   Variable-length pendulum

are $x(0) = \dot{x}(0) = 0$. The following are the functions R and x_c considered in the simulation:

$$R = 1 + 0.25 \sin 4t + 0.5 \cos t,$$

$$x_c = 0.5 \sin 0.5t + 0.5 \cos t.$$

In case chattering of the torque signal is unacceptable, $u = \dot{w}$ is considered as a new control. Define $\sigma = \dot{x} - \dot{x}_c + 2(x - x_c)$. The relative degree of the system is 2. Condition (6.3) holds here only locally: $\ddot{\sigma}|_{u=0}$ depends on $\dot{x}$ and is not uniformly bounded. Thus, the controllers are effective only in a bounded vicinity of the origin $x = \dot{x} = w = 0$. Their global application requires the standard techniques described at the end of Section 6.2, which are not implemented here for simplicity.

The applied controller of the form (6.18) is based on the twisting controller (6.9)

$$\dot{w} = u = -15 \, \mathrm{sign} \, z_0 + 10 \, \mathrm{sign} \, z_1, \tag{6.23}$$

$$\dot{z}_0 = -35 |z_0 - \sigma|^{1/2} \, \mathrm{sign} \, (z_0 - \sigma) + z_1, \quad \dot{z}_1 = -70 \, \mathrm{sign} \, (z_0 - \sigma),$$
$$\sigma = \dot{x} - \dot{x}_c + 2(x - x_c). \tag{6.24}$$

The angular velocity $\dot{x}$ is considered here to be directly measured. Otherwise, a 3-sliding controller can be applied together with a second order differentiator [18] producing both $\dot{x} - \dot{x}_c$ and $\ddot{x} - \ddot{x}_c$. In the case when discontinuous torque is acceptable, another option is to directly implement a 2-sliding controller considering $x - x_c$ as the output to be nullified. Indeed, the corresponding relative degree is also 2, and the appropriate discontinuous controller of form (6.18) is

$$w = -10 \, \mathrm{sign} \, z_0 + 5 \, \mathrm{sign} \, z_1, \tag{6.25}$$

$$\dot{z}_0 = -6 |z_0 - \sigma|^{1/2} \, \mathrm{sign} \, (z_0 - \sigma) + z_1, \quad \dot{z}_1 = -35 \, \mathrm{sign} \, (z_0 - \sigma),$$
$$\sigma = x - x_c. \tag{6.26}$$

Initial values $x(0) = \dot{x}(0) = 0$ were taken, $w(0) = 0$ is taken for controller (6.23) and (6.24), the sampling step $\tau = 0.0001$. The trajectories in the coordinates $x - x_c$ and $\dot{x} - \dot{x}_c$ in the absence of noise are shown for systems (6.22)–(6.24) and (6.22), (6.25) and (6.26) in Figs 6.5a and b respectively, the corresponding accuracies being $|x - x_c| \leq 1.6 \cdot 10^{-6}$, $|\dot{x} - \dot{x}_c| \leq 1.8 \cdot 10^{-5}$ and $|x - x_c| \leq 6.7 \cdot 10^{-6}$, $|\dot{x} - \dot{x}_c| \leq 0.01$.

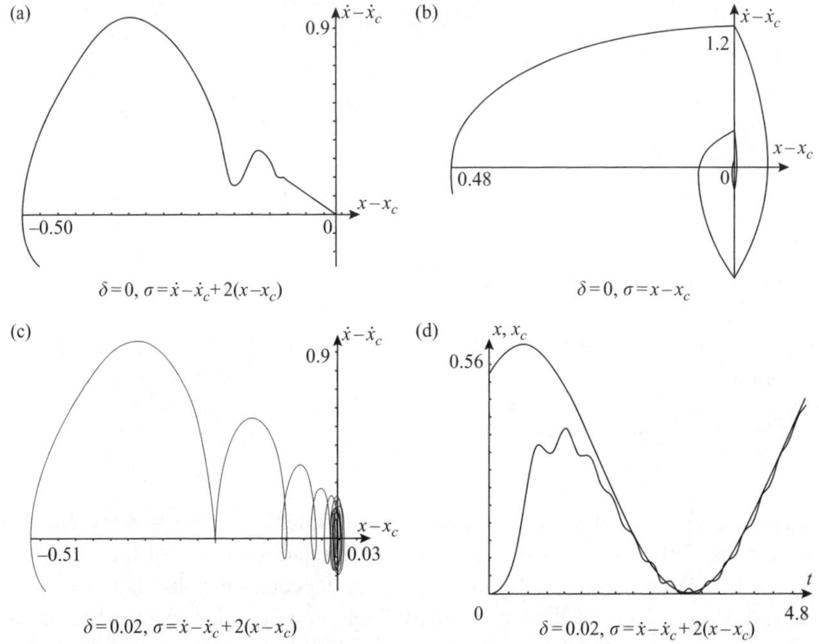

Figure 6.5   *Output-feedback 2-sliding control: simulation results*

The trajectory of (22)–(24) in the presence of noise with magnitude 0.02 in the σ-measurements is shown in Fig. 6.5c. The tracking results are shown in Fig. 5d, the tracking accuracy being $|x - x_c| \leq 0.018$, $|\dot{x} - \dot{x}_c| \leq 0.16$. The performance does not differ when the frequency of the noise changes from 10 to 10000.

6.8   Influence of the actuator dynamics

Real control systems contain fast actuators and sensors, whose dynamics are often not accounted for. Obviously, the resulting relative degree r is the sum of the relative degrees of the actuator, sensor and the plant. Thus, only an r-sliding mode, $r \geq 3$, is possible in the closed-loop system and the validity of the designed 2-sliding control is questionable. As a result, one may expect some motion to take place in a small vicinity of the 2-sliding manifold $\sigma = \dot{\sigma} = 0$.

6.8.1   *Instability of r-sliding modes, r > 2, generated by the twisting controller*

It is known that relay control systems are unstable when the relative degree exceeds 2 [34–36]. It is reasonable to expect that the same phenomenon occurs if the system is governed by the twisting 2-sliding mode algorithm. The idea of the proof of this follows. Consider a linear time invariant system of relative degree 3 governed by the twisting algorithm (6.9)

$$\dot{y}_1 = y_2, \quad \dot{y}_2 = y_3, \quad \dot{y}_3 = a_{31}y_1 + a_{32}y_2 + a_{33}y_3 + u, \tag{6.27}$$

$$u = -r_1 \operatorname{sign} y_1 - r_2 \operatorname{sign} y_2, \qquad r_1 > r_2 > 0.$$

Let the Lyapunov function be $V = y_1 y_3 - y_2^2/2$. Thus,

$$\dot{V} = y_1(a_{31}y_1 + a_{32}y_2 + a_{33}y_3 - r_1 \operatorname{sign} y_1 - r_2 \operatorname{sign} y_2) \le -a|y_1| \le 0,$$

$$0 < a < r_1 - r_2$$

and $\dot{V}$ is negative at least in a small neighbourhood of the origin $(0, 0, 0)$. This means that the zero solution of system (6.27) is unstable. It is simple to generalise this approach for systems with relative degree $n \ne 4k + 2, k \ge 1$.

6.8.2 *High relative-degree systems with fast actuators*

It is known [4] that in relay systems with relative degree 1, introduction of an actuator of the first relative degree can lead to the establishment of an asymptotically stable 2-sliding mode, which leads to spontaneous chattering elimination. On the contrary, as we have just seen, the higher order sliding mode is unstable in any system governed by the twisting controller and having relative degree 3 and more. This leads to an important conclusion: even being stable, inertial actuators do not suppress chattering in closed-loop systems.

One has to distinguish two main cases. In the first case, the controller is designed for systems with relative degree 2 and produces a discontinuous control. In that case, the actuator output will have some finite magnitude and high frequency, i.e., it chatters. In the second case, the chattering removal version (6.15) of the controller is applied with systems of relative degree 1, i.e., the controller is used as a filter. In such a case, the produced control is continuous and Lipschitzian. The actuator output will track its Lipschitzian input and only infinitesimal control chattering will be produced at the output of a fast stable actuator. In all cases, the trajectory evolves in a small vicinity of the 2-sliding manifold.

For investigation of the chattering phenomena in sliding mode systems, the averaging technique is used (6.19) and (6.18). Higher-order actuators may give rise to high frequency periodic solutions. The general model of systems with a fast actuator governed by the twisting algorithm has the form:

$$\dot{x} = h(x, y_1, y_2, z, u), \quad \dot{y}_1 = y_2, \quad \dot{y}_2 = g_2(x, y_1, y_2, z, u),$$

$$\mu \dot{z} = g_1(x, y_1, y_2, z, u), \tag{6.28}$$

where $z \in R^m, x \in R^n, y_1, y_2 \in R, z \in R^m, u = -r_1 \operatorname{sign} y_1 - r_2 \operatorname{sign} y_2$, and g_1, g_2 are smooth functions of their arguments. Variables x, y_1, y_2 may be considered as the state coordinates of the plant, z and μ being the fast-actuator coordinates and the actuator time constant respectively. Following [37, 38] consider the solutions of system (6.28) in a small neighbourhood of the second order sliding manifold $y_1 = y_2 = 0$.

Taking into account the homogeneity of the twisting algorithm, it is reasonable to consider the solutions of the system (6.28) with the initial conditions $y_1 = O(\mu^2)$, $y_2 = O(\mu)$. Then, introducing the new variables $y_1 = \mu^2\sigma_1, y_2 = \mu\sigma_2$, we can rewrite the system (6.28) in the form:

$$\dot{x} = h(x,\mu^2\sigma_1,\mu\sigma_2,z,u), \quad \mu\dot{\sigma}_1 = \sigma_2, \quad \mu\dot{\sigma}_2 = g_2(x,\mu^2\sigma_1,\mu\sigma_2,z,u),$$

$$\mu\dot{z} = g_1(x,\mu^2\sigma_1,\mu\sigma_2,z,u).$$

$$(6.29)$$

Suppose that the following conditions are true:

1.   The fast-motion system

$$\frac{d\sigma_1}{d\tau = \sigma_2}, \quad \frac{d\sigma_2}{d\tau} = g_2(x,0,0,z,u), \quad \frac{dz}{d\tau} = g_1(x,0,0,z,u), \tag{6.30}$$

has a $T(x)$- periodic solution $(\sigma_1^*(\tau,x),\sigma_2^*(\tau,x),z^*(\tau,x))$. System (6.30) generates a Poincare map $\Phi(\sigma_2,z)$ of the surface $\sigma_1 = 0$ into itself, which for any $x \in X$ has a fixed point $(\sigma_2^*(x),z^*(x))$ such that $(\sigma_2^*(x),z^*(x)) = \Phi(\sigma_2^*(x),z^*(x))$. Moreover, the Frechet derivative of $\Phi(\sigma_2,z)$ with respect to the variables σ_2,z calculated at $(\sigma_2^*(x),z^*(x))$ is a contractive matrix for any $x \in X$.
2.   The averaged system

$$\dot{x} = \frac{1}{T(x)} \int\limits_0^{T(x)} h(x,0,0,z^*(\tau,x),u(\sigma_1^*(\tau,x),\sigma_2^*(\tau,x)))d\tau$$

has a unique equilibrium point $x = x_0$. This equilibrium point is exponentially stable.

Theorem 6.   *Under conditions 1, 2, system (6.28) has an isolated orbitally asymptotically stable periodic solution with period $\mu(T(x) + O(\mu))$ near the closed curve $(x_0, \mu^2\sigma_1^*(t/\mu,x_0), \mu\sigma_2^*(t/\mu,x_0), z^*(t/\mu,x_0))$*

6.8.3   *Frequency domain analysis of chattering in 2-sliding mode systems with actuators*

In this subsection, we follow Reference 39. Taking into account that the introduction of an actuator increases the order of the system, the analysis of the corresponding Poincare maps becomes very complicated. The describing function (DF) method [24] seems to be a good choice in this case. However, the DF method provides only an approximate solution. There are two ways to use the twisting algorithm (6.9): control of systems with relative degree 2; or control of systems with relative degree 1 with the introduction of an integrator in the loop (twisting-as-a-filter). For the systems with relative degree two, it can be formulated as follows. The plant (or plant plus actuator) is described by the differential equations

$$\dot{x} = Ax + Bu,$$

$$y = Cx, \quad u = -r_1\,\text{sign}\,(y) - r_2\,\text{sign}\,(\dot{y}), \quad r_1 > r_2 > 0,$$

$$(6.31)$$

where A and B are matrices of appropriate dimensions; y can be treated as either the sliding variable or the output of the plant. The closed-loop system can be analysed by means of the DF method. Assume that a periodic motion occurs in the system with the twisting algorithm. According to the definition of the DF [24], find the DF q of the twisting algorithm as the first harmonic of the periodic control signal divided by the amplitude of $y(t)$,

$$q = \frac{\omega}{\pi A_1} \int_0^{2\pi/\omega} u(t) \sin \omega t \, dt + j \frac{\omega}{\pi A_1} \int_0^{2\pi/\omega} u(t) \cos \omega t \, dt, \tag{6.32}$$

where A_1 is the amplitude of the first harmonic and ω is the frequency of $y(t)$. However, the twisting algorithm can be analysed as the parallel connection of two ideal relays where the input to the first relay is the sliding variable and the input to the second relay is the derivative of the sliding variable. The DF for these nonlinearities are known. For the first relay, the DF is: $q_1 = 4r_1/\pi A_1$, and for the second relay it is: $q_2 = 4r_2/\pi A_2$, where A_2 is the amplitude of dy/dt. Take also into account the relationship between y and dy/dt in the Laplace domain, which gives the relationship between the amplitudes A_1 and A_2: $A_2 = A_1 \Omega$, where Ω is the frequency of the oscillation. As a result, taking into account the parallel connection of those relays, the DF of the twisting algorithm can be given as a sum of the DF of the first relay and the DF of the second relay multiplied by the Laplace operator:

$$q = q_1 + s q_2 = \frac{4r_1}{\pi A_1} + j\Omega \frac{4r_2}{\pi A_2} = \frac{4}{\pi A_1}(r_1 + jr_2). \tag{6.33}$$

Note that the DF of the twisting algorithm depends only on the amplitude value. This suggests a technique for finding the parameters of the limit cycle – via the solution of the complex equation [24]:

$$-\frac{1}{q(A_1)} = W(j\Omega), \tag{6.34}$$

where $W(j\omega)$ is the complex frequency response characteristic (Nyquist plot) of the plant and the function at the left-hand side is given by the equality $-1/q = \pi A_1(-r_1 + jr_2)/[4(r_1^2 + r_2^2)]$. Equation (6.34) is equivalent to the condition of the complex frequency response characteristic of the open-loop system intersecting the real axis in the point $(-1, j0)$. The graphical illustration of the solution technique for equation (6.34) is given in Fig. 6.6.

The function $-1/q$ is a straight line, the slope of which depends on the c_2/c_1 ratio. It is located in the second quadrant of the complex plane. The intersection point of the graph of this function and of the Nyquist plot $W(j\omega)$ provides the solution of the periodic problem. This point gives the frequency of the oscillation Ω and the amplitude A_1. Therefore, *if the transfer function of the plant (or plant plus actuator) has relative degree higher than 2, a periodic motion may occur in such a system.* For this reason, *if an actuator of first or higher order is added to the plant with relative degree 2 driven by the twisting controller, a periodic motion may occur in the system.*

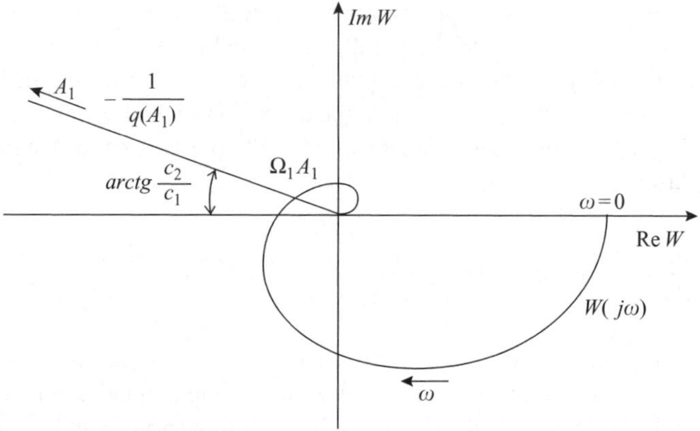

Figure 6.6   Finding the periodic solution

The *asymptotic second order SM relay controller* was studied [26, 34, 35]. The simplest scalar example of this controller has the form $\ddot{x} = -a\dot{x} - bx - k$ sign (x), $a > 0, k > 0$. It is shown in these references that this system is exponentially stable (no finite-time convergence). According to our analysis, it also follows from Fig. 6.6 that *the frequency of the periodic solution for the twisting algorithm is always higher than the frequency of the asymptotic second order sliding mode relay controller,* because the later is determined by the point of intersection of the Nyquist plot and the real axis.

Another modification of the twisting algorithm is its application to a plant with relative degree 1 with the addition of an integrator. This will be further referred to as the 'twisting as a filter' algorithm. The above reasoning is applicable in this case too. The introduction of the integrator in series with the plant makes the relative degree of this part of the system equal to 2. As a result, any actuator introduced in the loop increases the overall relative degree to at least 3 and the system becomes unstable, which may result in periodic solutions with small amplitudes. In this case, there always exists a point of intersection of the Nyquist plot of the serial connection of the actuator, the plant and the integrator and of the negative reciprocal of the DF of the twisting algorithm (Fig. 6.6). Thus, *if an actuator of first or higher order is added to a plant with relative degree 1, a periodic motion may occur in the system with the twisting as a filter algorithm.* The DF analysis provides proof of possible existence of a periodic solution in systems where the twisting algorithm is used and an actuator is introduced. However, the DF method is an approximate one and a more rigorous analysis would be desirable.

Consider first an example of analysis of the system with a relay feedback control. This will serve as a benchmark for the comparison with other types of control. Let the plant be given by $\dot{x}_1 = x_2$, $\dot{x}_2 = -x_1 - x_2 + u_a$ and the actuator by $\dot{u}_a + u_a = u$. Perform an analysis of periodic motions in systems with the asymptotic second order SM relay control and with the twisting control algorithm. Now carry out an analysis of periodic motions in the system with the twisting algorithm. Suppose the relay amplitudes are $r_1 = 0.8$ and $r_2 = 0.6$ (which provides the same amplitude of the fundamental frequency of the control signal as the unity amplitude). The simulations of the system with the given actuator-plant and the relay algorithm as well as with the twisting algorithm prove a very good match with the exact analysis. The DF analysis was also carried out and also provided a good match with the exact analysis and with the simulations.

The results of the analysis of a number of combinations of first/second order actuators, first/second relative degree plants, and twisting/twisting-as-a-filter algorithms are presented in Table 6.1. The review of the results shows good correlation between the DF analysis and the simulations. A periodic motion occurs if the relative degree of the actuator-plant system is higher than 2. The frequency of the periodic solution for the twisting control algorithm is always higher than the frequency for the second order SM asymptotic relay control (for the same actuator-plant system) – this fact was predicted by the DF analysis. Also, a comparison between the twisting-as-a-filter algorithm and the classical first order SM control is performed (in both cases, the plant is of first order). The frequency of chattering of the twisting-as-a-filter algorithm is always lower than the frequency of the first order SM control – this fact can also be explained by the above analysis if the plant is viewed as the original plant plus an integrator, with the twisting algorithm applied to that combined plant. The amplitudes of the oscillations are obtained only analytically. In all the examples, the obtained amplitude values reflect the relationship between the chattering frequency and the magnitude of the transfer function at this frequency.

6.9 Conclusions

The robustness aspects of 2-sliding mode control were studied. In particular, the influences of measurement noise and of unmodelled fast actuator dynamics were considered.

The measurement noise does not destroy the standard 2-sliding controller performance if both the sliding variable and its time derivative are available. A problem arises when the sliding variable derivative is not available and the first difference is used instead. It is proposed in this chapter to use a real-time robust exact differentiation together with standard 2-sliding controllers to provide for full SISO control based on input measurements only. The obtained controllers are locally applicable to general case SISO systems, and are also globally applicable if the boundedness restrictions (6.3) hold globally. In the absence of noise, the tracking accuracy proportional to τ^2 is provided, τ being a sampling period, which is the best possible accuracy obtainable with a discontinuous second output derivative. In the presence of bounded input noise, the tracking error is proportional to the maximal noise magnitude.

The proposed output-feedback controller is shown to be robust with respect to output noise. Thus, the differentiator is to be used whenever the sampling step can be taken small. At the same time, in the practically important case when the sampling step is sufficiently large with respect to the noise and the output derivative, the differentiator is successfully replaced by the first finite difference [6].

Table 6.1   Oscillation: computation and simulation results

	Twisting controller	Twisting controller	Asymptotic second order SM relay controller	Asymptotic second order SM relay controller
Plant $Wp(s)$	$W_p(s) = \dfrac{1}{s^2+s+1}$	$W_p(s) = \dfrac{1}{s^2+s+1}$	$W_p(s) = \dfrac{1}{s^2+s+1}$	$W_p(s) = \dfrac{1}{s^2+s+1}$
Actuator $Wa(s)$	$W_a(s) = \dfrac{1}{0.01s+1}$	$W_a(s) = \dfrac{1}{0.0001s^2+0.01s+1}$	$W_a(s) = \dfrac{1}{0.01s+1}$	$W_a(s) = \dfrac{1}{0.0001s^2+0.01s+1}$
$W(s)$	$W = W_a W_p$	$W = W_a W_p$	$W = W_a W_p$	$W = W_a W_p$
Ω (DF analysis)	77.05	54.64	10.05	10.00
Ω (simulations)	77.68	54.53	9.36	9.13
Plant output chattering amplitudes	$1.67e{-}4$	$4.83e{-}4$	0.0146	0.0155

Another robustness problem arises when an unaccounted-for actuator dynamics is introduced. It is shown that the corresponding higher order sliding mode is locally unstable and a periodic motion may occur in such systems, when the combined relative degree of the actuator and the plant is higher than 2. The parameters of this periodic motion are approximately calculated by means of the DF method. The oscillations reveal themselves in small vibrations of the sliding variable and its derivative near zero. The performed analysis shows that the frequency of the oscillations grows and their amplitude decreases due to the use of the twisting algorithm in comparison with the asymptotic 2-sliding mode controller. Also, the frequency of the oscillations of the twisting-as-a-filter algorithm is always lower than the frequency of the 1-sliding mode control.

6.10   References

1   UTKIN, V. I.: 'Sliding modes in optimization and control problems' (Springer Verlag, New York, 1992)

2   ZINOBER, A. S. I. (Ed.): 'Variable structure and Lyapunov control' (Springer Verlag, Berlin, 1994)

3   SLOTINE, J.-J. E. and LI, W.: 'Applied nonlinear control' (Prentice-Hall London, 1991)

4   FURUTA, K. and PAN, Y.: 'Variable structure control with sliding sector', *Automatica*, 2000, **36**, pp. 211–228

5   EMELYANOV, S. V., KOROVIN, S. K., and LEVANTOVSKY, L. V.: 'Higher order sliding regimes in the binary control systems', *Soviet Physics, Doklady*, 1986, **31**(4), pp. 291–293

6   LEVANT, A. (LEVANTOVSKY, L. V.): 'Sliding order and sliding accuracy in sliding mode control', *International Journal of Control*, 1993, **58**(6), pp. 1247–1263

7   BARTOLINI, G., PISANO, A., and USAI, E.: 'Second-order sliding-mode control of container cranes', *Automatica*, 2002, **38**, pp. 1783–1790

8   SIRA-RAMÍREZ, H.: 'On the dynamical sliding mode control of nonlinear systems', *International Journal of Control*, 1993, **57**(5), pp. 1039–1061

9   MAN Z., PAPLINSKI, A. P., and WU, H. R.: 'A robust MIMO terminal sliding mode control for rigid robotic manipulators', *IEEE Trans. Automat. Control*, 1994, **39**(12), pp. 2464–2468

10   BARTOLINI, G., FERRARA, A., and PUNTA, E.: 'Multi-input second-order sliding-mode hybrid control of constrained manipulators', *Dynamics and Control*, 2000, **10**, pp. 277–296

11   BARTOLINI, G., PISANO, A., PUNTA, E., and USAI, E.: 'A survey of applications of second-order sliding mode control to mechanical systems', *International Journal of Control*, 2003, **76**(9/10), pp. 875–892

12   FERRARA, A. and GIACOMINI, L.: 'Control of a class of mechanical systems with uncertainties via a constructive adaptive/second order VSC approach', *J. DYN SYST-T ASME*, 2000, **122**(1), pp. 33–39

13   FLOQUET, T., BARBOT, J.-P., and PERRUQUETTI, W.: 'Higher-order sliding mode stabilization for a class of nonholonomic perturbed systems', *Automatica*, 2003, **39**, pp. 1077–1083

14   LEVANT, A., PRIDOR, A., GITIZADEH, R., YAESH, I., and BEN-ASHER, J. Z.: 'Aircraft pitch control via second order sliding technique', *J. of Guidance, Control and Dynamics*, 2000, **23**(4), pp. 586–594

15   ORLOV, Y., AGUILAR, L., and CADIOU, J. C.: 'Switched chattering control vs. backlash/friction phenomena in electrical servo-motors', *International Journal of Control*, 2003, **76**(9/10), pp. 959–967

16   SIRA-RAMÍREZ, H.: 'Dynamic second-order sliding mode control of the hovercraft vessel', *IEEE Transactions On Control Systems Technology*, 2002, **10**(6), pp. 860–865

17   SHKOLNIKOV, I. A., SHTESSEL Y. B., LIANOS D., and THIES, A. T.: 'Robust missile autopilot design via high-order sliding mode control' Proceedings of AIAA *Guidance, Navigation, and Control Conference*, Denver, CO, 2000, AIAA paper no. 2000-3968

18   LEVANT, A.: 'Higher-order sliding modes, differentiation and output-feedback control', *International Journal of Control*, 2003, **76**(9/10), pp. 924–941

19   BARTOLINI, G., FERRARA, A., and USAI, E.: 'Chattering avoidance by second-order sliding mode control', *IEEE Trans. Automat. Control*, 1998, **43**(2), pp. 241–246

20   LEVANT, A.: 'Variable measurement step in 2-sliding control', *Kibernetica*, 2000, **36**(1), pp. 77–93

21   LEVANT, A.: 'Robust exact differentiation via sliding mode technique', *Automatica*, 1998, **34**(3), pp. 379–384

22   BACCIOTTI, A. and ROSIER, L.: 'Liapunov functions and stability in control theory', Lecture notes in control and information sciences 267 (Springer-Verlag, New-York, 2001)

23   FILIPPOV, A. F.: 'Differential equations with discontinuous right-hand side' (Kluwer, Dordrecht, The Netherlands, 1988)

24   ATHERTON, D. P.: 'Nonlinear control engineering – describing function analysis and design' (Van Nostrand, Workingham, Berks, UK, 1975)

25   ISIDORI, A.: 'Nonlinear control systems' (Springer Verlag, New York, 1989, 2nd edn)

26   EMELYANOV, S. V., KOROVIN, S. K., and LEVANT, A.: 'Higher-order sliding modes in control systems', *Differential Equations*, 1993, **29**(11), pp. 1627–1647

27   LEVANT, A.: 'Construction principles of output-feedback 2-sliding mode design'. Proceedings of the IEEE conference on *Decision and Control*, Las-Vegas, Nevada, December 10–13, 2002

28   ROSIER L.: 'Homogeneous Lyapunov function for homogeneous continuous vector field', *System and Control Letters*, 1992, **19**, pp. 467–473

29   ATASSI, A. N. and KHALIL, H. K.: 'Separation results for the stabilization of nonlinear systems using different high-gain observer designs', *Systems and Control Letters*, 2000, **39**, pp. 183–191

30   BARTOLINI, G., PISANO, A. and USAI E.: 'First and second derivative estimation by sliding mode technique', *Journal of Signal Processing*, 2000, **4**(2), pp. 167–176

31   KRUPP, D., SHKOLNIKOV, I. A., and SHTESSEL, Y. B.: '2-sliding mode control for nonlinear plants with parametric and dynamic uncertainties'. Proceedings of AIAA *Guidance, Navigation, and Control Conference*, Denver, CO, 2000, AIAA paper no. 2000-3965, 2000

32   KOBAYASHI, S., SUZUKI, S., and FURUTA, K.: 'Adaptive VS differentiator', *Advances in Variable Structure Systems*. Proceedings of the 7th VSS Workshop, July 2002, Sarajevo

33   YU, X. and XU, J. X.: 'An adaptive signal derivative estimator', *Electronic Letters*, 1996, **32**(16), pp. 1445–1447

34   TSYPKIN, Y. Z.: 'Relay control systems' (Cambridge University Press, Cambridge, 1984)

35   ANOSOV, D. V.: 'On stability of equilibrium points of relay systems', *Automation and Remote Control*, 1959, **2**, pp. 135–149 (in Russian)

36   FRIDMAN, L. and LEVANT, A.: 'Higher order sliding modes', in: BARBOT, J. P. and PERRUGUETTI, W. (Eds): 'Sliding mode control engineering' (Marcel Dekker, New York, 2002), pp. 53–102

37   FRIDMAN, L. M.: 'The problem of chattering: an averaging approach', in YOUNG, K. K. D. and OZGUNER, U. (Eds): 'Variable structure systems, sliding mode and nonlinear control, *Lecture Notes in Control and Information Sciences*, 247 (Springer-Verlag, Berlin, 1999), pp. 363–386

38   FRIDMAN, L.: 'An averaging approach to chattering', *IEEE Transactions of Automatic Control*, 2001, **46**, pp. 1260–1265

39   BOIKO, I., CASTELLANOS, M. I., and FRIDMAN, L.: 'Analysis of second order sliding mode algorithms in the frequency domain'. Proceedings of 42th conference on *Decision in Control*, Maui, Hawaii, 2003

Chapter 7

Sliding modes, delta-modulation and output feedback control of dynamic systems[*]

Hebertt Sira-Ramírez and Luis Iván Lugo Villeda

7.1 Introduction

In this chapter, we propose a sliding mode based algorithm for robust differentiation of reference signals with uniformly bounded rates which may also be subject to additive measurement noise. The proposed algorithm is based entirely in the reinterpretation of sliding mode features of *Delta Modulation* based signal tracking (see Steele [1] and Norsworthy et al. [2]), in combination with well known properties of the *Equivalent Control* method (Utkin [3]). We specifically show that an elementary reference signal tracking problem, with control decision inputs restricted to a discrete set, naturally yields a classic delta modulation tracking scheme consisting of a feed forward sign function nonlinearity in feedback connection with a pure integrator. The reference signal is only assumed to be differentiable with an absolutely bounded time derivative. The delta modulator output coincides, under ideal sliding conditions, with the equivalent control associated with the tracking problem. This 'equivalent' modulator output signal is just the time derivative of the exogenous reference input signal to the modulator, provided the switched gain is chosen in accordance with the (known) uniform absolute bound of the reference signal rate. Hence, using well known results of the equivalent control method, a first order low pass filtering of the modulator's output asymptotically converges to the time derivative of the input signal. Since the cut-off frequency of the low pass filter can be chosen to be relatively high, due to the ideal infinite switching frequency of the tracking feedback signal, the low pass

* This research was supported by the Centro de Investigación y Estudios Avanzados del IPN, (CINVESTAV-IPN) and by the Consejo Nacional de Ciencia y Tecnología (CONACYT) under Research Grant 42231-Y.

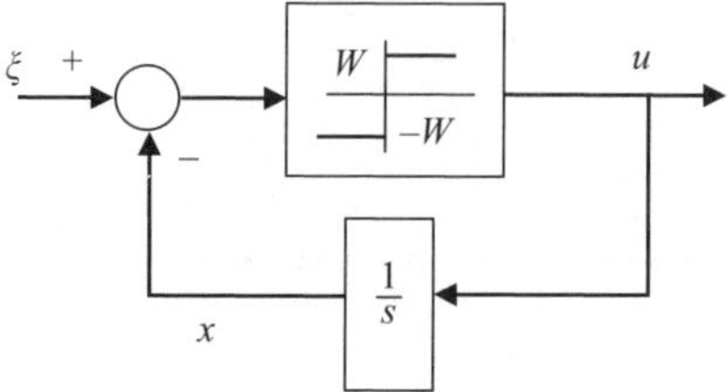

Figure 7.1   Classical analogue Δ-modulator

filter can be made quite fast and, hence, the asymptotic convergence of the filter output to the time derivative of the input signal can be made to occur very quickly. This differentiation result, already known from the work of Golembo et al. [4], is here put in the context of delta modulation and proofs of the basic facts are provided. A sliding mode approach to Sigma-Delta modulation can be found in recent articles [5, 6]. Earlier results in this area were known from Sira-Ramírez [7].

The scheme is also shown to be robust with respect to additive noise in the reference input signal. For smooth signals with unknown uniform absolute bounds on the time derivative, an adaptive scheme is proposed that automatically induces a sliding motion on the tracking error space by suitable adjustment of the switching gain. The resulting adaptation scheme is strikingly reminiscent of classical *syllabically companded* delta modulation schemes.

Section 7.2 presents the developments leading to a delta modulator based differentiator. In this section we also describe an experimental circuit for signal differentiation. Section 7.3 is devoted to presenting some illustrative examples of output feedback control of some nonlinear systems accomplishing non-trivial trajectory tracking manoeuvres. The illustrative examples are of SISO as well as MIMO nature. Section 7.4 is devoted to extending the delta modulation based differentiation results, using first order sliding motions, to one using second order sliding. For this we base our developments on a well-studied algorithm proposed in Fridman and Levant [8]. The last section presents the conclusions and suggestions for further research.

7.2   Delta-modulators and sliding modes

Consider the basic block diagram of Fig. 7.1 depicting a classical analogue Δ-modulator, traditionally used in the early stages of voice signal encoding systems. The following theorem summarises the relation of delta-modulators with sliding mode control and depicts the basic features of performance of this (forgotten) modulator[1].

---

[1] For interesting details about delta modulation and its many variations, devoid of sliding mode control considerations, the reader is referred to the classical book by Steele [1].

Theorem 1.   *Given a continuously differentiable signal, $\xi(t) \in C^1$, with absolutely uniformly bounded first order time derivative, $\dot{\xi}(t)$, there exists a strictly positive gain, W, such that for all initial values $x(t_0)$ of the feedback (locally decoded) signal $x(t)$, it is verified that $x(t) \to \xi(t)$ in a finite amount of time $t_h > t_0$, provided the following encoding condition is satisfied,*

$$W > \sup |\dot{\xi}(t)| \tag{7.1}$$

Moreover, from any arbitrary initial value of the tracking, or local encoding, error $e(t_0) = x(t_0) - \xi(t_0)$, a sliding motion exists on the perfect encoding condition $e = 0$ for all $t \geq t_h$, where the quantity t_h satisfies

$$t_h \leq t_0 + \frac{|e(t_0)|}{W - \sup |\dot{\xi}(t)|}$$

Proof.   From the figure, the variables in the Δ-modulator satisfy the following relations:

$$\dot{x} = u$$

$$u = W \text{sign}(\xi - x) \tag{7.2}$$

$$e = x - \xi$$

Clearly, $\dot{e} = -W \text{sign}(e) - \dot{\xi}(t)$ and since $\dot{\xi}(t)$ is assumed to be absolutely uniformly bounded, choosing $W > \sup |\dot{\xi}(t)|$ we have, for $|e| > 0$:

$$e\dot{e} = -W|e| - e\dot{\xi}(t) = -W|e| - |e|\dot{\xi}\text{sign}(e)$$

$$\leq -W|e| + |e| \sup |\dot{\xi}| = -(W - \sup |\dot{\xi}|)|e| < 0 \tag{7.3}$$

A sliding regime exists on $e = 0$ for all time t after the hitting time t_h (see [3]). Under ideal sliding, or encoding, conditions, $e = 0$, $\dot{e} = 0$, we have that $x = \xi(t)$ and the *equivalent* (average) value of the coded output signal u is given by $u_{eq} = \dot{\xi}(t)$ for all $t \geq t_h$.

Remark 1.   *Note that if $\xi(t_0)$ is known, by setting the initial conditions $x(t_0)$ to be arbitrarily close to $\xi(t_0)$, the sliding regime starts to exist in a correspondingly arbitrarily small time. Ideally, then, one could set to zero the sliding surface reaching time t_h.*

7.2.1   *The equivalent control method in time differentiation of signals*

It is easy to see that the Δ-modulator output u ideally differentiates the modulator input signal $\xi(t)$ in an *equivalent control* sense, or *average* sense (see Utkin [3]). Indeed, let $W > 0$ be a positive scalar, if one considers the elementary tracking problem of having the state x of a first order integrator system: $\dot{x} = u$, with $u \in \{-W, W\}$, track the C^1 signal $\xi(t)$ with $\sup_t |\dot{\xi}(t)| < \infty$, it readily follows from the developments above, that a *control input* switching policy exists of the form: $u = W \text{sign}(\xi - x)$. This induces, in finite time, a sliding motion on the zero level set of the tracking error signal $e = \xi - x$, provided $W > \sup_t |\dot{\xi}(t)|$, i.e. the condition $e = 0$ is achievable in finite time, and maintained thereafter, whenever W uniformly absolutely bounds the time derivative of the tracking signal $\xi(t)$. Clearly, under the ideal sliding surface invariance conditions $e = \dot{e} = 0$, it follows from the definition of e itself that the corresponding *equivalent control*, u_{eq}, is given by $u_{eq} = \dot{\xi}(t)$.

The following development is a restatement of a rather well known, and fundamental, result in sliding mode control, established by Utkin [3].

Ideal sliding motions require infinite switching frequency for the control input u, aside from other idealised behaviour of the switch defining the control input. *Real* sliding motions may be plagued by switch imperfections (small delays, parasitic dynamics, noisy inputs) as well as the natural limitation of a high, but finite, switching frequency. Ideal sliding motions are thus never achievable in practice and the sliding mode conditions $e = 0$, $\dot{e} = 0$ are not rigorously valid. In fact, only conditions of the form $\sup_t |e(t)| < \delta_0$, $\sup_t |\dot{e}(t)| < \delta_1$, for small positive scalar constants δ_0, δ_1, may be actually guaranteed, or enforced. Nevertheless, even under a large class of realistic imperfections of the sliding mode implementation, and non-ideal sliding mode conditions, the equivalent control signal has been shown to be approximately synthesised, in practice, by letting the actual high frequency switched control input, u, undergo the effects of a unit-gain low pass filter.

Let $\tau > 0$ be a constant. Consider the delta modulator signal tracking system of the previous theorem, with a first order low pass filter connected to the output of the modulator (see Fig. 7.2). In other words, consider the system

$$\dot{x} = u$$

$$u = W \mathrm{sign}(\xi - x)$$

$$e = x - \xi$$

$$\tau \dot{y} + y = u$$

$$(7.4)$$

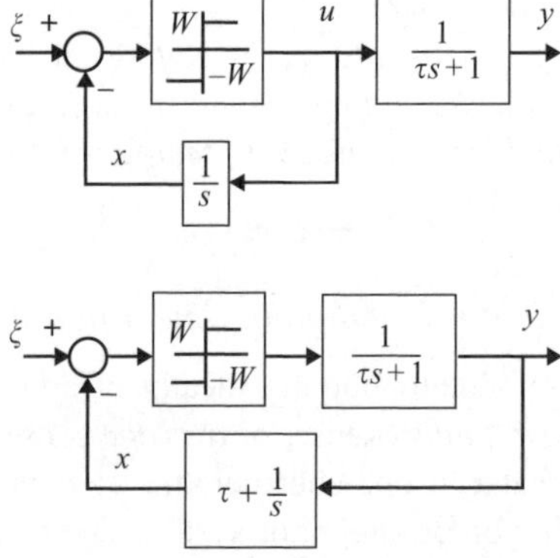

Figure 7.2   Classical analogue Δ-modulator with a low-pass filter; equivalent representations

Assume, that we set the filter initial condition $y(t_0)$ to zero. Under sliding mode conditions we have that

$$\dot{e} = -\dot{\xi} - W \operatorname{sign} e \qquad (7.5)$$

Hence, $W \operatorname{sign} e = -\dot{\xi} - \dot{e}$ and therefore

$$\tau \dot{y} + y = \dot{\xi} + \dot{e}$$

Integrating the filter differential equation we have, for any time $t > t_0$,

$$y(t) = \frac{e^{-t/\tau}}{\tau} \int_{t_0}^{t} e^{\sigma/\tau} \left(\dot{\xi}(\sigma) + \dot{e}(\sigma) \right) d\sigma$$

$$= \frac{e^{-t/\tau}}{\tau} \left[e^{t/\tau}(\xi(t) + e(t)) - (\xi(0) + e(0)) - \frac{1}{\tau} \int_{t_0}^{t} e^{\sigma/\tau}(\xi(\sigma) + e(\sigma))d\sigma \right]$$

$$= \frac{1}{\tau} \left[(\xi(t) + e(t)) - (\xi(t_0) + e(t_0))e^{-t/\tau} - \frac{e^{-t/\tau}}{\tau} \int_{t_0}^{t} e^{\sigma/\tau}(\xi(\sigma) + e(\sigma))d\sigma \right]$$

$$\qquad (7.6)$$

It can be shown from the above expressions, following the steps in Utkin [3], that, given an arbitrary small scalar quantity ϵ, there then exists a sufficiently small filter time constant, τ, and time instants $T(\epsilon, \tau) > 0$, and $\Delta(\epsilon, \tau)$, with $T(\epsilon, \tau) > \Delta(\epsilon, \tau)$, such that the $\sup_t |y(t) - \dot{\xi}(t)| < \epsilon$ for all $t_0 + \Delta(\epsilon, \tau) < t < T(\epsilon, \tau)$. In other words, the ideal equivalent control signal can be physically reproduced, in an approximate manner, during a certain time interval for a sufficiently fast filter which still behaves as a low pass filter for the high frequency switching inputs.

7.2.2   *An illustrative example with experimental results*

Consider the perfectly known signal $\xi(t) = A \sin(\omega t)$. From the results of the previous section, if a gain W_1 is chosen such that $W_1 > A\omega$, then the filtered output of the delta-modulator constitutes an approximation to the time derivative, $\dot{\xi}(t) = A\omega \cos(\omega t)$, of the input signal to the modulator $\xi(t)$.

Figure 7.3 depicts an electronic circuit synthesising the delta-modulation-low pass filter differentiator proposed in the previous section. The circuit uses commercially available operational amplifiers, a high speed buffer, a signal generator and standard passive and active elements. The tracking error signal activates a Schmidt trigger acting as the feedforward 'sign' function in the delta modulator. An octal transceiver, or digital buffer, ensures an output signal in the range 0–5 [V]. The switched output is passed through an 'adder' whose purpose is to adjust the signal to a bipolar range of $[-2.5, 2.5]$ [V]. This signal is then amplified to the required encoding range $[-W, W]$, here set to be $[-10.8, 10.8]$ [V], and fed back to the input comparator via an operational amplifier based integration circuit with time constant

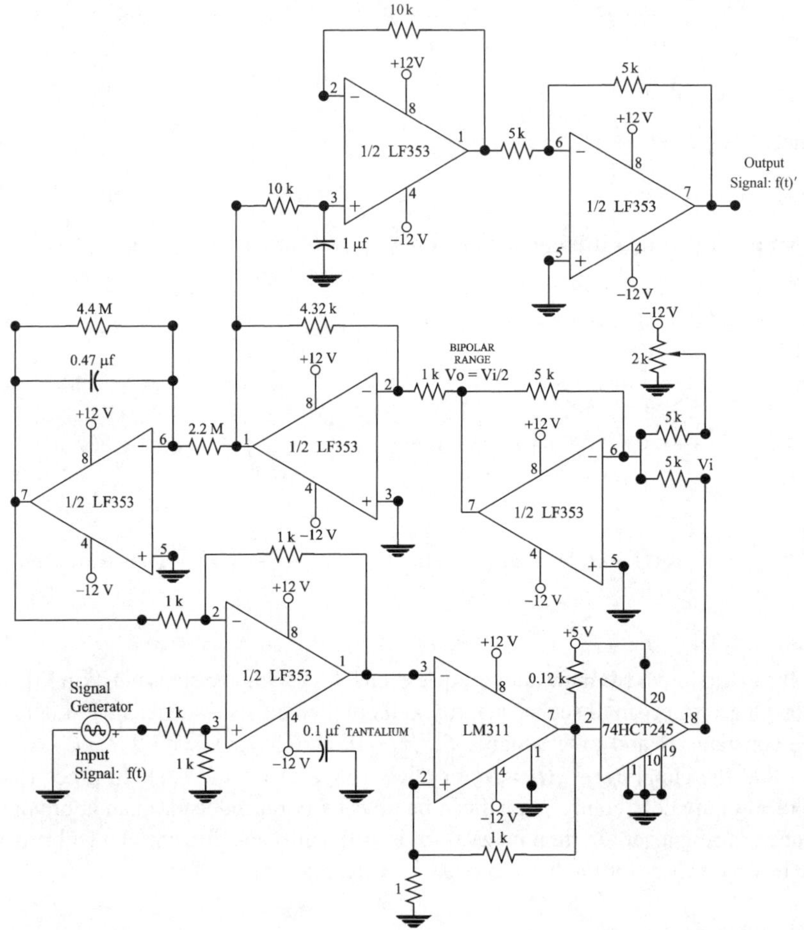

Figure 7.3   *Electronic circuit realisation of Δ-modulation-low pass filter based differentiator*

$R_i C_i = 0.9671$. The output low pass filter time constant was set to be $\tau = 0.01$ [s], which corresponds with a cut-off angular frequency of 100 [rad/s] or 15.91 [Hz]. The sinusoid input parameters were set to be

$$A = 0.75 \text{ [V]}, \quad f = 0.913 \text{ [Hz]}$$

Figure 7.4 shows the experimental performance of the differentiator when the input is represented by a low frequency sinusoid signal, of the form $y(t) = A \sin(\omega t + \phi)$, generated by a commercial wave generator. The figure shows the output of the low pass filter $\hat{y}_f$ which approximates, rather well, the time derivative function $A\omega \cos(\omega t + \phi)$. We also depict the behaviour of the sliding surface coordinate function (or encoding error function), $e = y - x$, with x being the integrated

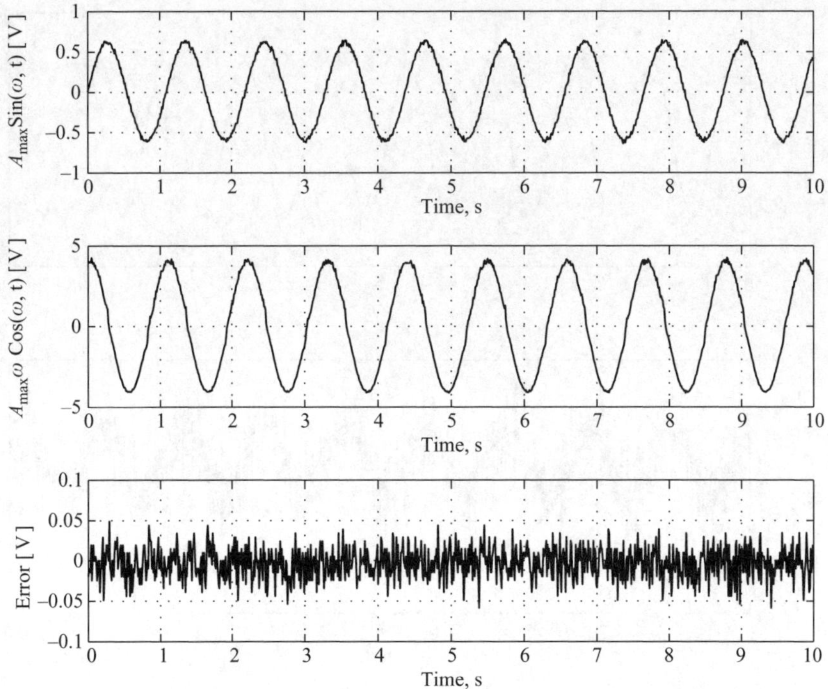

Figure 7.4   *Actual sinusoidal input, Δ-modulation based differentiated output and encoding error*

feedback signal of the modulator (locally decoded output), i.e. $\dot{x} = W \operatorname{sign} e$. Just for comparison purposes, Fig. 7.5 shows the same sinusoidal signal numerically differentiated, in the traditional backward difference scheme, through a 5 [ms] sampling interval.

7.3   Output feedback control of differentially flat systems

Consider a nonlinear SISO observable system, defined by the smooth drift vector field $f(x)$ and input vector field $g(x)$

$$\dot{x} = f(x) + g(x)u, \qquad x \in R^n, \quad u \in R$$
$$y = h(x), \qquad y \in R \tag{7.7}$$

We are primarily interested in SISO systems that exhibit a linearising, or flat, output. In this case, the system is known to be linearisable by means of static state feedback. In fact, if the system is not linearisable by means of static state feedback, then dynamic extension of the system does not yield a linearisable system either. For simplicity, we assume that y is the linearising, or flat, output. Then the system is easily shown to be

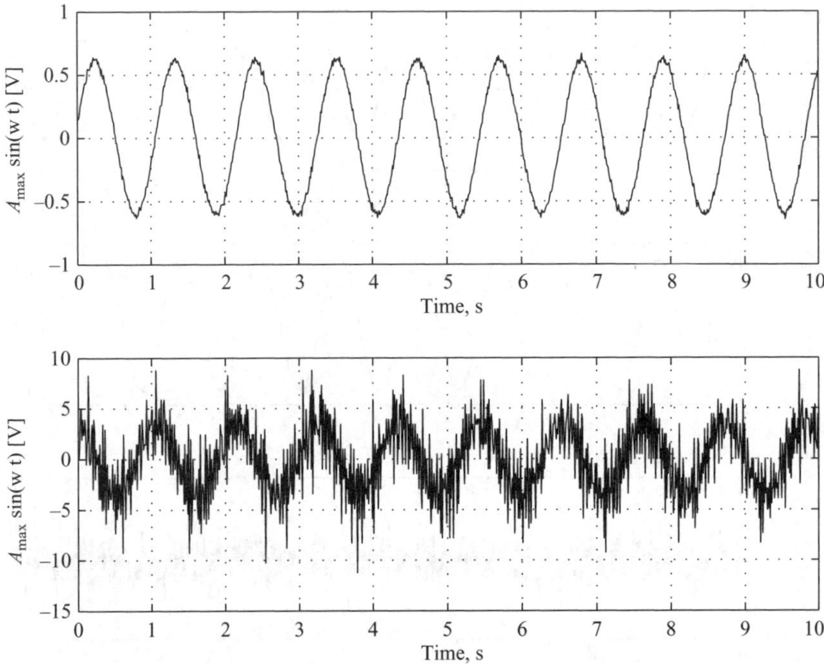

Figure 7.5   Numerically differentiated sinusoid input signal

locally (globally) observable from y and the following map, which is independent of the inputs, is locally (globally) invertible.

$$\begin{bmatrix} y \\ \dot{y} \\ \vdots \\ y^{(n-1)} \end{bmatrix} = \begin{bmatrix} h(x) \\ L_f h(x) \\ \vdots \\ L_f^{n-1} h(x) \end{bmatrix} \tag{7.8}$$

In other words, the state x is locally (globally) obtainable via a nonlinear (differential) vector function of a finite number of time derivatives of the output signal y, i.e. $x = \varphi(y, \dot{y}, \ldots, y^{(n-1)})$. Any suitable feedback control, which is synthesised on the basis of the state vector x, can then also be synthesised in terms of the flat output and a finite number of its time derivatives. This is the basis for a feedback control approach based on iterated time differentiation of the system output.

Consider the case of a nonlinear MIMO observable system with the same number of inputs and outputs (i.e. a *square* system) defined by the smooth drift vector field $f(x)$ and the smooth input matrix $G(x)$,

$$\dot{x} = f(x) + G(x)u, \qquad x \in R^n, \quad u \in R^m$$

$$y = h(x), \qquad y \in R^m \tag{7.9}$$

For simplicity, we assume that the system (7.9) is, in general, a suitable dynamic extension of an originally given $n - p$ dimensional system that becomes static feedback linearisable. Assume that the m-vector $z = (z_1, \ldots, z_m)$ qualifies as the set of flat outputs for the suitably extended system. In general, if the system is flat, we have that $x = \psi(z, \dot{z}, \ldots, z^{(\alpha)})$, $u = \vartheta(z, \dot{z}, \ldots, z^{(\gamma)})$ with $z = \varphi(x, u, \dot{u}, \ldots, u^{(\beta)})$ for some multi-index $\alpha = (\alpha_1, \ldots, \alpha_m)$ and $\beta = (\beta_1, \ldots, \beta_m)$ and $\gamma = (\gamma_1, \ldots, \gamma_m)$ with $z^{(\alpha)}$ meaning $z^{(\alpha)} = (z_1^{(\alpha_1)}, \ldots, z_m^{(\alpha_m)})$.

It is clear that a flatness based controller may be based on the auxiliary multi-input decoupled set of equations, representing in general a dynamic input coordinate transformation

$$z^{(\gamma)} = v, \qquad \left(z^{(\gamma_i)} = v_i, \qquad i = 1, \ldots, m, \qquad \sum_i \gamma_i = n \right) \tag{7.10}$$

A finite number of time derivatives of the flat outputs z are to be generated for any stabilising, or trajectory tracking, feedback controller based on exact linearisation and pole placement. We propose to use differentiators in the generation of such feedback signals.

7.3.1   A third order integrator system

Consider the third order integrator system

$$y^{(3)} = u \tag{7.11}$$

It is desired to track a given signal $y^*(t)$ smoothly rising from an initial value to a final constant value in a finite time interval $[t_0, T]$. A pole-placement based feedback controller for the tracking error signal is readily proposed to be

$$u = [y^*(t)]^{(3)} - k_2(\ddot{y} - \ddot{y}^*(t)) - k_1(\dot{y} - \dot{y}^*(t)) - k_0(y - y^*(t)) \tag{7.12}$$

Evidently, the time derivatives of the output signal need to be obtained from the measured output signal y. To this end, for the synthesis of the required time derivatives of the output signal, we advocate the use of cascade arrangements of low pass filtered outputs of delta modulation circuits. We denote these filtered outputs by $\dot{y}_e$, $\ddot{y}_e$ (see Fig. 7.6) to indicate the approximate, or estimated, nature of these derivative signals.

The delta modulator low pass filter combination for the synthesis of the ith time derivative of y ($i = 1, 2$) was realised as

$$y_{dm}^{(i)}(t) = W_i \mathrm{sign}(e_i(t))$$

$$e_i(t) = y_e^{(i-1)}(t) - x_i(t)$$

$$\dot{x}_i(t) = y^{(i)}(t) \tag{7.13}$$

$$\tau_i \dot{y}_e^{(i)}(t) = -y_e^{(i)}(t) + y^{(i)}(t)$$

with $y_e^{(0)}(t) = y(t)$.

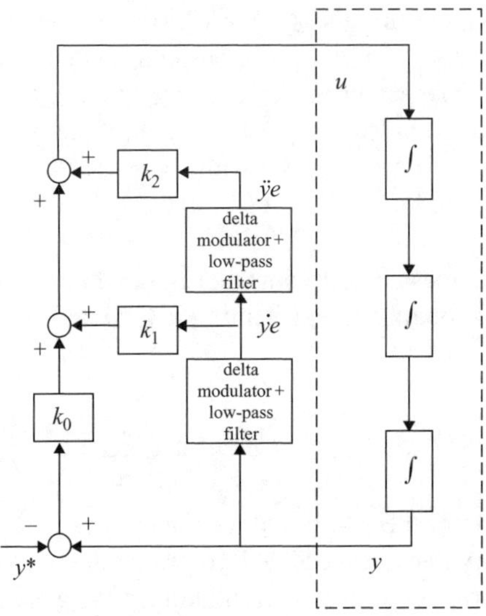

Figure 7.6   Output feedback control scheme for third order integrator

The gains $\{k_0, k_1, k_2\}$, for the closed loop linear system, can be chosen by identifying, term by term, the closed loop characteristic polynomial, $p(s) = s^3 + k_2 s^2 + k_1 s + k_0$, with a desired strictly stable polynomial of the form: $p_d(s) = (s^2 + 2\zeta\omega_n s + \omega_n^2)(s + p)$, i.e.

$$k_2 = p + 2\zeta\omega_n, \quad k_1 = \omega_n^2 + 2\zeta\omega_n p, \quad k_0 = \omega_n^2 p$$

where ζ, ω_n and p are chosen to be strictly positive design constants. Figure 7.7 depicts the performance of the feedback control strategy for the control input, u, the position variable, y, the velocity variable, $\dot{y}$, and the acceleration variable, $\ddot{y}$. The reference signal $y^*(t)$ was set to be a smooth polynomial function of the form

$$y^*(t) = y(t_0) + [y(T) - y(t_0)]\phi(t, t_0, T)$$

with $\phi(t_0, t_0, T) = 0$, $\phi(T, t_0, T) = 1$ and given by

$$\phi(t, t_0, T) = \left[\frac{t - t_0}{T - t_0}\right]^8 \left[r_1 - r_2\left(\frac{t - t_0}{T - t_0}\right) + \cdots - r_8\left(\frac{t - t_0}{T - t_0}\right)^7 \right.$$
$$\left. + r_9\left(\frac{t - t_0}{T - t_0}\right)^8\right]$$

with

$$r_1 = 12\,870, \quad r_2 = 91\,520, \quad r_3 = 288\,288, \quad r_4 = 524\,160, \quad r_5 = 600\,600,$$
$$r_6 = 443\,520, \quad r_7 = 205\,920, \quad r_8 = 54\,912, \quad r_9 = 6435$$

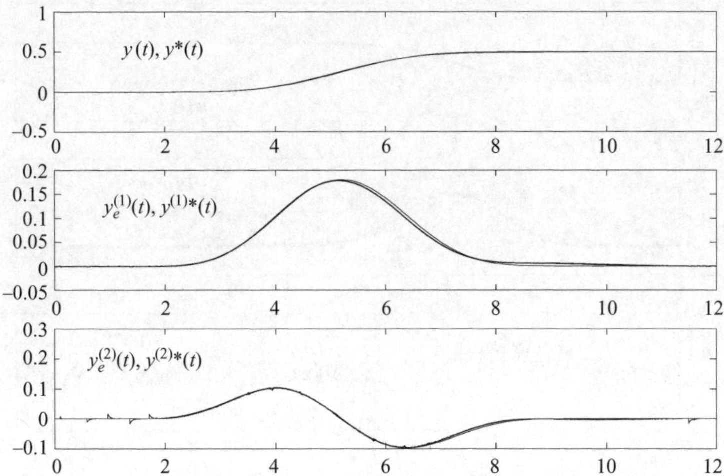

Figure 7.7   Performance of third order controlled system using output derivatives feedback

We set the following values for the reference signal $y^*(t)$ $t_0 = 1$, $T = 10$, $y(t_0) = 0$, $y(T) = 0.5$. The controller gains were set to be $p = 1$, $\zeta = 0.707$, $w_n = 1$ and the delta modulator gains were set to be $W_1 = 0.3$, $W_2 = 0.2$. The time constants of the low pass filters were set to be: $\tau_1 = \tau_2 = 25$.

In order to test the robustness of the proposed feedback control scheme with respect to input and measurement noise, we use the derived controller on the perturbed system

$$z^{(3)} = u + \eta(t)$$

$$y = z + \nu(t)$$

where $\eta(t)$ is an exogenous perturbation input modelled by a computer generated noise with an uniform rectangular probability distribution function at each instant of time t. The measurement noise $\eta(t)$ is also a rectangularly distributed random variable at each instant of time. Figure 7.8 represents the output signal and the two filtered time derivatives in comparison with the actual trajectories of these variables.

7.3.2   Flatness based control of the synchronous generator

The following model constitutes a popular representation of a single synchronous generator connected to an infinite bus (see Kundur [9]). The dynamic model is given by the following set of differential equations

$$\dot{x}_1 = x_2$$

$$\dot{x}_2 = -b_1 x_3 \sin(x_1) - b_2 x_2 + P \tag{7.14}$$

$$\dot{x}_3 = b_3 \cos(x_1) - b_4 x_3 + E + u$$

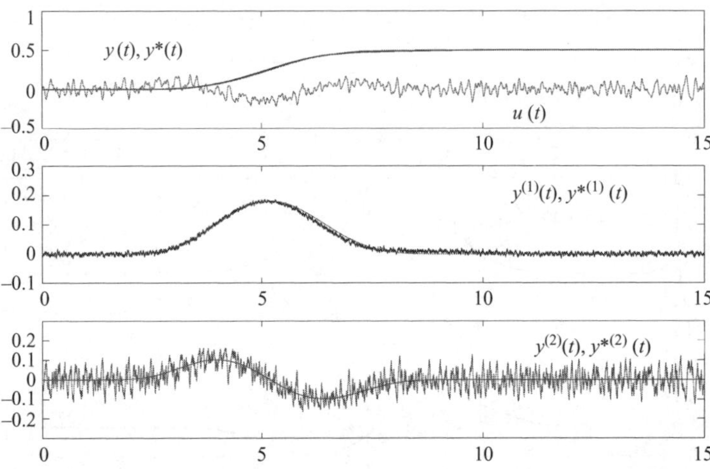

Figure 7.8   *Performance of perturbed third order controlled system using output derivatives feedback*

where x_1 is the load angle, x_2 is the velocity deviation of the rotor axis with respect to the synchronous velocity and x_3 is the internal voltage in the quadrature axis. The coefficients $b_1,\ldots,b_4$ are assumed to be known positive parameters. P represents the mechanical power delivered to the generator. The input field voltage is represented by the sum $u + E$ where E is a constant equilibrium voltage and u is a supplementary control input.

The system is clearly differentially flat, with flat output $y = x_1$. Indeed, the following differential parameterisation is clear from the system equations

$$x_1 = y$$

$$x_2 = \dot{y}$$

$$x_3 = \frac{1}{b_1 \sin(y)}[P - b_2\dot{y} - \ddot{y}]$$

$$u = -E + \frac{b_4}{b_1 \sin(y)}[P - b_2\dot{y} - \ddot{y}] - b_3\cos(y)$$

$$- \frac{1}{b_1 \sin^2(y)}\left[(b_2\ddot{y} + y^{(3)})\sin(y) + (P - b_2\dot{y} - \ddot{y})\dot{y}\cos(y)\right] \tag{7.15}$$

The following trajectory tracking controller, forcing the flat output y to track the desired trajectory $y^*(t)$, is usually proposed in flatness based control of similar third order systems:

$$v = [y^*(t)]^{(3)} - k_3(\ddot{y} - \ddot{y}^*(t)) - k_2(\dot{y} - \dot{y}^*(t)) - k_1(y - y^*(t))$$

$$- k_0 \int_0^t (y - y^*(\sigma))d\sigma \tag{7.16}$$

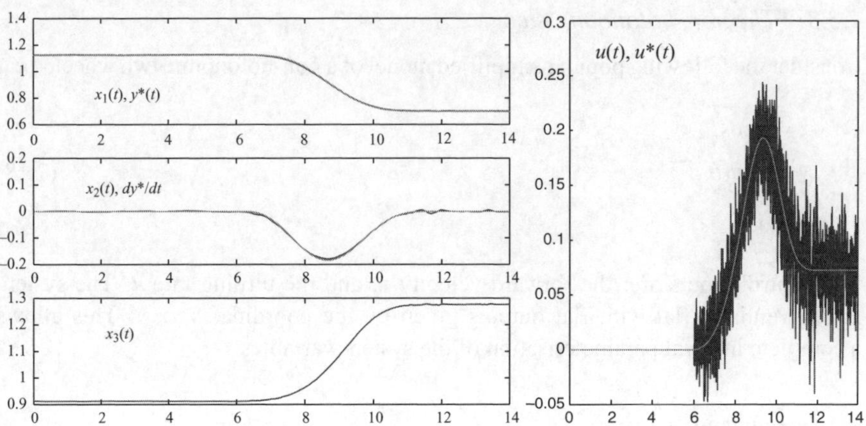

Figure 7.9   *Performance of controlled synchronous generator using flat output derivatives feedback*

where v is an auxiliary control input obtained on the basis of a flat output input coordinate transformation given by

$$u = -E + \frac{b_4}{b_1 \sin(y)}[P - b_2\dot{y} - \ddot{y}] - b_3\cos(y)$$

$$- \frac{1}{b_1\sin^2(y)}[(b_2\ddot{y} + v)\sin(y) + (P - b_2\dot{y} - \ddot{y})\dot{y}\cos(y)] \tag{7.17}$$

The proposed controller evidently requires, for both the linear tracking controller (7.16) and the input coordinate transformation (7.17), the online generation of the time derivatives of the flat output y up to a second order. We propose the use of a set of two cascaded delta-modulator based differentiation units, fed by the measured flat output y, for such a synthesis purpose.

Figure 7.9 depicts the simulated performance of the proposed feedback controller based on the generated output derivatives of the flat output. We used the following parameter values, taken from Espinoza-Pérez et al. [10], for the synchronous generator system,

$$b_1 = 34.29, \quad b_2 = 0, \quad b_3 = 0.1490, \quad b_4 = 0.3341, \quad P = 28.220, \quad E = 0.2405$$

We have also assumed that the measurement of the load angle x_1 undergoes an additive stochastic perturbation noise $\xi(t)$ represented by a computer generated pseudo random Gaussian noise (denoted by 'norm (t)') with an amplitude of 0.001, i.e. $\xi(t) = 0.01$ norm(t) and $y = x_1 + \xi(t)$. The velocity deviation dynamics and the internal voltage dynamics were also assumed to be additively perturbed by stochastic processes of similar nature to that affecting the angular deviation measurement (with an amplitude to 0.01 in both cases).

7.3.3   *The non-holonomic car*

Consider the following popular simplified model of a non-holonomic two wheeled car

$$\dot{x} = v\cos\theta$$
$$\dot{y} = v\sin\theta \qquad\qquad (7.18)$$
$$\dot{\theta} = u$$

The control inputs are: the forward velocity v, and the turning rate u. The system is differentially flat with flat outputs given by the coordinates x, y. This allows a complete integral parameterisation of the system variables.

$$\theta = \arctan\left(\frac{\dot{y}}{\dot{x}}\right)$$

$$u = \frac{\ddot{y}\dot{x} - \dot{y}\ddot{x}}{\dot{x}^2 + \dot{y}^2}$$

$$v = \sqrt{\dot{x}^2 + \dot{y}^2}$$

Given a set of desired position trajectories $x^*(t)$ and $y^*(t)$, it is desired to determine an output feedback tracking controller, based solely on the knowledge of the position coordinates x and y, so that the given trajectories $x^*(t)$, $y^*(t)$, are asymptotically tracked by the system coordinates x and y, respectively

The differential parameterisation of the control inputs u and v clearly reveals that v must undergo a first order extension in order to obtain an invertible relation between the flat output highest order derivatives and the control inputs. We obtain after one differentiation of v the following relation

$$\begin{pmatrix} u \\ \dot{v} \end{pmatrix} = \begin{bmatrix} -\dfrac{\dot{y}}{\dot{x}^2 + \dot{y}^2} & \dfrac{\dot{x}}{\dot{x}^2 + \dot{y}^2} \\[2ex] \dfrac{\dot{x}}{\sqrt{\dot{x}^2 + \dot{y}^2}} & \dfrac{\dot{y}}{\sqrt{\dot{x}^2 + \dot{y}^2}} \end{bmatrix} \begin{bmatrix} \ddot{x} \\ \ddot{y} \end{bmatrix}$$

The system is therefore equivalent, under dynamic feedback and a state dependent input coordinates transformation, to the set of decoupled linear systems:

$$\ddot{x} = \vartheta_1, \quad \ddot{y} = \vartheta_2$$

where

$$\begin{pmatrix} \vartheta_1 \\ \vartheta_2 \end{pmatrix} = \begin{bmatrix} -\dfrac{\dot{y}}{\dot{x}^2 + \dot{y}^2} & \dfrac{\dot{x}}{\dot{x}^2 + \dot{y}^2} \\[2ex] \dfrac{\dot{x}}{\sqrt{\dot{x}^2 + \dot{y}^2}} & \dfrac{\dot{y}}{\sqrt{\dot{x}^2 + \dot{y}^2}} \end{bmatrix}^{-1} \begin{bmatrix} u \\ v \end{bmatrix}$$

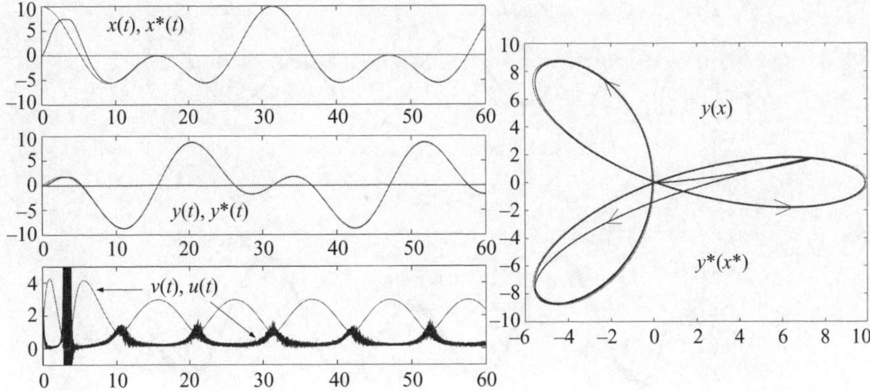

Figure 7.10   *Performance of dynamically controlled non-holonomic car using flat output delta modulation generated time derivatives feedback*

A multivariable feedback tracking controller, including integral error action, is immediately found to be,

$$\vartheta_1 = \ddot{x}^*(t) - k_2(\dot{x} - \dot{x}^*(t)) - k_1(x - x^*(t)) - k_0 \int_0^t (x - x^*(\sigma))d\sigma$$

$$(7.19)$$

$$\vartheta_2 = \ddot{y}^*(t) - \gamma_2(\dot{y} - \dot{y}^*(t)) - \gamma_1(y - y^*(t)) - \gamma_0 \int_0^t (y - y^*(\sigma))d\sigma$$

where the set of coefficients, $\{k_2, k_1, k_0\}$ and $\{\gamma_2, \gamma_1, \gamma_0\}$ are chosen so that the closed loop characteristic polynomials $p_x(s) = s^3 + k_2 s^2 + k_1 s + k_0$ and $p_y(s) = s^3 + \gamma_2 s^2 + \gamma_1 s + \gamma_0$ are Hurwitz polynomials.

The proposed nonlinear multi-variable tracking controller specified for the auxiliary control inputs, ϑ_1 and ϑ_2 requires the tracking error signals $x - x^*(t)$, $y - y^*(t)$ and their first order time derivatives, $\dot{x} - \dot{x}^*(t)$, $\dot{y} - \dot{y}^*(t)$. The actual control input signal u and the extended input v also require online knowledge of $\dot{x}$ and $\dot{y}$ in their nonlinear expressions. We synthesise the required derivatives by means of the proposed delta modulation-low pass filter scheme and evaluate the performance of the closed loop system. The results of the dynamically controlled trajectory tracking task with delta modulation generated time derivatives of the position variables are shown in Fig. 7.10.

The pole placement based controller, including integral action, set the closed loop poles for each independent second order integration chain at the roots of a characteristic polynomial of the form $(s^2 + 2\zeta\omega_n s + \omega_n^2)(s + p)$ with $\zeta = 0.8$, $\omega_n = 0.7$ and $p = 0.5$. Figure 7.11 depicts the actual and the generated time derivatives of the position variables x and y, denoted respectively by $(dx/dt)_e$ and $(dy/dt)_e$.

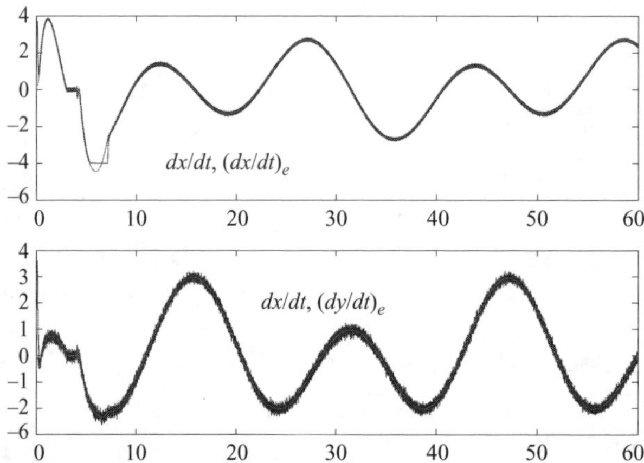

Figure 7.11   Flat outputs time derivatives and their delta modulation generated estimates

7.4 Delta modulation and higher order sliding mode differentiation

Here we propose a second order sliding based differentiator of the delta modulation type. For this, we use a typical second order sliding mode system [8]. By slightly reinterpreting the robustness features of this second order sliding mode system, we obtain a differentiator based on second order sliding modes which resembles a delta modulator in a loose sense. We first prove an auxiliary result regarding finite time reachability of the origin by the trajectories of a certain differential equation.

Theorem 2. *Consider the following nonlinear differential equation*

$$\dot{y} = -\sqrt{|y|}\,\operatorname{sign}(y) \tag{7.20}$$

Then for any initial condition, $y(t_0)$, the trajectory of the system reaches $y(t_h) = 0$ at a finite instant of time t_h, given by $t_h = t_0 + 2\sqrt{|y(t_0)|}$ and, $y(t)$ is identically zero for all times $t \geq t_h$. Moreover, the quantity: $\dot{y}(t)/2\sqrt{|y(t)|}$, evaluated along the solutions of the differential equation, remains constant and equal to $-\frac{1}{2}\operatorname{sign} y(t_0)$. In fact

$$\lim_{t \to t_h} \left(\frac{\dot{y}}{2\sqrt{|y(t)|}} \right) = -\frac{1}{2}\operatorname{sign}(y(t_0))$$

Proof. Consider first an initial condition $y(t_0) > 0$. We have,

$$\dot{y} = -\sqrt{y} \tag{7.21}$$

The solution of the differential equation (7.21), from an arbitrary initial condition $y(t_0) > 0$, is given by

$$y(t) = y(t_0) - (t - t_0)\sqrt{y(t_0)} + \tfrac{1}{4}(t - t_0)^2 = \left(\sqrt{y(t_0)} - \tfrac{1}{2}(t - t_0) \right)^2 \tag{7.22}$$

Evidently, at time $t_h = t_0 + 2\sqrt{y(t_0)}$, the solution of the differential equation (7.21) reaches the condition $y(t_h) = 0$. Since $\dot{y}(t)$ is also zero at time $t = t_h$, then the solution remains at zero for any $t \geq t_h$. Note, moreover, that the quantity $\dot{y}/2\sqrt{y(t)}$ exists for all time t. In fact this quantity is a constant of value $-\frac{1}{2}$ for all t. In particular, we have:

$$\lim_{t \to t_h} \frac{\dot{y}}{2\sqrt{y(t)}} = -\frac{1}{2}$$

As can be easily inferred from the expression for $y(t)$ in (7.22) and of its time derivative:

$$\dot{y}(t) = -\left(\sqrt{y(t_0)} - \tfrac{1}{2}(t - t_0)\right)$$

Note that this last statement can be also directly obtained from (7.21).

Consider now the case in which $y(t_0) < 0$. We have

$$\dot{y} = \sqrt{-y} \tag{7.23}$$

Similarly, for $y(t_0) < 0$, we have that $y(t_h) = 0$ for $t_h = t_0 + 2\sqrt{-y(t_0)}$ and, also, $y(t) = 0$ for any $t \geq t_f$. We also find that

$$\lim_{t \to t_f} \frac{\dot{y}}{2\sqrt{-y(t)}} = \frac{1}{2}$$

Thus, independently of the sign of the initial condition $y(t_0)$, the differential equation (7.20) exhibits a finite time reachability of the origin, at time $t = t_h$, given by $t_h = t_0 + 2\sqrt{|y(t_0)|}$, and the quantity $\dot{y}/2\sqrt{|y|}$ is constant, of value, $-\frac{1}{2}\mathrm{sign}\, y(t_0)$, for all t.

Let σ be a function defined by $\sigma = \dot{y} + \sqrt{|y|}\,\mathrm{sign}\, y$, from a certain time t_0 on. It is clear that if the quantity σ is driven to zero, say, within a finite time interval $[t_0, T_h]$, i.e. $\sigma(T_h) = 0$ and $\sigma(t)$ remains zero indefinitely for all later times $t > T_h$, then by the preceding theorem, y and $\dot{y}$, which at time $t = T_h$ exhibit the values $y(T_h)$, $\dot{y}(T_h)$, also converge to zero in an additional finite amount of time t_h, after the hitting instant T_h, given by $t_h = T_h + 2\sqrt{|y(T_h)|}$.

We present the following result.

Theorem 3. *Let $\bar{\Omega}$ be a strictly positive real number. Consider a compact set of the real line, Ω, containing the origin, given by $\Omega = [-\bar{\Omega}, +\bar{\Omega}]$. Let $\omega(t)$ be an absolutely continuous, scalar function of time, uniformly bounded within the set Ω and such that $\omega(t) \to 0$, in the finite time interval $[t_0, T_h]$. Suppose, furthermore, that $\omega(t)$ remains at zero for all times after time T_h. Then, the trajectories of the uncertain differential equation*

$$\dot{y} = -\sqrt{|y|}\,\mathrm{sign}\, y + \omega(t), \qquad \omega(t) \in \Omega \quad \forall t \tag{7.24}$$

remain bounded in the phase space $(y, \dot{y})$ for all times $t \in [t_0, T_h]$.

Proof.   From the hypothesis in the theorem it follows that the product signal $y(t)\omega(t)$ has a bounded integral in the interval $[t_0, T_h]$ and that such an integral becomes constant for all $t \geq T_h$. Let γ be a strictly positive constant that bounds the finite integral of the product signal $y(t)\omega(t)$ for all times. The following Lyapunov function candidate is positive definite and well defined

$$V(y) = \frac{1}{2}y^2 + \gamma - \frac{\bar{\Omega}}{4}\int_0^t \sqrt{|y(\sigma)|}\,d\sigma$$

The time derivative of $V(y)$, along the solutions of (7.24), is obtained as:

$$\dot{V}(y) = -\sqrt{|y|}|y| \leq 0$$

The set of trajectories compatible with $\dot{V}(y) = 0$ is given by the equilibrium point $y = 0$. It follows by virtue of LaSalle's theorem that $y = 0$ is an asymptotically stable equilibrium point. In particular, the signal $y(t)$ is absolutely bounded for any forcing signal $\omega(t)$ of the hypothesised form.

Based in the previous theorems, we have the following essential result on the robustness of the reachability of the origin in a second order sliding mode autonomous system (see [8]).

Theorem 4.   *Consider the following discontinuous second order differential equation*

$$\ddot{y} = \xi(t) - W\mathrm{sign}[\dot{y} + \sqrt{|y|}\mathrm{sign}\,y] \tag{7.25}$$

For any initial state $y(t_0)$, $\dot{y}(t_0)$, and any absolutely bounded signal $\xi(t)$, there exists a strictly positive real number W such that if $W > \sup_t |\xi(t)| + \frac{1}{2}$, then, ideally, $y(t) = 0$ and $\dot{y}(t) = 0$ for all $t \geq t_h + T_h$.

Proof.   Let, as before, $\sigma = \dot{y} + \sqrt{|y|}\,\mathrm{sign}\,y$. Consider a compact neighbourhood Ω of the real line, containing the origin, such that $\sigma(t_0) \in \Omega$. Similarly, consider a sufficiently large compact set N containing the origin in the phase space $(y, \dot{y})$ such that the solutions of the differential equation $\dot{y} + \sqrt{|y|}\mathrm{sign}\,y = \sigma(t)$ remain bounded in the cylinder $\Omega \times N$. Then, for all $|y| > 0$, we have

$$\dot{\sigma} = \xi(t) - W\mathrm{sign}\,\sigma + \frac{\dot{y}}{2\sqrt{|y|}}\mathrm{sign}\,y \tag{7.26}$$

Define the signal $\eta(t) = \xi(t) + (\dot{y}/2\sqrt{|y|})\mathrm{sign}\,y$. Then, for a sufficiently large value of W, the trajectories of σ decrease towards the origin, $\sigma = 0$, while the trajectories of σ are governed by

$$\dot{\sigma} = \eta(t) - W\mathrm{sign}\,\sigma$$

Since $\eta(t)$ satisfies $\sup_t |\eta(t)| < W$, clearly, a sliding regime exists on $\sigma = 0$ within a finite amount of time, say $T_h - t_0$. Thus, the invariance conditions $\sigma = 0$, $\dot{\sigma} = 0$ become ideally valid after a finite time interval has elapsed. As a result y and $\dot{y}$ are forced to satisfy the second order differential equation (7.20). The phase variables y and $\dot{y}$ converge to the origin in an additional finite time t_h, after the instant T_h.

7.5 References

1   STEELE, R.: 'Delta modulation systems' (London and Halsted Press, New York, 1975)

2   NORSWORTHY, S. R., SHREIER, R., and TEMES, G. C.: Delta-sigma data converters: theory, design and simulation (John Wiley and Sons, IEEE Press, 1996)

3   UTKIN, V. I.: 'Sliding modes and their applications in variable structure systems' (Mir Publishers, Moscow, 1978)

4   GOLEMBO, B., EMELYANOV, S. V., UTKIN, V. I., and SHUBLADE, A. M.: 'Applications of piecewise continuous dynamic systems to filtering problems', *Automation and Remote Control*, 1976, **73**(3), Part I, pp. 369–377

5   PLEKHANOV, S., SHKOLNIKOV, Y., and SHTESSEL, Y. B.: 'High order sigma-delta modulator design via sliding mode control', Proceedings of the American Control Conference, Denver, 2003

6   SHKOLNIKOV, I. A., SHSTESSEL, Y. B., and PLEKHANOV, S. V.: 'Analog-to-digital converters: sliding mode observer as a pulse modulator', Proceedings of the conference on *Decision and Control*, Orlando, FL, 2001

7   SIRA-RAMÍREZ, H.: 'Sliding regimes in analog signal encoding and delta modulation circuits', Proceedings 25th Annual Allerton Conference on *Communications, Control and Computing*, Monticello, Il, 1987, pp. 78–87

8   FRIDMAN, L. and LEVANT, A.: 'Higher order sliding modes', in BARBOT, J. P. and PERRUQUETTI, W. (Eds): 'Sliding mode in engineering' (Marcel Dekker, New York, 2002)

9   KUNDUR, P.: 'Power system stability and control' (McGraw Hill, New York, 1994)

10 ESPINOZA-PÉREZ, G., GODOY-ALCANTARA, M., and GUERRERO-RAMIREZ, G.: 'Passivity based control of synchronous generator', Proceedings of the 1997 IEEE International Symposium on *Industrial Electronics*, Guimaraes, Portugal, 1997, pp. SS101–SS106

Chapter 8

Analysis of sliding modes in the frequency domain

Igor Boiko

8.1 Introduction

Sliding modes are usually studied in the state space domain (see [1] and references therein). The term *sliding* itself is used as an illustration of the character of the system motion (trajectory) in the state space. A frequency domain analysis normally deals with a characteristic that represents a system response to a periodic input. By using a frequency domain approach we, therefore, imply that a periodic motion occurs in the system being studied. However, in linear system analysis due to the validity of the superposition principle, a frequency domain characteristic can be used for assessment of the system behaviour, which is not a periodic motion (i.e. analysis of a transient process). In nonlinear systems, a frequency domain approach can be used for analysis of a periodic motion or a complex behaviour that includes a periodic motion (i.e. describing function analysis of a transient process in a limit cycling system). In sliding mode (SM) control, there are a few phenomena related to the periodicity of the motion, which are more convenient to analyse in the frequency domain. They are considered below.

It is known that the presence of an actuator, sensor or switching imperfections results in the convergence of the transient process in a SM system to a steady state that is not an equilibrium point but a periodic motion, which is usually referred to as chattering. As stated [2, 3], chattering can be associated with the lag effect of the transitions across the sliding surface. It occurs in both the steady state and in the transient process where it exists together with the motion along the sliding surface. Therefore, real sliding represents a complex behaviour, which can be considered as one consisting of a motion component along the sliding surface (sliding) and a high frequency periodic component of the motion across the sliding surface (chattering). Analysis of chattering as an undesired component of the system motion is important in practice. Development of models of chattering and obtaining associated parameters open the way to chattering reduction and elimination. There also exist applications where chattering is a normal operating mode (i.e. control of a DC motor). In such applications, the parameters of the chattering motion are the design objectives. In both types of applications, analysis of chattering is an important theoretical and especially practical problem.

Another manifestation of real SM compared to the ideal behaviour is the distinction of the averaged motions (the system output in particular) in the real SM from the motions in the reduced order system. This distinction becomes especially significant when an external disturbance (static load) is applied to the plant. The ideal SM system would totally reject this disturbance, and the reduced order model reflects the property of ideal disturbance rejection. However, the real SM system does not totally reject the disturbance. It only attenuates its impact to a certain degree. The reduced order model cannot handle this effect, and this is going to be considered below with the use of a frequency domain approach.

One of the features of real SM that complicates its analysis is that SM control can be implemented via the use of different algorithms. The most popular algorithms are: relay feedback control; relay control with state dependent amplitudes; and linear state dependent feedback control with switched gains, which is a classical variable structure approach. However, in many publications the SM is simply studied as a SM in the relay feedback system. The relay feedback system is particularly important in SM control theory. It is known that SM control is essentially a relay feedback control with the sliding variable being the input to the relay. This property is realised exactly for the ideal SM and approximately for the real SM if the control is not designed as a relay control. This observation allows analysis of chattering in a SM system as oscillations in a relay feedback system.

The fundamental approach to the analysis of periodic motions in relay systems is based on the Poincare maps, which is reflected in publications [4–6]. However, the direct use of those maps is not always convenient, and methods where the fixed points of the Poincare maps are expressed in the form more convenient for analysis and design, or approximate methods, are normally used in engineering practice. In respect to the application of the relay systems theory to SM analysis, two frequency domain methods should be mentioned, as they can furnish the solution of some important aspects of the analysis problem indicated above. These are Tsypkin's method [7] and the describing function (DF) method [8]. Yet, the DF method is an approximate one, and Tsypkin's method cannot provide a solution of the input-output problem, which would not allow us to analyse the effect of external disturbances on the system motion. An approach called the locus of a perturbed relay system (LPRS), within which all fixed points of the Poincare maps are given explicitly in the form of a function of the frequency, is going to be considered in the present chapter. This approach provides exactness of the periodic problem solution and also the solution of the input-output problem.

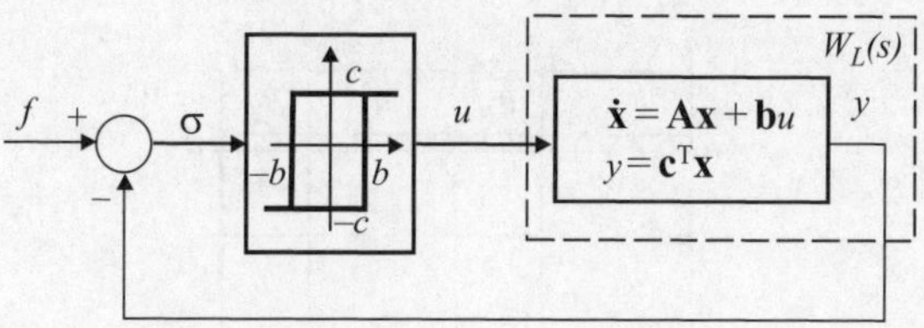

Figure 8.1   Relay server system

8.2   Introduction to the locus of a perturbed relay system (LPRS)

It is known that due to its discontinuous nature, SM control (in the case of ideal sliding) is essentially a relay feedback control with the sliding variable being the input to the relay (Fig. 8.1). In Fig. 8.1, f is a cumulative input (disturbance) to the SM system transposed to the relay input, u is the control, y is the output (the sliding variable), σ is the error signal (the sliding variable affected by the disturbance), c is the amplitude of the relay, $2b$ is the hysteresis of the relay function, $u = u(\sigma)$, and $W_L(s)$ is the transfer function of the linear part (of the actuator, plant and the sliding surface). Normally the hysteresis of the relay in a SM control is zero. However, since the subject of the present study is mainly non-ideal sliding, the hysteresis of the relay function must be considered. Alternatively, the actuator, plant and sliding surface can be given by matrix differential equations, which is also reflected in Fig. 8.1.

Let us call the system Fig. 8.1 a relay servo system emphasising the fact that an external input or a disturbance is applied to the system and an effect of this external signal is one of the subjects of the study. The describing function method provides a simple and often fairly precise approach to the problem of input-output analysis (within the framework of the assumption about a sinusoidal input to the relay). The motions are normally analysed as motions in two separate dynamic subsystems: the 'slow' subsystem and the 'fast' subsystem. The 'slow' subsystem deals with a non-zero initial conditions component of the motion and the forced motions caused by an input signal or by a disturbance. The 'fast' subsystem pertains to the self-excited oscillations or chattering analysis. The two dynamic subsystems interact with each other via a set of parameters: the results of the solution of the 'fast' subsystem are used by the 'slow' subsystem. This decomposition of the dynamics is possible if the external input is much slower than the self-excited oscillations, which is normally the case. Exactly like within the DF method, we shall proceed from the assumption that the external signals applied to the system are slow in comparison with the oscillations. By comparatively slow, we shall understand that the signals meet the following condition: the external signal can be considered constant over the period of the oscillations without significant loss of accuracy of the oscillations estimation. Although this is not a rigorous definition, it outlines a framework for the subsequent analysis. Assume that the input to the system is a constant signal f_0: $f(t) \equiv f_0$. Then an asymmetric periodic motion occurs in the system (Fig. 8.2), so that each signal now has a periodic and a constant term: $u(t) = u_0 + u_p(t)$, $y(t) = y_0 + y_p(t)$, $\sigma(t) = \sigma_0 + \sigma_p(t)$, where

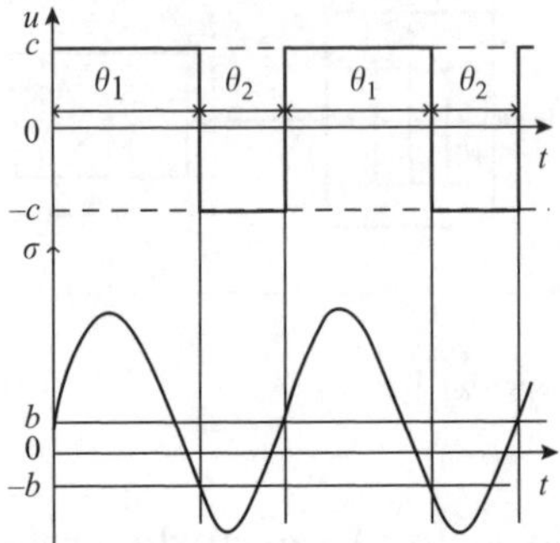

Figure 8.2   Asymmetric oscillations at unequally spaced switches

subscript '0' refers to the constant term in the Fourier series, and subscript '*p*' refers to the periodic term of the function (the sum of periodic terms of the Fourier series).

The constant term is the mean or averaged value of the signal on the period. If we slowly slew the input from a certain negative value to a positive value and measure the values of the constant term of the control (mean control) and the constant term of the error signal (mean error) we can determine the constant term of the control signal as a function of the constant term of the error signal, which would not be discontinuous but a smooth function: $u_0 = u_0(\sigma_0)$.

Let us call it the *bias function*. The described effect is known as the 'chatter smoothing' phenomenon, which was studied with the use of the DF method [8]. The derivative of the mean control with respect to the mean error taken in the point of zero mean error $\sigma_0 = 0$ (corresponding to zero constant input) provides the so-called *equivalent gain* of the relay k_n or the *incremental gain* at zero argument [8], which is used as a local approximation of the *bias function*:

$$k_n = \left.\frac{du_0}{d\sigma_0}\right|_{\sigma_0=0} = \lim_{f_0 \to 0}\left(\frac{u_0}{\sigma_0}\right).$$

Since for the slow inputs the relay servo system behaves similarly to a linear system, finding the *equivalent gain* value is the main point of the input-output analysis. Once it is found, all subsequent analysis of the slow motions can be carried out exactly as for a linear system with the relay replaced with the *equivalent gain*. The model obtained via the replacement of the relay with the *equivalent gain* would represent the model of the averaged (on the period of the oscillations) motions in the system. This is especially pertinent to a SM analysis because the deviations of the sliding variable from the zero value are usually small, and the *equivalent gain,* being a local approximation of the *bias function,* usually provides good accuracy. The model obtained as described above would not be a reduced order model. It would retain the order of the original system. The reduced order model can be obtained from the non-reduced model via setting the equivalent gain to infinity. The same approach is applicable to the analysis of a free motion caused by non-zero initial conditions, i.e. the motion along the sliding surface (this is considered below).

The following development is aimed at designing an analysis tool that would allow us to find the parameters of the oscillations and the *equivalent gain* and through the linearisation of the relay to build a model of the slow motions in the oscillatory system. To understand the meaning of the function, which will be defined below, consider first the DF analysis of the relay feedback system (Fig. 8.1). The DF of the hysteresis relay with a biased sine input is represented by the following well-known formula:

$$N(a,\sigma_0) = \frac{2c}{\pi a}\left[\sqrt{1-\left(\frac{b+\sigma_0}{a}\right)^2}+\sqrt{1-\left(\frac{b-\sigma_0}{a}\right)^2}\right]-j\frac{4cb}{\pi a^2},$$

$$(a \geq b+|\sigma_0|), \tag{8.1}$$

where a is the amplitude of the oscillations. The mean control as a function of a and σ_0 is given by the following formula:

$$u_0(a,\sigma_0) = \frac{c}{\pi}\left(\arcsin\frac{b+\sigma_0}{a}-\arcsin\frac{b-\sigma_0}{a}\right). \tag{8.2}$$

From (8.1) and (8.2), we can obtain the DF of the relay and the derivative of the mean control with respect to the mean error for the case of the symmetric sine input:

$$N(a) = \frac{4c}{\pi a}\sqrt{1-\left(\frac{b}{a}\right)^2}-j\frac{4cb}{\pi a^2}, \qquad (a \geq b), \tag{8.3}$$

$$\left.\frac{\partial u_0}{\partial \sigma_0}\right|_{\sigma_0=0} = \frac{2c}{\pi a}\frac{1}{\sqrt{1-(b/a)^2}} = k_{n(DF)}. \tag{8.4}$$

We denote the right-hand side of (8.4) as $k_{n(DF)}$, which is the value of the *equivalent gain* computed with the use of the DF method. The periodic solution in the relay feedback system can be found from the equation:

$$W_L(j\Omega) = -\frac{1}{N(a)}, \tag{8.5}$$

which can be transformed to the following form via the replacement of $N(a)$ with its respective formula:

$$W_L(j\Omega) = -\frac{1}{2}\frac{\sqrt{1-(b/a)^2}}{2c/\pi a}-j\frac{\pi b}{4c}. \tag{8.6}$$

We note that the fraction in the real part of (8.6) is the reciprocal of the equivalent gain, and also that the condition of the switch of the relay from minus to plus (defined as zero time) is the equality of the system output to the negative half hysteresis $(-b)$. Taking this into account, we can rewrite formula (8.6) as the following expression:

$$W_L(j\Omega) = -\frac{1}{2}\frac{1}{k_{n(DF)}}+j\frac{\pi}{4c}y_{(DF)}(0). \tag{8.7}$$

Now let us define a certain function J exactly as the expression in the right-hand side of formula (8.7) but require from it that the *values of the equivalent gain and the output at the zero time should be exact values*. As a result, we can write the following definition of this function:

$$J(\omega) = -0.5 \lim_{f_0 \to 0} \left(\frac{\sigma_0}{u_0} \right) + j\frac{\pi}{4c} \lim_{f_0 \to 0} y(t)|_{t=0}, \tag{8.8}$$

where $t = 0$ is the time of the switch of the relay from '$-c$' to '$+c$'. Formula (8.8) is a definition and involves the parameters of the oscillations in the closed-loop system. To obtain a function of frequency, there must be some means of varying the frequency in the system that does not involve the parameters of formula (8.8). However, the frequency cannot be varied by manipulating a parameter of the plant either. The only parameter that fits these requirements is the hysteresis $2b$. Therefore, ω is the frequency of the self-excited oscillations varied by changing the hysteresis $2b$ while all other parameters of the system are considered constant; σ_0, u_0 and $y(t)|_{t=0}$ are, therefore, functions of ω. In the definition (8.8), an assumption is made that the limit cycle becomes symmetric if the input f_0 tends to zero. Thus, $J(\omega)$ is defined as a characteristic of the response of the linear part to the unequally spaced pulse input $u(t)$ subject to $f_0 \to 0$ as the frequency ω is varied. The real part of $J(\omega)$ contains information about the gain k_n, and the imaginary part of $J(\omega)$ comprises the condition of the switching of the relay and, consequently, contains information about the frequency of the oscillations. The meaning of the above definition is that with the function $J(\omega)$ computed, we will be able to apply the existing techniques of the DF method to the analysis and design of relay servo systems. However, unlike in the DF analysis, we will be able to obtain exact values of the frequency of the oscillations and of the equivalent gain.

Let us call the function $J(\omega)$ defined above as well as its plot on the complex plane (with the frequency ω varied) the locus of a perturbed relay system (LPRS). Suppose we have computed the LPRS of a given system. Then (like in the DF analysis) we are able to determine the frequency of the oscillations (as well as the amplitude) and the equivalent gain k_n (Fig. 8.3). The point of an intersection of the LPRS and of the straight line, which lies at the distance $\pi b/(4c)$ below (if $b > 0$) or above (if $b < 0$) the horizontal axis and parallel to it (line '$-\pi b/4c$'), allows the frequency of the oscillations and the equivalent gain k_n of the relay to be computed.

According to (8.8), the frequency Ω of the oscillations can be computed via solving the equation:

$$\text{Im}\,J(\Omega) = -\frac{\pi b}{4c}, \tag{8.9}$$

(i.e. $y(0) = -b$ is the condition of the relay switch) and the gain k_n can be computed as:

$$k_n = -\frac{1}{2\text{Re}\,J(\Omega)}. \tag{8.10}$$

Formula (8.9) is, therefore, a necessary condition for the existence of the periodic solution. Formula (8.2) is only a definition and not intended for the purpose of computing of the LPRS $J(\omega)$. It is shown below that although $J(\omega)$ is defined via the parameters of the oscillations in a closed-loop system, it can be easily derived from the parameters of the linear part without employing the variables of formula (8.8).

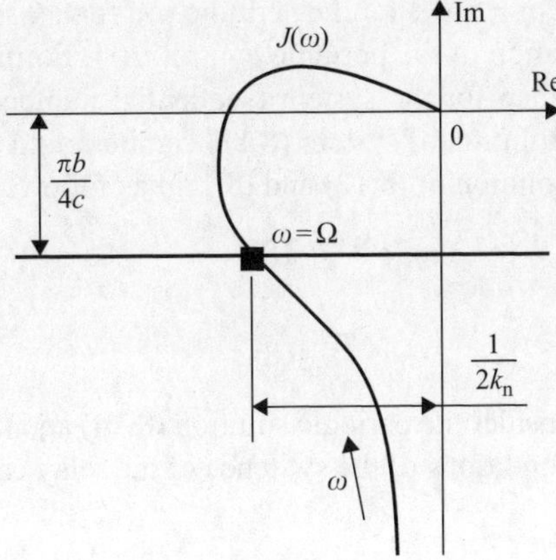

Figure 8.3   The LPRS and oscillations analysis

8.3   Computation of the LPRS for a non-integrating plant

8.3.1   Matrix state space description approach

The computational formula of the LPRS involves only the parameters of the linear part for the case of the non-integrating (self-regulating) linear part given by matrix differential equations. Let the system Fig. 8.1 be described by the following equations:

$$\dot{\mathbf{x}} = \mathbf{A}\mathbf{x} + \mathbf{b}u,$$

$$y = \mathbf{c}^{\mathrm{T}}\mathbf{x},$$

$$u = \begin{cases} +c & \text{if } \sigma = f_0 - y > b \quad \text{or} \quad \sigma > -b, \dot{\sigma} < 0, \\ -c & \text{if } \sigma = f_0 - y < -b \quad \text{or} \quad \sigma < b, \dot{\sigma} > 0, \end{cases} \tag{8.11}$$

where $\mathbf{A} \in R^{n \times n}, \mathbf{b} \in R^{n \times 1}, \mathbf{c}^{\mathrm{T}} \in R^{1 \times n}$ are matrices and $\mathbf{A}$ is nonsingular.

Let us find the periodic solution of system (8.11) at unequally spaced switching caused by a nonzero input signal. A common way to find a periodic solution is to use a Poincare map. Since the control switches are unequally spaced and the oscillations are not symmetric, a Poincare return map must be considered. Suppose that an asymmetric periodic process with period T exists in the system. Then, considering the solution for the constant control $u = \pm 1$ (it will be shown below that the LPRS is a characteristic of the linear part only and we can assume without loss of generality $c = 1$):

$$\mathbf{x}(t) = e^{\mathbf{A}t}\mathbf{x}(0) + \mathbf{A}^{-1}(e^{\mathbf{A}t} - \mathbf{I})\mathbf{b}u,$$

the periodic solution of system (8.10) and (8.11) can be written as:

$$\vec{\eta} = e^{\mathbf{A}\theta_1}\vec{\rho} + \mathbf{A}^{-1}(e^{\mathbf{A}\theta_1} - \mathbf{I})\mathbf{b}, \tag{8.12}$$

$$\vec{\rho} = e^{\mathbf{A}\theta_2}\vec{\eta} - \mathbf{A}^{-1}(e^{\mathbf{A}\theta_2} - \mathbf{I})\mathbf{b}, \tag{8.13}$$

where $\vec{\rho} = \mathbf{x}(0) = \mathbf{x}(T)$, $\vec{\eta} = \mathbf{x}(\theta_1)$ for the periodic solution, θ_1, θ_2 are the positive and the negative pulse duration of the periodic control $u(t)$. Formulas (8.12) and (8.13) are a return Poincare map for the system (sequential numbers of switches are not shown). The periodic solution of system (8.11) can be found as a fixed point of the Poincare return map (solution of (8.12) and (8.13)) as follows:

$$\vec{\rho} = (\mathbf{I} - e^{\mathbf{A}T})^{-1}\mathbf{A}^{-1}[e^{\mathbf{A}T} - 2e^{\mathbf{A}\theta_2} + \mathbf{I}]\mathbf{b}, \tag{8.14}$$

$$\vec{\eta} = (\mathbf{I} - e^{\mathbf{A}T})^{-1}\mathbf{A}^{-1}[2e^{\mathbf{A}\theta_1} - e^{\mathbf{A}T} - \mathbf{I}]\mathbf{b}. \tag{8.15}$$

We now need to consider the periodic solution (8.14) and (8.15) as a result of the feedback action. The conditions of the switches of the relay can be written as:

$$\begin{aligned} f_0 - y(0) &= b, \\ f_0 - y(\theta_1) &= -b. \end{aligned} \tag{8.16}$$

Having solved the set of equations (8.16) for f_0 we can obtain: $f_0 = (y(0) + y(\theta_1))/2$. Hence, the constant term of $\sigma(t)$ is:

$$\sigma_0 = f_0 - y_0 = \frac{(y(0) + y(\theta_1))}{2} - y_0 \tag{8.17}$$

and the real part of the LPRS definition formula can be transformed into:

$$\operatorname{Re} J(\omega) = -0.5 \lim_{\gamma \to \frac{1}{2}} \frac{0.5[y(0) + y(\theta_1)] - y_0}{u_0}, \tag{8.18}$$

where $\gamma = \theta_1/(\theta_1 + \theta_2) = \theta_1/T$.

Then $\theta_1 = \gamma T$, $\theta_2 = (1 - \gamma)T$, $u_0 = 2\gamma - 1$ and (8.18) can be rewritten as:

$$\operatorname{Re} J(\omega) = -0.5 \lim_{\gamma \to \frac{1}{2}} \frac{0.5\mathbf{c}^{\mathrm{T}}[\vec{\rho} + \vec{\eta}] - y_0}{2\gamma - 1},$$

where $\vec{\rho}$ and $\vec{\eta}$ are given by (8.14) and (8.15) respectively. The imaginary part of the formula of $J(\omega)$ can be transformed into:

$$\operatorname{Im} J(\omega) = \frac{\pi}{4}\mathbf{c}^{\mathrm{T}} \lim_{\gamma \to \frac{1}{2}} \vec{\rho}.$$

Finally, the state space description based formula of the LPRS can be derived on the basis of the previous two formulas and (8.14), (8.15) as follows:

$$J(\omega) = -0.5\mathbf{c}^{\mathrm{T}}\left[\mathbf{A}^{-1} + \frac{2\pi}{\omega}(\mathbf{I} - e^{(2\pi/\omega)\mathbf{A}})^{-1}e^{(\pi/\omega)\mathbf{A}}\right]\mathbf{b}$$
$$+ j\frac{\pi}{4}\mathbf{c}^{\mathrm{T}}(\mathbf{I} + e^{(\pi/\omega)\mathbf{A}})^{-1}(\mathbf{I} - e^{(\pi/\omega)\mathbf{A}})\mathbf{A}^{-1}\mathbf{b}. \tag{8.19}$$

Therefore, if the system under consideration is given in state-space form, (8.19) can be used for the LPRS computation. The LPRS computed as (8.19) comprises all possible periodic solutions and equivalent gain values for a given linear part. For that reason, the LPRS can be considered a relatively universal frequency domain characteristic of the linear part of a relay servo system. An actual periodic solution for given linear part and parameters of the relay can be found from equation (8.9).

8.3.2   Partial fraction expansion technique

Now consider deriving the LPRS formula if the description of the linear part is given in the form of the transfer function expanded into partial fractions. At first prove a property of the LPRS $J(\omega)$.

Theorem 1 (additivity property).   *If the transfer function $W_L(s)$ of the linear part is a sum of n transfer functions: $W_L(s) = W_1(s) + W_2(s) + \cdots + W_n(s)$ then the LPRS $J(\omega)$ can be calculated as a sum of n LPRS: $J(\omega) = J_1(\omega) + J_2(\omega) + \cdots + J_n(\omega)$, where $J_i(\omega)$ $(i = 1, \ldots, n)$ is the LPRS of the relay system with the transfer function of the linear part being $W_i(s)$.*

Proof.   Prove the property for $n = 2$. It is obvious that if the property is true for $n = 2$ it is true for any n. Consider the steady asymmetric oscillations in the system when $f(t) \equiv f_0 \neq 0$. Assume that a unimodal asymmetric limit cycle occurs (Fig. 8.2). Suppose we know the frequency Ω and the amplitude c of the oscillations, and the pulse duration (θ_1 and θ_2) of the periodic control $u(t)$. If $W_L(s) = W_1(s) + W_2(s)$ then the output $y(t) = y_1(t) + y_2(t)$, where $y_{1,2}(t)$ is the output of the linear part, which has the transfer function $W_{1,2}(s)$ with its input $u(t)$ being as determined above. Substitute $y_1(t) + y_2(t)$ for $y(t)$ in (8.17) and obtain: $\sigma_0 = \sigma_{01} + \sigma_{02}$, where $\sigma_{01} = (y_1(0) + y_1(\theta_1))/2 - y_{01}$, $\sigma_{02} = (y_2(0) + y_2(\theta_1))/2 - y_{02}$, y_{01} and y_{02} are the constant terms of $y_1(t)$ and $y_2(t)$ respectively. Thus, when the parameters of $u(t)$ are as specified above, the constant term of $\sigma(t)$ is equal to the sum of the constant terms of $\sigma_1(t)$ and $\sigma_2(t)$ where $\sigma_1(t)$ and $\sigma_2(t)$ are the errors in two different relay systems with the transfer functions $W_1(s)$ and $W_2(s)$ respectively. Since the additivity property is true for σ_0, it is also true for σ_0/u_0 because $u_0 = const$ and, consequently, this is true for $\lim(\sigma_0/u_0)$. It is also obvious that $y(0) = y_1(0) + y_2(0)$. Thus, according to (8.8): $J(\omega) = J_1(\omega) + J_2(\omega)$.

The proved property offers a way of computing the LPRS $J(\omega)$ via expanding $W_L(s)$ into the sum of first and second order dynamic elements (partial fractions), calculating the component LPRS $J_i(\omega)$ for each of them and summation of the LPRS $J_i(\omega)$. Analytical formulas have been derived for $J(\omega)$ of first and second order dynamic elements and are presented in Table 8.1.

8.3.3   Transfer function description approach

Another formula for $J(\omega)$ can now be derived for the case of the linear part given by a transfer function. Suppose the linear part does not have integrators. Write the

Table 8.1　Formulas of the LPRS $J(\omega)$

Transfer function $W(s)$	The LPRS $J(\omega)$
K/s	$0 - j\pi^2 K/(8\omega)$
$K/Ts+1$	$0.5K(1 - \alpha\,\mathrm{cosech}\,\alpha) - j0.25\pi K\,\mathrm{th}\,(\alpha/2),\ \alpha = \pi/(T\omega)$
$K/[(T_1 s + 1)(T_2 s + 1)]$	$0.5K[1 - T_1/(T_1 - T_2)\alpha_1\,\mathrm{cosech}\,\alpha_1 - T_2/(T_2 - T_1)\alpha_2\,\mathrm{cosech}\,\alpha_2)]$ $\quad - j0.25\pi K/(T_1 - T_2)[T_1\,\mathrm{th}\,(\alpha_1/2) - T_2\,\mathrm{th}\,(\alpha_2/2)],$ $\quad \alpha_1 = \pi/(T_1\omega),\ \alpha_2 = \pi/(T_2\omega)$
$K/(s^2 + 2\xi s + 1)$	$0.5K[(1 - (B + \gamma C)/(\sin^2\beta + \mathrm{sh}^2\alpha)]$ $\quad - j0.25\pi K(\mathrm{sh}\,\alpha - \gamma\sin\beta)/(\mathrm{ch}\,\alpha + \cos\beta),$ $\quad \alpha = \pi\xi/\omega,\ \beta = \pi(1 - \xi^2)^{1/2}/\omega,\ \gamma = \alpha/\beta,$ $\quad B = \alpha\cos\beta\,\mathrm{sh}\,\alpha + \beta\sin\beta\,\mathrm{ch}\,\alpha,\ C = \alpha\sin\beta\,\mathrm{ch}\,\alpha - \beta\cos\beta\,\mathrm{sh}\,\alpha$
$Ks/(s^2 + 2\xi s + 1)$	$0.5K[\xi(B + \gamma C) - \pi/\omega\cos\beta\,\mathrm{sh}\,\alpha]/(\sin^2\beta + \mathrm{sh}^2\alpha)$ $\quad - j0.25K\pi(1 - \xi^2)^{-1/2}\sin\beta/(\mathrm{ch}\,\alpha + \cos\beta),$ $\quad \alpha = \pi\xi/\omega,\ \beta = \pi(1 - \xi^2)^{1/2}/\omega,\ \gamma = \alpha/\beta,$ $\quad B = \alpha\cos\beta\,\mathrm{sh}\,\alpha + \beta\sin\beta\,\mathrm{ch}\,\alpha,\ C = \alpha\sin\beta\,\mathrm{ch}\,\alpha - \beta\cos\beta\,\mathrm{sh}\,\alpha$
$Ks/(s+1)^2$	$0.5K[\alpha(-\mathrm{sh}\,\alpha + \alpha\,\mathrm{ch}\,\alpha)/\mathrm{sh}^2\alpha - j0.25\pi\alpha/(1 + \mathrm{ch}\,\alpha)],\ \alpha = \pi/\omega$
$Ks/[(T_1 s + 1)(T_2 s + 1)]$	$0.5K/(T_2 - T_1)[\alpha_2\,\mathrm{cosech}\,\alpha_2 - \alpha_1\,\mathrm{cosech}\,\alpha_1]$ $\quad - j0.25K\pi/(T_2 - T_1)\,[\mathrm{th}\,(\alpha_1/2) - \mathrm{th}\,(\alpha_2/2)],$ $\quad \alpha_1 = \pi/(T_1\omega),\ \alpha_2 = \pi/(T_2\omega)$
$K\exp(-\tau s)/(Ts + 1)$	$(K/2)(1 - \alpha e^\gamma\,\mathrm{cosech}\,\alpha) + j(\pi/4)K(2e^{-\alpha}e^\gamma/(1 + e^{-\alpha}) - 1),$ $\quad \alpha = \pi/T\omega,\ \gamma = \tau/T$

Fourier series expansion of the signal $u(t)$ (Fig. 8.2):

$$u(t) = u_0 + 4c/\pi \sum_{k=1}^{\infty} \sin(\pi k\theta_1/(\theta_1 + \theta_2))/k$$

$$\times \left\{ \cos\left(\frac{k\omega\theta_1}{2}\right)\cos(k\omega t) + \sin\left(\frac{k\omega\theta_1}{2}\right)\sin(k\omega t) \right\},$$

where $u_0 = c(\theta_1 - \theta_2)/(\theta_1 + \theta_2)$, $\omega = 2\pi/(\theta_1 + \theta_2)$. Therefore, $y(t)$ as a response of the linear part with the transfer function $W_L(s)$ can be written as:

$$y(t) = y_0 + 4c/\pi \sum_{k=1}^{\infty} \sin(\pi k\theta_1/(\theta_1 + \theta_2))/k \times \left\{ \cos\left(\frac{k\omega\theta_1}{2}\right)\cos[k\omega t + \varphi_L(k\omega)] \right.$$

$$\left. + \sin\left(\frac{k\omega\theta_1}{2}\right)\sin[k\omega t + \varphi_L(k\omega)] \right\} A_L(k\omega), \tag{8.20}$$

where $\varphi_L(k\omega) = \arg W_L(jk\omega)$, $A_L(k\omega) = |W_L(jk\omega)|$, $y_0 = u_0|W_L(jD)|$. The conditions of the switches of the relay have the form of equations (8.16) where $y(0)$ and

$y(\theta_1)$ can be obtained from (8.20) if we set $t = 0$ and $t = \theta_1$ respectively:

$$y(0) = y_0 + 4c/\pi \sum_{k=1}^{\infty} [0.5 \sin(2\pi k\theta_1/(\theta_1 + \theta_2))\mathrm{Re}\, W_L(jk\omega)$$

$$+ \sin^2(\pi k\theta_1/(\theta_1 + \theta_2))\mathrm{Im}\, W_L(jk\omega)]/k, \tag{8.21}$$

$$y(\theta_1) = y_0 + 4c/\pi \sum_{k=1}^{\infty} [0.5 \sin(2\pi k\theta_1/(\theta_1 + \theta_2))\mathrm{Re}\, W_L(jk\omega)$$

$$- \sin^2(\pi k\theta_1/(\theta_1 + \theta_2))\mathrm{Im}\, W_L(jk\omega)]/k. \tag{8.22}$$

Differentiating (8.16) with respect to f_0 (and taking into account (8.21) and (8.22)) we obtain the formulas containing the derivatives in the point $\theta_1 = \theta_2 = \theta = \pi/\omega$. Having solved those equations for $d(\theta_1 - \theta_2)/df_0$ and $d(\theta_1 + \theta_2)/df_0$ we shall obtain: $d(\theta_1 + \theta_2)/df_0|_{f_0=0} = 0$, which corresponds to the derivative of the frequency of the oscillations, and:

$$\frac{\mathrm{d}(\theta_1 - \theta_2)}{\mathrm{d}f_0}\bigg|_{f_0=0} = \frac{2\theta}{c(|W_L(0)| + 2\sum_{k=1}^{\infty} \cos(\pi k)\mathrm{Re}\, W_L(\omega k))}. \tag{8.23}$$

Considering the formula of the closed-loop system transfer function we can write:

$$\frac{\mathrm{d}(\theta_1 - \theta_2)}{\mathrm{d}f_0}\bigg|_{f_0=0} = k_n/(1 + k_n|W_L(0)|)2\theta/c. \tag{8.24}$$

Solving together equations (8.23) and (8.24) for k_n we obtain the following expression:

$$k_n = \frac{0.5}{\sum_{k=1}^{\infty} (-1)^k \mathrm{Re}\, W_L(k\pi/\theta)}. \tag{8.25}$$

Taking into account formula (8.25) and the definition of the LPRS (8.8), we obtain the final form of expression for $\mathrm{Re}\, J(\omega)$. Similarly, having solved the set of equations (8.16) where $\theta_1 = \theta_2 = \theta$ and $y(0)$ and $y(\theta_1)$ have the form (8.21) and (8.22) respectively, we obtain the final formula of $\mathrm{Im}\, J(\omega)$. Having put the real and the imaginary parts together, we can obtain the final formula of the LPRS $J(\omega)$ for servo systems with non-integrating plants:

$$J(\omega) = \sum_{k=1}^{\infty} (-1)^{k+1}\mathrm{Re}\, W_L(k\omega) + j\frac{\sum_{k=1}^{\infty} \mathrm{Im}\, W_L[(2k-1)\omega]}{2k-1}. \tag{8.26}$$

8.4   Computation of the LPRS for an integrating plant

8.4.1   *Matrix state space description approach*

If the plant contains an integrator, the linear part of the SM system (which includes the actuator, the plant, and the sliding surface) will also contain an integrator. For an integrating linear part, the formulas derived above cannot be used without some modifications. Although the solution $\mathbf{x}(t)$ of the system is well-defined even if the matrix $\mathbf{A}$ does not have an inverse and, therefore, the above results might seem to be applied to an integrating linear part, in the case of unequally spaced switches the system that has a conventional form of description, strictly speaking, cannot have a periodic process even if a ramp signal is applied to the input of the system Fig. 8.1. The motion exhibited by such a system would be a combination of a periodic and a ramp motion – due to unlimited integration. To enable the system to have an asymmetric periodic motion, transpose the constant input signal to the integrator input (Fig. 8.4). The balance of the constant terms of the signals in the various points of the system must be achieved for the periodic motion to occur.

Similarly, derive the formulas of $J(\omega)$ for the case of an integrating linear part. The state space description of the system (Fig. 8.4) has the following form:

$$\dot{\mathbf{x}} = \mathbf{A}\mathbf{x} + \mathbf{b}u, \tag{8.27}$$

$$\dot{y} = \mathbf{c}^{\mathrm{T}}\mathbf{x} - f_0, \tag{8.28}$$

$$u = \begin{cases} +c & \text{if } \sigma = -y > b \quad \text{or} \quad \sigma > -b, \dot{\sigma} < 0 \\ -c & \text{if } \sigma = -y < -b \quad \text{or} \quad \sigma < b, \dot{\sigma} > 0 \end{cases}$$

where $\mathbf{A} \in R^{(n-1)\times(n-1)}, \mathbf{b} \in R^{(n-1)\times 1}, \mathbf{c}^{\mathrm{T}} \in R^{1\times(n-1)}$, $\mathbf{A}$ is nonsingular, f_0 is a constant input to the system, σ is the error signal. A separate consideration of the variable $y(t)$ from the other state variables is possible due to the integrating property of the linear part.

This allows us at first to find a periodic solution for $\mathbf{x}(t)$ (for a given unequally spaced switching), and after that to determine a periodic solution for the system output. The periodic solution for $\mathbf{x}(t)$ before the integrator was given above (formulas (8.14) and (8.15)). The periodic output $y(t)$ can be obtained via integrating equation (8.11) from the initial states determined by formulas (8.14) and (8.15). As a result, for the control $u = 1$ the system output can be written as:

$$y_1(t) = y_1(0) - \mathbf{c}^{\mathrm{T}}\mathbf{A}^{-1}\mathbf{b}t - f_0 t + \mathbf{c}^{\mathrm{T}}\mathbf{A}^{-1}[(e^{\mathbf{A}t} - \mathbf{I})\vec{\rho} + \mathbf{A}^{-1}(e^{\mathbf{A}t} - \mathbf{I})\mathbf{b}] \tag{8.29}$$

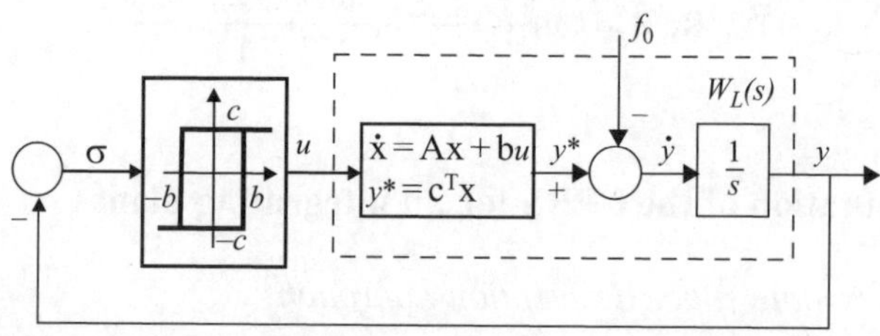

Figure 8.4   Relay server system with integrating linear part

and for the control $u = -1$ as the following formula:

$$y_2(t) = y_1(\theta_1) + \mathbf{c}^{\mathrm{T}}\mathbf{A}^{-1}\mathbf{b}t - f_0 t + \mathbf{c}^{\mathrm{T}}\mathbf{A}^{-1}[(e^{\mathbf{A}t} - \mathbf{I})\vec{\eta} - \mathbf{A}^{-1}(e^{\mathbf{A}t} - \mathbf{I})\mathbf{b}],$$

$$(8.30)$$

where $y_1(t) = y(t)$, $y_2(t) = y(t + \theta_1)$.

The time t in formulas (8.29) and (8.30) is independent and $t = 0$ in formula (8.29) is the time of the switch from minus to plus, and in formula (8.30) $t = 0$ is the time of the switch from plus to minus. For a periodic motion, the following equations should hold, which represents a return Poincare map:

$$y(\theta_1) = y(0) - (\mathbf{c}^{\mathrm{T}}\mathbf{A}^{-1}\mathbf{b} + f_0)\theta_1 + \mathbf{c}^{\mathrm{T}}\mathbf{A}^{-1}[(e^{\mathbf{A}\theta_1} - \mathbf{I})\vec{\rho} + \mathbf{A}^{-1}(e^{\mathbf{A}\theta_1} - \mathbf{I})\mathbf{b}],$$

$$(8.31)$$

$$y(0) = y(\theta_1) + (\mathbf{c}^{\mathrm{T}}\mathbf{A}^{-1}\mathbf{b} - f_0)\theta_2 + \mathbf{c}^{\mathrm{T}}\mathbf{A}^{-1}[(e^{\mathbf{A}\theta_2} - \mathbf{I})\vec{\eta} - \mathbf{A}^{-1}(e^{\mathbf{A}\theta_2} - \mathbf{I})\mathbf{b}].$$

$$(8.32)$$

Analysis of equations (8.31) and (8.32) shows that the set of equations may have a solution if and only if:

$$f_0 = -\mathbf{c}^{\mathrm{T}}\mathbf{A}^{-1}\mathbf{b}(2\gamma - 1),$$

$$(8.33)$$

where

$$\gamma = \frac{\theta_1}{\theta_1 + \theta_2} = \frac{\theta_1}{T},$$

which corresponds to the situation when the constant term of the signal $y^*(t)$ is equal to f_0 and, therefore, the constant term at the integrator input is zero – the only possibility for the system to have a periodic process. Furthermore, equations (8.31) and (8.32) are equivalent and have an infinite number of solutions. An explanation of this fact can be as follows. If a periodic signal with zero constant term is applied to the integrator input, its output signal is not uniquely determined but depending on the initial value can represent an infinite number of biased periodic signals. To define a unique solution introduce an additional condition:

$$y(\theta_1) = -y(0).$$

$$(8.34)$$

The solution of equations (8.31) and (8.34) results in

$$y(0) = \mathbf{c}^{\mathrm{T}}\mathbf{A}^{-1}\mathbf{b}\gamma(1 - \gamma)T + \tfrac{1}{4}\mathbf{c}^{\mathrm{T}}\mathbf{A}^{-2}\{(\mathbf{I} - e^{\mathbf{A}T})^{-1}[6e^{\mathbf{A}T} - 3(e^{\mathbf{A}\theta_1} + e^{\mathbf{A}\theta_2})$$

$$- e^{\mathbf{A}T}(e^{\mathbf{A}\theta_1} + e^{\mathbf{A}\theta_2}) + 2\mathbf{I}] - (e^{\mathbf{A}\theta_1} + e^{\mathbf{A}\theta_2}) + 2\mathbf{I}\}\mathbf{b}.$$

$$(8.35)$$

The output at $t = \theta_1$ is a negative value of the same formula. Thus, the periodic solution of system (8.27) and (8.28) is found. The LPRS formula can be derived from the analysis of the closed-loop system with an unequally spaced switching control having an infinitesimally small asymmetry. The constant term y_0 of the output $y(t)$

can be determined as the sum of integrals of functions (8.29) and (8.30) divided by the period T.

$$y_0 = \frac{1}{T} \left\{ \int_0^{\theta_1} y_1(\tau) d\tau + \int_0^{\theta_2} y_2(\tau) d\tau \right\}, \tag{8.36}$$

where $y_1(\tau)$ is given by (8.29) and $y_2(\tau)$ is given by (8.30). The formula of the real part of $J(\omega)$ can be transformed into:

$$\mathrm{Re}\, J(\omega) = 0.5 \lim_{\gamma \to \frac{1}{2}} \frac{y_0}{c(2\gamma - 1)}, \tag{8.37}$$

where expression (8.36) can be used for y_0. The formula of the imaginary part of $J(\omega)$ is determined by (8.35) with a coefficient, which follows from the LPRS definition. Finally, the LPRS for the case of an integrating linear part can be written as the following formula:

$$J(\omega) = 0.25 \mathbf{c}^{\mathrm{T}} \mathbf{A}^{-2} \left\{ (\mathbf{I} - \mathbf{D}^2)^{-1} \left[\mathbf{D}^2 - \left(\mathbf{I} + \frac{4\pi}{\omega} \mathbf{A} \right) \mathbf{D} + \mathbf{D}^3 - \mathbf{I} \right] + \mathbf{D} - \mathbf{I} \right\} \mathbf{b}$$
$$+ j \frac{\pi}{8} \mathbf{c}^{\mathrm{T}} \mathbf{A}^{-1} \left\{ \frac{\pi}{\omega} + \mathbf{A}^{-1} [(\mathbf{I} - \mathbf{D}^2)^{-1} \cdot (3\mathbf{D}^2 - 3\mathbf{D} - \mathbf{D}^3 + \mathbf{I}) - \mathbf{D} + \mathbf{I}] \right\} \mathbf{b}, \tag{8.38}$$

where $\mathbf{D} = e^{(\pi/\omega)\mathbf{A}}$. Therefore, the state space description based LPRS formula for the case of an integrating linear part has been derived above.

8.4.2 Transfer function description approach

Derive the LPRS formula for the case of an integrating linear part given by a transfer function. The model suitable for the following analysis is given in Fig. 8.4. One notices that the periodic terms of the signals of the system Fig. 8.4 are the same as the periodic terms of respective signal of the system Fig. 8.1. For that reason, we can use some results of the above analysis for the case of a non-integrating linear part. The constant input f_0 causes an asymmetry in the periodic motion. In a steady periodic motion, the constant term of the input signal to the integrator is zero. Yet, the input $\sigma(t)$ to the relay has two terms: the constant term σ_0 and the periodic term $\sigma_p(t)$. The periodic term $\sigma_p(t)$ coincides with the one of formula (8.20) (negative value of the latter). The constant term σ_0 can be expressed as: $\sigma_0 = 0.5(\sigma_p(0) + \sigma_p(\theta_1))$, which with (8.21) and (8.22) taken into account results in:

$$\sigma_0 = \frac{2c}{\pi} \sum_{k=1}^{\infty} \sin\left(\frac{2\pi k \theta_1}{\theta_1 + \theta_2} \right) \mathrm{Re}\, W_L(jk\omega).$$

The equivalent gain k_n can be obtained as a reciprocal of the derivative $d\sigma_0/du_0$ at $\theta_1 = \theta_2 = \pi/\omega$. The imaginary part of the LPRS remains the same for the case of an integrating linear part. Finally, a formula for the LPRS can be constructed on the basis of the definition (8.8) and of the above analysis. The final formula for the LPRS, which also incorporates the results obtained above for the case of a non-integrating linear part, is given as follows:

$$J(\omega) = \sum_{k=1}^{\infty} k^m (-1)^{k+1} \operatorname{Re} W_L(k\omega) + j \frac{\sum_{k=1}^{\infty} \operatorname{Im} W_L[(2k-1)\omega]}{2k-1}, \qquad (8.39)$$

where $m = 0$ for a non-integrating linear part and $m = 1$ for an integrating linear part. For an accurate calculation of a point of $J(\omega)$, consideration of a few initial terms in the series (8.39) is often sufficient. It can be shown that the series (8.39) always converges for strictly proper transfer functions. Formula (8.39) can also be used for the LPRS calculation from a frequency response characteristic (Bode plot, Nyquist plot) of the linear part.

Naturally, the LPRS method overlaps with other existing methods and produces the same results under certain circumstances. In this respect, to compare the LPRS with the DF method and Tsypkin's method is interesting.

The describing function method. Since the DF method is based upon the filtering hypothesis, it might be expected that the LPRS method should provide the same result if this hypothesis is accepted. It can be better illustrated if the series form of the LPRS (8.39) is considered. Indeed, if only the first terms of the series of the real and imaginary parts are used (this corresponds to acceptance of the filtering hypothesis) this formula would coincide with that of the DF method (see formula (8.7)). The LPRS method, therefore, provides a more precise model of the oscillations in a relay system compared to the DF method. In particular, it takes into account the non-sinusoidal shape of the output signal and the precision enhancement is due to this.

The Tsypkin's method. The main similarity between Tsypkin's method and the LPRS is in the imaginary parts of the two loci. The imaginary part of the Tsypkin's locus is defined as the output value in a periodic motion at the time of the relay switch from minus to plus. The imaginary part of the LPRS is essentially the same: the difference is only in the coefficient. However, the real part of Tsypkin's locus is defined as a derivative of the output at the time of the switch and is intended for verifying the condition of the proper direction of the switch. The real part of the LPRS is defined as a ratio of the two infinitesimally small constant terms of the signals caused by the infinitesimally small asymmetry of the switching in a closed loop system. As a result, Tsypkin's locus is a method of analysis of possible periodic motions only. The LPRS is intended for a complex analysis, the solution of the periodic problem and the input-output analysis (disturbance rejection and external signal propagation).

8.5   Frequency domain conditions of sliding mode existence

The LPRS is a function of the frequency and contains all possible periodic solutions for a given plant, including the solution of infinite frequency corresponding to the ideal SM. Since a periodic solution is found as a point of intersection of the LPRS and the real axis, the location of the high-frequency segment of the LPRS can be very informative with respect to whether the ideal SM or chattering will occur in the system. If, for example, the high-frequency segment of the LPRS is located in the upper half-plane, and, therefore, the LPRS must have an intersection with the real axis at a finite frequency, chattering normally occurs (there may be situations when both finite and infinite periodic solutions occur).

Let us now consider the location of the high-frequency segment of the LPRS of an arbitrary order linear plant. Let the transfer function $W_L(s)$ of the linear plant be given as a quotient of two polynomials of degrees m and n:

$$W_L(s) = \frac{B_m(s)}{A_n(s)} = \frac{b_m s^m + b_{m-1} s^{m-1} + \cdots + b_1 s + b_0}{a_n s^n + a_{n-1} s^{n-1} + \cdots + a_1 s + a_0}. \tag{8.40}$$

The relative degree of the transfer function $W_L(s)$ is $(n-m)$. Then the following statements hold (given without proof).

Lemma 1.   If function $W_L(s)$ is strictly proper $(n > m)$ there exists ω^* corresponding to any given $\varepsilon > 0$ such that for every $\omega \geq \omega^*$:

$$\left| \operatorname{Re} W_L(j\omega) - \operatorname{Re} \left(\frac{b_m}{a_n \cdot (j\omega)^{n-m}} \right) \right| \leq \varepsilon \left(\frac{\omega^*}{\omega} \right)^{(n-m)}, \tag{8.41}$$

$$\left| \operatorname{Im} W_L(j\omega) - \operatorname{Im} \left(\frac{b_m}{a_n \cdot (j\omega)^{n-m}} \right) \right| \leq \varepsilon \left(\frac{\omega^*}{\omega} \right)^{(n-m)}. \tag{8.42}$$

Lemma 2 (monotonicity of high-frequency segment of the LPRS).   If $\operatorname{Re} W_L(j\omega)$ and $\operatorname{Im} W_L(j\omega)$ are monotone functions of the frequency ω and $|\operatorname{Re} W_L(j\omega)|$ and $|\operatorname{Im} W_L(j\omega)|$ are decreasing functions of the frequency ω for every $\omega \geq \omega^*$, then the real and imaginary parts of the LPRS $J(\omega)$ corresponding to that transfer function are monotone functions of the frequency ω and magnitudes of the real and imaginary parts are also monotone functions of the frequency ω within the range $\omega \geq \omega^*$. The proof can be based on formula (8.26) and finding a dominating series. Consider the following statement.

Theorem 2.   *If the transfer function $W_L(s)$ is a quotient of two polynomials $B_m(s)$ and $A_n(s)$ of degrees m and n respectively (8.40) then the high-frequency segment (where the above Lemma 1 holds) of the LPRS $J_L(\omega)$ corresponding to the transfer function $W_L(s)$ is located in the same quadrant of the complex plane where the high-frequency segment of the Nyquist plot of $W_L(s)$ is located with either the real axis (if the relative degree $(n-m)$ is even) or the imaginary axis (if the relative degree $(n-m)$ is odd) being an asymptote of the LPRS. Again, the proof can be based on formula (8.26), Lemma 2 and a dominating series.*

Theorem 3.   *If the transfer function $W_L(s)$ is a quotient of two polynomials $B_m(s)$ and $A_n(s)$ of degrees m and n respectively with nonnegative coefficients and the relative degree $(n-m)$ being one or two, then a periodic motion of infinite frequency may occur (subject to initial conditions) in the relay feedback system with the plant being $W_L(s)$. Note: this does not, however, concern the case of the plant that has* *two or more imaginary poles (integrators). Such a system may not have a periodic solution at all. The proof can be based on Theorem 2 and the property considered in [7], which in terms of the LPRS can be reformulated as a necessity for the LPRS to intersect the real axis from below for the periodic solution to be a stable limit cycle (a necessary condition in a general case).*

This theorem provides one more proof of a well-known property. The considered theorems provide a foundation for the analysis of possible modes in a relay system. With the LPRS computed or only a transfer function available, one can easily see if either the ideal SM or chattering may occur in the SM system being analysed.

8.6   Example of chattering and disturbance attenuation analysis

Consider an example that illustrates the proposed approach to the analysis of chattering and the static load (disturbance) propagation. The equations of the spring-loaded cart with viscous output damping on the inclined plane can be written as follows:

$$\dot{x}_1 = x_2,$$

$$\dot{x}_2 = -x_1 - x_2 + u_a + d,$$

where x_1 is the linear displacement of the cart, x_2 is the linear velocity, u_a is the force developed by the actuator and d is the disturbance (projection of the gravity onto the inclined plane). The goal is to stabilise the cart in the point corresponding to zero displacement. Let us design the switching surface (line) as follows: $x_1 + x_2 = 0$ and the control as a relay control that can make the point $x = 0$ an asymptotically stable equilibrium point of the closed-loop system under the applied disturbance $d = -1$: $u = -4 \, \text{sign} \, (x_1 + x_2)$. Suppose that the force u_a is developed by a fast actuator with the second order dynamics:

$$T_a^2 \ddot{u}_a + 2 \xi_a T_a \dot{u}_a + u_a = u,$$

where $T_a = 0.01 \, \text{s}^{-1}$, $\xi_a = 0.5$. Clearly, the system should exhibit oscillations due to the actuator presence. Finding the frequency and the amplitude of those oscillations is one of the goals of this analysis. Another goal is an assessment of the disturbance effect. In the case of ideal sliding, even if the disturbance is applied the trajectory tends to the origin. In the case of non-ideal sliding (due to the actuator presence) the trajectory does not tend to the origin. Write an expression for the transfer function of the linear part: $W_L(s) = (s + 1) \cdot W_a(s) \cdot W_p(s)$, where $W_a(s) = 1/(T_a^2 s^2 + 2 \xi_a T_a s + 1)$, $W_p(s) = 1/(s^2 + s + 1)$. Compute the LPRS for $W_L(s)$ as per (8.39) and plot it on the complex plane (Fig. 8.5).

Find the point of intersection of the LPRS and the real axis. This point corresponds to the frequency $\Omega = 99.27 \, \text{s}^{-1}$, which is the frequency of chattering in the system. The real part of the LPRS in this point is $\text{Re} \, J(\Omega) = -0.009 \, 46$ and the equivalent gain

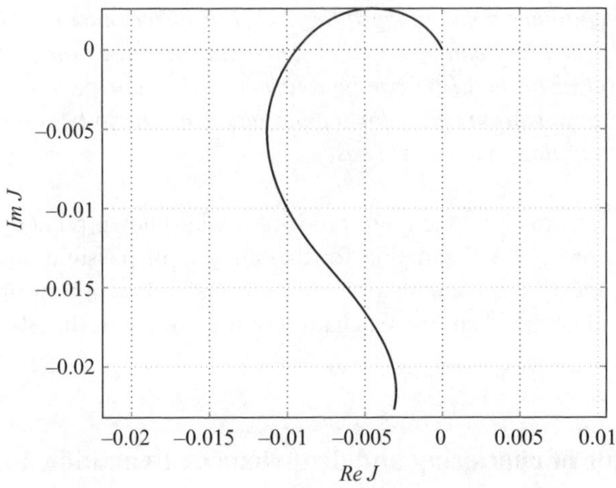

Figure 8.5   LPRS of the linear part (actuator, plant and sliding surface)

of the relay (according to formula (8.10)) is $k_n = 52.8$. As a result, the non-reduced order model of the slow motions can be written as follows (subscript '0' denotes the slow component of respective variables):

$$\dot{x}_{01} = x_{02},$$

$$\dot{x}_{02} = -x_{01} - x_{02} + u_{0a} + d,$$

$$\ddot{u}_{0a} = \frac{u_0 - 2\xi_a T_a \dot{u}_{0a} - u_{0a}}{T_a^2}, \qquad (8.43)$$

$$u_0 = -k_n \sigma_0,$$

$$\sigma_0 = x_{01} + x_{02}.$$

The reduced order model can be obtained from (8.43) as a limiting case: if the equivalent gain is set to infinity: $k_n \to \infty$ (that would result in $\sigma_0 = 0$ and, consequently, in $x_{01} = x_{02}$ – the condition of ideal sliding). Note that the actual value of the equivalent gain is finite. For that reason, in the analysed case, the non-reduced order model provides an additional accuracy in comparison with the reduced order model. Since the transient processes in both the reduced model and the non-reduced model look alike, the advantage of the non-reduced order model can be best demonstrated, if an external disturbance is applied to the system, and the effect of this disturbance is of interest.

In the example being considered, the equivalent gain k_n does not vary. For that reason, the effect of the applied disturbance is identical in the transient and the steady state modes, and the analysis of disturbance attenuation can be carried out with the use of the techniques relevant to linear systems. Analyse the disturbance attenuation. In a steady state, there exists a periodic motion of frequency Ω with the centre $(x_{01}, 0)$ where $x_{01} = d/(1 + k_n) = -0.018$, which can be considered a disturbance rejection measure. This means that in a steady state, the cart exhibits oscillations around the point $x_{01} = -0.018$, with the frequency $\Omega = 99.27\,\text{s}^{-1}$ and the amplitude of the fundamental frequency component: $A_{x1} = 4c/\pi\,|W_a(j\Omega) \cdot W_p(j\Omega)| = 5.19 \cdot 10^{-4}$. The simulations of the original equations provide the following results. The frequency of chattering is $\Omega_{sim} = 99.21\,\text{s}^{-1}$, and the output averaged steady state value is $x_{01sim} = -0.017$, which closely match the frequency domain analysis.

8.7 Conclusion

The considered frequency domain methodology of analysis of SM control systems is based on the notion of the LPRS and an approach that involves substitution of the relay element with the equivalent gain, and analysis of the obtained linearised system. The LPRS comprises both: the oscillatory and the transfer properties of a relay system and succeeds even if the filtering hypothesis fails, and, therefore, can be used as a relatively universal characteristic of a relay system. It is proved that despite the fact that the LPRS is defined via the parameters of the periodic motion in the closed-loop system, it is actually a characteristic of the linear part only. Three different formulas of the LPRS for both non-integrating and integrating linear parts are derived and a methodology of analysis that involves the LPRS is presented. An illustrative example of the frequency-domain analysis of a SM system is considered.

8.8 References

1 UTKIN, V.: 'Sliding modes in control and optimization' (Springer Verlag, Berlin, 1992)

2 YOUNG, K. D., UTKIN, V. I., and OZGUNER, U.: 'A control engineer's guide to sliding mode control', *IEEE Trans. Control System Technology*, 1999, **7**, pp. 328–342

3 FRIDMAN, L.: 'Singularly perturbed analysis of chattering in relay control systems', *IEEE Transactions on Automatic Control*, 2002, **47**(12), pp. 2079–2084

4 BROMBERG, P. V.: 'Stability and self-existed oscillations of impulse control systems' (Oborongiz, Moscow, 1953)

5 ASTROM, K. J.: 'Oscillations in systems with relay feedback', *The IMA Volumes in Mathematics and its Applications: Adaptive Control, Filtering and Signal Processing*, 1995, **74**, pp. 1–25

6 VARIGONDA, S. and GEORGIOU, T. T.: 'Dynamics of relay relaxation oscillators', *IEEE Trans. on Automatic Control*, 2001, **46**(1), pp. 65–77

7 TSYPKIN, Y. Z.: 'Relay control systems' (Cambridge University Press, Cambridge, 1984)

8 ATHERTON, D. P.: 'Nonlinear control engineering – describing function analysis and design' (Workingham, Berks, UK: Van Nostrand Company, 1975)

Chapter 9

Output tracking in causal nonminimum-phase systems using sliding modes

Yuri B. Shtessel and Ilya A. Shkolnikov

9.1 Introduction

Nonminimum-phase output tracking is a challenging, real life control problem that has been extensively studied recently [1, 2]. Much consideration has been given to this problem in the area of nonlinear aircraft and missile control [3–5]. One of the most interesting aspects, from the theoretical and especially from the practical point of view, is to provide robust nonlinear nonminimum-phase output tracking in causal systems. A nonlinear control system will be recognised as nonminimum-phase if its internal or zero dynamics [1] are unstable. The nonminimum-phase nature of a plant restricts the application of powerful nonlinear control techniques such as feedback linearisation control [1] and sliding mode control [6–8]. Restrictions on tracking have been extensively investigated for linear systems [9] and nonlinear systems [2, 10, 11]. In general, exact tracking in causal nonlinear nonminimum-phase systems seems to be impossible for arbitrary reference inputs even in the absence of plant uncertainties and external disturbances. All existing approaches to the problem can be divided into two categories: either methods to modify the system model (or to redefine the output) in order to get a minimum-phase system; or methods to restrict the class of input signals to be tracked.

For tracking in causal nonminimum-phase systems, a variety of approximate solutions have been established in the literature. A radical design of a piecewise dynamical unstable controller with state resetting [12] has been employed for power electronic systems. For flight control systems that are slightly nonminimum-phase, meaning that the right half plane zeros of the linearised system have a large real part, different solutions are available; e.g., techniques that involve neglecting small parasitic coupling between the mechanisms of lift and pitch [3–5]. They typically involve either redefining the output or increasing the relative degree of the plant model.

Restriction on the class of signals to be tracked is another method to tackle the problem. In the work [10] the problem of tracking a class of signals given by a known nonlinear exosystem is reduced to solving a first order partial differential algebraic equation. Approximate solutions to this equation have been proposed in References 13 and 14 for a special class of systems and given trajectories. Tracking a very narrow class of trajectories, slowly varying around the given trajectory, is addressed in Reference 15. Exact tracking of a known trajectory via stable nonlinear noncausal inverse is achieved in Reference 16.

Some specific plant models allow for particular nonminimum-phase tracking techniques. For instance, a VTOL aircraft model near hover allows for some indirect feedback regulation using the differentially flat outputs [17], which are indeed minimum-phase outputs, as proposed in Reference 18. The same problem is addressed [19] using dynamical variable structure control and Fliess' Generalised Observability Canonical Form.

Due to the general limitations [2] of exact nonminimum-phase output tracking, all the aforementioned methods explore particular cases and impose their own restrictions on the problem statement. Many methods, discussed above, use linearisation of the internal dynamics, a certain type of canonical representation of a plant model, and an exogenous system for the reference input. In this work, similar techniques are applied to a nonlinear feedback linearisable plant model presented in the normal canonical form with linearised internal dynamics [14, 22] and in the regular form [6, 21]. Also, unmatched disturbances presented by an exosystem and plant uncertainties multiplicative in the control are accounted for [21, 22].

The structure of this chapter is as follows. In Section 9.2 a motivating example, nonminimum-phase aircraft flight control, is considered and discussed. The method of the stable system centre design [20, 21], which is presented in Section 9.3, addresses nonminimum-phase tracking via feedforward/feedback control for a plant given in a normal canonical form. The dynamic sliding manifold technique [23–25] that is used to design a feedback sliding mode control for nonminimum-phase tracking is discussed in Section 9.4. The conclusions to the toolbox of design methods developed in Sections 9.3 and 9.4 for nonminimum-phase output tracking are given in Section 9.5.

9.2 Motivational example: consideration of a nonminimum-phase plant

9.2.1 *Aircraft flight path angle tracking in the pitch plane*

The main purpose of this section is to introduce the reader to the phenomenon of a nonminimum-phase output in control systems. For aircraft control in the pitch plane, the difficulties and specific features of command tracking by a nonminimum-phase output of the controlled plant are discussed. In the paradigm of sliding mode control

(SMC) design, it is shown that standard application of the SMC method cannot solve the problem of tracking a nonminimum-phase output in a causal closed-loop system.

An approximate model of F-16 jet fighter, which is taken at one point of the flight envelope and linearised around a constant trim condition in the pitch plane, is considered

$$
\begin{bmatrix} \dot{\theta} \\ \dot{\alpha} \\ \dot{q} \end{bmatrix} = \begin{bmatrix} 0 & 0 & 1 \\ 0 & -1 & 1 \\ 0 & 4 & -1.2 \end{bmatrix} \cdot \begin{bmatrix} \theta \\ \alpha \\ q \end{bmatrix} + \begin{bmatrix} 0 \\ -0.2 \\ -20 \end{bmatrix} \delta,
\tag{9.1}
$$

where θ is the pitch angle [rad], α is the angle of attack [rad], q is the pitch rate [rad/sec], δ is a control input, the elevator deflection [rad].

The commanded output of the plant (9.1) is

$$
\gamma = \theta - \alpha,
\tag{9.2}
$$

the flight path angle (the angle of the vehicle velocity vector in the vertical plane with respect to the space datum).

The airframe dynamics (9.1) are completely controllable in the state space and open-loop unstable, having the set of eigenvalues $\{0, -3.102, 0.902\}$. The input-output transfer function of system (9.1) and (9.2) is identified as

$$
H(s) = \frac{\gamma(s)}{\delta(s)} = \frac{0.2 \cdot (s + 10.816)(s - 9.616)}{s \cdot (s + 3.102)(s - 0.902)}.
\tag{9.3}
$$

A linear single-input-single-output (SISO) plant is nonminimum-phase if its transfer function has either poles or zeros in the right half of the complex plane. However, it is the right half zeros that cause problems in tracking an arbitrary signal. The nonminimum-phase property of a linear system will be further understood in a narrow sense as having only 'unstable' zeros. Moreover, the extension of this definition for nonlinear multiple-input-multiple-output (MIMO) systems [1] corresponds to the narrow sense. Our system is obviously of nonminimum-phase, having one zero at 9.616. Following the approach [1], for system (9.1) and (9.2) one can identify input-output (I/O) dynamics of first order

$$
\dot{\gamma} = -\gamma + \theta + 0.2\delta
\tag{9.4}
$$

and a residual part of second order, which is not directly involved in the I/O relation

$$
\begin{bmatrix} \dot{\theta} \\ \dot{q} \end{bmatrix} = \begin{bmatrix} 0 & 1 \\ 4 & -1.2 \end{bmatrix} \cdot \begin{bmatrix} \theta \\ q \end{bmatrix} + \begin{bmatrix} 0 \\ 1 \end{bmatrix} (-4\gamma - 20\delta).
\tag{9.5}
$$

Consider the output stabilisation problem for the flight path angle γ using the sliding mode control method and the relative degree approach [6, 26, 27]. For system (9.4), the sliding surface is introduced as $\sigma = \gamma = 0$, and the 'equivalent control' providing system motion on the surface $\sigma = 0$ is identified from (9.4), (9.5) as

$$
\delta = u_{eq} = -5\theta.
\tag{9.6}
$$

The sliding mode dynamics on the surface $\sigma = 0$ are obtained by substituting (9.6) into (9.5)

$$\begin{bmatrix} \dot{\theta} \\ \dot{q} \end{bmatrix} = \begin{bmatrix} 0 & 1 \\ 104 & -1.2 \end{bmatrix} \cdot \begin{bmatrix} \theta \\ q \end{bmatrix}. \tag{9.7}$$

This has the set of eigenvalues $\{-10.816, 9.616\}$.

System (9.7) is called the *zero dynamics* of the plant (9.1) with respect to the output (9.2). The concept of zero dynamics for nonlinear plants was introduced and used to define a nonminimum-phase plant [1].

Definition.   *The plant output is of nonminimum-phase if the corresponding zero dynamics are unstable.*

The instability of the zero dynamics causes the equivalent control (9.6) to grow unbounded. Thus, no bounded sliding mode control can keep $\sigma = 0$ indefinitely when the direct lift force represented by the aerodynamic coefficient $+0.2$ in (9.4) is actually used to stabilise γ.

The direct lift force can be considered as parasitic since another more powerful control input represented by the aerodynamic coefficient -20 in (9.5), lagging in its effect on γ, creates a lift force of opposite sign. Consider a negative control deflection $\delta < 0$, and then if initially $\gamma = \theta = 0$, we have $\dot{\gamma} \sim 0.2\delta$ and γ will grow negative.

However, if we consider the dynamic relation between flight path angular rate $\dot{\gamma}$ and pitch rate q, neglecting the parasitic lift force,

$$\dot{\gamma}(s) = \frac{1}{T_i s + 1} q(s), \qquad T_i = 1 \text{ s}.$$

This is the so-called incidence lag; we observe from (9.5) that $\dot{q} \sim -20\delta$, and q as well as $\dot{\gamma}$ eventually, will grow positive. One can see from (9.4) that when $1.0(\theta - \gamma) > |0.2\delta|$, we have $\dot{\gamma} > 0$, and γ will increase, compensating for the initial decrease due to the opposite action of the direct lift force. This peculiar struggle of forces creates the nonminimum-phase phenomenon for aircraft dynamics from the physical standpoint.

Many practical approaches to nonminimum-phase output-tracking, especially when there exists a set-point regulation task only, exclude a 'fast unstable zero' from the plant model (methods for so-called slightly nonminimum-phase systems [4, 5]). In this case the modified plant model (9.1) is of relative degree equal to 2 and can be stabilised via a traditional PD controller. The disadvantage of this method, which is limited to slightly nonminimum-phase systems, is that it is not developed for the problem of tracking an arbitrary profile, and the solution to the regulation problem is not robust to plant parameter variations and external disturbances. In the next sections, novel techniques that address the problem discussed above are developed in the framework of sliding mode control.

9.2.2 The normal form and the inverse dynamics of a nonminimum-phase plant: the feedforward/feedback control approach

In this section, the question of how the instability of the zero dynamics complicates the tracking of an arbitrary reference profile is discussed with regard to a feedforward control action.

A feedforward control that uses plant dynamic inversion may provide tracking of an arbitrary reference profile in combination with a stabilising control. For the pitch plane motion of the F-16 in (9.1), the causal dynamic inversion is obviously unstable (see (9.3)) making the feedforward control unbounded. It is convenient to study the causal dynamic inversion of the plant (9.1) given in state-variable format using the normal form ([1] p. 144), which will be used later for a general class of nonlinear MIMO systems.

Using the nonsingular linear transformation

$$\begin{bmatrix} \eta_1 \\ \eta_2 \\ \xi \end{bmatrix} = \begin{bmatrix} -2.201 & 2.435 & -0.0243 \\ 2.437 & -2.199 & 0.022 \\ 1 & -1 & 0 \end{bmatrix} \cdot \begin{bmatrix} \theta \\ \alpha \\ q \end{bmatrix}, \tag{9.8}$$

one can transform system (9.1) to the form

$$\begin{bmatrix} \dot{\eta}_1 \\ \dot{\eta}_2 \end{bmatrix} = \begin{bmatrix} -10.816 & 0 \\ 0 & 9.616 \end{bmatrix} \cdot \begin{bmatrix} \eta_1 \\ \eta_2 \end{bmatrix} - \begin{bmatrix} 23.806 \\ 23.428 \end{bmatrix} \xi, \tag{9.9}$$

$$\dot{\xi} = 2.01\eta_1 + 2.226\eta_2 - \xi + 0.2u, \tag{9.10}$$

where the commanded output is $y = \gamma = \xi$ and the control input is $u = \delta$.

The representation (9.9), (9.10) is called *normal form*, where the system (9.10) represents the I/O dynamics, and system (9.9) represents the internal dynamics, which is not affected by the input u.

The homogeneous part of system (9.9) represents the zero dynamics in uncoupled form. One should note that in general, obtaining the normal form is not easy, and form (9.4), (9.5) is used instead of the representation (9.9), (9.10) for control design.

The *problem of following the reference profile* $y_R(t)$ (exact tracking) can be solved for the system (9.9), (9.10), using the control input

$$u(t) = \frac{1}{0.2}(\dot{y}_R(t) + y_R(t) - [2.01 \quad 2.226]\eta_R(t)), \tag{9.11}$$

$$\dot{\eta}_R = \begin{bmatrix} -10.816 & 0 \\ 0 & 9.616 \end{bmatrix} \eta_R - \begin{bmatrix} 23.806 \\ 23.428 \end{bmatrix} y_R(t), \tag{9.12}$$

with the initial conditions $\xi(0) = y_R(0), \forall \eta_R(0)$.

System (9.11), (9.12), where $y_R(t)$ is the input, $\eta_R(t)$ is the state, u is the output, can be interpreted as the inverse of the original system in (9.9), (9.10).

The inverse dynamics (9.12) is unstable, and the feedforward control (9.11) will be unbounded. Thus, the asymptotic stability of the zero dynamics is essential to provide exact tracking of an arbitrary reference input via feedforward control.

However, if one can identify a bounded solution $\eta_R^o(t)$ to the inverse dynamics (9.12), which has been called the *ideal internal dynamics* (IID) [14], then a bounded feedforward control is possible. In this case, a sliding control can provide asymptotic tracking with the error dynamics robust to uncertainties, initial conditions, and disturbances in the I/O dynamics.

For instance, the bounded solution to system (9.12) can be identified, if the reference profile $y_R(t)$ is known in advance (noncausal inverse solution).

$$
\eta_R^o(t) =
\begin{bmatrix}
e^{-10.816t}\eta_1^o(0) - 23.806 \displaystyle\int_0^t e^{-10.816(t-\tau)}y_R(\tau)\,d\tau \\[2em]
-23.428 \displaystyle\int_t^\infty e^{9.616(t-\tau)}y_R(\tau)\,d\tau
\end{bmatrix},
\qquad \forall \eta_1^o(0).
$$

The problem of noncausal stable inversion for nonlinear systems has been studied [16, 28]. The application of sliding mode control to tracking in noncausal systems is presented in Reference 28. This work concentrates on sliding mode tracking control in causal systems where the application of traditional sliding mode control could lead to unbounded control as was discussed earlier. An SMC design technique based on computing and following bounded state-tracking profiles for the causal nonminimum-phase tracking system that is presented in normal canonical form is discussed in Section 9.3.

9.2.3   Asymptotic output tracking problem: the state-feedback approach

Revisiting the original system in the normal form (9.9), (9.10), we consider the output (tracking-error) $e_y = y_R(t) - \xi$ as a state-variable. The problem is to ensure $e_y \to 0$ asymptotically. To design a state-feedback SMC control, one has to consider the following state model of the plant in the vector-space $(\eta_1, \eta_2, e_y) \in \Re^3$

$$
\begin{bmatrix} \dot{\eta}_1 \\ \dot{\eta}_2 \\ \dot{e}_y \end{bmatrix} =
\begin{bmatrix}
-10.816 & 0 & 23.806 \\
0 & 9.616 & 23.428 \\
0 & -2.226 & -1
\end{bmatrix}
\cdot
\begin{bmatrix} \eta_1 \\ \eta_2 \\ e_y \end{bmatrix}
+
\begin{bmatrix} -23.806 \\ -23.428 \\ 0 \end{bmatrix} y_R(t)
$$

$$
+ \begin{bmatrix} 0 \\ 0 \\ 1 \end{bmatrix} (\dot{y}_R(t) + y_R(t) + 2.01\eta_1 + 0.2u). \tag{9.13}
$$

As has been seen in Section 9.2.1, the SMC with a sliding surface designed in accordance with the relative degree approach cannot stabilise system (9.1) due to instability of the zero dynamics. Thus, the manifold $\sigma = e_y = 0$ providing for output regulation does not achieve the closed-loop bounded-input-bounded-state stability of the system. The question remains as to if there exists another manifold that can provide internal stability and asymptotic convergence of e_y in the sliding mode. Analysing system (9.9), we can conclude that an unstable subspace of inverse dynamics is described by the η_2 dynamics, and the η_1 dynamics are bounded-input-bounded-state stable, provided ξ is bounded. Thus, for regulating $e_y \to 0$, the (η_2, e_y) dynamics must be considered. This is

$$\begin{bmatrix} \dot{\eta}_2 \\ \dot{e}_y \end{bmatrix} = \begin{bmatrix} 9.616 & 23.428 \\ -2.226 & -1 \end{bmatrix} \cdot \begin{bmatrix} \eta_2 \\ e_y \end{bmatrix} + \begin{bmatrix} -23.428 \\ 0 \end{bmatrix} y_R(t)$$

$$+ \begin{bmatrix} 0 \\ 1 \end{bmatrix} (\dot{y}_R(t) + y_R(t) + 2.01\eta_1 + 0.2u), \tag{9.14}$$

For the reduced-order system (9.14), the input η_1 can be considered as a disturbance input, which is matched to the control. System (9.14) is written in the *regular form* [6], where the η_2-dynamics are driven by the 'unmatched disturbance' $y_R(t)$, and the matched e_y-dynamics.

The sliding mode control cannot cancel the effect of unmatched disturbances exactly. For static state-feedback, it is possible to eliminate the effect of disturbances whose action is orthogonal to the selected sliding surface, which corresponds to the condition that some designated output (the sliding quantity) is zero. The surface that is orthogonal to the unmatched disturbances in the state space of (η_2, e_y) is defined by $e_y = 0$. However, the motion of the system (9.13) on this surface is unstable. Conversely, any stable one-dimensional manifold is affected by the 'disturbance' $y_R(t)$. Thus, the problem is that of designing a manifold in the dynamically extended state space of the system (9.13), where the tracking-error, e_y, dynamics are autonomous and asymptotically stable. It is possible to decouple the disturbances modelled by a known exogenous model from the manifold of asymptotically stable e_y dynamics using ideas of servocompensation [10, 11] and the internal model principle [9]. This approach to the so-called *dynamic sliding manifold* design will be presented in Section 9.4.

9.2.4   *Conclusions*

In Section 9.2, an overview of various SMC design approaches has been presented with application to the tracking problem for a linearised nonminimum-phase F-16 jet fighter pitch plane model. It has been shown that application of traditional sliding mode control does not solve the nonminimum-phase tracking problem. It has been identified that in order to design a sliding surface for a feedforward/feedback control, one has to find a stable solution to the unstable inverse (internal) dynamics of the plant. Having identified bounded reference profiles for the internal states, one can enforce robust asymptotic state-tracking via the corresponding SMC. In order to design a feedback control treating all the unknown inputs (including the reference one) as disturbances, one has to consider the design of a sliding surface for the system with unmatched disturbances. The goal here is to decouple all non-decaying modes of the disturbance input vector from the manifold of asymptotically stable output tracking-error dynamics in the dynamically extended state space. The corresponding new SMC design methods will be presented in Sections 9.3 and 9.4.

9.3 Stable system centre design for feedforward/feedback tracking control for systems in a normal canonical form

9.3.1 Problem formulation

A nonlinear plant model, presented in a normal canonical form [1] with the internal dynamics linearised at some selected point, is considered.

Consider the following Input/Output (I/O) dynamics

$$\begin{pmatrix} y_1^{(r_1)} \\ y_2^{(r_2)} \\ \vdots \\ y_m^{(r_m)} \end{pmatrix} = \varphi(\xi, \eta, t) + \gamma(\xi, \eta)u, \tag{9.15}$$

where $\xi = [y_1, \dot{y}_1, \ldots, y_1^{(r_1-1)}, \ldots, y_m, \dot{y}_m, \ldots, y_m^{(r_m-1)}]^T \in \Re^r$ is a state-vector of the I/O dynamics, $[r_1, \ldots, r_m] = $ vector relative degree, $r_1 + r_2 + \cdots + r_m = r$ is the total relative degree, $r \leq n$, $u \in \Re^m$ is the control input, $y \in \Re^m$ is the commanded output, $\varphi(\cdot)$ is a partially uncertain but bounded (may be non-smooth) function for all time and in any compact bounded set of its arguments, $\gamma(\xi, \eta) = \gamma_o(\xi, \eta) + \Delta\gamma(\xi, \eta)$, $\gamma_o(\xi, \eta)$ is a known bounded function in any compact bounded set of its arguments, and $\Delta\gamma(\xi, \eta)$ is a bounded uncertainty.

The Internal Dynamics (ID) are given by

$$\dot{\eta} = Q_1\eta + Q_2\xi + \psi(\xi, \eta) + f(t), \tag{9.16}$$

where $\eta \in \Re^{n-r}$ is a vector of internal states, $Q_1 \in \Re^{(n-r)\times(n-r)}$ is a known non-Hurwitz matrix, $Q_2 \in \Re^{(n-r)\times r}$ is a known matrix, the pair (Q_1, Q_2) is completely controllable, $\psi(\cdot): \Re^n \to \Re^{(n-r)}$ is a partially uncertain, smooth vector field representing higher order terms of the linearised ID, and $f(t)$ is a smooth external disturbance. The output y and the internal state-vector η are accessible for measurement. The problem is to provide tracking of a smooth reference (command) profile, $y \to y^c(t)$, in real time, in the presence of system model uncertainties and external disturbances. If $Q_1 \in \Re^{(n-r)\times(n-r)}$ is a non-Hurwitz matrix, then the linear part of the ID (9.16) is unstable, and the formulated problem is a *nonminimum-phase output-tracking* problem.

9.3.2 Replacing output-tracking by state-tracking

If one can identify state-reference profiles such that state-tracking produces output tracking at least asymptotically for any output-reference profile, then the state-tracking-error stabilisation problem can be solved using a conventional SMC method [6]. Following this approach, one has to use a dynamic inverse of the plant to build the state-reference profile generator. For a nonminimum-phase output, this inverse is unstable. In particular, for the system (9.15), (9.16) one has to identify the reference profiles for the ID states, $\eta^c \in \Re^{n-r}$.

For a system with known ID and the output-reference profile, $y^c(t)$, defined by a known exosystem, the problem of finding a bounded η^c can be reduced to solving a partial differential algebraic equation (determining the centre manifold) [10]. A bounded solution to the unstable ID driven by $y^c(t)$ has been called the ideal internal dynamics (IID) [14]. A method to get the bounded profile that converges to a solution on the centre manifold asymptotically, for a class of systems and desired trajectories, has been developed [14]. In addition, we assume some uncertainty in the ID (9.16), $\psi(\xi,\eta) = \psi_o(\xi,\eta) + \varepsilon\psi_\Delta(\xi,\eta)$ ($\psi_o(\xi,\eta) =$ nominal nonlinear term, $\psi_\Delta(\xi,\eta) =$ smooth uncertain term, $\varepsilon =$ small number), and the additive external disturbance term, $f(t)$, which is assumed to be given by a known exosystem. An alternative method to get the IID asymptotically for the system (9.16) and the output-reference profile given by a known exosystem will be presented next. However, some preliminary steps must first be taken in a similar way to those in Reference 14.

9.3.2.1 Reduction of vector relative degree to $[1,\dots,1]$

Following [14], we define a coordinate transformation as follows. For each $\xi_i = [y_i, \dot{y}_i, \dots, y_i^{(r_i-1)}]^T$, $i = \overline{1,m}$,

$$\begin{pmatrix} z^i \\ S_i \end{pmatrix} = \begin{bmatrix} 1 & 0 & \cdots & 0 \\ 0 & 1 & \cdots & 0 \\ \vdots & & \ddots & 0 \\ a_0^i & a_1^i & \dots a_{r_i-1}^i & 1 \end{bmatrix} \cdot \xi_i, \qquad z^i \in \Re^{r_i-1}, \quad S_i \in \Re^1, \tag{9.17}$$

such that for each new output $i = \overline{1,m}$

$$S_i = y_i^{(r_i-1)} + a_{r_i-1}^i y_i^{(r_i-2)} + \cdots + a_1^i \dot{y}_i + a_0^i y_i, \tag{9.18}$$

we define a new output-tracking profile

$$S_i^c = y_i^{c(r_i-1)} + a_{r_i-1}^i y_i^{c(r_i-2)} + \cdots + a_1^i \dot{y}_i^c + a_0^i y_i^c. \tag{9.19}$$

If we achieve $S = S^c$, then $y \to y^c$ asymptotically with the eigenvalue placement defined by Hurwitz polynomials

$$\lambda_i^{r_i-1} + a_{r_i-1}^i \lambda_i^{r_i-2} + \cdots + a_1^i \lambda_i + a_0^i, \qquad i = \overline{1,m}. \tag{9.20}$$

Now, we have the output, $S \in \Re^m$, with a vector relative degree of $[1,\dots,1]$.

9.3.2.2 Secondary coordinate transformation

Relabel the variables [14] as

$$\zeta = \begin{pmatrix} \eta \\ z \end{pmatrix}, \qquad \zeta \in \Re^{n-m}. \tag{9.21}$$

Then, in the (S, ζ) coordinates, the dynamics of the system (9.15), (9.16) can be written as

$$\dot{S} = \tilde{\varphi}(S, \zeta, t) + \tilde{\gamma}(S, \zeta)u, \tag{9.22}$$

$$\dot{\zeta} = \tilde{Q}_1 \zeta + \tilde{Q}_2 S + \tilde{\psi}(S, \zeta) + \tilde{f}(t), \tag{9.23}$$

where the functions $\tilde{\varphi}(\cdot)$, $\tilde{\gamma}(\cdot)$, $\tilde{\psi}(\cdot)$, $\tilde{f}(t)$ have the same properties as $\varphi(\cdot)$, $\gamma(\cdot)$, $\psi(\cdot)$, $f(t)$, respectively. We can present $\tilde{\psi}(\cdot)$ as $\tilde{\psi}(S, \zeta) = \tilde{\psi}_o(S, \zeta) + \varepsilon \tilde{\psi}_\Delta(S, \zeta)$ where the term $\varepsilon \tilde{\psi}_\Delta(S, \zeta)$ will represent the uncertainty in the new ID (9.23). The ζ-dynamics are defined by the original ID (9.16) and the dynamics characterised by the roots of polynomials (9.20). The pair $(\tilde{Q}_1, \tilde{Q}_2)$ is completely controllable. Obviously, the matrix $\tilde{Q}_1$ is still non-Hurwitz.

9.3.2.3 Replacing output tracking by state-tracking

Now, we have an output tracking problem for the nonminimum-phase system (9.22), (9.23), which is in the normal form [1] and in the regular form [6] simultaneously. Further, we consider two cases:

Case I: the ID (9.23) is known, i.e., $f(t) \equiv 0, \varepsilon = 0$ and (9.23) can be presented as

$$\dot{\zeta} = \tilde{Q}_1 \zeta + \tilde{Q}_2 S + \tilde{\psi}_o(S, \zeta), \tag{9.24}$$

Case II: the ID (9.23) is partially uncertain, $\tilde{\psi}(S, \zeta) = \tilde{\psi}_o(S, \zeta) + \varepsilon \tilde{\psi}_\Delta(S, \zeta)$, and driven by $\tilde{f}(t)$.

In Case I, a bounded state-reference profile ζ^c for the system (9.24), which satisfies

$$\dot{\zeta}^c = \tilde{Q}_1 \zeta^c + \tilde{Q}_2 S^c + \tilde{\psi}_o(S^c, \zeta^c), \tag{9.25}$$

is the ideal internal dynamics (IID) for system (9.22), (9.24) (see (52) in Reference 14). Once the IID ζ^c is identified, the problem of providing state-tracking in the system (9.22), (9.24) can be solved using sliding mode control [6] as follows.

Introducing $e_\zeta = \zeta^c - \zeta, e_S = S^c - S$, the internal state-tracking-error dynamics is written as

$$\dot{e}_\zeta = \tilde{Q}_1 e_\zeta + \tilde{Q}_2 e_S + q_3(e_\zeta, e_S, t), \tag{9.26}$$

where $q_3(e_\zeta, e_S, t) = \tilde{\psi}_o(S^c, \zeta^c) - \tilde{\psi}_o(S^c - e_S, \zeta^c - e_\zeta)$.

Defining the sliding surface $\sigma \in \Re^m$ as

$$\sigma = e_S + K e_\zeta = 0, \qquad K \in \Re^{m \times (n-m)} \tag{9.27}$$

and considering e_S as a virtual control in the sliding mode on the surface (9.27), $e_S = -K e_\zeta$, (9.26) is rewritten in the closed loop as

$$\dot{e}_\zeta = (\tilde{Q}_1 - K\tilde{Q}_2)e_\zeta + q_3(e_\zeta, -K e_\zeta, t). \tag{9.28}$$

Since the pair $(\tilde{Q}_1, \tilde{Q}_2)$ is completely controllable, then by selecting the eigenvalues of $(\tilde{Q}_1 - K\tilde{Q}_2)$ to lie sufficiently far in the left-half plane, local asymptotic stability for the system (9.28) is ensured (compare with (66) in Reference 14). Then, the existence of the sliding mode on surface (9.27) under a standard SMC control law can be established [6].

9.3.3 Stable system centre design
   (a method to obtain the IID asymptotically)

The method of system centre has been developed [20] for the plant model presented in the regular form [6], which is convenient for SMC design. As a result, a system of differential-algebraic equations has been obtained to generate the state-reference vector profile, the *system centre*.

For a piecewise output reference profile defined by polynomial splines, a method to generate a stable system centre, which provides output tracking in systems with a linear unstable internal dynamics, has been developed [21]. In this work, we generalise the stable system centre design for system (9.22), (9.23) and for any output reference profile that satisfies a linear exosystem, and consider both Case I and Case II.

Given the exosystems for the output reference profile S^c and for the unmatched disturbance $\tilde{f}(t)$, we assume that the stable closed-loop behaviour of the term $\tilde{\psi}_o(S,\zeta) + \varepsilon\tilde{\psi}_\Delta(S,\zeta) + \tilde{f}(t)$ in (9.23) can be characterised by a known linear exosystem. Let the 'cumulative' characteristic polynomial for this exosystem, which can describe each component of $\tilde{\psi}_i(S,\zeta) + \tilde{f}_i(t)$, $i = \overline{1, n-m}$, be

$$P_k(\lambda) = \lambda^k + p_{k-1}\lambda^{k-1} + \cdots + p_1\lambda + p_0, \tag{9.29}$$

where 'k' is the order of this exosystem, and $p_{k-1}, \ldots, p_1, p_0$ are specified numbers.

Assuming one can measure/estimate the state-vector $[S, \zeta]^T \in \Re^n$, the estimate for the uncertain part in (9.23) can be calculated as

$$\hat{f} = \hat{\dot{\zeta}} - \tilde{Q}_1\zeta - \tilde{Q}_2 S - \tilde{\psi}_o(S,\zeta) \approx \varepsilon\tilde{\psi}_\Delta(S,\zeta) + \tilde{f}(t). \tag{9.30}$$

Using the characteristic polynomial (9.29) and the estimate (9.30), we define a stable system centre for the system (9.22), (9.23). A bounded internal state-reference profile $\tilde{\zeta}^c$ (the system centre) for system (9.23), which asymptotically converges in Case I to the IID (9.25) and in Case II to the IID

$$\dot{\zeta}^c = \tilde{Q}_1\zeta^c + \tilde{Q}_2 S^c + \tilde{\psi}(S^c,\zeta^c) + f_1(t) \tag{9.31}$$

is defined by

$$\dot{\tilde{\zeta}}^c = \tilde{Q}_1\tilde{\zeta}^c + \theta^c(\tilde{\zeta}^c, S^c, \hat{f}) + g^c, \tag{9.32}$$

where $\theta^c = \tilde{Q}_2 S^c + \tilde{\psi}_o(S^c, \tilde{\zeta}^c) + \hat{f}$, and g^c should converge to zero asymptotically with any desired eigenvalue placement.

Conditions and a set of algorithms to generate the internal state-reference profile and to transform the nonminimum-phase tracking problem for system (9.22), (9.23) to the problem of stabilisation of the system (9.26) to zero using a conventional sliding mode control can be summarised in the following theorem.

Theorem 1.   *Given the nonminimum-phase system (9.22), (9.23) with the measurable state-vector (S, ζ) and the following set of conditions:*

i.   *The matrix $\tilde{Q}_1$ in (9.23) is nonsingular.*
ii.   *The output reference profile $S^c(t)$ (9.19), the unmatched disturbance $\tilde{f}_1(t)$, and the nonlinear partially uncertain term $\tilde{\psi}(S, \zeta)$ can be piecewise presented by known linear exosystems.*

Then:

1.   *The output tracking in real time of a bounded reference profile, $S^c \in \mathfrak{R}^m$, can be replaced by tracking the state-reference profile $(S^c, \zeta^c)^T \in \mathfrak{R}^n$, such that $(S, \zeta)^T \to (S^c, \zeta^c)^T$ asymptotically with given eigenvalue placement.*
2.   *The internal state-reference profile $\zeta^c \in \mathfrak{R}^{n-m}$ is generated by the matrix differential equation*

$$\tilde{\zeta}^{c(k)} + c_{k-1} I \cdot \tilde{\zeta}^{c(k-1)} + \cdots + c_1 \dot{\tilde{\zeta}}^c + c_0 \tilde{\zeta}^c$$
$$= -(P_{k-1}\theta^{c(k-1)} + \cdots + P_1\dot{\theta}^c + P_0\theta^c), \tag{9.33}$$

where the numbers $c_{k-1}, \ldots, c_1, c_0$ are chosen to provide any desired eigenvalue placement, and matrices $P_{k-1}, \ldots, P_1, P_0 \in \mathfrak{R}^{(n-m) \times (n-m)}$ are given by

$$P_{k-1} = (I + c_{k-1}\tilde{Q}_1^{-1} + \cdots + c_0\tilde{Q}_1^{-k})$$
$$\times (I + p_{k-1}\tilde{Q}_1^{-1} + \cdots + p_0\tilde{Q}_1^{-k})^{-1} - I,$$
$$P_{k-2} = c_{k-2}\tilde{Q}_1^{-1} + \cdots + c_0\tilde{Q}_1^{-(k-1)} - (P_{k-1} + I)$$
$$\times (p_{k-2}\tilde{Q}_1^{-1} + \cdots + p_0\tilde{Q}_1^{-(k-1)}),$$
$$\vdots \tag{9.34}$$
$$P_1 = c_1\tilde{Q}_1^{-1} + c_0\tilde{Q}_1^{-2} - (P_{k-1} + I)\cdot(p_1\tilde{Q}_1^{-1} + p_0\tilde{Q}_1^{-2})$$
$$P_0 = c_0\tilde{Q}_1^{-1} - (P_{k-1} + I)\cdot p_0\tilde{Q}_1^{-1}.$$

3.   *The uncertainty in system (9.23) is estimated as $\hat{f}$ given by (9.30), where each component of the vector $\dot{\zeta}$ is estimated via an exact differentiator [29].*

Proof.   See Reference 22.

9.3.4   Conclusion

A complete constructive algorithm to address the nonlinear nonminimum-phase output tracking problem for a causal system written in a normal canonical form is obtained. A sliding mode controller has been designed to provide robust tracking with matched as well as unmatched nonlinear uncertain terms and disturbances, using the method of stable system centre and a second-order SMC-based observer. Such a controller is insensitive to matched disturbances and nonlinearities, and can accommodate unmatched terms as well. The proposed control scheme allows the tracking-error to be cancelled from a causal reference input piecewise defined by a known linear exosystem.

9.4 Asymptotic output tracking by state-feedback: dynamic sliding manifold technique

9.4.1 *Dynamic sliding manifold (DSM) of full order*

Consider the nonminimum-phase plant with time-varying uncertainties

$$\dot{x} = Ax + B\left((I + \Delta B_2(x,t))u + f_2(x,t)\right) + F \cdot f_1(x,t),$$

$$y = Cx,$$

(9.35)

where $A \in \Re^{n \times n}$, $B \in \Re^{n \times m}$, $C \in \Re^{m \times n}$ are known matrices, $\Delta B_2(x,t)$, $f_2(x,t)$ are matched multiplicative and additive disturbances respectively; and $F = B_\perp^T$ is any matrix such that the column-range space $\mathbf{R}[B, B_\perp^T] \equiv \Re^n$ is a basis in $\Re^n$, and $B_\perp B \equiv [0]_{(n-m) \times m}$, such that $f_1(x,t)$ is the unmatched time-varying uncertainty with respect to the control distribution B.

In the feedback design approach, an output tracking problem is transformed into an output regulation problem, where the designated output

$$e_y = y_R(t) - Cx. \tag{9.36}$$

The output tracking-error should be robustly regulated to zero in the presence of model uncertainties and external disturbances including the reference input $y_R(t)$, which is now treated as a disturbance. For a nonminimum-phase plant, this problem is amended with the requirement of internal stability. As we have seen in the example of Section 9.2, the plant motion on the manifold $\sigma = e_y = 0$ is unstable for a nonminimum-phase plant. On the other hand, any 'stable' manifold in $\Re^n$ is exposed to unmatched disturbances and cannot ensure $e_y \to 0$.

The problem of decoupling all of the uncertainties, modelled by the states of an exogenous system, from the manifold of asymptotically stable output error dynamics can be solved using the ideas of servocompensation (dynamic extension of the state space) and the internal model principle [9, 30, 31]. In this case, designing the dynamic state-feedback, the SMC approach can provide enhanced robustness to matched uncertainties. The solution is a two-loop cascade structure, where the inner-loop SMC controller enforces some nominal plant behaviour on the full order or reduced-order system, and the outer-loop dynamic compensator provides for asymptotic output tracking with overall stability, such that the unmatched disturbances modelled by an exosystem are decoupled from the asymptotically stable e_y-dynamics.

Considering that full state-variable feedback is available for the system (9.35), the full-order nominal plant dynamics can be enforced in the integral-type *dynamic sliding manifold* [32]

$$\sigma = B^T(x - x(0)) - B^T \int_0^t (Ax + Bu_c)\,d\tau = 0, \tag{9.37}$$

where u_c is the outer-loop servocompensator control. One can calculate the equivalent control differentiating (9.37) and using (9.35) as follows

$$\dot{\sigma} = B^T(B(I + \Delta B_2(\cdot))u_{eq} + f_2(\cdot) - u_c + B_\perp^T f_1(x,t)) = 0 \tag{9.38}$$

or, since $B^T B_\perp^T \equiv 0$, and $|B^T B| \neq 0$, assuming B is of full rank, and certain conditions on ΔB_2, we obtain

$$u_{eq} = (I + \Delta B_2)^{-1}(u_c(\cdot) - f_2(\cdot)). \tag{9.39}$$

Substituting (9.39) into (9.35) we have the nominal closed-loop behaviour in the sliding mode $\sigma = 0$ on the dynamic surface (9.37)

$$\dot{x} = Ax + Bu_c + F \cdot f_1(x,t),$$
$$y = Cx, \tag{9.40}$$

where without loss of generality we assume $F = B_\perp^T$, i.e., the column-range space of matrix F is the orthogonal complement of the column-range space of B. The sliding mode $\sigma = 0$ can be maintained via a traditional discontinuous SMC in the format

$$u = u_c + R \cdot \text{SIGN}(\sigma), \tag{9.41}$$

where $R = \text{diag}\{\rho_1, \ldots, \rho_m\}$, $\text{SIGN}(\sigma) = [\text{sgn}(\sigma_1), \ldots, \text{sgn}(\sigma_m)]^T$.

The solution to the output tracking problem for the system (9.40) can be obtained using servocompensator control (see the review [30]), if the unmatched disturbance $f_1(x,t)$ and the reference input $y_R(t)$ can be modelled by the states of a known exosystem.

Not all modes of the internal dynamics of the plant can be unstable. In this case feedback on a particular subvector of the internal states can be enough to provide overall stability of the output tracking-error e_y, as was discussed in Section 9.2. This observation calls for development of the SMC design technique, which can provide the same unmatched disturbance accommodation and cancellation of matched uncertainties using a dynamic sliding manifold of lower order than the full-order integral-type dynamic sliding manifold (9.37). This technique is presented next.

9.4.2 *Dynamic sliding manifold of reduced order*

Assume that the vector relative degree of system (9.35) is equal to $[1, 1, \ldots, 1]$. If this is not true, the technique of Section 9.3 can transform the problem to this condition using output redefinition. Thus, assume that after suitable transformation our system is given in form (9.35), such that $|CB| \neq 0$. Without loss of generality we assume $CB = I_{m \times m}$, and transform the system (9.35) once more to state-variables $[z_0, z_1, \xi] = Mx$ where the subvector $\xi \in \Re^m$ describes the matched subspace, the subvector $z_0 \in \Re^{n-2m}$ is the state-vector of a stable manifold of the unmatched subspace and the subvector $z_1 \in \Re^m$ of the subspace includes the unstable manifold of the system (9.35). System (9.35) in these new coordinates will have the form

$$
\begin{aligned}
\dot{z}_0 &= A_{00}z_0 + A_{01}z_1 + A_{02}\xi + f_{01}(\cdot,t), \\
\dot{z}_1 &= A_{11}z_1 + A_{12}\xi + f_{11}(\cdot,t), \\
\dot{\xi} &= A_{20}z_0 + A_{21}z_1 + A_{22}\xi + f_2(\cdot,t) + (I + \Delta B_2(\cdot))u, \\
e_y &= -\xi + y_R(t),
\end{aligned}
\tag{9.42}
$$

where the z_0-dynamics are bounded-input-bounded-state (BIBS) stable provided bounded inputs z_1,ξ; the matrix A_{11} is non-Hurwitz. We assume additionally that $\det(A_{11}) \neq 0$, and the pair A_{11}, A_{12} is completely controllable.

The problem is to ensure $e_y \to 0$ making the z_1-dynamics stable simultaneously.

Remark.   In accordance with the structure of the system (9.35), one can observe that the state vector of the zero dynamics (z_0, z_1) consists of stable (z_0) and unstable (z_1) parts. Instability of z_1 makes the output tracking problem nonminimum-phase.

A sliding mode control u can collapse the ξ-dynamics in finite time and enforce the state ξ or the output e_y to follow any smooth trajectory. Considering e_y as a virtual control, we have the following output stabilisation problem of reduced order for part of the unmatched dynamics (the z_0-dynamics are BIBS stable and can be excluded from the feedback design)

$$
\dot{z}_1 = A_{11}z_1 - A_{12}e_y + (A_{12}y_R(t) + f_{11}(\cdot,t)),
\tag{9.43}
$$

where e_y is considered as a 'control' and as a regulated output. The goal is to achieve $e_y \to 0$, $|z_1| < \infty$. There is a direct feed-through of the control input to the regulated output in the system (9.43).

For the sake of clarity in the design algorithm, we accept that the exosystem model, which should describe the behaviour of the term $A_{12}y_R(t) + f_{11}(\cdot,t)$, has a characteristic polynomial with all eigenvalues at zero. In other words, the uncertainty in system (9.43) is supposed to be presented by piecewise polynomial splines. Generalisation of this approach to any arbitrary linear exosystem model can be made similar to the design technique presented in Section 9.3.

The following theorem gives a solution to the stabilisation problem.

Theorem 2.   *For the nonminimum-phase system (9.42), where*

 i)   *$\det(A_{11}) \neq 0$, and A_{11}, A_{12} is a completely controllable pair;*
 ii)   *under condition $e_y \to 0$, the behaviour of the term $A_{12}y_R(t) + f_{11}(\cdot,t)$ can be piecewise modelled by an exosystem $\theta^{(k)} = 0$, k is a specified number;*
 iii)   *the sliding mode on the dynamic sliding manifold*

$$
\sigma = z_1 + P_k e_y + \int \left(P_{k-1}e_y + \int \left(P_{k-2}e_y + \cdots + \int (P_0 e_y)\, d\tau \right) d\tau \right) d\tau = 0,
\tag{9.44}
$$

1)   *Provides for the uncoupled asymptotically stable tracking-error dynamics with given eigenvalue placement*

$$e_y^{(k+1)} + c_k I e_y^{(k)} + \cdots + c_1 I \dot{e}_y + c_0 I e_y = 0, \tag{9.45}$$

where $(c_k, \ldots, c_1, c_0)$ are specified numbers.

2)   *Given the set of Hurwitz polynomial coefficients $(c_k, \ldots, c_1, c_0)$, the set of matrices $(P_k, \ldots, P_1, P_0)$ in (9.44) is calculated as*

$$
\begin{aligned}
P_0 &= -c_0 A_{11}^{-1} P_k, \\
P_1 &= -[c_0 A_{11}^{-2} + c_1 A_{11}^{-1}] P_k, \\
&\ \vdots \\
P_{k-1} &= -[c_0 A_{11}^{-k} + \cdots + c_{k-1} A_{11}^{-1}] P_k, \\
P_k &= -[c_0 A_{11}^{-k} + \cdots + c_{k-1} A_{11}^{-1} + A_{11} + c_k I]^{-1} A_{12}.
\end{aligned}
\tag{9.46}
$$

Proof.   See Reference 25.

A standard SMC that provides existence of a sliding mode $\sigma = 0$ on the dynamic sliding manifold (9.44) can be designed in the form

$$u = \hat{u}_{eq} + R \cdot \mathrm{SIGN}(\sigma), \tag{9.47}$$

where $R = \mathrm{diag}\{\rho_1, \ldots, \rho_m\}$, $\mathrm{SIGN}(\sigma) = [\mathrm{sgn}(\sigma_1), \ldots, \mathrm{sgn}(\sigma_m)]^T$, and $\hat{u}_{eq}$ is the best estimate of the actual equivalent control.

9.4.3   Case study: The flight path angle tracking in a pitch plane of F-16 jet fighter

The DSM technique developed above (feedback control only) is illustrated by a SMC design for flight path angle tracking in the pitch plane of the F-16 jet fighter discussed in Section 9.2. The pitch dynamics of the F-16 jet fighter originally described by (9.1) and (9.2) have been transformed into the regular form (9.9), (9.10)

$$
\begin{bmatrix} \dot{\eta}_1 \\ \dot{\eta}_2 \end{bmatrix} = \begin{bmatrix} -10.816 & 0 \\ 0 & 9.616 \end{bmatrix} \cdot \begin{bmatrix} \eta_1 \\ \eta_2 \end{bmatrix} - \begin{bmatrix} 23.806 \\ 23.428 \end{bmatrix} \xi,
$$

$$\dot{\xi} = 2.01\eta_1 + 2.226\eta_2 - \xi + 0.2u,$$

where the commanded output is $y = \gamma = \xi$ and the control input is $u = \delta$.

Define the output tracking-error $e_y = y_R(t) - \xi$. One can observe the η_1-dynamics are BIBS stable, and the η_2 dynamics are unstable. Thus, the internal state η_2 must be BIBS stabilised while the output tracking-error reaches zero asymptotically.

Therefore, the problem is to achieve $e_y \to 0$, while $|\eta_2| < \infty$.

Taking into account $e_y = y_R(t) - \xi$, the unstable part of the internal dynamics (9.9) is rewritten as follows

$$\dot{\eta}_2 = 9.616\eta_2 + 23.428 e_y - 23.428 y_R(t).$$

Assume $\dddot{y}_R = 0$ almost everywhere, i.e., the reference input can be described in a piecewise manner by second-order polynomials. Then, according to the theorem, the output tracking-error dynamics will be of fourth order.

Select the asymptotic behaviour of e_y according to an ITAE criterion with $\omega = 2$

$$e_y^{(4)} + 2.1\omega e_y^{(3)} + 3.4\omega^2 \ddot{e}_y + 2.7\omega^3 \dot{e}_y + \omega^4 e_y = 0,$$

then the set of matrices (P_3, P_2, P_1, P_0), which are just numbers in this case, is obtained as

$$P_3 = 1.513, \quad P_2 = -2.521, \quad P_1 = -3.661, \quad P_0 = -2.518.$$

The dynamic sliding manifold is calculated as

$$\sigma = \eta_2 + P_3 e_y + \int \left(P_2 e_y + \int \left(P_1 e_y + \int (P_0 e_y)\, d\tau \right) d\tau \right) d\tau = 0, \qquad (9.48)$$

where $\eta_2 = 2.437\theta - 2.199\alpha + 0.022q$.

The SMC that stabilises the DSM (9.48) is taken in *saturation function* format that approximates the sign function

$$u = SAT_{0.3}(100\sigma)$$

providing convergence to a small domain around the DSM.

The results of a simulation are presented in Figs 9.1–9.4.

9.4.3.1   Discussion of the simulation results

The command profile for the output is selected to be $y_R(t) = 0.2\sin(0.3t)$. In Fig. 9.1, we see that after a typical transient of the nonminimum-phase plant, the output, in this case the flight path angle, follows the reference command accurately, even though the controller is tuned to follow asymptotically only piecewise parabolic signals.

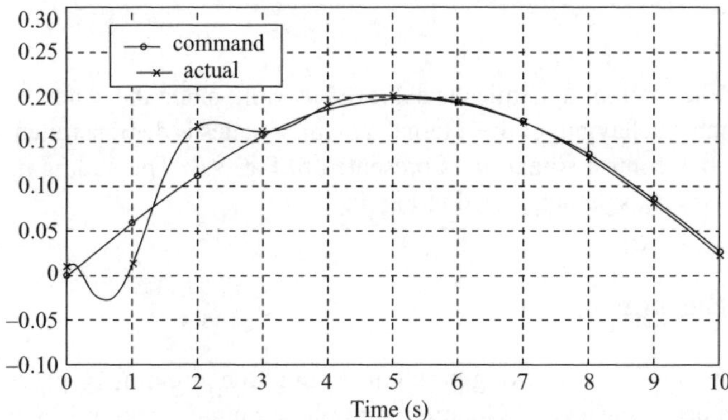

Figure 9.1   Flight path angle γ and its command, $y_R(t)$, versus time

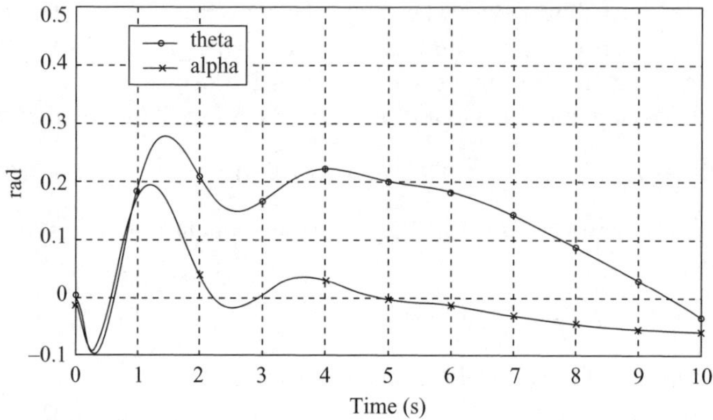

Figure 9.2   The states θ and α versus time

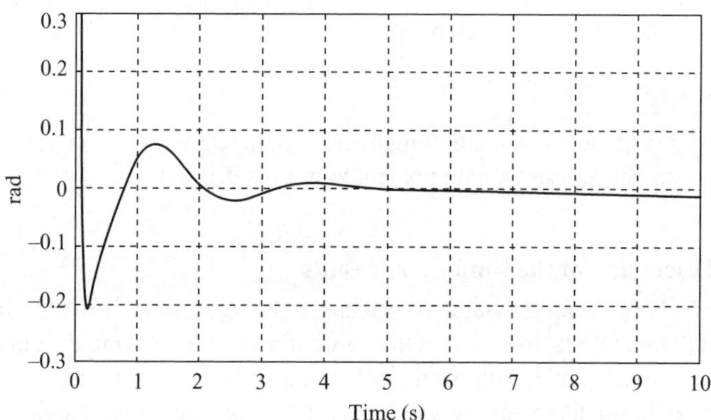

Figure 9.3   Control u versus time

Stability of the internal dynamics is achieved as well, since the states, θ and α, are bounded. Stable behaviour of the internal dynamics states is demonstrated in Fig. 9.2. The continuous control signal, u, is presented in Fig. 9.3. The sliding performance on the DSM $\sigma = 0$ is demonstrated in Fig. 9.4.

9.5   Conclusions

In this chapter an output tracking problem relating to nonminimum-phase nonlinear systems has been considered. Nonminimum-phase output tracking is a challenging, real life control problem that restricts the use of powerful control techniques such as sliding mode control and feedback linearisation. Taking into consideration uncertain

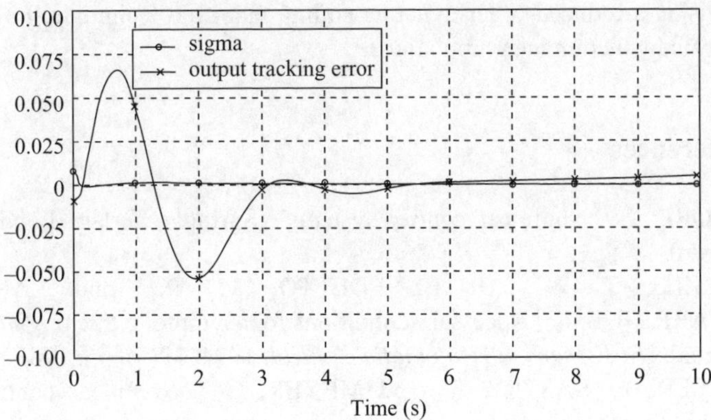

Figure 9.4   Sliding quantity σ and tracking-error e_y versus time

causal systems that have to follow real-time reference profiles only complicates the problem further. In this chapter, the output-tracking problem for causal nonminimum-phase systems with uncertainties and disturbances has been tackled by means of a robust nonlinear control technique, sliding mode control.

A toolbox has been presented: a set of fully constructive design algorithms to develop sliding mode controllers (SMC). One SMC design technique is based on a stable system centre approach. The idea of this approach is to replace the output reference profile tracking by state-reference profile tracking. If one identifies the state-reference profiles (*the system centre*), such that state-tracking yields output tracking at least asymptotically for arbitrary choice of real-time output reference profile, the goal is to design the SMC to stabilise the state-tracking-error at zero in the sliding mode. The key task is to properly build the *stable* state-reference profile generator that produces bounded state-reference profiles (*the stable system centre*) given a bounded real-time output reference profile and bounded uncertainties and disturbances. This non-trivial nonminimum-phase tracking problem is addressed by SMC design based on the properly built stable system centre. An SMC design technique of feed forward/feedback type is developed for systems presented in normal canonical form.

Two other SMC design techniques have been developed based on a dynamic sliding manifold (DSM) approach. Considering the output tracking-error as a virtual control, the DSM is designed such that in the sliding mode, the output tracking-error tends to zero asymptotically while the unstable zero dynamics are stabilised. One SMC was designed based on full-order integral type DSM in combination with a traditional servocompensator that compensates for unmatched disturbances and uncertainties. A second SMC was designed based on a reduced-order DSM that incorporates the exosystem for the output reference profile and unmatched disturbances/uncertainties. Both SMCs are of a pure feedback type.

A case study, flight path angle tracking in the pitch plane of a F-16 jet fighter with unstable zero dynamics, was considered. The pure feedback-type SMC was designed using a reduced-order dynamic sliding manifold technique. Its efficiency was confirmed via computer simulations.

9.6   References

1   ISIDORI, A.: 'Nonlinear control systems' (Springer Verlag, London, 1995, 2nd edn)

2   GRIZZLE, J. W., DI BENEDETTO, M. D., and LAMNABHI-LAGARRIGUE, F.: 'Necessary conditions for asymptotic tracking in nonlinear systems', *IEEE Trans. on Automatic Control*, 1994, **39**(9), pp. 1782–1794

3   HAUSER, J., SASTRY, S., and MEYER, G.: 'Nonlinear control design for slightly nonminimum phase systems: application to V/STOL aircraft', *Automatica*, 1992, **28**(4), pp. 665–679

4   BENVENUTI, L., DI BENEDETTO, M. D., and GRIZZLE, J. W.: 'Approximate output tracking for nonlinear nonminimum phase systems with an application to flight control', *Journal of Nonlinear and Robust Control*, 1994, **4**, pp. 397–414

5   AZAM, M. and SINGH, S. N.: 'Invertibility and trajectory control for nonlinear maneuvers of aircraft', *Journal of Guidance, Control, and Dynamics*, 1998, **17**(1), pp. 192–200

6   UTKIN, V., GULDNER, J., and SHI, J.: 'Sliding modes in electromechanical systems' (Taylor and Francis, London, 1999)

7   DeCARLO, R., ZAK, S. H., and MATHEWS, G. P.: 'Variable structure control of nonlinear multivariable systems: a tutorial', *Proc. IEEE*, 1988, **76**, pp. 212–232

8   HUNG, J. Y., GAO, W. B., and HUNG, J. C.: 'Variable structure control: a survey', *IEEE Trans. on Industrial Electronics*, 1993, **40**, pp. 2–22

9   FRANCIS, B. A. and WONHAM, W. M.: 'The internal model principle of control theory', *Automatica*, 1976, **12**, pp. 457–465

10   ISIDORI, A. and BYRNES, C. I.: 'Output regulation of nonlinear systems', *IEEE Trans. on Automatic Control*, 1990, **35**(2), pp. 131–140

11   ISIDORI, A. and MOOG, C. H.: 'On the nonlinear equivalent of the notion of transmission zeros', in BYRNES, C. I. and KURZHANSKY, A. B. (Eds): 'Modeling and adaptive control' (Springer Verlag, Berlin, 1991)

12   LLANES-SANTIAGO, O. and SIRA-RAMÍREZ, H.: 'A controller resetting strategy for the stabilization of DC-to-DC power converters towards nonminimum phase equilibria', *Proceedings on the 33rd Conference on Decision and Control*, Florida, 1994, pp. 2920–2925

13   HUANG, J. and RUGH, W. J.: 'On a nonlinear multivariable servomechanism problem', *Automatica*, 1990, **26**(6), pp. 963–972

14   GOPALSWAMY, S. and HEDRICK, J. K.: 'Tracking nonlinear non-minimum phase systems using sliding control', *International Journal of Control*, 1993, **57**(5), pp. 1141–1158

15 HUANG, J. and RUGH, W. J.: 'Approximate noninteracting control with stability for nonlinear systems', *IEEE Transactions on Automatic Control*, 1991, **36**, pp. 295–304

16 DEVASIA, S., CHEN, D., and PADEN, B.: 'Nonlinear inversion-based output tracking', *IEEE Transactions on Automatic Control*, 1996, **47**(7), pp. 930–942

17 FLIESS, M., SIRA-RAMÍREZ, H., and MÁRQUEZ, R.: 'Regulation of nonminimum-phase outputs: a flatness based approach', in NORMAND-CYROT, D. (Ed.): 'Perspectives in control' (Springer Verlag, London, 1998)

18 MARTIN, P., DEVASIA, S., and PADEN, B. E.: 'A different look at output tracking: control of VTOL aircraft', *Automatica*, 1996, **32**, pp. 101–107

19 LU, X. Y., SPURGEON, S. K., and POSTLETHWAITE, I.: 'Robust variable structure control of a PVTOL aircraft', *International Journal of Systems Science*, 1997, **28**(6), pp. 547–558

20 SHTESSEL, Y. B.: 'Nonlinear output tracking in conventional and dynamic sliding manifolds', *IEEE Transactions on Automatic Control*, 1997, **42**(9), pp. 1282–1286

21 SHTESSEL, Y. B. and SHKOLNIKOV, I. A.: 'Tracking controller design for nonlinear nonminimum phase systems via method of system centre', *IEEE Transactions on Automatic Control*, 2001, **46**(10), pp. 1639–1643

22 SHTESSEL, Y. B. and SHKOLNIKOV, I. A.: 'Tracking a class of nonminimum phase systems with nonlinear internal dynamics via sliding mode control using method of system centre', *Automatica*, 2002, **38**(5), pp. 837–842

23 SHTESSEL, Y. B.: 'Nonlinear nonminimum phase output tracking via dynamic sliding manifolds', *Journal of the Franklin Institute*, 1998, **335**B(5), pp. 841–850

24 SHKOLNIKOV, I. A. and SHTESSEL, Y. B.: 'Nonminimum phase tracking in MIMO systems with square input-output dynamics via dynamic sliding manifolds', *Journal of the Franklin Institute*, 2000, **337**(1), pp. 43–56

25 SHKOLNIKOV, I. A. and SHTESSEL, Y. B.: 'Aircraft nonminimum phase control in dynamic sliding manifolds', *AIAA Journal of Guidance, Control, and Dynamics*, 2001, **24**(3), pp. 566–572

26 SIRA-RAMÍREZ, H.: 'Sliding regimes in general non-linear systems: a relative degree approach', *International Journal of Control*, 1989, **50**(4), pp. 1487–1506

27 FERNÁNDEZ, B. R. and HEDRICK, J. K.: 'Control of multivariable non-linear systems by the sliding mode method', *International Journal of Control*, 1987, **46**(3), pp. 1019–1040

28 JEONG, H.-S. and UTKIN, V. I.: 'Sliding mode tracking control of systems with unstable zero dynamics', in YOUNG, K. D. and ÖZGUNER, Ü. (Eds): 'Variable structure systems, sliding mode and nonlinear control', *Lecture Notes in Control and Information Sciences*, no. 247 (Springer Verlag, London, 1999), p. 303

29 LEVANT, A.: 'Robust exact differentiation via sliding mode technique', *Automatica*, 1998, **34**(3), pp. 379–384

30   BYRNES, C. I. and ISIDORI, A.: 'Output regulation for nonlinear systems: an overview', *International Journal of Robust and Nonlinear Control*, 2000, **10**(5), pp. 323–337
31   JOHNSON, C. D.: 'A new approach to adaptive control', in LEONDES C. T. (Ed.): 'Advances in control and dynamic systems' (Academic Press, New York, 1988)
32   ACKERMANN, J. and UTKIN, V.: 'Sliding mode control design based on Ackermann's formula', *IEEE Transaction on Automatic Control*, 1998, **43**(2), pp. 234–237

Chapter 10

Sliding mode control and chaos

Xinghuo Yu and Guanrong Chen

10.1 Introduction

Chaos refers to a type of complex dynamical behaviours of some nonlinear systems that possess such features as extreme sensitivity to initial conditions, boundedness of trajectories while having positive Lyapunov exponents, continuous power spectra, fractional dimensions, etc. Chaos has been shown to be a common phenomenon in nature.

The question that should be asked is: What does chaos have to do with SMC? An answer may not be straightforward but we will consider two aspects of this question in this chapter: the first is whether SMC can cause chaos, and the second considers what SMC can offer to chaos research.

Sliding mode control is about regulating dynamical behaviours using some sort of 'disruptive/discontinuous' control actions to achieve fast reactions (see Chapter 1 for an introduction). One may wonder if such control actions, by nature, would cause any possible chaotic behaviours at all. Indeed, if the SMC is ideal, that is, the switching frequency used for any variable structure control actions is infinite, the controlled system should behave as desired. However, there is a gap between 'theory' and 'practice'. Nowadays, practical SMC is commonly implemented via digital computers or microprocessors. Digitised control is implemented by 'freezing' the control force during the sampling period. This very feature may deteriorate the elegant invariance property enjoyed by most (if not all) continuous-time SMC systems. The deterioration of SMC performance due to digitisation was observed a long time ago. Various techniques were developed to specifically address the problems associated with digitising SMC by using a relatively low switching frequency (see Chapter 5 for some detailed discussion). However, the 'micro-behaviours' of SMC after digitisation were relatively under-studied. In this chapter, we first report some of our recent research on discretisation chaos of a popular SMC scheme – the equivalent control based SMC systems.

Differing from Chapter 5, the purpose here is to investigate how the SMC performance changes with respect to the increase of sampling period when no principles for discrete SMC are imposed at the design stage.

Second, we discuss the use of SMC for chaos control. Controlling chaos for engineering applications has emerged as a new and attractive field within the scientific community, and many new theories and methodologies have been developed to date [1, 2]. Chaos control refers to purposefully manipulating chaotic dynamical behaviours of some complex nonlinear systems. Chaos control is particularly useful for time- and energy-critical engineering applications. Examples include data traffic congestion control on the Internet, encryption and secure communication, high-performance circuits and devices (e.g., delta-sigma modulators and power converters), liquid mixing, chemical reactions, power systems collapse prediction and prevention, oscillators design, biological systems modelling and analysis (e.g., the brain and the heart), crisis management (e.g., jet-engine surge and stall), and so on. There are many practical reasons for controlling or ordering chaos. In a system where a chaotic response is undesired or harmful, it should be reduced as much as possible, or totally suppressed. Examples of this include avoiding fatal voltage collapse in power networks, eliminating deadly cardiac arrhythmias, guiding disordered circuit arrays (e.g., multi-coupled oscillators and cellular neural networks) to reach a certain level of desirable pattern formation, regulating dynamical responses of mechanical and electronic devices (e.g., diodes, laser machines, and machine tools), and organising a multi-agency corporation to achieve optimal performance.

SMC has been recently used in chaos control [3–5]. A particular reason is that chaos control usually involves 'small control energy' to 'direct' or 'induce' a desired dynamical behaviour, e.g., stabilising an inherently unstable periodic orbit (UPO) or directing the system trajectory from one orbit to another in the state space. Unlike conventional control systems, chaotic systems are bounded hence there is no global instability issue. The system can afford to wait and let the chaotic orbit evolve and eventually come close to a neighbourhood of a desired orbit by the ergodicity property before taking a small local control action. The principles of SMC seem ideal for chaos control as switching control is often preferred for the drastic control actions needed to achieve fast and effective control. We discuss the use of SMC in chaos control for two tasks: one is time-delayed feedback control to stabilise UPOs and the other is a generalisation of a well-known model-free chaos control method.

This chapter is organised as follows. Section 10.2 discusses discretisation chaos in SMC systems. Section 10.3 presents some results on SMC for controlling chaos to their UPOs. Section 10.4 outlines an extension of a well-known chaos control method from the literature using SMC. Some conclusions are drawn in Section 10.5.

10.2   Discretisation chaos in SMC

In this section, we investigate discretisation chaos in SMC systems. We focus on discretisation chaos in an equivalent control based SMC for linear systems.

10.2.1   Discretisation of an equivalent control based SMC system

Consider the following controllable single-input linear SMC system with a switching manifold s:

$$\dot{x} = Ax + bu, \tag{10.1}$$

$$s(x) = c^{\top}x, \tag{10.2}$$

where $x \in R^n$, $u, s \in R^1$, A is an $n \times n$ matrix, and b and c are n-dimensional vectors, respectively. The switching manifold s is predefined to represent some desirable asymptotically stable dynamics. Its corresponding equivalent control based SMC is

$$u = u_{eq} + u_s, \tag{10.3}$$

$$u_{eq} = -(c^{\top}b)^{-1}c^{\top}Ax, \tag{10.4}$$

$$u_s = -\alpha(c^{\top}b)^{-1}\,\mathrm{sgn}\,s(x), \tag{10.5}$$

with $\alpha > 0$ being a constant control gain, sgn the sign function, and $c^{\top}b \neq 0$. Without loss of generality, we assume that $c^{\top}b = 1$.

Note that the equivalent control u_{eq} is derived by solving $\dot{s} = 0$ subject to (10.1). It can be easily verified that for the Lyapunov function $V = \frac{1}{2}s^2$, the time derivative of V along the dynamics (10.1) with (10.3), (10.4) and (10.5) yields $\dot{V} = -\alpha|s| = -\alpha V^{1/2}$, which indicates the finite-time attainability and global stability of $s = 0$.

The control law (10.3) is by far the most popular SMC structure. For simplicity, and without loss of generality, we assume that the system is in the controllable canonical form.

An interesting question to ask is how discretisation affects the control performance of this class of SMC if u is implemented (digitised) via, for example, a zero-order holder (ZOH) at discrete moments, i.e., $u = u_k$ over the time interval $[kh, (k+1)h)$, where h is a sampling period.

To study the discretisation behaviours, we first convert the continuous-time system (10.1) in the controllable form under the ZOH into the discrete form

$$x(k+1) = e^{Ah}x(k) + \int_0^h e^{A\tau}\,d\tau\,bu_k, \tag{10.6}$$

where

$$u_k = u_{eq}(k) + u_s(k) = -c^{\top}Ax(k) - \alpha\,\mathrm{sgn}\,s(x(k)), \qquad k = 0, 1, \ldots. \tag{10.7}$$

During the evolution of the system state $x(k)$, the function $\mathrm{sgn}\,s(x(k))$ generates a sequence of binary values of -1 and $+1$, which can be considered as a *symbolic sequence* of the underlying dynamics. In the following, for simplicity, we denote $\mathrm{sgn}\,s(x(k))$ as σ_k, hence the symbolic sequence, denoted as σ, can be represented by $\sigma = (\sigma_0, \sigma_1, \sigma_2, \ldots)$. If a symbolic sequence has a minimal period L, we name the sequence as a period-L sequence. With these definitions, the discrete system becomes

$$x(k+1) = \Phi x(k) - \alpha \Gamma \sigma_k, \tag{10.8}$$

$$\Phi = e^{Ah} - \left\{ \int_0^h e^{A\tau} d\tau \right\} (bc^\top A), \tag{10.9}$$

$$\Gamma = \left\{ \int_0^h e^{A\tau} d\tau \right\} b. \tag{10.10}$$

In fact, the dynamic system (10.8) can be considered as two separate affine maps:

$$F_s(x) = \Phi x - \alpha \Gamma \sigma, \quad \text{for } \sigma = -1, +1. \tag{10.11}$$

Let us consider a trajectory starting from the initial point $x(0)$, and assume that $x(k)$ corresponds to the symbolic sequence σ. First, observe that the kth-iteration of the system starting from $x(0)$ can be computed as

$$x(k) = (F_{\sigma_{k-1}} \circ \cdots \circ F_{\sigma_1} \circ F_{\sigma_0})(x(0)) = \Phi^k x(0) - \alpha \sum_{i=0}^{k-1} \Phi^i \Gamma \sigma_{k-1-i}, \quad k \geq 1. \tag{10.12}$$

One can see that for a fixed symbolic sequence, $F_{\sigma_{k-1}} \circ \cdots \circ F_{\sigma_1} \circ F_{\sigma_0}$ is an affine map. There is a close association between the orbits of system (10.8) and its corresponding symbolic sequence $\sigma = (\sigma_0, \sigma_1, \ldots, \sigma_{k-1})$. The sequence σ can be used to describe the mapping between the phase plane and the sequence [6]. Equation (10.12) can be rewritten in the following form:

$$x(k) = \Phi^k x(0) - \alpha \Psi_k (\sigma_0, \sigma_1, \ldots, \sigma_{k-1})^\top, \tag{10.13}$$

where

$$\Psi_k = (\Phi^{k-1}\Gamma, \ldots, \Phi^2\Gamma, \Phi\Gamma, \Gamma), \quad k = 1, 2, \ldots. \tag{10.14}$$

We now present several interesting properties of this discretised single-input SMC system.

Lemma 1 [7].   *The matrix Φ has the following form:*

$$\Phi = \begin{bmatrix} 1 & v^\top(h) \\ \bar{0} & D(h) \end{bmatrix}, \tag{10.15}$$

where $v(h)$ is an $(n-1)$-dimensional vector, $\bar{0}$ is an $(n-1)$-dimensional zero vector, and $D(h)$ is an $(n-1) \times (n-1)$ matrix.

Lemma 1 implies that

$$\Phi^j = \begin{bmatrix} 1 & v^\top \sum_{i=0}^{j-1} D^i \\ \bar{0} & D^j \end{bmatrix} = \begin{bmatrix} 1 & v^\top (I-D)^{-1}(I-D^j) \\ \bar{0} & D^j \end{bmatrix}.$$

For convenience of the analysis, we rewrite the discrete dynamical system (10.8) using Lemma 1 as

$$\begin{bmatrix} x_1(k+1) \\ z(k+1) \end{bmatrix} = \begin{bmatrix} 1 & v^\top \\ 0 & D \end{bmatrix} \begin{bmatrix} x_1(k) \\ z(k) \end{bmatrix} - \alpha \begin{bmatrix} \Gamma_1 \\ \Gamma_2 \end{bmatrix} \sigma_k,$$

where $z = [x_2, \ldots, x_n]^\top \in R^{(n-1)}$, Γ_1 is a scalar and Γ_2 is an $(n-1)$-dimensional vector. Hence, given the upper block-triangular structure of the matrix Φ, the system in fact can be decomposed into two subsystems:

$$x_1(k+1) = x_1(k) + v^\top z(k) - \alpha \Gamma_1 \sigma_k, \tag{10.16}$$

$$z(k+1) = Dz(k) - \alpha \Gamma_2 \sigma_k. \tag{10.17}$$

It is well known that discretised SMC systems, even with moderate sampling rates, may exhibit chattering/zigzagging and sometimes chaotic motions [8, 9]. The questions of interest for the discretised equivalent control based SMC systems are:

1. When does bifurcation occur from stable to unstable motion with respect to the sampling period h?
2. What kinds of chattering/zigzagging bebaviours will appear?

These questions are addressed in the following.

10.2.2   *Discretisation behaviours analysis*

For convenience of the analysis, the general nth-order SMC system (10.8) with (10.9) and (10.10) is alternatively expressed as

$$x_1(k+1) = x_1(k) + v^\top z(k) - \alpha \Gamma_1 \sigma_k, \tag{10.18}$$

$$z(k+1) = Dz(k) - \alpha \Gamma_2 \sigma_k, \tag{10.19}$$

where $z(k) \in R^{n-1}$, and all notations are as defined above.

Theorem 1.   *The system (10.18) and (10.19) is stable in the sense of Lyapunov if*

$$\|D\| < 1, \tag{10.20}$$

$$|\Gamma_1| > \frac{\|v\| \|\Gamma_2\|}{1 - \|D\|}, \tag{10.21}$$

where $\|\cdot\|$ is the spectral norm. Furthermore,

$$|x_1(\infty)| < \alpha|\Gamma_1| + \alpha\|(v^\top - c_1^{-1}\bar{c}^\top)\Gamma_2(I-D)^{-1}\|, \tag{10.22}$$

$$\|z(\infty)\| < \frac{\alpha\|\Gamma_2\|}{1-\|D\|}, \tag{10.23}$$

where $\bar{c} = (c_2, c_3, \ldots, 1)^\top$.

Proof.   First, it follows from (10.19) that

$$\|z(k+1)\| \leq \|D\|\|z(k)\| + \alpha\|\Gamma_2\|, \tag{10.24}$$

so that iterating n times on (10.24) yields

$$\|z(n)\| \leq \|D\|^n \|z(0)\| + \alpha\|\Gamma_2\| \sum_{i=0}^{n-1} \|D\|^{n-1-i}$$

$$= \|D\|^n \|z(0)\| + \alpha\|\Gamma_2\|(1-\|D\|^n)(1-\|D\|)^{-1},$$

since $\|D\| < 1$. Then, as $n \to \infty$, we have

$$\|z(\infty)\| \leq \alpha\|\Gamma_2\|(1-\|D\|)^{-1}.$$

This completes the proof of inequality (10.23).

The switching line for the nth-order system, $s(x) = c^\top x = c_1 x_1 + \bar{c}^\top z$, can be decomposed into 'two' variables (x_1, z'), where $z' = \bar{c}^\top z$ is a scalar variable and the switching line can be viewed on the 'plane' x_1–z'. First, in the limiting case,

$$z \to \alpha\Gamma_2(I-D)^{-1}\sigma,$$

for a fixed sign σ. From (10.18),

$$x_1(k+1) = x_1(k) + v^\top z(k) - \alpha\Gamma_1\sigma_k.$$

As far as the scalar variable z' is concerned, its effect with respect to the switching line on the plane x_1–z' is the same as the scalar $x_2 = z$ for the second-order case [7]. Hence, similar reasoning leads to

$$|x_1(k+1)| < \alpha|\Gamma_1| + \alpha\|(v^\top - c_1^{-1}\bar{c}^\top)\Gamma_2(I-D)^{-1}\|.$$

Tedious details somewhat repeat the second-order case [7] and therefore are omitted here.

Remark 1.   *Theorem 1 indicates some basic features of the boundedness of linear higher-order systems under the equivalent control. The most interesting behaviours are those within the boundaries given in Theorem 1. It is known that within these boundaries, some intriguing behaviours, such as 'fast' chattering and 'slow' periodic zigzagging, occur [7]. Numerous simulations have shown a general pattern that the system trajectory travels between a finite set of points and exhibits periodic behaviours. The periods of trajectories depend on the initial conditions and the system parameter setting such as the sampling period h. Note that the period of the* *symbolic sequence may not be the same as the period of the system periodic trajectory, evidenced by the results of another class of discrete systems [10], where discontinuity is involved.*

We want to know the relationship between the periodic trajectories of the system and their symbolic sequences. The result is the following.

Theorem 2.   *For $\|D\| < 1$, if system (10.18) and (10.19) exhibits a periodic behaviour with a period-L symbolic sequence, then the system trajectory will eventually converge to a set of L fixed states. Furthermore:*

1. *the z-coordinates of the L fixed states are uniquely determined by*

$$z(i) = -\alpha(1 - D^L)^{-1}\Psi_L\sigma^i,$$

   for $i = 0, \ldots, L - 1$, where

$$\sigma^i = (\sigma_i, \sigma_{i+1}, \ldots, \sigma_{L-1}, \sigma_0, \ldots, \sigma_{i-1})^\top,$$

$$\Psi_L = (D^{L-1}\Gamma_2, D^{L-2}\Gamma_2, \ldots, D\Gamma_2, \Gamma_2);$$

2. *the following equality holds:*

$$\sum_{i=0}^{L-1} \sigma_i = 0.$$

Proof.   First, from Theorem 1, we know that the trajectory will eventually be confined within the boundaries defined by (10.22) and (10.23), therefore we only need to consider the behaviours within the boundaries.

Given a period-L symbolic sequence, we denote a set of L states as $x(0), x(1), \ldots, x(L - 1)$. From the periodicy and (10.19), we have

$$z(1) = Dz(0) - \alpha\Gamma_2\sigma_0,$$
$$\vdots$$
$$z(L - 1) = Dz(L - 2) - \alpha\Gamma_2\sigma_{L-2},$$
$$z(0) = dz(L - 1) - \alpha\Gamma_2\sigma_{L-1},$$

$$(10.25)$$

where $z(i) \neq z(j)$ for $i \neq j$. Straightforward algebraic manipulation on (10.25) yields

$$(I - D^L)z(0) = -\alpha\Psi_L\sigma^0,$$
$$\vdots$$
$$(I - D^L)z(L - 1) = -\alpha\Psi_L\sigma^{L-1}.$$

Break the number of iterations from $z(0)$ into a number of finite sets of length L. A trajectory from $z(i + jL)$ has a periodic symbolic sequence σ^i, where $i = 0, 1, \ldots, L - 1$ and $j \geq 0$. Denote $y_i(j) = z(i + jL) - z(i)$. Using similar arguments as in the proof of Theorem 2 in Reference 7, we get

$$y_i(j + 1) = D^L y_i(j).$$

$$(10.26)$$

Iterating (10.26) m times yields

$$y_i(j+m) = D^{mL} y_i(j).$$

If $m \to \infty$, then starting from any $j \geq 0$ and for any $i = 0, 1, \ldots, L-1$, we have

$$y_i(\infty) = 0,$$

which indicates that the z-coordinates of all trajectories with period L will converge to the uniquely determined $z(0), z(1), \ldots, z(L-1)$.

Although the uniqueness of the z-coordinates of the period L states has been determined, the solutions of x_1 are not unique, as can be seen from (10.18), which could be any value within the boundaries defined by (10.22) and (10.23) depending upon the initial conditions. Summing up the L equations in (10.25), using $\sum_{i=0}^{L-1} D^i = (1 - D^L)(I - D)^{-1}$, yields

$$\sum_{i=0}^{L-1} z(i) = -\alpha(I - D)^{-1}\Gamma_2 \sum_{i=0}^{L-1} \sigma_i. \tag{10.27}$$

From (10.18), it can also be observed that

$$x_1(L) = x_1(0) + v^{\top} \sum_{j=0}^{L-1} z(j) - \alpha\Gamma_1 \sum_{j=0}^{L-1} \sigma_j.$$

We can then prove that

$$v^T \sum_{j=0}^{L-1} z(j) = \alpha\Gamma_1 \sum_{j=0}^{L-1} \sigma_j, \tag{10.28}$$

by using similar arguments as in the proof of Theorem 2 in Reference 7. From (10.28) and (10.27) and the fact that

$$\Gamma_1(h) \neq -v^{\top}(h)(I - D(h))^{-1}\Gamma_2(h),$$

we can infer that $\sum_{i=0}^{L-1} \sigma_i = 0$.

Remark 2.   *Note that because σ_k can only take binary values $\{-1, 1\}$, the only possibility that makes $\sum_{i=0}^{L-1} \sigma_i = 0$ in Theorem 2 is that L is an even integer and there are equal numbers of 1 and -1 on both sides. This is perhaps quite special to the discretised SMC. In general, the numbers of 1 and -1 may not be equal to each other, e.g., in digital filters [10].*

10.2.3 An example

We now show some simulations to verify the theoretical results presented above.

We simulate a third-order system to validate the results for higher-order systems. For the third-order system, we first choose $a_1 = -10$, $a_2 = 9$, $a_3 = -4$, $\alpha = 1$, $c_1 = 1$, $\bar{c} = (1, 1)$. Let $h = 0.1$ and $x(0) = (-2, 1, 1)$. The resulting eigenvalues are

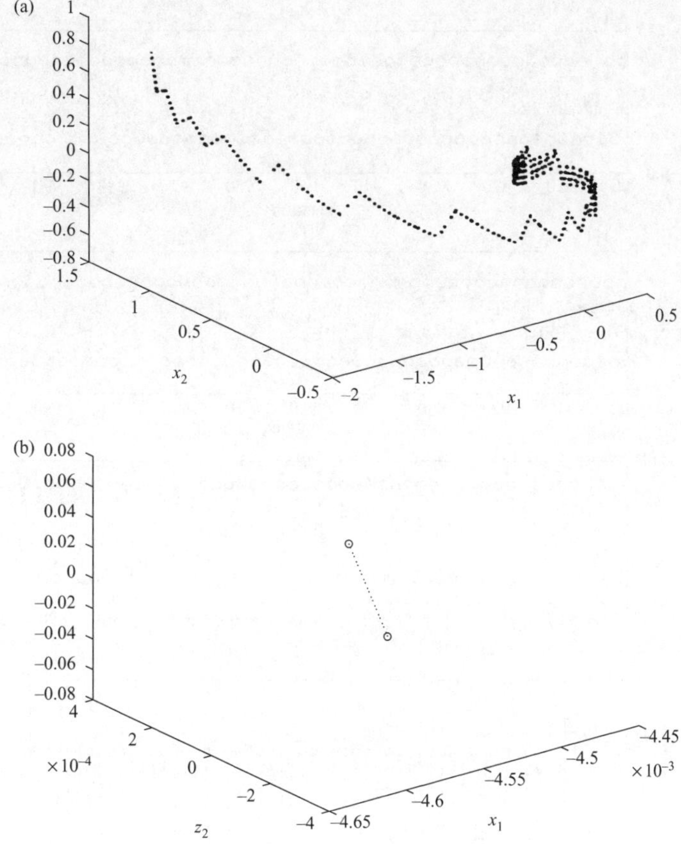

Figure 10.1   a) 3D trajectory; b) system states of the last 100 iterations

1.0000, 0.8812, 0.9444. From Figs 10.1(a)–(c), it can be seen that the system trajectory converges to two fixed points:

$$(-0.0045, 0.0002, -0.0662) \quad \text{and} \quad (0.0046, -0.0002, 0.0662).$$

By Theorem 1, the theoretical values of the boundaries are

$$|x_1(\infty)| < 3.8401 \quad \text{and} \quad \|z(\infty)\| < 2.1806. \tag{10.29}$$

The converged fixed points are well within the boundaries (see Fig. 10.1(e)). From Fig. 10.1(d), one can see that the symbolic sequence is period-2, and $s = (+1, -1)$. Hence, from Theorem 2, the trajectory will converge to these 2 fixed points, which is confirmed by Fig. 10.1(b).

We now set $a_1 = -20$, $a_2 = 14$, $a_3 = -4$, $\alpha = 1$, $c_1 = 1$, $\bar{c} = (1, 1)$ with the same $h = 0.1$. Let $x(0) = (-2, 1, 1)$. The resulting eigenvalues are 1.0000, 0.8093, 0.9947. From Figs 10.2(a)–(c), it is observed that the trajectory converges to 38 fixed points

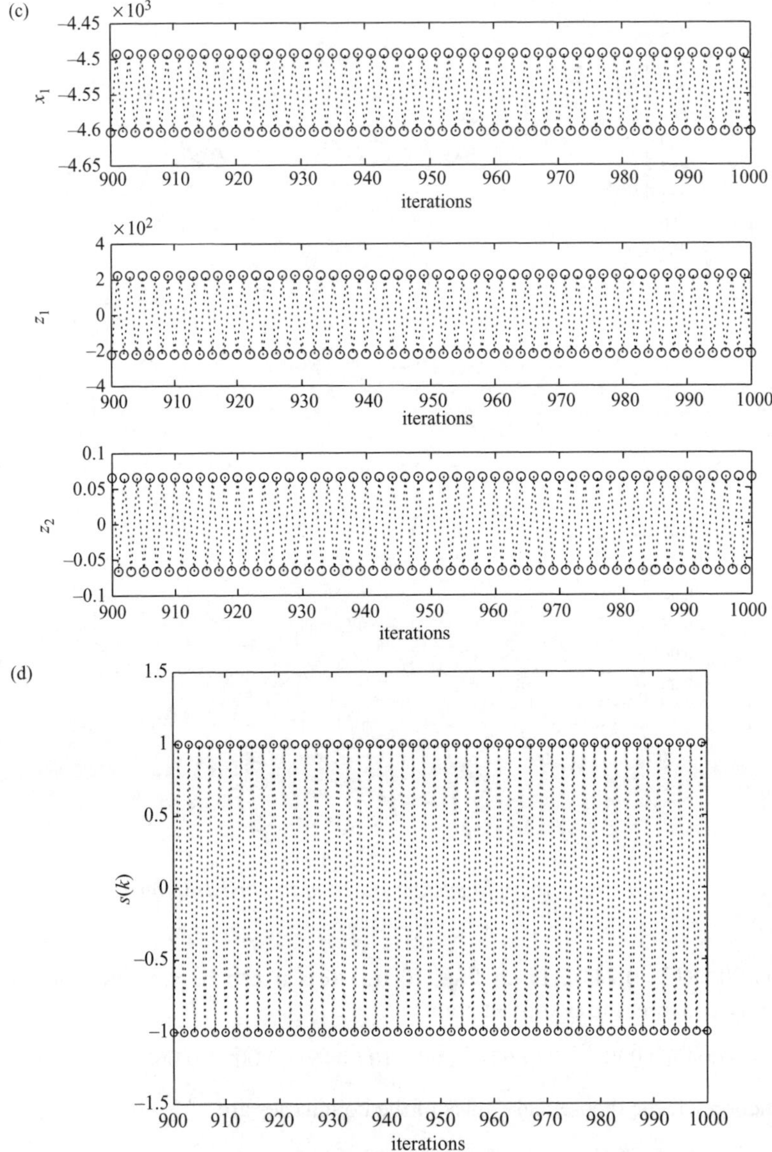

Figure 10.1   Continued. c) System states of the last 50 iterations; d) symbolic sequence of the last 100 iterations

(the actual symbolic sequence is omitted since it is too long). By Theorem 1, the theoretical values of the boundaries are

$$|x_1(\infty)| < 26.6456 \quad \text{and} \quad \|z(\infty)\| < 22.8821. \tag{10.30}$$

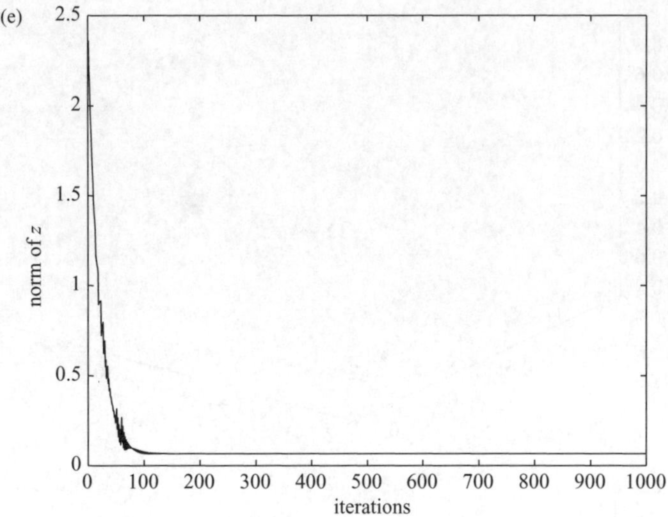

Figure 10.1   Continued. e) Norm function of z

The converged fixed points are well within the boundaries. From Fig. 10.2(d), one can see that the symbolic sequence is a strange long period-38 sequence, which is constructed by two leading $(-1, -1)$ and 17 repeating $(+1, -1)$ subsequences and two trailing $(+1, +1)$. This sequence would be very easily mistaken as period-2 sequence if one does not look at a long enough data record. From Theorem 2, the trajectory is supposed to converge to these 38 fixed points, which is confirmed by Fig. 10.2(a).

10.3   Time-delayed chaos control with SMC

Recently, stabilising UPOs of chaotic systems has become an active and focussing direction in the field of chaos control [11]. This problem can be formulated as a (target) tracking problem in classical control theory. Therefore, the rich literature of conventional tracking control theory is readily applicable for the tasks of stabilising UPOs, provided that the UPOs as reference signals are available for use. In practice, it is very difficult to obtain exact and analytic formulas for UPOs, except the degenerate case of unstable equilibria, and is extremely difficult (if not impossible) to implement UPOs by physical means such as circuitry due to the unstable nature of such orbits. There is a time-delayed feedback control (TDFC) method in classical control theory [12], which receives a renewal of great interest spurred by Pyragas' paper [13] for stabilisation of UPOs in chaotic systems. The novel idea in this methodology is to use the current as well as past system states for feedback, thereby avoiding a direct use of the target UPO in the controller. In this section, we discuss a SMC based TDFC method for chaos control.

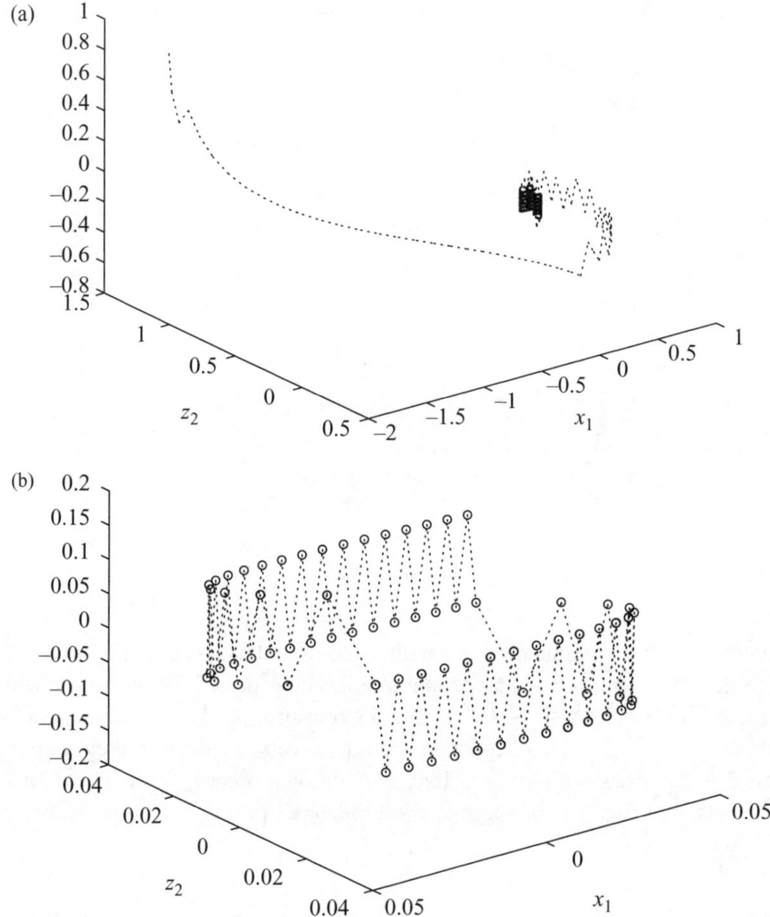

Figure 10.2   a) 3D trajectory; b) system states of the last 100 iterations

10.3.1   Time-delayed feedback control based on SMC

The main idea of the TDFC based SMC strategy is to perturb some parameters in the chaotic system to create a local attraction region (usually, a neighbourhood of the target UPO). A global control strategy is applied to let the chaotic system freely evolve until it enters the local region due to the ergodicity of chaotic dynamics, and then to engage a control so that the trajectory will stay in the region (hence the UPO) thereafter. SMC principles are ideal for this kind of control.

Chaos control does not need full information about the system states. Furthermore, stabilisation of chaotic systems does not require global stability. This is due to the boundedness of trajectories and the ergodicity of chaotic dynamics. To control a chaotic system, we may select one or more appropriate system parameters to manipulate chaos. Without loss of generality, we discuss a scalar control using

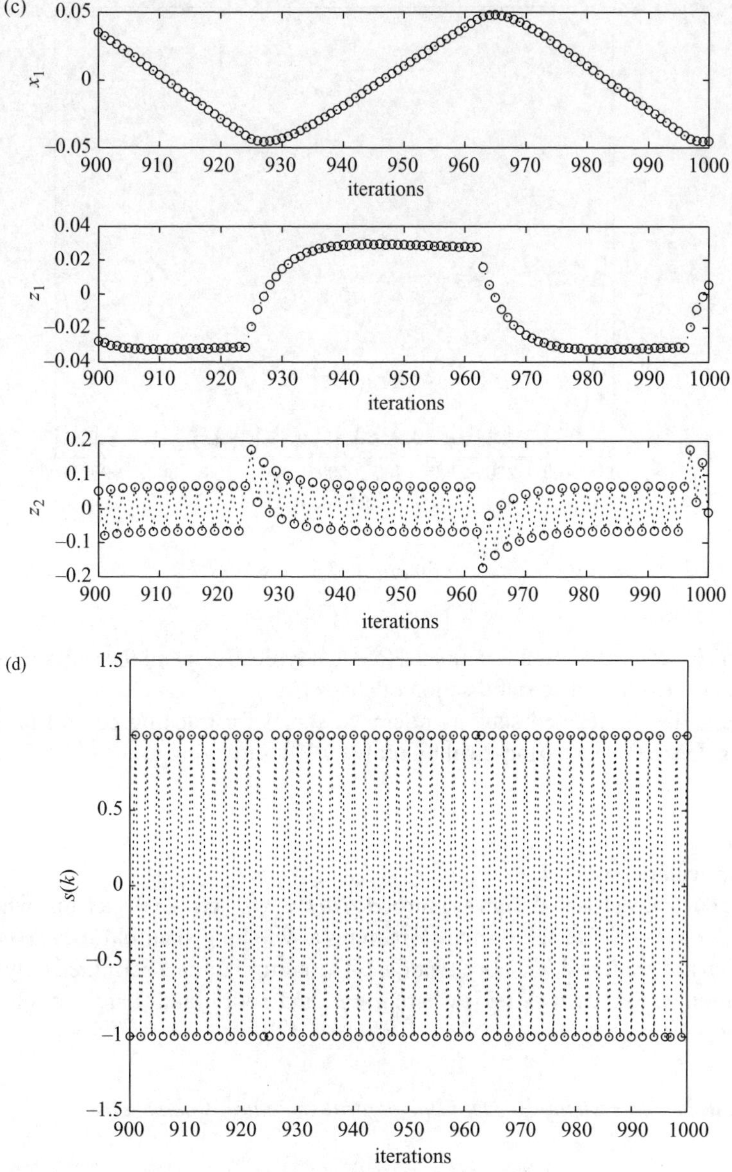

Figure 10.2   *Continued. c) System states of the last 50 iterations; d) symbolic sequence of the last 50 iterations*

one parameter (named as u for consistency) such that the controlled chaotic system becomes

$$\dot{x} = f(x) + b(x)u, \tag{10.31}$$

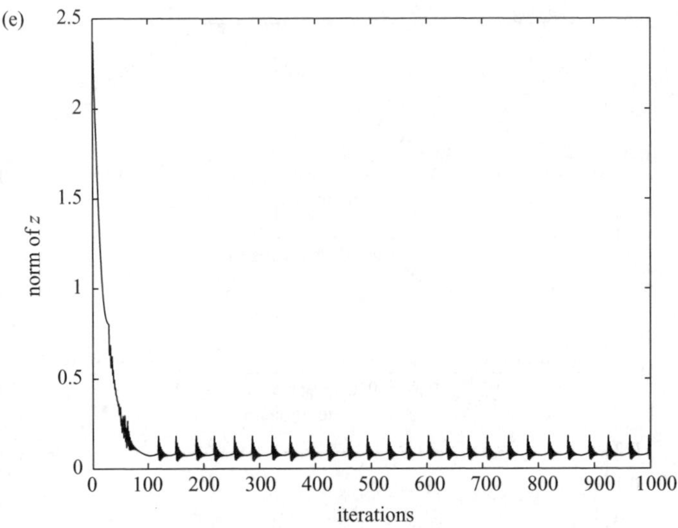

Figure 10.2   Continued. e) Norm function of z

where $f, b \in R^n$ are smooth functions. Note that typically not all the states need to be fully controlled to realise stabilisation of chaos [3].

TDFC uses a delayed state as reference signal for tracking control in chaotic systems. Let $\bar{x}(t)$ be a period-τ solution of (10.31), satisfying

$$\dot{\bar{x}}(t) = f(\bar{x}) + b(\bar{x})u, \quad \bar{x}(t) = \bar{x}(t - \tau). \tag{10.32}$$

for a particular constant u.

The control task is to design a control $u = u(t, \tau, x(t), x(t - \tau))$ such that when $t \to \infty$, $x(t) \to x(t - \tau)$. Designing a SMC requires a switching manifold $s(t, \tau, x(t), x(t - \tau))$ which has relative degree one (w.r.t. u). The perturbation is permitted only within a small range, that is, $u \in [\underline{u}, \bar{u}]$, where $\underline{u} < \bar{u}$, and both $\underline{u}$ and $\bar{u}$ are very close. We have the following result.

Theorem 3.   *In system (10.31), if the control structure is chosen as*

$$u(x) = \begin{cases} \bar{u} & s(t, \tau, x(t), x(t - \tau)) > 0 \\ \underline{u} & s(t, \tau, x(t), x(t - \tau)) < 0, \end{cases} \tag{10.33}$$

where $s(t, \tau, x(t), x(t - \tau))$ is an asymptotically stable switching manifold, then there always exists an attraction region defined by

$$\underline{u} < -\left(\frac{\partial s}{\partial x} b(x)\right)^{-1} \frac{\partial s}{\partial x} f(x) < \bar{u}. \tag{10.34}$$

Moreover, the controlled system orbits are always bounded outside the attraction region. Once the system trajectory enters this attraction region, the target UPO will be stabilised.

Proof.   It is well known from the SMC design principle (see Chapter 1) that to ensure sliding on the manifold $s(t, \tau, x(t), x(t - \tau)) = 0$, the following condition is needed within a neighbourhood of $s(t, \tau, x(t), x(t - \tau)) = 0$:

$$s\dot{s} < 0. \tag{10.35}$$

When sliding,

$$s(t, \tau, x(t), x(t - \tau)) = 0 \quad \text{and} \quad \dot{s}(t, \tau, x(t), x(t - \tau), u) = 0. \tag{10.36}$$

The equivalent control, denoted by $u_{eq}(t, \tau, x(t), x(t - \tau))$ and calculated by solving the following equation:

$$\dot{s} = \frac{\partial}{\partial x}\dot{x} = \langle \nabla s, f(x) + b(x)u_{eq} \rangle = 0, \tag{10.37}$$

satisfies (10.36). The attraction region can then be constructed as [14]

$$\underline{u} < u_{eq} < \bar{u}. \tag{10.38}$$

The boundedness of the chaotic dynamics is maintained, even outside the attraction region, thanks to the SMC mechanism. Since we have assumed $u \in [\underline{u}, \bar{u}]$, then small variation between $\bar{u}$ and $\underline{u}$ does not change the chaotic nature of the system (which may only change slightly the orientation of the dynamical chaos flows). For any chaotic system, there always exists a Lyapunov function $V(x, u)$ and a positive constant M such that

$$V(x, \underline{u}) < M \quad \text{and} \quad V(x, \bar{u}) < M.$$

When the sliding mode is not reached, the perturbation switches between $\underline{u}$ and $\bar{u}$, which does not change the boundedness nature of the chaotic system. Hence, $V(x, u) < M$. When the sliding mode is reached, that is, $\dot{s}(t, \tau, x(t), x(t - \tau)) = 0$, the equivalent control, u_{eq}, plays an equivalent role in guiding the system to remain in the sliding mode. This means $\underline{u} \leq u_{eq} \leq \bar{u}$. Hence, $V(x, u_{eq}) < M$. This implies that the SMC strategy does not cause instability even if the controlled system orbit is outside the attraction region.

Remark 3.   *With Theorem 3, one can design a switching manifold with relative degree one (w.r.t. u) for stabilisation of an UPO, that is, $x(t) \rightarrow x(t - \tau)$ for $t \rightarrow \infty$, for example, with a Hurwitz polynomial of $e(t) = x_i(t) - x_i(t - \tau)$ as the error state.*

Remark 4.   *One interesting phenomenon observed is that even if the time delay constant is slightly different from the period of the target UPO, the boundedness of the controlled trajectory is still maintained under the SMC. This demonstrates the robustness of the SMC based chaos control method. The effectiveness of TDFC based* *on SMC relies on the accuracy of the estimation of the delay constant τ, an important topic to be further addressed below.*

10.3.2   Estimation of the delay time τ

One common assumption about the TDFC strategies for chaos control is that the delay time τ has to be known a priori, in order to deliver an effective control performance. The acquisition of τ is equally difficult compared to acquiring the analytic solution of UPOs.

We now introduce an iterative algorithm based on the gradient descent approach for searching an accurate estimate of τ [15]. The performance index for search is defined as

$$E = \int_{t_0}^{T+t_0} \|(y(t) - y(t-\tau))\| \, dt, \tag{10.39}$$

where $y(t)$ represents either a full system state or some manifold of partial system states, and T is a large enough instant that can cover a sufficiently long length of time period for estimation. The adaptive search of τ is carried out as follows:

1.   Set a tolerance error ϵ and large T. Simulation starts from $t_0 = 0$. For a given initial condition, $\tau = \tau_0$, let the chaotic system run freely for a period of time, τ_0.
2.   Enable the control $u(t)$ and let the system run for a period of time T. Set $i = 1$ and let $\tau_1 = \tau_0$.
3.   Compute the adjustment to τ_i:

$$\tau_{i+1} = \tau_i - \beta \frac{\partial E}{\partial \tau_i}, \tag{10.40}$$

where

$$\frac{\partial E}{\partial \tau_i} = \int_{iT+t_0}^{(i+1)T+t_0} (y(t) - y(t-\tau_i))\dot{y}(t-\tau_i) \, dt \tag{10.41}$$

with β being a proper adaptation parameter. Set $i = i+1$.
4.   If

$$E > \epsilon,$$

go to Step 3; otherwise, stop.

10.3.3   An example

We now present a simulation on the Rössler system to show the effectiveness of the TDFC method. The Rössler system is described by

$$\begin{bmatrix} \dot{x}_1 \\ \dot{x}_2 \\ \dot{x}_3 \end{bmatrix} = \begin{bmatrix} 0 & -1 & -1 \\ 1 & a & 0 \\ x_3 & 0 & -c \end{bmatrix} \begin{bmatrix} x_1 \\ x_2 \\ x_3 \end{bmatrix} + \begin{bmatrix} 0 \\ 0 \\ b \end{bmatrix},$$

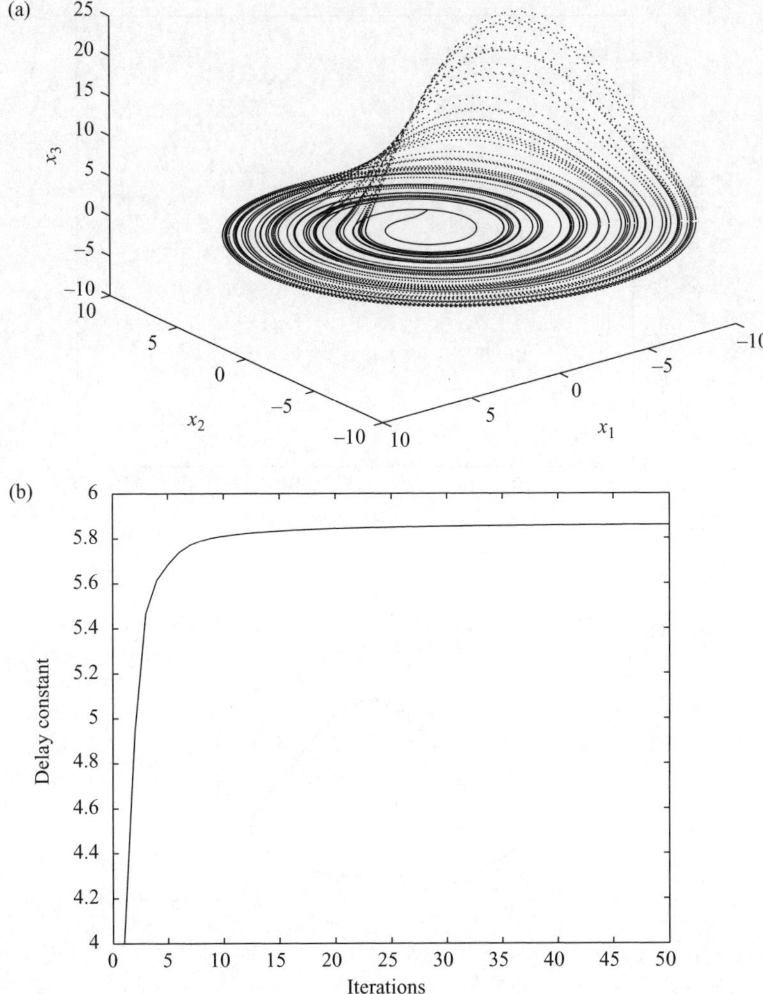

Figure 10.3   a) Rössler chaotic attractor; b) approximating the delay constant τ

where the parameters were taken as $a = b = 0.2$ and $c = 5.7$, respectively. The Rössler chaotic attractor is shown in Fig. 10.3(a).

We used the SMC-based TDFC method for chaos control by using the x_2-component to form the one-dimensional manifold, chosen as $s = x_2(t) - x_2(t - \tau)$. The control signal $u(t) = -k_0 \operatorname{sign}(x_2(t)s)$ was added to the second equation (x_2) of the Rössler system, as discussed above, forming a local attraction region. In this simulation, we set the initial condition as $x(0) = (0\ 0\ 0)^\top$, $\tau_0 = 4$, $\beta = 5$ and $k_0 = 0.2$.

Figure 10.3(b) shows the convergence of the time delay constant τ. Clearly, τ converges to 5.861, which is the period of the target UPO. Figure 10.3(c) depicts the convergence of function E, showing that the estimation error converges to zero.

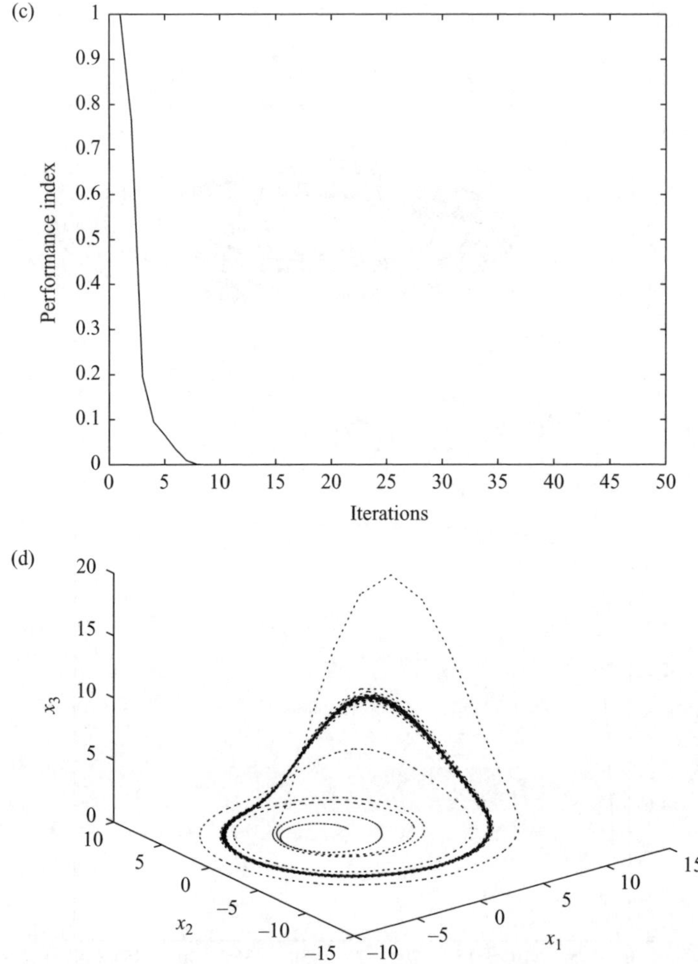

Figure 10.3   Continued. c) Approximation performance; d) controlled Rössler chaotic attractor

The system behaviour when $\tau = 5.860$ is shown in Fig. 10.3(d), in which the period-1 UPO is stabilised as required.

10.4   Generalising the OGY method using SMC

Among the existing chaos control methodologies, the model-free chaos control approach has attracted a great deal of attention due to the difficulty of obtaining a faithful model for a chaotic system from data in many real applications. The first model-free chaos control method was suggested by Ott, Grebogi and Yorke

[16], known as the OGY method. It has lately been extended, analysed, and applied widely [1, 17]. Essentially, this kind of control technique requires identification of the stable and unstable manifolds from available time-series data and, based on that, a suitable control action is developed to bring the system orbit to the stable manifold leading to the target. This type of control method, although with basic features of classical control, exploits some particular properties of chaos such as the ergodicity and structural stability of chaos. It is now known that the OGY-type of chaos control is effective only for controlling lower-dimensional chaotic systems because it utilises the prominent feature of saddle type of fixed points that have both stable and unstable manifolds. From an algebraic point of view, the difficulty lies in the situation where the system Jacobian at the target fixed point has complex eigenvalues or multiple unstable eigenvalues. Even with distinct real eigenvalues, the construction of stable and unstable manifolds for higher-dimensional chaotic systems is a technical challenge.

In this section, we use the SMC methodology to extend the OGY chaos control method, which not only preserves the spirit of the original method to direct the system orbit to designated stable manifold(s), but also can deal with higher-dimensional chaotic systems fairly easily.

10.4.1   *SMC-based OGY method for MIMO systems*

The essence of the multi-input and multi-output (MIMO) SMC-based OGY method involves two steps. First, it is necessary to construct suitable manifolds, independent of the system Jacobian eigenvalues and eigenvectors, which are selected to present a desired dynamics. Once these manifolds are made invariant, the desired dynamics, usually containing the desired fixed point corresponding to an ideal inherent periodic orbit to be stabilised, will take any state in the manifolds to the equilibrium asymptotically. Secondly, a switching control must be designed to force the system state to reach and lie on these selected invariant manifolds, so that whenever the system state lies on it, the system orbit will be guided to the designated fixed point that corresponds to the originally targeted unstable periodic orbit. It should be noted that the above procedure is a typical SMC type of approach.

The basics of the OGY control method is to restrict the orbital movement along the direction perpendicular to the tangent of the stable manifold, so that sooner or later the system orbit is confined to the stable manifold on which the subdynamics of the manifold will drive the orbit to the target fixed point. Such a stable manifold is invariant because in a neighbourhood of the manifold, the orbit will always be attracted to it under the control, which will stay within the manifold thereafter.

In the two-dimensional Poincaré section case, the stable and unstable manifolds are constructed by using the system Jacobian eigenvalues and eigenvectors as in the OGY method. For higher-dimensional chaotic systems, this construction becomes very difficult, if not impossible, because the system Jacobian may have complex eigenvalues, perhaps with multiplicity, so it is very difficult to determine the stable and unstable manifolds.

Here, we make use of the SMC concept to extend the OGY method to deal with the higher-dimensional case. First, we construct a set of invariant manifolds for the higher-dimensional chaotic system, independent of the system Jacobian eigenvalues and eigenvectors. This implies that the manifolds, on which the system orbit will eventually stay, can be prescribed independently of the system Jacobians. We then design an OGY-type of control so that in the neighbourhood of the invariant manifolds, the system orbit is guided to reach and then stay on a stable manifold. Note that the parameters of the switching manifolds can be chosen beforehand so that, if needed, the eigenvalues of the resulting dynamics can be assigned desirable values.

To present the approach formally, we assume that the locally linearised model is obtained as

$$x(k+1) = Ax(k) + Bu(k), \tag{10.42}$$

where $x(k) \in R^{n-1}$, and the system is controllable. Note that here the dimension of the discrete model is $n-1$. This represents the reduction of dimension by one due to the Poincaré mapping.

For system (10.42), we choose a set of manifolds, which can be independent of the system Jacobian, represented by

$$s(k) = s(x(k)) = Cx(k) = 0 \in R^m, \tag{10.43}$$

where $C \in R^{m \times (n-1)}$. Assuming that the $m \times m$ matrix CB is nonsingular. We want to find a control action, $u(k)$, such that

$$s(k+1) = Cx(k+1) = CAx(k) + CBu(k) = 0 \qquad \forall k \geq k_0,$$

so that $s(k) = 0$ defines the invariant manifolds, i.e., the system orbit will lie on the intersection of these manifolds. Note that the idea here is similar to the conventional discrete sliding mode control, which requires $\Delta s(k) = s(k+1) - s(k) = 0$, but here we require $s(k+1) = 0$.

When this is achieved, we have

$$s(k) = Cx(k) = [C_1 \ C_2] \begin{bmatrix} x_1(k) \\ x_2(k) \end{bmatrix}, \tag{10.44}$$

where $x_1 \in R^{n-1-m}$, $x_2 \in R^m$, and C_1 and C_2 are $m \times (n-1-m)$ and $m \times m$ matrices, respectively. From (10.44), we have

$$x_2(k) = -C_2^{-1} C_1 x_1(k). \tag{10.45}$$

This, although in the discrete-time form, appears to be similar to the continuous-time sliding mode setting.

The question now is how to guarantee that the dynamical system subject to the constraint $s(k) = 0$ is asymptotically stable toward the desired fixed point. This can be ensured by choosing suitable matrices C_1 and C_2 in the above formula [18]. Thus, once the system orbit reaches these stable manifolds, it will converge to the intersection point; subsequently, the controlled system orbit converges to the desired fixed point asymptotically, which corresponds to the desired periodic orbit of the given system.

It would be interesting to see the inherent dynamical and geometrical properties of the system when the controlled orbit stays in the invariant manifolds $s(k)=0$.

On the invariant manifolds, the equivalent control, denoted $u_{eq}(k)$, may be chosen as

$$u_{eq}(k) = -(CB)^{-1}CAx(k), \tag{10.46}$$

which is obtained by solving the manifold equation $g(k+1)=0$. The resulting dynamics on the manifolds is described by

$$x(k+1) = Ax(k) + Bu_{eq}(k) = (I - B(CB)^{-1}C)Ax(k). \tag{10.47}$$

It can be proved by using the same techniques for continuous-time sliding mode control [19] that such dynamics are invariant with respect to C. We can conclude that the chaotic system, when confined on the invariant manifold so constructed, will have m zero eigenvalues along with $n-1-m$ eigenvalues which are the transmission zeros of the 'equivalent system' [19]. These $n-1-m$ eigenvalues can be arbitrarily assigned by properly chosen C, for example, by choosing all stable eigenvalues for the purpose of constructing the desired invariant manifolds. For a single stable manifold, one can simply use the pole assignment technique; for multiple stable manifolds, one can use some well-known algorithms [20] (but assigning eigenvalues within a unit disk rather than to the left hand side of the complex plane).

10.4.2 An example

We now present a simulation study of chaos control in a fourth-order double-rotor map. The fourth-order discrete kicked double-rotor map is described [21]

$$\begin{bmatrix} Y(k+1) \\ Z(k+1) \end{bmatrix} = \begin{bmatrix} MZ(k)+Y(k) \\ LZ(k)+G(Y(k+1)) \end{bmatrix}, \tag{10.48}$$

where $Y = (y_1, y_2)^\top \in S^1 \times S^1$ is the angular position of the rods at the instant of the kth kick, $Z = (z_1, z_2)^\top$ is the angular velocity of the rods immediately after the kth kick, $G(Y(K+1)) = (q_1 \sin x_1(k+1),\ q_2 \sin x_2(k+1))^\top$, S^1 is a circle (mod 2π), $q_i = f_0 l_i / I$, f_0 is the constant strength of the period impulse kicks, and l_i is the length of the rods ($i=1,2$).

Using the same values given in Reference 21, the matrices are obtained as

$$M = \begin{bmatrix} 0.4860 & 0.2134 \\ 0.2134 & 0.6993 \end{bmatrix}, \quad L = \begin{bmatrix} 0.2414 & 0.2726 \\ 0.2726 & 0.5140 \end{bmatrix}.$$

To stabilise its unstable periodic orbit of interest, we perturbed the strength of the period impulse kicks f, such that $f = f_0 + \Delta f$ with $f_0 = 9$. The map (10.48) was linearised about the unstable periodic orbit x^*, as

$$x(k+1) = Ax(k) + B\Delta f(k), \tag{10.49}$$

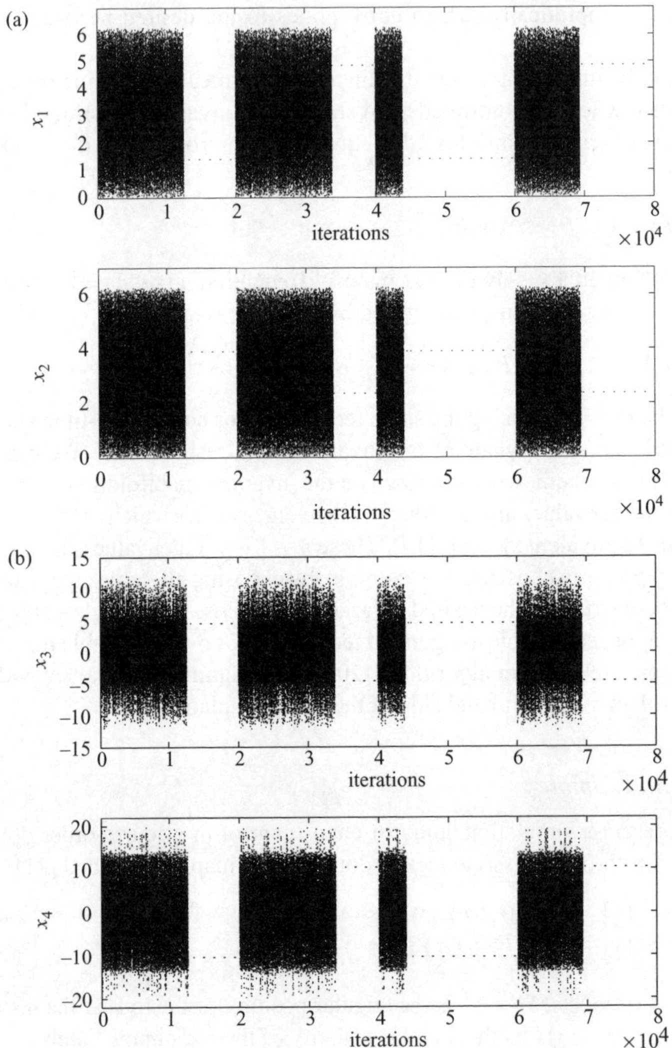

Figure 10.4   a) Time responses of x_1 and x_2; b) time responses of x_3 and x_4

where $x(k) = (X(k) - X^*\ Y(k) - Y^*)^\top$, $B = (0\ 0\ (l_1/I)\sin(x_1^*)\ (l_2/I)\sin(x_2^*))^\top$ and

$$A = \begin{bmatrix} I_2 & M \\ H(x^*) & L + H(x^*)M \end{bmatrix}, \quad H = \begin{bmatrix} q_1 \cos x_1^* & 0 \\ 0 & q_2 \cos x_2^* \end{bmatrix}.$$

We selected two different unstable period-one fixed points from the control targets [18], which are

$$x_a^* = (1.4113, 3.9144, 4.5547, -10.3743)^\top$$

and

$$x_b^* = (4.8719, 2.3688, -4.5547, 10.3743)^\top.$$

The control task was to force the system orbit to settle at these two fixed points alternatively. Using the algorithm in Reference 20, the following matrices were obtained:

$$C_a = (1.1171 \ -2.5878 \ 0.2284 \ -1.2040)$$

$$C_b = (-1.1171 \ 2.5878 \ -0.2284 \ 1.2040),$$

where C_a is for stabilising x_a^* and C_b for x_b^*. These two matrices give the four desired eigenvalues of $(I - B(CB)^{-1}C)A$, i.e., -0.15, 0, 0.11, and 0.08.

A controller capable of accomplishing the task was then constructed using (10.46), as

$$\Delta f_a = (-1.3480 \ -5.1702 \ -1.4851 \ -3.3466)^\top \Delta x$$
$$\Delta f_b = (1.3480 \ 5.1702 \ 1.4851 \ 3.3466)^\top \Delta x. \tag{10.50}$$

The eigenvalues of the manifold $h(k)$ are then computed, as 0.11, -0.15, and 0.08. The simulation result is shown in Fig. 10.4 where, as one can see, the angular positions and velocities of the two rods are settled at the two target fixed points alternatively, indicating the success of the control.

10.5   Conclusions

In this chapter, we have examined the relationship between SMC and chaos. We have shown that digitising SMC in practice may cause some micro-level 'chaotic' behaviours, such as different periodic behaviours due to different initial conditions, an aspect of sensitivity to initial conditions. An interesting correlation between the periodic trajectories and their symbolic sequences has been explored. We have also discussed two SMC-based chaos control methods: one is the TDFC control and the other is a generalised OGY method. Their effectiveness has also been verified by computer simulations.

10.6   References

1 CHEN, G. and DONG, X.: 'From chaos to order: methodologies, perspectives and applications' (World Scientific, Singapore, 1998)
2 CHEN, G. and YU, X.: 'Chaos control: theory and applications', *Lecture Notes in Control and Information Sciences*, 292 (Springer Verlag, Berlin, 2003)
3 YU, X.: 'Controlling Lorenz chaos', *International Journal of Systems Science*, 1995, **27**, pp. 355–361
4 YU, X.: 'Variable structure control approach for controlling chaos', *Chaos, Solitons and Fractals*, 1997, **8**(9), pp. 1577–1586

5   YU, X.: 'Tracking inherent periodic orbits using time delayed variable structure self-control', *IEEE Transactions on Circuits and Systems – Part I*, 1999, **46**(11), pp. 1408–1411

6   CHUA, L. O. and LIN, T.: 'Chaos in digital filters', *IEEE Transactions on Circuits and Systems – Part I*, 1998, **35**, pp. 648–652

7   YU, X. and CHEN, G.: 'Discretization behaviours of equivalent control based variable structure systems', *IEEE Transactions on Automatic Control*, 2003, **48**(9), pp. 1641–1646

8   YU, X.: 'Bifurcation and chaotic behaviours in the variable structure system with unlimited control magnitudes', *International Journal of Bifurcation and Chaos*, 1997, **7**(8), pp. 1897–1993

9   YU, X.: 'Discretization effect on a sliding mode control system with bang-bang type switching', *International Journal of Bifurcation and Chaos*, 1998, **8**(6), pp. 1245–1257

10   YU, X. and GALIAS, Z.: 'Periodic behaviors in a digital filter with two's complement arithmetic', *IEEE Transactions on Circuits and Systems – Part I*, 2001, **48**(10), pp. 1177–1190

11   YAMAMOTO, S. and USHIO, T.: 'Odd number limitation in delayed feedback control', Chaos Control: Theory and Applications, *Lecture Notes in Control and Information Sciences*, 292 (Springer Verlag, Berlin, 2003), pp. 71–88

12   BELLMAN, R. and COOKE, K. L.: 'Differential-difference equations' (Academic Press, New York, 1963)

13   PYRAGAS, K.: 'Continuous control of chaos by self-controlling feedback', *Physical Letters A*, 1992, **170**, pp. 421–428

14   SIRA-RAMÍREZ, H.: 'A relative degree approach for the control in sliding mode of nonlinear systems of general type,' Proceedings of International Workshop on *Variable Structure Systems and Their Applications*, 1990, Sarajevo, p. 29

15   CHEN, G. and YU, X.: 'On time delayed feedback control of chaos', *IEEE Transactions on Circuits and Systems – Part I*, 1999, **46**(6), pp. 767–772

16   OTT, E., GREBOGI, C., and YORKE, J. A.: 'Controlling chaos', *Physical Review Letters*, 1990, **64**(11), pp. 1196–1199

17   DITTO, W. L., SPANO, M. L., and LINDNER, J. F.: 'Techniques for the control of chaos', *Physica D*, 1995, **86**, pp. 198–211

18   YU, X., CHEN, G., XIA, Y., SONG, Y., and CAO, Z.: 'An invariant manifold approach for controlling higher order chaotic systems', *IEEE Transactions on Circuits and Systems – Part I*, 2001, **48**(8), pp. 930–937

19   EL-GHEZAWI, O. M. E., ZINOBER, A. S. I., and BILLINGS, S. A.: 'Analysis and design of variable structure systems using a geometric approach', *International Journal of Control*, 1983, **38**, pp. 657–671

20   HARVEY, C. A. and STEIN, G.: 'Quadratic weights for asymptotic regulator properties', *IEEE Transactions on Automatic Control*, 1978, **23**(3), pp. 378–387

21   GREBOGI, C., KOSTELICH, E., OTT, E., and YORKE, J. A.: 'Multidimensional intertwined basin boundaries: basin structure of the kicked double rotor,' *Physica D*, 1987, **25**, pp. 347–360

Part III

Applications of sliding mode control

Chapter 11

Sliding modes in fuzzy and neural network systems

Kemalettin Erbatur, Yildiray Yildiz and Asif Sabanovic

11.1 Introduction

It is a well known fact that sliding mode control (SMC) is a powerful control methodology for both linear and nonlinear systems because of its robustness to parameter changes, external disturbances and unmodelled dynamics. Besides its power, the design of sliding mode controllers needs the information of the system's state, which makes the design relatively austere in some applications where the mathematical modelling of the system is very hard and where the system has a large range of parameter variations together with unexpected and sudden external disturbances. For those applications, a controller that will provide predicted performance even if the model of the system is not very well known, is needed. That controller should also adapt itself to large parameter variations and to unexpected external disturbances. These types of controllers are generally called 'intelligent' controllers, mainly working on the principles of fuzzy logic, neural networks, genetic algorithms and other technologies derived from artificial intelligence. The idea of combining these intelligent control structures with the power of sliding mode control approach has attracted much research. A recent survey on the combination of SMC and intelligent control can be found in Reference 1. In this chapter, the union of sliding mode with neural networks and fuzzy logic is examined with examples from literature, and then a new technique combining neural networks and sliding mode control is presented.

11.2 Sliding mode control and intelligence

In this section, a brief review of sliding mode control with discontinuous control is given and in the following sections, the merging of the SMC method with computational intelligence is presented.

11.2.1   Sliding mode control design

Consider the n-dimensional dynamical system linear with respect to the control.

$$\frac{dx}{dt} = f(x,t) + B(x,t)u + d(x,t)$$

$$x \in R^n, \quad u \in R^m, \quad B = \bar{B} + \Delta B(x,t), \quad \text{rank}(B) = m \tag{11.1}$$

Here d is an external disturbance, ΔB represents an input matrix uncertainty and $\bar{B}$ is an estimated input matrix of the actual B.

The goal is to find a control u such that the motion of the system (11.1) is restricted to the manifold S in the state space.

$$S(x,t) = \{x : \sigma(x,t) = G(x^d(t) - x(t)) = \phi(t) - S_a(x) = 0\} \tag{11.2}$$

where $\sigma(x,t) \in R^m$ is the sliding variable and $S_a(x) = Gx(t)$. x^d represents the desired state vector and G is an $(m \times n)$ matrix. $\phi(t)$ denotes the time dependent part of the sliding variable.

The control should be selected such that a candidate Lyapunov function satisfies Lyapunov stability criteria. The control that will ensure existence of a sliding mode in the manifold $\sigma(x,t) = 0$ can be determined as,

$$u(t) = u_{eq}(t) + u_c(t) \tag{11.3}$$

where $u_{eq}(t)$ is the equivalent control given by

$$u_{eq}(t) = -(GB)^{-1}\left(Gf(x,t) + Gd(x,t) - \frac{d\phi(t)}{dt}\right) \tag{11.4}$$

and $u_c(t)$ is the corrective control term computed as

$$u_c(t) = (GB)^{-1} D\,\text{sign}(\sigma) \tag{11.5}$$

If $u_c(t)$ is selected in the form

$$u_c(t) = (GB)^{-1} D\sigma \tag{11.6}$$

the solution $\sigma(x,t) = 0$ is asymptotically stable but the control action is smooth. After reaching the manifold S, the control signal (11.3) will be equal to the equivalent control signal (11.4).

11.2.2   Intelligence in action

In practical applications, sliding mode control suffers from some difficulties. First, there is the chattering problem, which can be described by high frequency oscillations of the controller output. Second, sliding mode control is sensitive to measurement noise since the control depends on the sign of a measured variable, $\sigma(x,t)$, which is very close to zero [2]. Finally, calculation of the equivalent control is problematic because it needs the exact model.

The concept of intelligence is used to refer to self-adaptation. Mainly, self-adaptation can be achieved using neural network and fuzzy logic technologies.

The merging of the intelligent control schemes with the sliding mode control would be a candidate to alleviate at least some of the difficulties in the application of sliding mode methods.

Neural networks have many definitions including 'computer representations of the mammalian brain' or 'simulation of central neural nervous systems' or 'software or hardware that can learn on the basis of the functioning of the human brain', etc. The definition that best fits control purposes may be the following. A neural network is a structure that consists of scalar elements called perceptrons which perform a nonlinear transformation $R \to R$. They can also be seen as 'universal approximators' [3–5]. Figure 11.1 shows a typical feedforward neural network.

With sliding mode control, neural networks have two main areas of use. In one of them, neural networks are used in the feedback or feedforward loop, functioning parallel to a sliding mode controller or calculating the equivalent control. In the other case, they are used for the online adaptation of the sliding mode parameters.

In the literature, one of the earliest works [6] concerning the calculation of the equivalent control appeared in 1995. In that paper, a design method using an online NN estimator to estimate part of the equivalent control – containing the system's nonlinear part f in (11.1), input uncertainty ΔBu and external disturbance d – as one linearly combined nonlinear function, is presented. To achieve this goal, first, the linearly combined nonlinear function $f(x,t) + d(x,t)$ is replaced by a mapping of a three layers neural network N, and second, the actual input matrix B is replaced by the estimated one, $\bar{B}$. Then, the control signal (11.3) has the form below:

$$u = -(G\bar{B})^{-1}\left(GN(x,t) - \frac{d\phi}{dt}\right) - (G\bar{B})^{-1}D\sigma \tag{11.7}$$

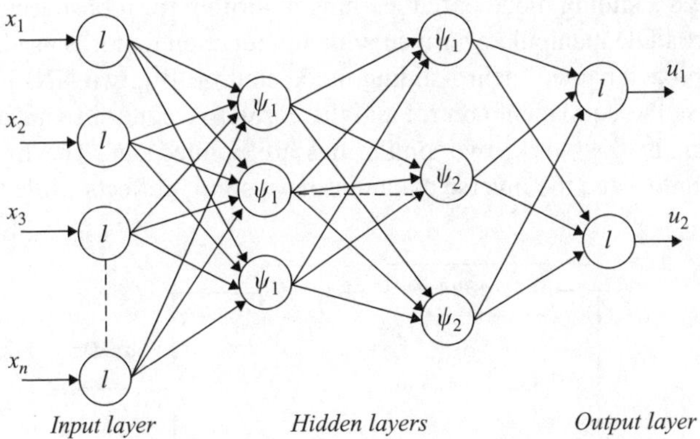

Figure 11.1   *A typical feedforward neural network.* ψ_is *represent nonlinear activation functions, l represents linear activation functions and arrows represent network weights*

Using (11.7) as a control input, the time derivative of Lyapunov function $V = \frac{1}{2}\sigma^T\sigma$ along the system trajectories is calculated as

$$\frac{dV}{dt} = \sigma^T \frac{d\sigma}{dt}$$

$$= \sigma^T(G(f(x,t) + \Delta B(x,t)u + d(x,t) - N(x,t))) - \sigma^T D\sigma \qquad (11.8)$$

From the equation above, suppose that the neural network can be trained to satisfy the following condition:

$$|\sigma^T G(f(x,t) + \Delta B(x,t)u + d(x,t) - N(x,t))| < |\sigma^T D\sigma| \qquad (11.9)$$

Then $dV/dt < 0$ holds and consequently the convergence of σ to zero is assured.

With the control input in (11.7), the estimation error, denoted by J, can be calculated from (11.1) and (11.2) as follows:

$$J = G(f(x,t) + \Delta B(x,t)u + d(x,t) - N(x,t)) = D\sigma + \dot{\sigma} \qquad (11.10)$$

From (11.9) and the estimation error obtained by (11.10), the neural network is to be trained to minimise the function below:

$$E = \frac{(\dot{\sigma} + D\sigma)^2}{2} \qquad (11.11)$$

When (11.11) tends to zero using the online estimator realised by the neural network, the mapping $GN \to G(f + \Delta Bu + d)$ is accomplished and consequently the stability condition is satisfied. The structure of the neural network estimator is shown in Fig. 11.2.

In Reference 7 an online NN estimator to control a 3 DOF PUMA type direct drive (DD) robot system is used. To avoid the chattering effect the equivalent control is estimated and this estimate is used in the sliding mode control algorithm. The estimation of the equivalent control was carried out using an online NN estimator. In Reference 8 a sliding mode based learning algorithm for robust accurate tracking of a single axis DD manipulator driven with an induction motor is used.

In an approach named 'neuro-sliding mode control' [9], two NNs in parallel are used to realise the equivalent control and the corrective control terms of the sliding mode control. In this work, two similarities are pointed out. The first is that the equivalent control and the inverse dynamics have similar effects while the system is

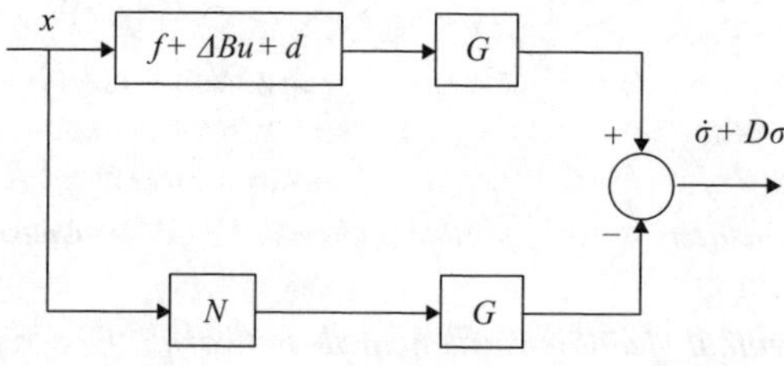

Figure 11.2   Online estimator realised by NN [4]

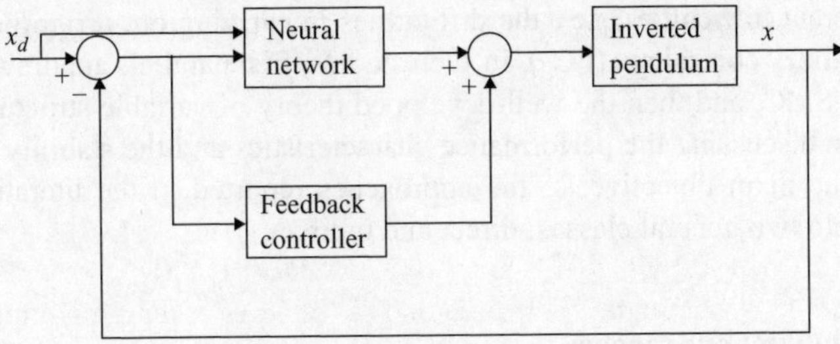

Figure 11.3   System structure proposed by Reference 10

in the sliding mode. The second similarity is between the corrective term of the sliding mode control and a proposed neuro-control structure. Based on the first, a two-layer feedforward NN is proposed to compute the equivalent control and the weights are adapted to minimise the square of the corrective term. This adaptation is based on the fact that if the NN learns the equivalent control, the corrective control term goes to zero when the system is on the sliding surface and any difference between the equivalent control and the NN output is reflected as a nonzero corrective control.

Calculating the equivalent control by means of a NN has been also presented [10]. The architecture of the overall control system is shown in Fig. 11.3. In this approach, a variable structure type feedback controller [11] is used in parallel with a NN controller for the control of an inverted pendulum. A gradient-descent learning algorithm is used for weight updates. The main goal of the algorithm is to minimise the function $J = (\dot{\sigma} + D\sigma)^2/2$, which results from the Lyapunov design. The algorithm is verified experimentally.

Also, an approach proposed by Ramirez and Morles [12] for robust adaptive learning in analogue adaptive linear elements (ADALINES) and a new online learning algorithm based on sliding mode control [13] can be found in the literature. Moreover, studies utilising variable structure system (VSS) theory in the training of computationally intelligent structures [14–16] and the use of NNs for computation of the dynamics and inverse dynamics [17] are reported.

Fuzzy Logic (FL) is mainly concerned with imprecision and approximate reasoning while NNs are mainly associated with learning and curve fitting. Generally these two approaches are not competitive but complementary and there is much to be gained by using them in a combined manner. For example, integration of fuzzy logic and neuro-computing has become very popular with many diverse applications, ranging from chemical process control to customer goods. Simply, a NN accomplishes what a person does with data and fuzzy logic realises what a person does with language. One of the important contributions on the integration of fuzzy logic and control can be found in Reference 18. In the following paragraphs, the union of fuzzy logic and VSS systems is described.

The integration of a FL system with SMC is seen in many examples where an attempt to relieve the implementation difficulties of the SMC is made via the addition of the FL system. On the other hand, some significant research work has originated due to different difficulties, i.e., the difficulties in carrying out a rigorous stability analysis of fuzzy controllers (FCs). In such studies, first parallels are drawn between the FC and SMC, and then the well-developed theory of variable structure systems is utilised in discussing the performance characteristics and the stability of the FC. Whatever the main objective is, the approaches reported in the literature can be separated into two general classes: direct and indirect.

11.2.2.1   Indirect approaches

In indirect approaches, the basic design and implementation philosophy of SMC is followed to a great extent and FL systems are used to fulfil a secondary function. It may be there either in order to adapt the controller parameters or to eliminate chattering problems and the consequent difficulty in the calculation of the equivalent control u_{eq}.

FL in a smoothing filter. One of the earliest works seen in the literature on the integration of VSS theory with FCs proposes smoothing the control input in a VSS and, thus, prevent chattering by the use of a low pass filter based on a fuzzy set of rules [19]

The filter equation was given as $\dot{u}_f = \lambda(u_f - u)$ where u_f is the filtered control, u is the unfiltered control and finally λ is the bandwidth of the filter.

To prevent abrupt changes in u, λ should be small. On the other hand, when λ is too small, u and u_f might be very different from each other, resulting in a deviation from the ideal sliding mode. To provide appropriate performance, λ should be small when the state is in the vicinity of the sliding surface and it can be made large to obtain the advantages of VSS with a sliding mode. The algorithm needed to accomplish this procedure can be obtained using a fuzzy set of rules [19].

SMC parameter tuning and FL. Consider the SMC as defined by (11.3)–(11.6). The parameter G determines the slope of the sliding line and therefore the larger it is, the faster will be the system response. However, a too large value for G can cause overshoot, or even instability. It would therefore be advantageous to adaptively vary the slope in such a way that the slope is increased as the magnitude of the error gets smaller.

Another adaptation rule can be found for the parameter D in (11.6) as follows. When the parameter D is large, the system states reach the sliding line in a short time, yet overshoot it by a considerable amount. When D is small, system response becomes sluggish. Neither of these two cases is desirable. An optimal value can be found using a fuzzy adaptation algorithm in which the parameter D is increased only when the states are close to the sliding line. Such an approach was followed for the trajectory control of robotic manipulators in Reference 20.

FL for modelling uncertainties. One of the main difficulties in the design of a SMC is the fact that exact knowledge of the plant is rarely available. Even the bounds of the uncertainties may not be known. This may result in an over conservative design. To solve this problem, a number of researchers have proposed the use of adaptive FL identifiers for the uncertainties. For example, a fuzzy system architecture was employed to adaptively model the plant nonlinearities that have unknown uncertainties [21].

In the proposed scheme, the bound of the modelling error, which results from the error between the fuzzy system and the actual nonlinear plant (an inverted pendulum system), is identified adaptively. Using this bound, a sliding control input is calculated.

The approach proposed in Reference 22 is similar. A nonlinear system is first linearised around a number of operating points and then FL principles are used to aggregate each locally linearised model into a global model representing the nonlinear system. Finally, a robust SMC is proposed that guarantees the asymptotic stability of the system.

11.2.2.2   Direct approaches

A vast amount of work is seen in the literature in which a computational intelligence methodology is used directly in the design of a VSS theory based scheme (generally, for control purposes) or, conversely, the VSS theory is used in the computationally intelligent architecture for parameter adaptation or for a robust and stable design. In this chapter, these schemes are classified as direct schemes. Some representative works from the many papers in the literature are outlined below.

Reference 23 draws parallels between FC and SMC by considering a single input, single output fuzzy controller with a set of rules as given below:

$$R^{(1)} : \text{IF } x = \text{NB THEN } y = \text{BIGGER}$$
$$R^{(2)} : \text{IF } x = \text{NS THEN } y = \text{BIG}$$
$$R^{(3)} : \text{IF } x = \text{Z THEN } y = \text{MEDIUM}$$
$$R^{(4)} : \text{IF } x = \text{PS THEN } y = \text{SMALL}$$
$$R^{(5)} : \text{IF } x = \text{PB THEN } y = \text{SMALLER}$$

$$(11.12)$$

In (11.12) x is the input, y is the output and NB, NS, Z, PS and PB are the labels of fuzzy sets, which are *negative big, negative small, zero, positive small* and *positive big* respectively. Let the universe of discourses of x and y be partitioned as shown in Figs 11.4 and 11.5, respectively.

The fuzzy inference performs a mapping from the fuzzy sets in X to the fuzzy sets in Y, based on the rule base and compositional rule of inference for fuzzy reasoning. According to the sup-min compositional rule of inference, a fuzzy set F'_y is generated by the rule base and the *centre-of-area* principle is used for defuzzification. Under these conditions, the input-output relation of the fuzzy system can be described by the curve [23] as shown in Fig. 11.6. The shape of this curve is very much like the

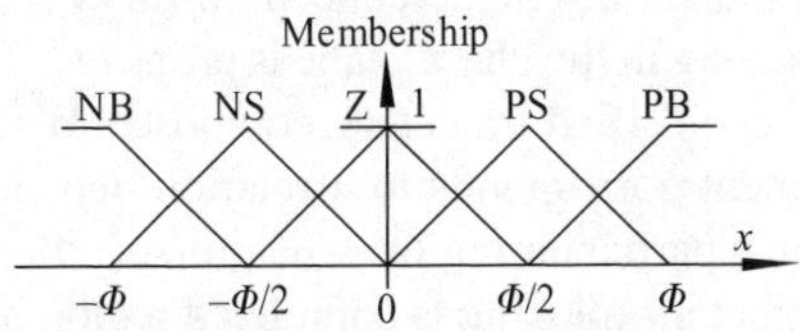

Figure 11.4   Universe of discourse for x

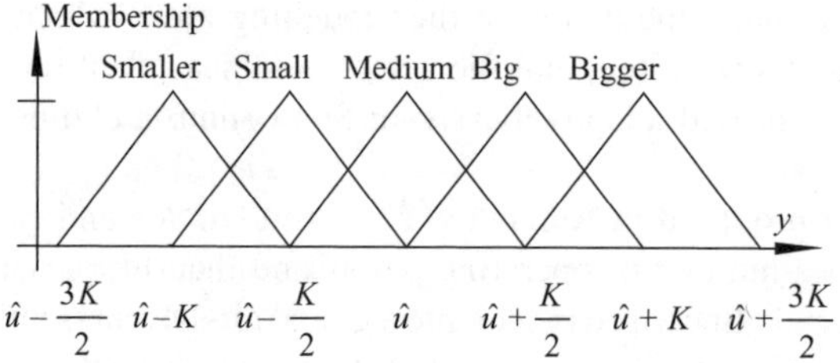

Figure 11.5   *Universe of discourse for y*

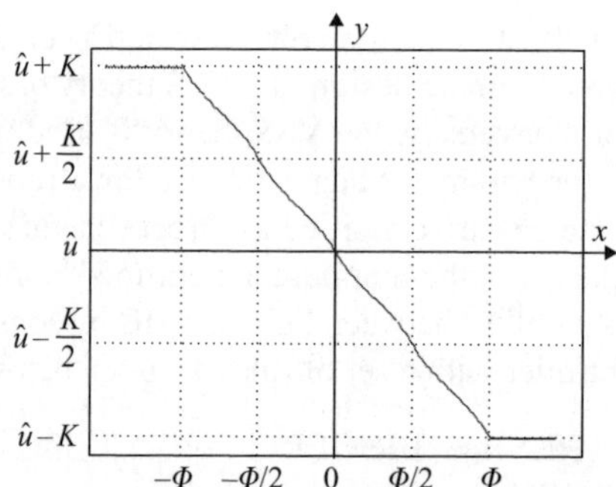

Figure 11.6   *Nonlinear operating line in a fuzzy SMC resembling the saturation function used in conventional SMC*

saturation function used with SMC systems. The fuzzy sliding mode control (FSMC) is therefore very much like SMC and theorems that are developed for the latter have corresponding ones in FSMC. This makes the performance and the stability analysis of FC possible.

The similarity of sliding mode systems and FCs has also been addressed [24], where a fuzzy compensator scheme for stick-slip friction is developed considering the effects of the fuzzy rules in the phase plane. The phase plane is divided into regions and this partitioning results in a switching line passing through the origin. Eight rules are used in the controller corresponding to the eight regions. Using the tuning parameters, the slope of this switching line is adjusted to obtain the desired dynamics.

The shape of the sliding surface can be used as a guide to design fuzzy controllers. In Reference 25, using the sliding mode concept, a fuzzy logic control for a linear system for trajectory tracking in the phase plane is proposed. The system's state $(e, \dot{e})$ is forced to track the pre-specified trajectory composed of several segments in the phase plane. Each segment corresponds to a relation between the tracking error e and the error change $\dot{e}$ in a particular region. The pre-specified trajectory is regarded as a sliding surface. Trajectory tracking is completed region by region. A systematic design procedure is formulated that makes trial and error unnecessary. The example of the cutting motion of an x–y table is given to demonstrate the merits of the controller.

11.3   A sliding mode neuro-controller

In addition to the control systems reviewed in Section 11.2, in this section, a new approach [26] based on the combination of Sliding Mode Control and Neural Networks is presented in detail.

Assume a SISO, nonlinear dynamical system that has the following dynamics:

$$\frac{dx}{dt} = f(x,t) + B(x,t)u + d(x,t) + f(z^r, \dot{z}^r) \tag{11.13}$$

Here $f(x,t)$ represents the unknown dynamics, $B(x,t)$ is the actual input matrix, $d(x,t)$ is the external disturbance and z^r is the reference input. The system state x is constructed as

$$x = \left[e \quad \frac{de}{dt} \right]^T \tag{11.14}$$

where e is the difference between the actual output and the reference input ($e =$ actual output-reference input). It is desired to find a control input u, such that the system will be stable and robust to parameter changes and to the variations in the disturbance.

The manifold for the states is defined as

$$S = \{x : \sigma(x) = Gx = 0\} \tag{11.15}$$

where $G = [C \quad 1]$ (C is a positive constant).

A Lyapunov function candidate is selected as

$$V = \frac{\sigma^2}{2} \tag{11.16}$$

By selecting the control input that will force the derivative of (11.16) to have the predefined structure

$$\dot{V} = -D\sigma^2 \tag{11.17}$$

where D is a positive constant, the asymptotic stability of the solution $\sigma(x) = Gx = 0$ will be enforced.

Now, the goal is to build an NN control that will provide the control input that may fulfil the above requirement. The structure of the controller is given in Fig. 11.7.

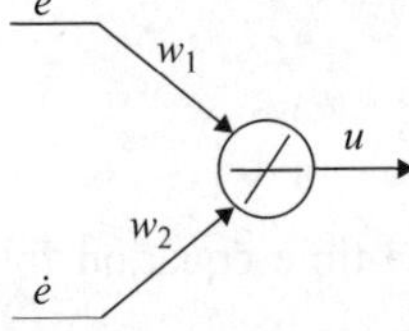

Figure 11.7   The structure of the controller

This structure is ADALINE, one of the simplest types of neural network structures. There is an input layer and only one output node, which has a linear activation function. The resulting control input u is given below:

$$u = W^T x \tag{11.18}$$

Here, $W = [w_1 \quad w_2]^T$ and $x = [e \quad \dot{e}]^T$.

Taking the derivative of (11.16) and inserting it into (11.17) it can be seen that stability can be obtained by satisfying $\dot{\sigma} + D\sigma = 0$. To achieve this, least square minimisation by introducing the square of the function '$\dot{\sigma} + D\sigma$' as the error function E for the NN controller training is employed. Namely, the NN is trained by minimising the error function defined by

$$E = \frac{(\dot{\sigma} + D\sigma)^2}{2} \tag{11.19}$$

11.3.1   Finding the weight updates

To find the weight updates the sensitivity (dE/dw_i) for weights should be found. The weights are updated as follows:

$$\dot{w}_i = -\eta \cdot \frac{dE}{dw_i} \tag{11.20}$$

Using the chain rule, the following equation can be written:

$$\frac{dE}{dw_i} = \left(\frac{dE}{du}\right)\left(\frac{du}{dw_i}\right). \tag{11.21}$$

From (11.18) and (11.19)

$$\frac{dE}{dw_i} = \frac{1}{2}\left(\frac{d(\dot{\sigma} + D\sigma)^2}{du}\right)x_i$$

$$= (\dot{\sigma} + D\sigma)\left(\frac{d(\dot{\sigma} + D\sigma)}{du}\right)x_i \tag{11.22}$$

Hence,

$$\frac{dE}{dw_i} = (\dot{\sigma} + D\sigma)\left(\frac{d(G\dot{x}_i + DGx_i)}{du}\right)x_i \tag{11.23}$$

and

$$\frac{dE}{dw_i} = (\dot{\sigma} + D\sigma)GB(x,t)x_i \tag{11.24}$$

Then, rewritten as a discrete-time equation for computer implementation, the weight update rule is

$$w_i^{new} = w_i^{old} - \eta(\dot{\sigma} + D\sigma)GB(x,t)x_i \tag{11.25}$$

Note that, if the system is linear, the equation reduces to

$$w_i^{new} = w_i^{old} - \eta(\dot{\sigma} + D\sigma)Gbx_i \tag{11.26}$$

Since ηGb is a constant, η, G and b can altogether be included in a constant $\bar{\eta}$ defined by $\bar{\eta} = \eta Gb$ and (11.26) can be reduced to the following equation:

$$w_i^{new} = w_i^{old} - \bar{\eta}(\dot{\sigma} + D\sigma)x_i \tag{11.27}$$

For a nonlinear system with constant input matrix B, the same result, (11.27), can be obtained.

11.3.2   Disturbance rejection

The error function (11.19) can be examined in more detail as follows:

$$\dot{\sigma} + D\sigma = G\dot{x} + DGx$$
$$= Gf(x,t) + GB(x,t)W^T x + Gd(x,t) + Gf(z^r, \dot{z}^r) + DGx \tag{11.28}$$

Taking the common terms into parenthesis, the following expression can be obtained:

$$\dot{\sigma} + D\sigma = Gf(x,t) + (GB(x,t)W^T + DG)x + Gd(x,t) + Gf(z^r, \dot{z}^r) \tag{11.29}$$

It can be seen from (11.29) that, when the error x goes to zero, the weights w_i have to go to infinity to compensate for the external disturbance. To avoid this, it is logical to add another term to the controller structure that will deal with the disturbance compensation. This additional term copes with the disturbance without being multiplied with the state x. The proposed controller is shown in Fig. 11.8.

With this controller structure, the expression for the control input u is

$$u = W^T x + 1w_3 \tag{11.30}$$

Using the same weight update procedure as described above, w_3 can be updated as

$$w_3^{new} = w_3^{old} - \bar{\eta}(\dot{\sigma} + D\sigma)GB(x,t) \tag{11.31}$$

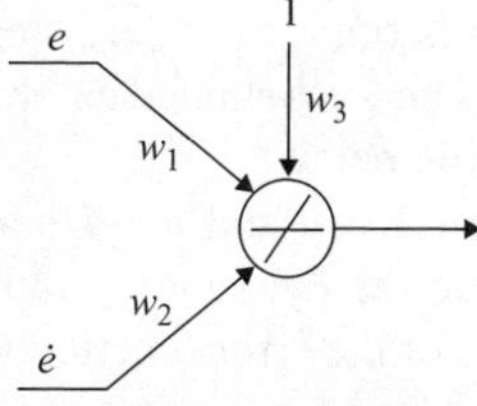

Figure 11.8   The improved controller structure

11.3.3   Stability and robustness analysis

It is well known that one of the most significant problems associated with the back-propagation algorithm is sticking to the local minima. In this part of the chapter it is shown that the proposed algorithm does not have this problem.

11.3.3.1   The error surface

It can be shown by elementary calculus that, if a function's second derivative does not change sign with respect to one of its arguments, then the function at hand does not have more than one minimum over that variable. Now, if the second derivative of the error function (11.19) is examined w.r.t. its variables (weights), the following equations hold.

First, substituting (11.14) into (11.24) and writing the result as two scalar equations we obtain:

$$\frac{dE}{dw_1} = (\dot{\sigma} + D\sigma)GB(x,t)e \tag{11.32}$$

$$\frac{dE}{dw_2} = (\dot{\sigma} + D\sigma)GB(x,t)\dot{e} \tag{11.33}$$

Taking one more derivative w.r.t. the weights the following expressions are found.

$$\frac{d^2E}{dw_1^2} = (GB(x,t)e)^2 \tag{11.34}$$

$$\frac{d^2E}{dw_2^2} = (GB(x,t)\dot{e})^2 \tag{11.35}$$

Furthermore, the second derivative of the error function w.r.t w_3 is

$$\frac{d^2E}{dw_3^2} = (GB(x,t))^2 \tag{11.36}$$

From (11.34), (11.35) and (11.36) it is seen that the curvature of the error surface through each weight variable is always positive. This tells us that the error surface does not have a local minimum. In addition, for proper values of the weight variables it is possible to reach the $E=0$ point. Hence, using the backpropagation algorithm with a proper learning rate, it is guaranteed that E converges to zero, without sticking to local minimums. In other words, for a bounded disturbance and unknown dynamics, it is guaranteed that the system is stable with zero steady state error and is robust to any bounded parameter changes and to bounded external disturbances. A proof of the convergence of E to zero is given below.

From (11.19), it is seen that $E > 0$ and $\dot{\sigma} + D\sigma = 0$ when $E = 0$. Therefore, to complete the proof, according to the Lyapunov stability criteria, what remains is to show that $\dot{E} < 0$. The derivative of the error function with respect to time is given by

$$\frac{dE}{dt} = \frac{\partial E}{\partial w_1}\frac{\partial w_1}{\partial t} + \frac{\partial E}{\partial w_2}\frac{\partial w_2}{\partial t} + \frac{\partial E}{\partial w_3}\frac{\partial w_3}{\partial t} \tag{11.37}$$

The expression for continuous changes of the weights is

$$\frac{dw_i}{dt} = -\eta \frac{dE}{dw_i} \tag{11.38}$$

Substituting (11.38) into (11.37)

$$\frac{dE}{dt} = -\eta \left(\frac{\partial E}{\partial w_1}\right)^2 - \eta \left(\frac{\partial E}{\partial w_2}\right)^2 - \eta \left(\frac{\partial E}{\partial w_3}\right)^2 \tag{11.39}$$

Note that (11.39) is a negative definite function, which completes the proof.

11.3.4 Simulation results

For the verification of the results obtained above, simulation studies are carried out with the model of a DC motor used to drive a toothed belt servo system located in the mechatronics laboratory of Sabanci University. Figure 11.9 shows a simplified model of this system.

The governing system equations of this system are as follows:

$$\frac{d\theta}{dt} = \omega \tag{11.40}$$

$$\frac{d\omega}{dt} = \frac{T}{J} - \frac{T_L(\omega, \theta)}{J} - \frac{r}{JG}(F_B(\theta, x) + F_D(\omega, v)) \tag{11.41}$$

$$\frac{dx}{dt} = v \tag{11.42}$$

$$\frac{dv}{dt} = \frac{F_B(\theta, x) + F_D(\omega, v)}{m} - \frac{F_L(x, v)}{m} \tag{11.43}$$

$$F_{Belt} = F_B(\theta, x) + F_D(\omega, v) \tag{11.44}$$

The variables in these equations are defined as follows.

θ	The angular position of the servomotor's shaft.
ω	The angular velocity of the servomotor's shaft.
x	The longitudinal position of the load.
v	The longitudinal velocity of the load.
$T = K_T i$	The torque developed by the servomotor.
$T_L(\omega, \theta)$	The friction torque at the servomotor side.

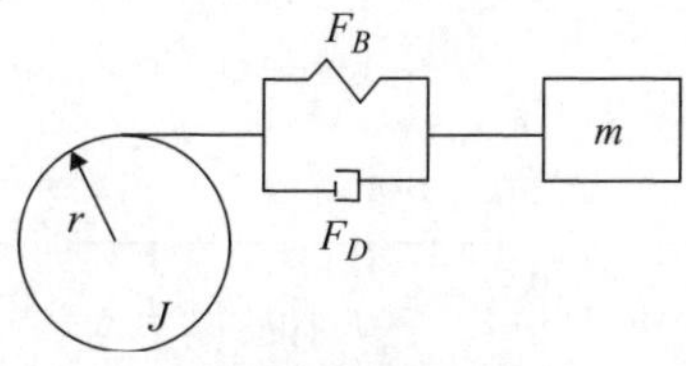

Figure 11.9   Simplified model of the simulated system

$F_B(\theta, x)$ The belt elasticity force proportional to the belt stretch.
$F_D(\omega, v)$ The damping force developed of the belt proportional to the derivative of stretch.
$F_L(x, v)$ The friction force at the load side.

In this system, the disturbance that affects the motor is the changing belt reaction. The employed controller used to control the position of the motor, has the following parameters: $D = 100$, $C = 5$, $\eta = 0.5$.

The simulation results are shown in Figs 11.10–11.17. The applied position reference is in the form of a smooth s-shaped curve realised as a linear segment preceded and followed by parabolic segments. This corresponds to a trapezoidal velocity reference curve. Figure 11.10 shows the actual position curve when the proposed controller scheme and the s-shaped position reference are applied to the motor. As can be seen, the actual position is very close to the position reference and the two curves cannot be distinguished in the figure. Indeed, the position error is very small as presented in Fig. 11.11. Figures 11.12–11.14 display the evolution of the three NN weights through time. The initial values are taken as zero for all the three parameters. It can be concluded from the figures that the training algorithm accomplishes a quick convergence of the weights to their final values. After a very short transient, the weights remain stable at their converged values and this indicates a stable behaviour of the backpropagation weight update. The control signal presented in Fig. 11.15 is smooth, as is desirable in the control of mechanical systems. The comparison of this figure with Fig. 11.14, indicates that the control signal is dominated by the contribution from w_3. Figure 11.16 shows the phase plane trajectory. Because the reference is smooth, the error and the derivative of the error start from zero and move in the vicinity of zero. Hence, the reaching phase to the sliding manifold can only be distinguished by

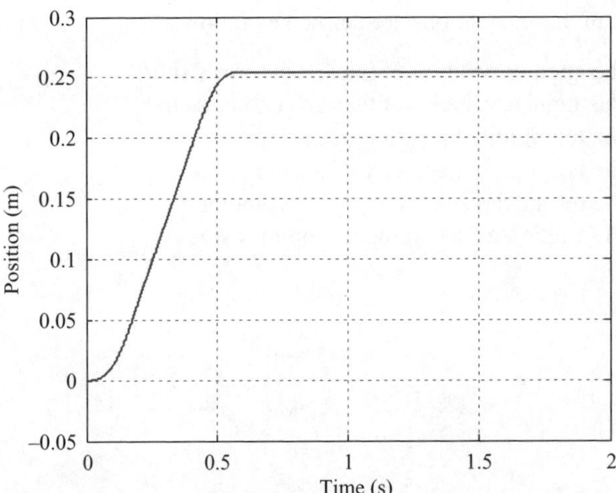

Figure 11.10   The reference and actual position curves

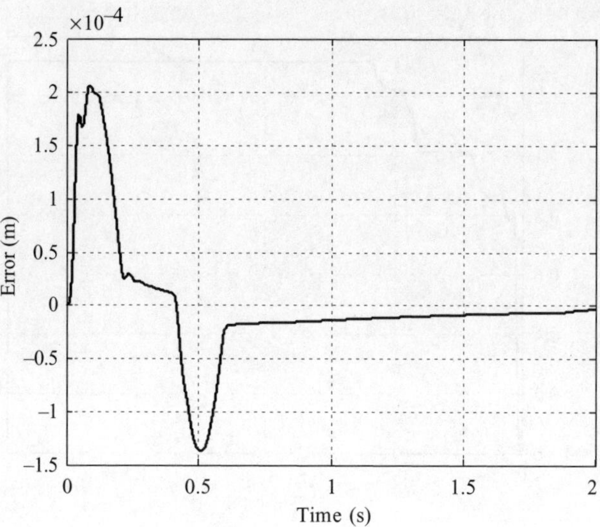

Figure 11.11   *The position tracking error*

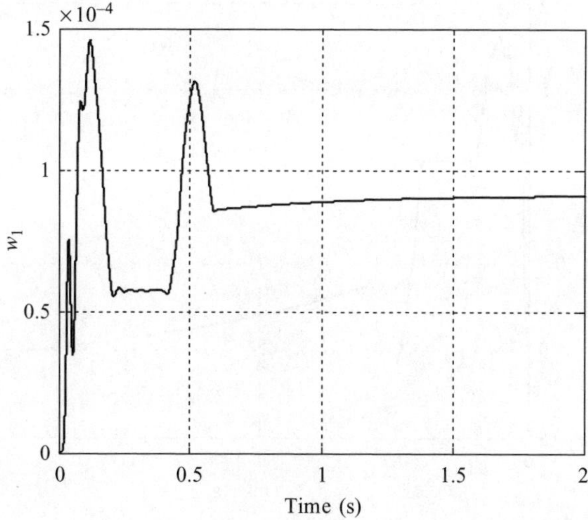

Figure 11.12   *Evolution of the NN weight w_1 in time*

zooming the plot to the origin of the plane. In order to investigate the reaching characteristics in more detail, simulations are carried out with step position references too. In this case, the state trajectory starts with a large position error giving a good opportunity to observe the system entering the sliding manifold. In Fig. 11.17, the phase plane is presented for a step reference input. This figure indicates a quick reaching

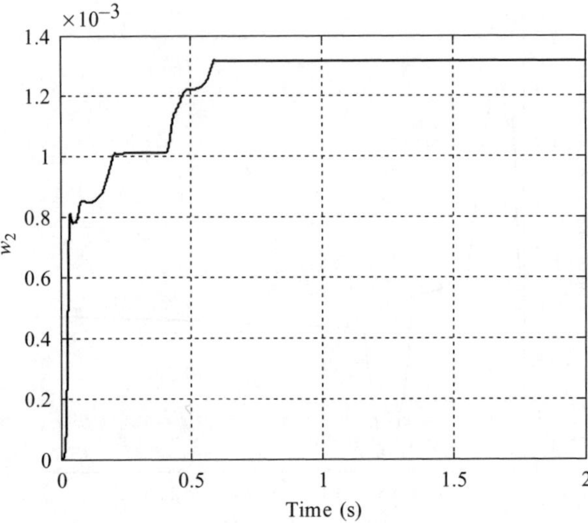

Figure 11.13   Evolution of the NN weight w_2 in time

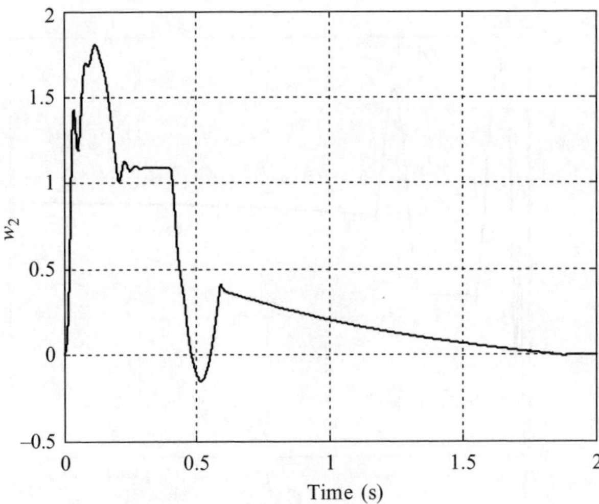

Figure 11.14   Evolution of the NN weight w_3 in time

phase without significant overshoot over the sliding line When the line is reached, the sliding behaviour is observed and the position error decays to zero with the error dynamics dictated by the parameter C of the SMC controller (sliding line slope is $-5=-C$). The simulation outputs are in harmony with the theoretical results and they indicate that the control system proposed is successful.

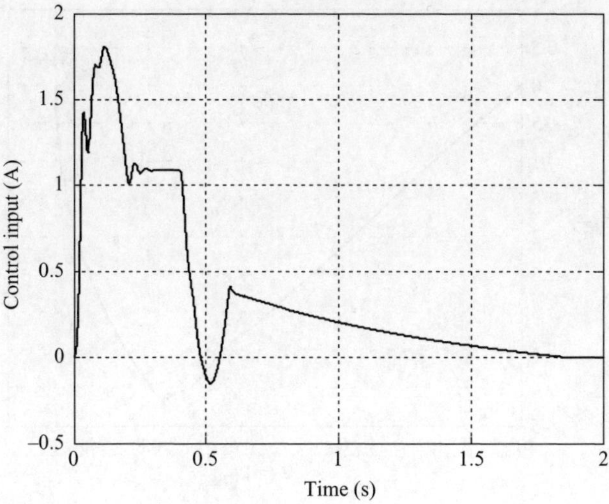

Figure 11.15   The control input signal

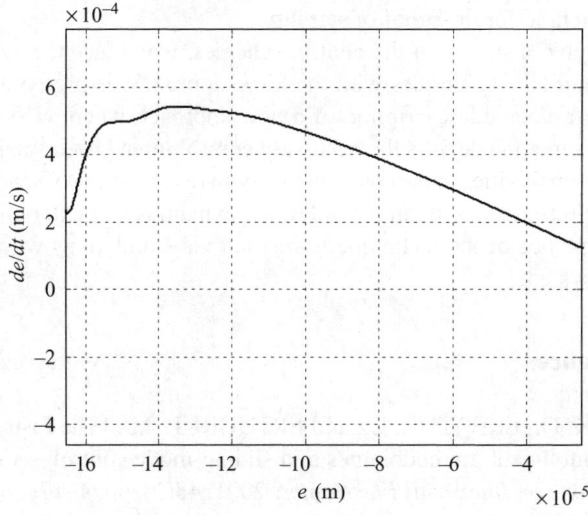

Figure 11.16   The phase plane for smooth reference

11.4   Conclusion

Merging FL and NN systems with SMC has been an attractive idea for many researchers. The use of intelligent computation techniques can solve problems of SMC such as chattering and alleviate difficulties in the computation of the equivalent

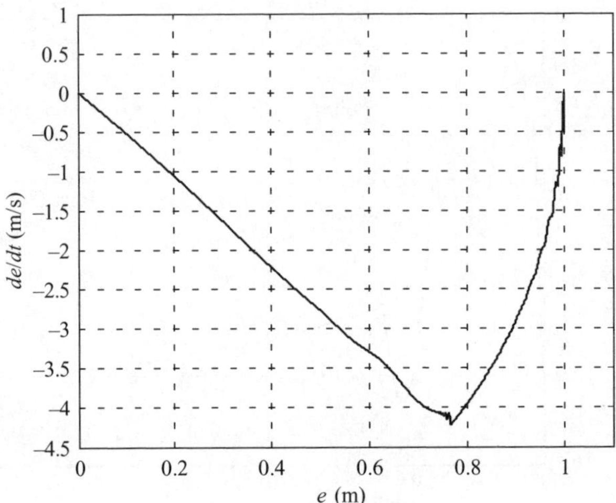

Figure 11.17   The phase plane for step reference

control. Considering another way of combination, the well-established results relating to SMC techniques can be applied to intelligent computation schemes, which lack a systematic method for the proof of stability.

In this chapter, a survey of the control schemes, which use the synergic combination of FL and NN techniques with the SMC approach, is presented. The survey is followed by a detailed description of a new approach where NNs and SMC are combined. This method derives the necessary control input based solely on the computation of the sensitivities of the cost function with respect to the NN weights. A proof of stability is presented for bounded unknown dynamics and external disturbances, and the performance of the technique is verified via simulations with the model of a mechanical system.

11.5   References

1   KAYNAK, O., ERBATUR, K., and ERTUGRUL, M.: 'The fusion of computationally intelligent methodologies and sliding-mode control – a survey', *IEEE Transactions on Industrial Electronics*, 2001, **48**(1), pp. 4–17

2   BARTOSZEWICZ, A.: 'On the robustness of variable structure systems in the presence of measurement noise' in *Proc. IEEE IECON'99*, Aachen, Germany, August 31–September 4, 1998, pp. 1733–1736

3   FUNAHASHI, K.: 'On the approximate realization of continuous mappings by neural networks', *Neural Networks*, 1989, **2**, pp. 183–192

4   CYBENKO, G.: 'Approximation by superpositions of a sigmoidal function', *Mathematics of Control, Signals and Systems*, 1989, **2**, pp. 303–314

5   HORNIK, K.: 'Multilayer feedforward networks are universal approximators', *Neural Networks*, 1989, **2**, pp. 359–366

6   MORIOKA, H., WADA, K., SABANOVIC, A., and JEZERNIC, K.: 'Neural network based chattering free sliding mode control', *Proceedings SICE'95*, Sapporo, Japan, 1995

7   JEZERNIK, K., RODIC, M., SAFARIC, R., and CURK, B.: 'Neural network sliding mode robot control', *Robotica*, 1997, **15**, pp. 23–30

8   RODIC, M., JEZERNIC, K., SABANOVIC, A., and SAFARIC, R.: 'Sliding mode based neural network learning procedure', in Proceedings, 5th International Workshop *on Robotics*, Budapest, 1996, Hungarian Robotics Association, pp. 547–552

9   ERTUGRUL, M. and KAYNAK, O.: 'Neuro sliding mode control of robotic-manipulators' *Mechatron.*, 2000, **10**(1–2), pp. 243–267

10   JEZERNIK, K., RODIC, M., SAFARIC, R., and CURK, B.: 'Neural network sliding mode robot control', *Robotica*, 1997, **15**, pp. 23–30

11   SABANOVIC, A., JEZERNIK, K., and RODIC, M.: 'Neural network application in sliding mode control systems', in *Proc. IEEE VSS'96*, 1996, pp. 143–147

12   RAMIREZ, H. S. and MORLES, E. C.: 'A sliding mode strategy for adaptive learning in adalines', *IEEE Trans. Circuits Syst. I*, 1995, **42**, pp. 1001–1012

13   PARMA, G. G., MENEZES, B. R., and BRAGA, A. P.: 'Sliding mode algorithm for training multilayer artificial neural networks', *Electron. Lett.*, 1998, **34**(1), pp. 97–98

14   SANNER, R. N. and SLOTINE, J. J. E.: 'Gaussian networks for direct adaptive control', *IEEE Trans. Neural Networks*, 1992, **3**, pp. 837–863

15   HSU, L. and REAL, J. A.: 'Dual mode adaptive control using Gaussian neural networks', *Proc. 36th Conf. Decision and Control*. New Orleans, LA, 1997, 1, pp. 4032–4037

16   HSU, L. and REAL, J. A.: 'Dual mode adaptive control', *Proc. IFAC World Congr.*, Beijing, China, 1999, vol. K, pp. 333–337

17   GE, S. S., LEE, T. H., and HARRIS, C. J.: 'Adaptive neural network control of robotic manipulators' (World Scientific, Singapore, 1998)

18   LEE, C. C.: 'Fuzzy logic in control systems: fuzzy logic controller – Part I and II', *IEEE Transactions on Systems, Man and Cybernetics*, 1990, **20**(2), pp. 404–435

19   HWANG, Y. R. and TOMIZUKA, M.: 'Fuzzy smoothing algorithms for variable structure systems', *IEEE Trans. Fuzzy Syst.*, 1994, **2**, pp. 277–284

20   ERBATUR, K., KAYNAK, O., SABANOVIC, A., and RUDAS, I.: 'Fuzzy adaptive sliding mode control of a direct drive robot', *Robot. Auton. Syst.*, 1996, **19**(2), pp. 215–227

21   CHEN, C. S. and CHEN, W. L.: 'Robust adaptive sliding-mode control using fuzzy modeling for an inverted-pendulum system', *IEEE Trans. Ind. Electron.*, 1998, **45**, pp. 297–306

22   YU, X. H., MAN, Z. H., and WU, B. L.: 'Design of fuzzy sliding-mode control systems', *Fuzzy Sets Syst.*, 1998, **95**(3), pp. 295–306

23   KIM, S. W. and LEE, J. J.: 'Design of a fuzzy controller with fuzzy sliding surface', *Fuzzy Sets Syst.*, 1995, **71**, pp. 359–367

24   CAO, C. T.: 'Fuzzy compensator for stick-slip friction', *Mechatronics*, 1993, **3**(6), pp. 783–794
25   WANG, W.-J. and LIN, H.-R.: 'Fuzzy control design for the trajectory tracking in phase plane', *IEEE Trans. Syst., Man, Cybern. A*, 1998, **28**, pp. 710–719
26   YILDIZ, Y., SABANOVIC, A., and ABIDI, K. S.: 'A novel approach to neuro-sliding mode controllers for systems with unknown dynamics', *Proceedings Electrical Drives and Power Electronics (EDPE'03)*, Slovakia, 2003

Chapter 12

SMC applications in power electronics

Domingo Biel Solé and Enric Fossas Colet

Power converters are widely used in applications where it is desired to obtain a totally regulated electric signal from a non-regulated one, keeping optimum energy efficiency in the conversion. These converters can be linear or switched, the latter being the most common due to their better energy efficiency. As will be seen in this chapter, switching converters can be modelled as variable structure systems. They therefore constitute a natural field of application of Sliding Mode Control techniques. The most usual conversion types, namely DC-DC, DC-AC and AC-DC, will be considered here. SMC controllers will be designed and several aspects involving the electronic implementation of the controllers will be discussed.

12.1   DC-DC power conversion

The aim of DC-DC power conversion is to obtain a regulated, continuous voltage (or current) at the load terminals. The power regulator consists of a power stage composed of semiconductors, inductors and capacitors, and a control stage commonly based on the processing of an error signal (the difference between a reference and an output voltage) and a voltage-time conversion through a Pulse-Width-Modulator (PWM). The control objective is to achieve a regulated robust output voltage with good dynamic performance from the switching converter.

12.1.1   Electrical and state-space models

The ideal[1] buck, boost and buck-boost topologies feeding a resistive load are depicted in Figs 12.1 and 12.2. The converter dynamics is modelled by two state-variables, i,

---

[1] Without semiconductor, capacitor and inductor losses.

$$\begin{cases} L\dfrac{di}{dt} = -v + Eu \\[2mm] C\dfrac{dv}{dt} = i - \dfrac{v}{R} \end{cases}$$

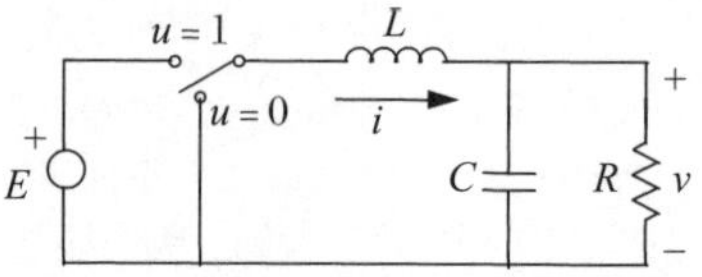

Figure 12.1   Buck converter topology

$$\begin{cases} L\dfrac{di}{dt} = -v(1-u) + E \\[2mm] C\dfrac{dv}{dt} = i(1-u) - \dfrac{v}{R} \end{cases}$$

$$\begin{cases} L\dfrac{di}{dt} = -(1-u)v + Eu \\[2mm] C\dfrac{dv}{dt} = (1-u)i - \dfrac{v}{R} \end{cases}$$

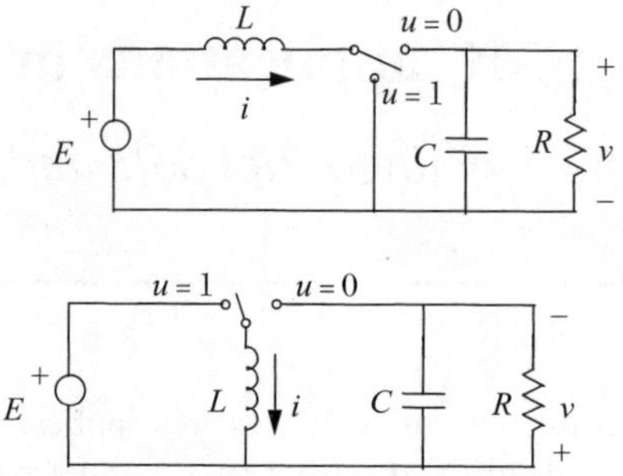

Figure 12.2   Boost and buck-boost converter topologies

the inductor current and v, the capacitor voltage[2], and by the control input $u \in \{0, 1\}$, which describes the position of a bidirectional switch. The state equations of the converter are listed below, where E is the DC-input voltage, L and C, the inductor and the capacitor value, respectively and R, the resistive load. Note that, in the case of the buck-boost converter, there exists an output voltage polarity inversion with respect to the input voltage.

A general model for buck, boost and buck-boost converters is given in (12.1). Specific models can be obtained by selecting the parameters λ and γ as follows: $\lambda = 0$, $\gamma = 1$ for the buck converter, $\lambda = 1$, $\gamma = 0$ for the boost converter and $\lambda = 1$, $\gamma = 1$ for the buck-boost converter:

$$\frac{d}{dt}\begin{pmatrix} Li \\ Cv \end{pmatrix} = \begin{pmatrix} 0 & -1 + \lambda u \\ 1 - \lambda u & -\dfrac{1}{R} \end{pmatrix} \begin{pmatrix} i \\ v \end{pmatrix} + \begin{pmatrix} E(1 + \gamma(u - 1)) \\ 0 \end{pmatrix} \tag{12.1}$$

Therefore, DC-DC switching converters can be modelled as bilinear systems, that is variable structure systems.

---

[2] The capacitor voltage coincides with the output voltage because the resistor losses associated to the capacitor are not considered.

12.1.2   *Sliding mode control analysis and design*

Output voltage regulation is the general control objective in DC-DC power conversion. A naive approach would design the action of the switch, the control action, based uniquely on the output voltage error (direct control). This approach will not be successful in general. An indirect approach, based on both the output voltage and the inductor current, is needed to achieve robust regulation.

SMC strategies for the DC-DC conversion problem via direct and indirect control will be considered here. Starting from a switching surface, the transversality condition is checked and the equivalent control is derived. The latter is used to obtain the ideal sliding dynamics and, when the ideal sliding dynamics are stable, to deduce the sliding domain. This subsection follows References 1 and 2 and Chapter 11 of Reference 3 partially.

12.1.2.1   Direct output voltage control

First, let us consider a direct output voltage control, which implies the use of the switching surface

$$\sigma_v = v - V_{ref} \tag{12.2}$$

where $V_{ref} > 0$ is a constant output voltage reference. Note that the transversality condition is not fulfilled in the buck converter case ($\lambda = 0$). For the other cases, ($\lambda = 1$ and $i \neq 0$), the equivalent control and the ideal sliding dynamics are given by

$$u_{eq} = \frac{i - (V_{ref}/RC)}{i} \tag{12.3}$$

$$\begin{cases} v = V_{ref} \\ \frac{di}{dt} = \frac{1}{Li}\left(Ei - \frac{V_{ref}}{R}\left(E\gamma + V_{ref}\right)\right) \end{cases} \tag{12.4}$$

The ideal sliding dynamics has an equilibrium point at $((E\gamma + V_{ref})V_{ref}/RE, V_{ref})$. Its stability is analysed by the first linear approximation, namely

$$\frac{d\hat{i}}{dt} = \frac{R}{L}\frac{E^2}{V_{ref}(E\gamma + V_{ref})}\hat{i} \tag{12.5}$$

where $\hat{i} = i - i^*$ and, $i^* = (E\gamma + V_{ref})V_{ref}/RE$.
Since $(R/L)(E^2/V_{ref}(E\gamma + V_{ref})) > 0$, the equilibrium point is unstable; hence direct voltage regulation results in instability of the inductor current.

12.1.2.2   Indirect output voltage control

Now the proposed switching surface is

$$\sigma_i = i - i_{ref} \tag{12.6}$$

where i_{ref} denotes a constant inductor current reference. Then, the equivalent control and the ideal sliding dynamics are given by

$$u_{eq} = \frac{v - E(1 - \gamma)}{E\gamma + \lambda v} \tag{12.7}$$

$$\begin{cases} i = I_{ref}, \\ \frac{dv}{dt} = \frac{1}{C}\left(I_{ref} - \lambda I_{ref}\left(\frac{v - E(1-\gamma)}{E\gamma + \lambda v}\right) - \frac{v}{R}\right) \end{cases} \tag{12.8}$$

The geometric locus defined by the equilibrium points is described in coordinates (I_{ref}, v^*) by $v^{*2} + E\gamma v^* - EI_{ref}R = 0$ for $\lambda = 1$ and $v^* = I_{ref}R$ for $\lambda = 0$. Linearising (12.8) around the equilibrium point (I_{ref}, v^*) yields

$$\frac{d\hat{v}}{dt} = \frac{1}{C}\left(-I_{ref}E\left(\frac{\lambda}{E\gamma + \lambda v^*}\right)^2 - \frac{1}{R}\right)\hat{v} \tag{12.9}$$

where $\hat{v} = v - v^*$. Thus, the indirect control results in a stable ideal sliding dynamics. The sliding domain on $i = I_{ref}$ resulting from $0 < u_{eq} < 1$, assuming $E > 0$, gives the following converter characteristics

	Characteristics	Sliding domain
Buck	Step-down	$i = I_{ref}$ and $0 < v < E$
Boost	Step-up	$i = I_{ref}$ and $0 < E < v$
Buck-boost	Step-up/Step-down	$i = I_{ref}$ and $0 < v$

Finally, the switching strategy is defined so that σ_i^2 qualifies as a Lyapunov function. From (12.7)

$$\frac{1}{2}\frac{d\sigma_i^2}{dt} = \sigma_i\frac{d\sigma_i}{dt} = \sigma_i(\lambda v + E\gamma)(u - u_{eq}) \tag{12.10}$$

Then, since $0 < u_{eq} < 1$ is assumed,

$$u = \begin{cases} u = 0 & \text{if } \sigma_i(\lambda v + E\gamma) > 0 \\ u = 1 & \text{if } \sigma_i(\lambda v + E\gamma) < 0 \end{cases} \tag{12.11}$$

In summary, the indirect output voltage control provides output voltage regulation presuming the converter states meet the sliding domain conditions. However, the output voltage depends on the load resistance; therefore, these controllers do not produce systems that are robust with respect to load variations.

12.1.2.3   Robustness

Two strategies to robustify indirect output voltage control are given here. The first is a specific method for linear systems of relative degree greater than 1 and the second is PI-type strategy.

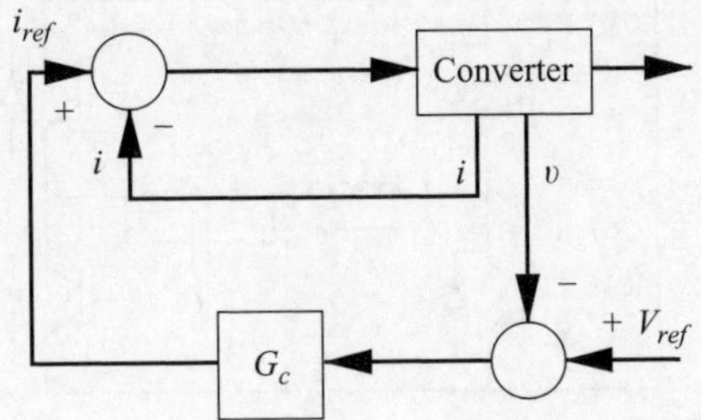

Figure 12.3   Classical two loops DC-DC regulator diagram

As for linear systems, by adding higher derivatives of the error, the relative degree decreases. For the buck converter [2–4], the switching surface considered is

$$\sigma_{vr} = e + k \cdot \frac{de}{dt} \tag{12.12}$$

where $e = V_{ref} - v$. This strategy cannot be applied to the boost or to the buck-boost converters because these systems have relative degree 1 and the derivative of the control input would appear in the expression of the equivalent control. Note, in addition, that the derivative of the output voltage (a discontinuous signal) should be processed in the nonlinear converters case. This makes it impossible for designers to use this switching scheme. To avoid processing discontinuities, a high frequency filtering (averaging) is used [5]. A linearising process can alternatively be used [6]. The switching surface

$$\sigma_{ir} = i + k_e e + k_{v_a} v_a \tag{12.13}$$

where $e = V_{ref} - v$ and v_a satisfies $(dv_a/dt) = e$, is a robust alternative. Both switching surfaces σ_{vr} and σ_{ir} can be represented by the block diagram in Fig. 12.3 which highlights two control loops. A fast inner control loop corresponding to the current dynamics, and a slow outer control loop which processes the output voltage error. The G_c block is a PD and a PI controller for σ_{vr} and σ_{ir} respectively.

The analysis of the controlled systems (buck, boost and buck-boost), namely checking the transversality condition, computing the equivalent control, the sliding domain, the equilibrium points and their stability is left to the reader as an exercise.

12.2   DC-AC power conversion

Electrical energy is carried through electric lines as Alternating Current (AC) since this format is easy to generate (there exist primary energy sources providing energy in this format). DC-AC is used in isolated applications when energy must be supplied by primary DC-voltage sources. Uninterruptible Power Systems and AC voltage sources are examples of DC-AC conversion. A classical circuit structure for this class of electric conversion is presented in this section, as well as an associated SMC design.

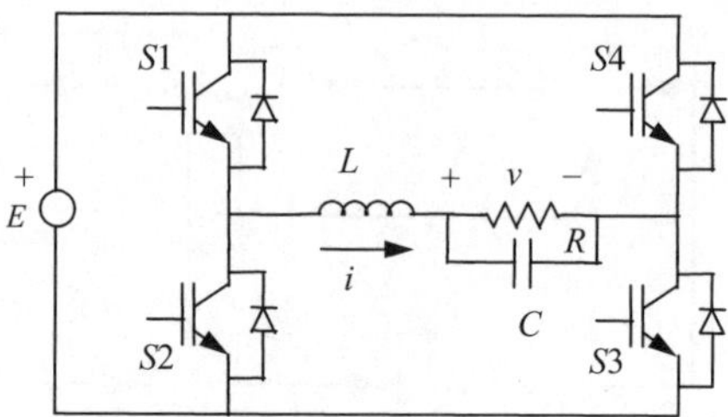

Figure 12.4   Full-bridge converter topology

Table 12.1   Switching sequence

	S1	S2	S3	S4
$u = 1$	ON	OFF	ON	OFF
$u = -1$	OFF	ON	OFF	ON

12.2.1   Full-bridge power converter

The full-bridge converter circuitry scheme is depicted in Fig. 12.4. It consists of reactive elements (L, C) and semiconductors, which, given two switching levels, are controlled according to Table 12.1 with the aim of achieving an AC output voltage on the load R from a DC input voltage E. Assuming a resistive load, the power stage can be modelled as a buck converter. In this case, the control input u takes values in the set $\{-1, 1\}$.

12.2.2   Tracking signal sliding mode control

12.2.2.1   Switching surface and sliding mode behaviour

As in Subsection 12.1.2, the switching scheme

$$\sigma = \lambda e(t) + \frac{de(t)}{dt} \tag{12.14}$$

is considered. This scheme can be found in References 4 and 7; $e(t)$ stands for the voltage error defined as $e(t) = v(t) - V_{ref}(t)$, $v(t)$ being the output voltage and $V_{ref}(t)$ the reference signal voltage. Sliding motion on the switching surface provides a first order dynamics response leading to the desired steady-state behaviour $v(t) = V_{ref}(t)$. This results in a robust output voltage performance with respect to load variations and source perturbations.

The equivalent control resulting from the application of the invariance condition to σ is

$$u_{eq}(t) = \frac{LC}{E}\left(\frac{d^2 V_{ref}}{dt^2} + \lambda\frac{dV_{ref}}{dt} + \left(\frac{1}{RC} - \lambda\right)\frac{dv}{dt} + \frac{v}{LC}\right) \qquad (12.15)$$

whereas the sliding domain can be obtained by imposing

$$\sigma = 0 \quad \text{and} \quad -1 < u_{eq} < 1 \qquad (12.16)$$

This leads to restrictions on the values of the converter parameters E, L, C, R, the desired output signal $V_{ref}(t)$ and the desired transient dynamics represented by the time constant $1/\lambda$. These restrictions must be taken into account in the design procedure.

For a fast transient response, the constant λ must be as large as possible. However, the greater the value of λ, the faster the transient response, but the greater the equivalent control value [4].

12.2.2.2 Design procedure

A useful design procedure can be derived from the equivalent control evaluated on the nominal trajectory. Replacing $v(t) = V_{ref}(t) = A\sin(\omega t)$ in (12.15), inequality (12.16) reads as

$$-E < A\left(\frac{\omega L}{R}\cos(\omega t) + (1 - LC\omega^2)\sin(\omega t)\right) < E \qquad \forall t \in [0, +\infty)$$

$$(12.17)$$

or, equivalently

$$A < \frac{E}{LC}\frac{1}{\sqrt{(\omega^2/(RC)^2) + (\omega^2 - (1/LC))^2}} \qquad (12.18)$$

which, in turn, can be written as

$$A < E\gamma(\omega) \qquad (12.19)$$

where

$$\gamma(\omega) = \frac{1}{LC}\frac{1}{\sqrt{(\omega^2/(RC)^2) + (\omega^2 - (1/LC))^2}} \qquad (12.20)$$

$\gamma(\omega)$ is the frequency response output filter gain. It is depicted plotted against the output signal frequency for several load values in Fig. 12.5.

The sliding domain can thus be expressed as a function of the frequency response of the converter output filter and the output signal parameters (amplitude and frequency). This suggests the following design procedure: given a value of R, a sliding regime is ensured for values of (A/E) lower than the output filter frequency response. Additionally, if an output load R meets inequality (12.19), then any $R' > R$ satisfies the inequality, too. For this reason, the design must take into account the minimum load value for which a sinusoidal output signal is desired. It is worth noting

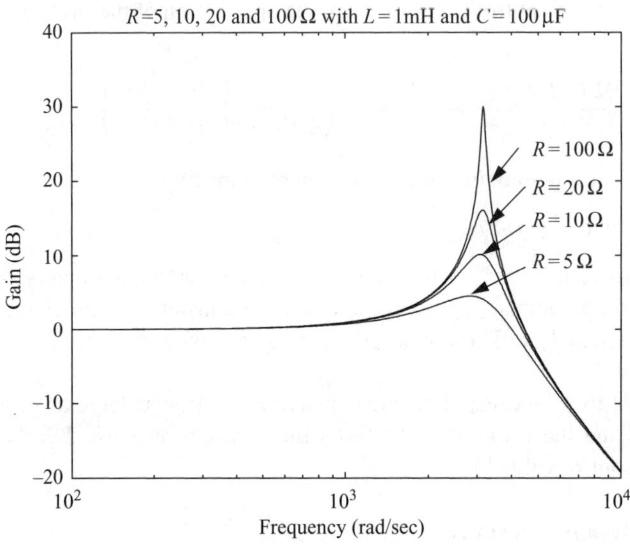

Figure 12.5 $\gamma(\omega)$ versus the output signal frequency

that an output voltage amplitude greater than the DC input voltage can be obtained for a small frequency range in the vicinity of the output filter resonant frequency.

12.2.2.3   Switching control law

Considering sliding mode control principles, the trajectories reach the switching surface σ provided that σ^2 qualifies as a Lyapunov function. From (12.14) and the definition of the equivalent control, we have

$$\frac{d\sigma^2}{dt} = 2\sigma\left(\frac{E}{LC}(u - u_{eq})\right) \tag{12.21}$$

Thus, assuming $-1 < u_{eq} < 1$ and $E > 0$, the previous expression leads to the switching strategy

$$u = \begin{cases} u = -1 & \text{if } \sigma > 0 \\ u = 1 & \text{if } \sigma < 0 \end{cases} \tag{12.22}$$

12.3   AC-DC power conversion

Traditionally, low power electronic equipment needs a DC power supply. As has been mentioned above, the electric utility grid has a sinusoidal waveform and, as a consequence, an AC-DC power conversion is required. For appropriate AC-DC electrical quality, the input current of the electronic equipment should be phased with the AC utility grid voltage (which is known as unity power factor ratio).

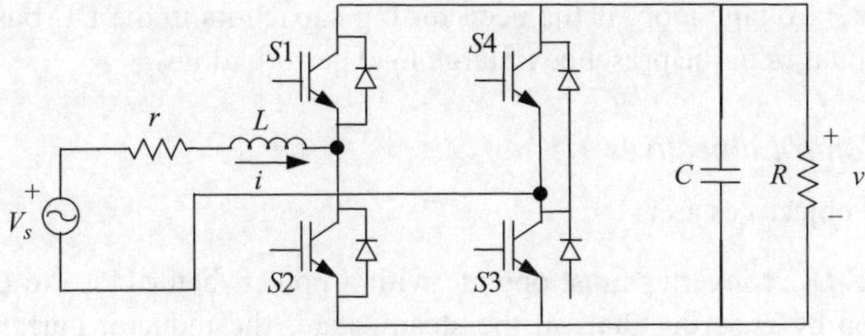

Figure 12.6   Bidirectional boost active rectifier converter

Additionally, the output voltage of the AC-DC switching power converter must be regulated and robust with respect to load or utility grid variations. The most popular switched-mode power converter for this AC-DC application is the boost rectifier power converter. This section deals with a complete SMC design of this power rectifier.

12.3.1   Rectifier power converter

The rectifier power converter depicted in Fig. 12.6 can be modelled as

$$L\frac{di}{dt} = -uv + v_s \tag{12.23}$$

$$C\frac{dv}{dt} = ui - \frac{1}{R}v \tag{12.24}$$

where i and v are the input inductor current and the output capacitor voltage variables, respectively; $v_s = E\sin(\omega_r t)$ is the ideal sinusoidal source that represents the AC-line source; R is the DC-side resistive load and L and C are the inductance and the capacitance of the converter, respectively. For simplicity, inductor losses, modelled by r in the picture, will not be considered here. The control variable u takes values in the set $\{-1, 1\}$.

In the following analysis, it will be interesting to deal with the DC component[3] of some variables that will be denoted as $\langle \cdot \rangle_0$. It is important to note that the system described by (12.23) and (12.24) can be seen as the interconnection of two subsystems with different time constants. In particular, the dynamics of (12.24) are much slower than the dynamics of (12.23). This fact has led to the development of classical control schemes for these systems consisting of two concentric control loops: the inner (fast) for shaping the inductor current, and the outer (slow) for regulating the output capacitor voltage. In this control architecture, the output of the outer loop controller acts as the modulating signal in an AM modulator, with carrier v_s, whose output is the reference for the inner loop. The disadvantage of this control topology, caused by

---

[3] The DC component, or averaged function, of a T-periodic signal $f(t)$ is defined by $\langle f(t) \rangle_0 \triangleq (1/T)\int_{t-T}^{t} f(\tau)d\tau.$

the slow outer voltage loop, is the need for big capacitors in the DC bus to prevent large overvoltages in the presence of large load perturbations.

12.3.2   Control objectives

The control objectives are:

1.   The AC-DC converter must operate with a power factor close to one. This is achieved by ensuring that, in the steady-state, the inductor current i follows a sinusoidal signal with the same frequency and phase as the AC-line voltage source v_s. The amplitude, I_d, of this sinus should be calculated by the controller in order to accomplish the following objective.
2.   The DC component of the output capacitor voltage $\langle v \rangle_0$ should be driven to the constant reference value $\langle v \rangle_{0d}$, where $\langle v \rangle_{0d} > E$ in order to have boost behaviour.
3.   The value of the DC bus capacitor must be as low as possible for cost reasons. This requirement implies that the controller should be able to reject large perturbations in the load with short transients to prevent overvoltages on the bus.

12.3.3   Ideal sliding dynamics

If the state vector of the system (12.23) and (12.24) is fixed assuming perfect control action at the desired values $(i_d = I_d \sin(\omega_r t), v_d = V_d = \langle v \rangle_{0d})$ and neglecting the higher order harmonics, an input-output active power balance [8] is performed resulting in

$$P_i = \langle i_d v_s \rangle_0 = \frac{E I_d}{2} \tag{12.25}$$

$$P_o = \frac{v_d^2}{R} = \frac{V_d^2}{R} \tag{12.26}$$

$$P_i = P_o \Rightarrow I_d = \frac{2 V_d^2}{E R} \tag{12.27}$$

As for the DC-DC boost converter, the bidirectional boost rectifier has relative degree 1 regardless of the output, i or v. As in the regulation problem, if the output is v, the system has a nonminimum-phase behaviour. For this reason, it is usually controlled through the current i, this being particularly appropriate in this case because of the shape specification for the input current.

Let us consider the dynamics evaluated on the nominal trajectory in order to define a dynamics for I_d which meets the output voltage specification. We replace $i = i_d = I_d \sin(\omega_r t)$ in (12.23) and (12.24). We have

$$\bar{u} = \frac{E \sin(\omega_r t) - \omega_r L I_d \cos(\omega_r t)}{\bar{v}} \tag{12.28}$$

$$C \frac{d\bar{v}}{dt} = -\frac{I_d{}^2 \sin(\omega_r t) \cos(\omega_r t) \omega_r L}{\bar{v}} + \frac{I_d (\sin(\omega_r t))^2 E}{\bar{v}} - \frac{\bar{v}}{R} \tag{12.29}$$

where $\bar{u}$ and $\bar{v}$ are the control variable and the capacitor voltage in the ideal sliding dynamics, respectively. Then, (12.29) describes the behaviour of the zero-dynamics of the system. This equation is a Bernoulli ODE, but multiplying each side of (12.29) by $\bar{v}$ and taking $\bar{z} = \frac{1}{2}\bar{v}^2$, we obtain the following linear ODE

$$\frac{d\bar{z}}{dt} = -\frac{I_d{}^2 \sin(\omega_r t)\cos(\omega_r t)\omega_r L}{C} + \frac{I_d\,(\sin(\omega_r t))^2\,E}{C} - \frac{2\bar{z}}{CR} \tag{12.30}$$

whose solution is $\bar{z}(t) = f(t) + p(t) + K$, where $f(t) = \frac{1}{2}C_1 \exp(-2t/RC)$ is the vanishing term corresponding to the first order linear dynamics, $p(t) = A\sin(2\omega_r t) + B\cos(2\omega_r t)$ is the oscillating term (at frequency $2\omega_r$), and $K = V_d^2/2$ is the constant term. It is worth noting that the DC value of $\bar{z}(t)$ in the steady-state is $\langle\bar{z}\rangle_0 = K = V_d^2/2$, i.e., averaging $\bar{z}(t)$ with period $T = \pi/\omega_r$ in steady-state results in the mean value of the DC capacitor bus squared and divided by 2. The same result can be obtained averaging (12.30):

$$\frac{d\langle\bar{z}\rangle_0}{dt} = \frac{EI_d}{2C} - \frac{2\langle\bar{z}\rangle_0}{RC} = \frac{V_d^2}{RC} - \frac{2\langle\bar{z}\rangle_0}{RC} \tag{12.31}$$

whose solution is $\langle\bar{z}\rangle_0 = V_d^2/2 + C_1 \exp(-2t/RC)$.

12.3.4   Control design

This subsection is devoted to the design of both the control u and I_d since the latter operates as a control in a linear equation describing the dynamics of $\langle v^2/2\rangle_0$. The control objectives can be written as follows:

1. $i(t) = I_d \sin(\omega_r t)$, and
2. $\langle z\rangle_0 = 0.5V_d^2$,

where $z = 0.5v^2$ and both requirements must be met in the steady-state.

As far as the first objective is concerned, $\sigma = i - I_d \sin(\omega_r t) = 0$ is considered as a switching surface. Following the standard procedure, we have

$$u_{eq} = \frac{E\sin(\omega_r t) - \omega_r L I_d \cos(\omega_r t)}{v}$$

$$u = \begin{cases} -1 & \text{if } \sigma < 0 \\ +1 & \text{if } \sigma > 0 \end{cases}$$

A necessary condition for a sliding mode is $v \neq 0$; note that the dot product of the gradient of σ and the control vector is $-v/L$ which, in turn, will be assumed negative. Furthermore, $-1 \leq u_{eq} \leq +1$ defines the subset of $\sigma = 0$, where sliding motion occurs, as in Subsection 12.2.2. The substitution of the steady-state ideal sliding dynamics in these inequalities results in the necessary conditions that must be satisfied by the plant parameters. This is left to the reader as an exercise.

With regard to the second objective, the variable $\langle z\rangle_0$ is regulated to $V_d^2/2$ applying classical linear control design to (12.31), where I_d acts as the control variable. This ordinary differential equation describes the zero dynamics, i.e., the Ideal Sliding

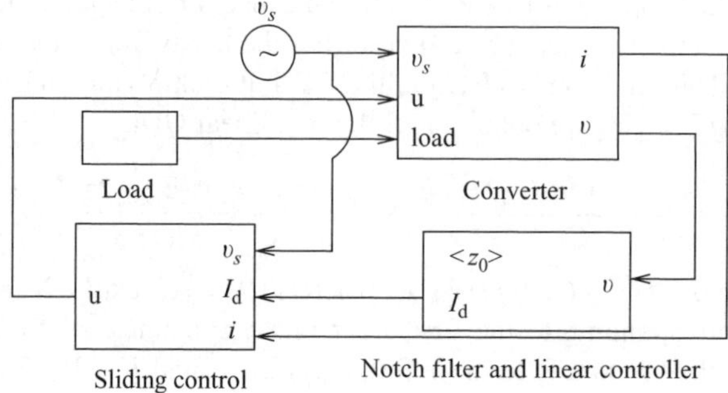

Figure 12.7   Control scheme block diagram

Dynamics. Taking the zero dynamics as the dynamics of $z = 0.5v^2$ makes sense because the current loop is much faster than the voltage one, as has already been pointed out. In addition, $z(t)$ has a DC component and a fundamental harmonic at $2\omega_r$ which is removed through the linear notch filter

$$H(s) = \frac{s^2 + 4\omega_r^2}{s^2 + 4\xi\omega_r s + 4\omega_r^2} \tag{12.32}$$

Simulation of the entire control strategy is left to the reader. Suggested plant parameter values are $L = 1\,\text{mH}$, $C = 4.7\,\text{mF}$, $v_s(t) = 220\sqrt{2}\sin(\omega_r t)\,\text{V}$, $\omega_r = 100\pi\,\text{rad/s}$, $V_d = 400\,\text{V}$. As a load, consider a pulsating function taking values $R = 100\,\Omega$ and $R = 10\,\Omega$. The control scheme block diagram is depicted in Fig. 12.7.

12.4   Control implementation

Several SMC strategies for electric power conversion were presented in this chapter. It is worth noting that SMC theory presumes an infinite switching frequency when the system operates in the sliding mode, and actual components cannot switch at infinite frequency. Further, higher switching frequencies become harmful in some applications. In power electronics, for instance, the higher the switching frequency, the higher the losses in the converter. Consequently, actual sliding mode controls operate at high, finite, possibly variable frequency which results in a chattering around the sliding surface.

Appropriate SMC implementations in switching systems must be considered at this stage. The switching frequency is required to be stable and synchronous for this type of system; this is a difficult requirement for non-standard implementation strategies. These problems have been tackled through fixed and variable bandwidth hysteresis comparators, by the addition of an external synchronous signal and by the use of the equivalent control as a duty cycle (with and without Zero Order Hold). These strategies are considered here and compared to the Zero Average Dynamics (ZAD) control strategy.

12.4.1 Sliding mode control implementation in switching converters

Let $\dot{x} = f(x) + g(x)u$ be a single input single output, autonomous system controlled through a sliding surface $S(x,t) = 0$ and an appropriate control law. Let us assume that the system behaviour is given by the ideal sliding dynamics

$$\begin{cases} S(x,t) = 0 \\ \quad \dot{x} = f(x) + g(x)u_{eq} \end{cases} \tag{12.33}$$

where u_{eq} is the equivalent control and, in this case, the switching frequency is assumed to be infinite.

As the system cannot reach the ideal sliding dynamics, the dynamics is characterised by

$$\begin{cases} S(x,t) \approx 0 \\ \quad \dot{x} = f(x) + g(x)\mu(x) \end{cases} \tag{12.34}$$

where, in the particular case of having a fixed switching frequency,

$$\mu(x) = \begin{cases} u^+ & \text{if } kT \leq t < (k+d_k)T \\ u^- & \text{if } (k+d_k)T \leq t < (k+1)T \end{cases} \tag{12.35}$$

The duty cycle d_k, or in general $d_k(x,t)$, defines the control action. It is usually obtained by Pulse-Width-Modulation of a processed system output. There are other control strategies providing fixed frequency switching. In Reference 9, for example, the duty cycle is defined as the equivalent control evaluated at the beginning of the control period:

$$d_k = \frac{u_{eq}(kT) - u^-}{u^+ - u^-}$$

The weak point of this strategy lies in the need to know the system parameters, which results in a loss of system robustness. For instance, in the Buck converter regulation problem, $u_{eq} = v/E$, which depends on the input voltage.

The use of a Zero Order Hold (ZOH) to synchronise signal control changes does not seem to be an appropriate option since commutations are gradually lost as the sampling period increases, as sketched in Fig. 12.8.

Authors in References 7, 10 and 11 propose the addition of a hysteresis cycle to the sliding mode control comparator, as shown in Fig. 12.9. The switching frequency

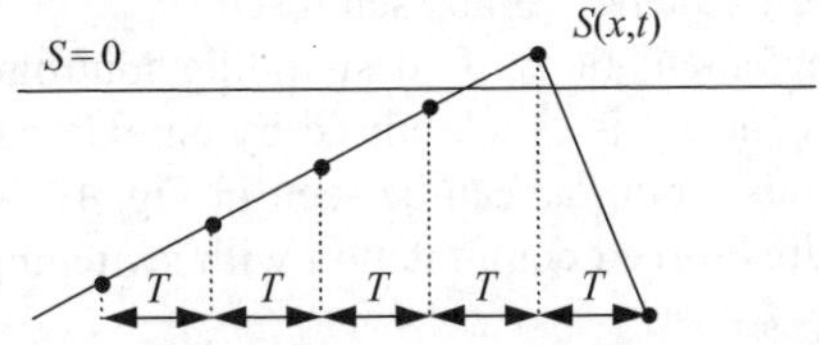

Figure 12.8   Losing switching opportunities due to the use of a T-period ZOH

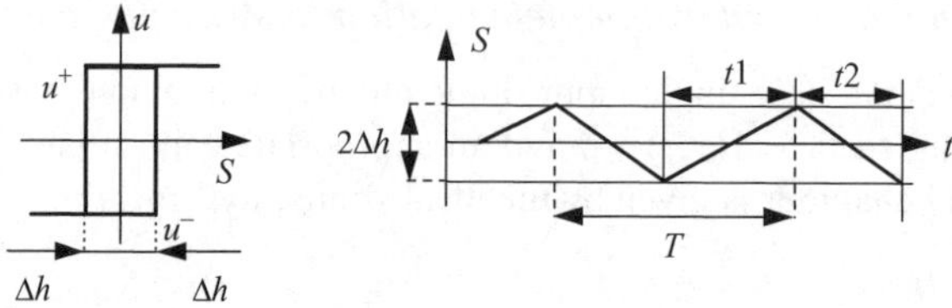

Figure 12.9   Hysteresis cycle and sliding surface dynamics

f_s can be estimated as follows:

$$
\begin{cases}
\dfrac{dS}{dt}\bigg|_{u=u^+} = \dfrac{2\Delta h}{t1} \\[4mm]
\dfrac{dS}{dt}\bigg|_{u=u^-} = -\dfrac{2\Delta h}{t2}
\end{cases}
\quad ; \quad
\begin{cases}
t1 = \dfrac{2\Delta h}{(\partial S/\partial x)g(x)(u^+ - u_{eq})} \\[4mm]
t2 = \dfrac{2\Delta h}{(\partial S/\partial x)g(x)(u^- - u_{eq})}
\end{cases}
\tag{12.36}
$$

As a conclusion, we have

$$
f_s = \frac{1}{t1+t2} = \frac{\partial S}{\partial x}\frac{g(x)}{2\Delta h}\frac{(u^- - u_{eq})(u_{eq}-u^+)}{(u^- - u^+)}
\tag{12.37}
$$

Note that the switching frequency is bounded but variable (not fixed) for time dependent equivalent controls.

It is worth noting that the processing time of the analogue or digital processor subsystem has not been taken into account. This processing time will affect the resulting dynamics if it is not sufficiently small when compared with the switching period.

Several approaches, [12] and [13], consider a variable bandwidth hysteresis cycle, the implementation of which depends on the system parameters and is complex. Taking

$$
\Delta h = \mu \frac{(u^- - u_{eq})(u_{eq}-u^+)}{(u^- - u^+)}
\tag{12.38}
$$

in (12.37) yields

$$
f_s = \frac{1}{2\mu}\frac{\partial S}{\partial x}g(x)
\tag{12.39}
$$

and hence the switching frequency can be stabilised.

Other electronic implementations of quasi-sliding controls are reported [14, 15]. The fixed switching frequency is synchronised by an external signal d defined by a T_d-periodic bipolar pulse train, as can be seen in Fig. 12.10. In this approach, a successful design requires forced commutation with switching frequency f_d. Then the inequalities

$$
D > \Delta h, \quad f_d > f_{s\,\text{max}}
\tag{12.40}
$$

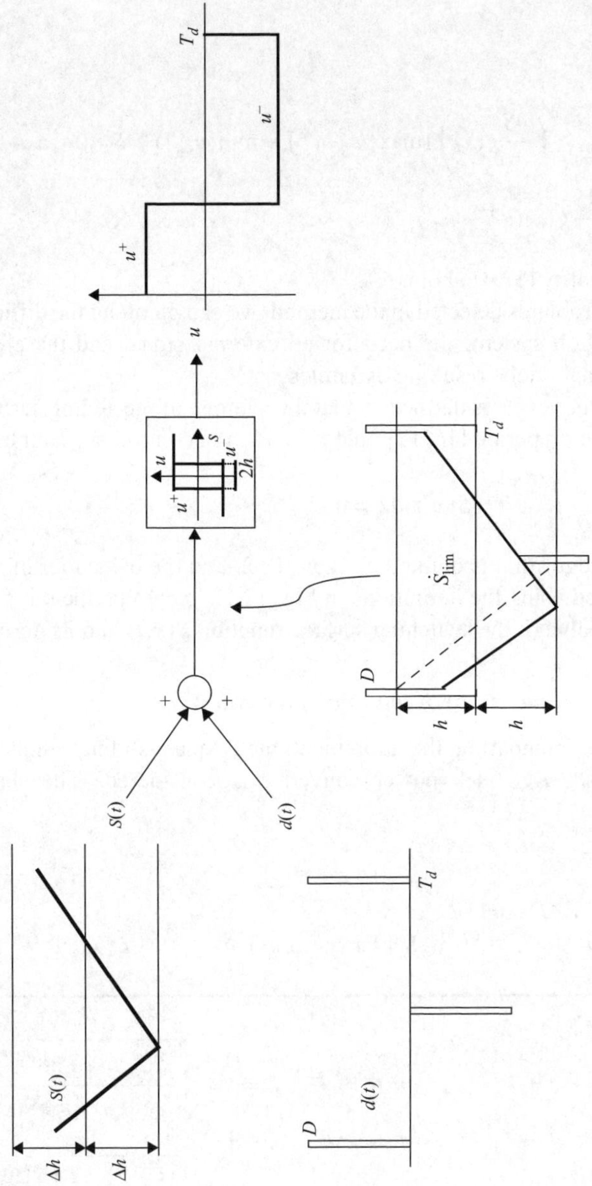

Figure 12.10   Sliding surface, external signal and switching signal

where D is the amplitude of the external signal and $f_{s\,max}$ is the maximum switching frequency achieved using a $2\Delta h$-width hysteresis cycle; otherwise, the switching frequency will be steered by the hysteresis cycle. The design must also avoid double commutations; a sufficient condition is

$$\dot{S}_{\lim} \geq \dot{S}_{\max}. \tag{12.41}$$

Since

$$\frac{\Delta h f_d}{4} = \dot{S}_{\lim} \geq \left(\frac{\partial S}{\partial x} g(x)\right)(\max\{u^-, u^+\} - \min\{u_{eq}\}) \geq \dot{S}_{\max} \tag{12.42}$$

$$\Delta h \geq \frac{1}{4 f_d}\frac{\partial S}{\partial x} g(x)(u^+ - u^-)$$

guarantees inequality (12.41) holds.

Among the problems detected in the method, we can point out the difficult tuning of the commutation system, the need for an external signal and the effect of the synchronism signal on the resulting dynamics.

Finally, the duty cycle is defined so that the average of the sliding surface is zero in each commutation period [16, 17]; that is to say, the controller guarantees

$$\langle S(x,t)\rangle = \frac{1}{T}\int_{KT}^{(K+1)T} S(x,\tau)d\tau = 0 \tag{12.43}$$

The control algorithm is defined in Table 12.2, and the behaviour of the sliding surface is outlined using the definitions in Fig. 12.11. Note that the duty cycle only depends on the value of the switching surface function $S(x,t)$ and its derivatives.

12.4.2   Comparative study of the implementation methods

With the aim of comparing the aforementioned 'quasi-sliding' implementation strategies, the DC-AC buck power converter is considered. The signal to be

Table 12.2   ZAD control algorithm

$S_1 = S(x(t_k), t_k),\ S_2 = S(x(t_k), t_k) + \frac{T}{2}\dot{S}|_{(k,u^+)},\ S_3 = S(x(t_k), t_k) + \frac{T}{2}\dot{S}|_{(k,u^-)}$

$S_1 \geq 0$ and $S_2 \geq 0$	$u(t_k) = u^+;\ d_k = 1$
$S_1 \geq 0$ and $S_2 < 0$	$u(t_k) = u^+;\ d_k = 1 - \sqrt{\dfrac{\|\dot{S}\|_{(k,u^+)}\| - 2\frac{\|S(x(t_k),t_k)\|}{T}}{\|\dot{S}\|_{(k,u^+)}\| + \|\dot{S}\|_{(k,u^-)}\|}}$
$S_1 \leq 0$ and $S_3 \leq 0$	$u(t_k) = u^-;\ d_k = 1$
$S_1 \leq 0$ and $S_3 > 0$	$u(t_k) = u^-;\ d_k = 1 - \sqrt{\dfrac{\|\dot{S}\|_{(k,u^-)}\| - 2\frac{\|S(x(t_k),t_k)\|}{T}}{\|\dot{S}\|_{(k,u^+)}\| + \|\dot{S}\|_{(k,u^-)}\|}}$

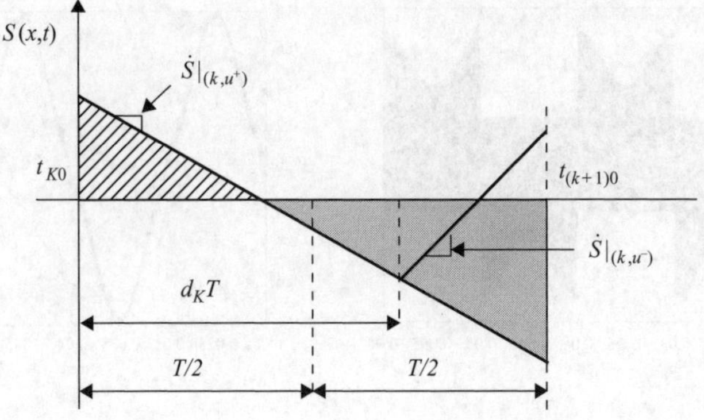

Figure 12.11   Zero Average Dynamics (ZAD) control

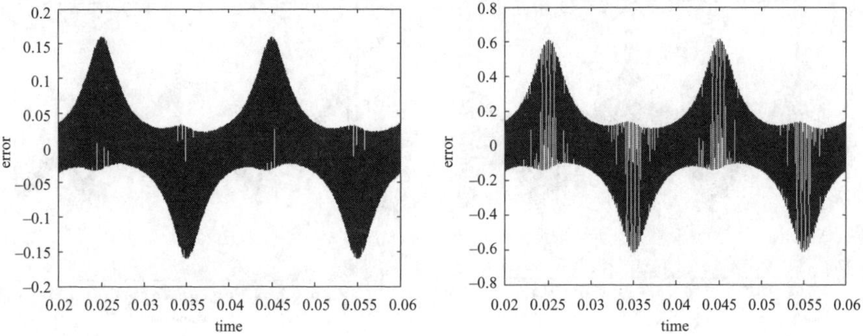

Figure 12.12   Voltage error with a fixed hysteresis cycle. $\Delta h = 0.5$ and $\Delta h = 1$

tracked is a sine wave. Simulation results are obtained for each implementation strategy using Matlab-Simulink and then compared. The switching surface is defined as $S = 0.5(V_{ref} - v) + 0.8 \cdot 10^{-4}(dV_{ref} - v)/dt$. The simulation parameters are $E = 50$ V, $L = 1.5$ mH, $C = 60\,\mu$F, $R = 20\,\Omega$, $f_r = 50$ Hz, the integrator is a $5 \cdot 10^{-8}$ fixed step Runge-Kutta 4-5.

Figure 12.12 shows the signal errors[4] when the sliding control law is implemented using a hysteresis cycle. $\Delta h = 0.5$ and $\Delta h = 1$ have been considered in the simulations, corresponding to a maximum switching frequency of 44 kHz and 22.5 kHz, respectively.

Figure 12.13 depicts the performance of the sliding surface $S(x, t)$ and the voltage error when an external signal of frequency 20 kHz and amplitude 1.2 is added to $S(x, t)$

---

[4] $error = V_{ref}(t) - v(t)$.

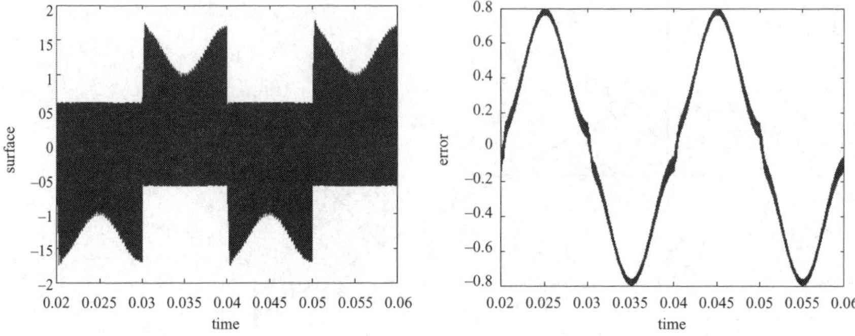

Figure 12.13 $S(x,t)$ *and voltage error with a fixed hysteresis cycle and a 20 kHz*
frequency external signal

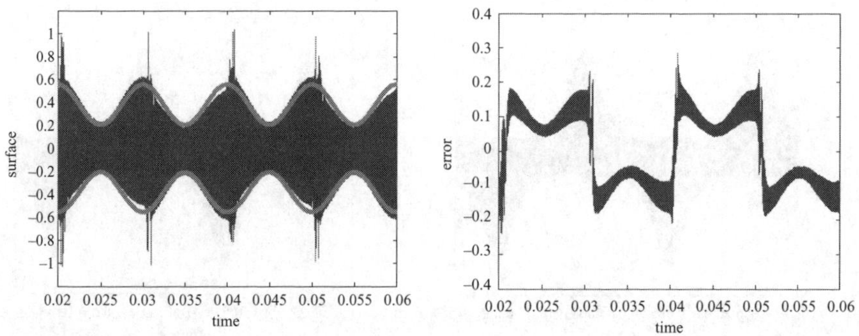

Figure 12.14 $S(x,t)$ *and voltage error with ZAD (20 kHz)*

and the hysteresis cycle is $\Delta h = 0.6$, which satisfies inequalities (12.40) and (12.41). As can be seen, the error is high compared to the previous case.

Results for a controller based on Zero Average Dynamics at 20 kHz do not differ much from the ones obtained by a hysteresis cycle. The performance of the function $S(x,t)$ and the error dynamics are displayed in Fig. 12.14. As can be seen in this figure, the envelope of the function $S(x,t)$ almost coincides with the values of the hysteresis cycle Δh obtained by defining $f_s = 20$ kHz in (12.39), solving for μ and replacing the result in (12.38). The best results are obtained with Pulse-Width-Modulation (see Fig. 12.15) which defines the duty cycle by comparing the equivalent control with a triangular signal. Specifically, the switching time is given by the signals intersection $0.5(1 + u_{eq}(kT + t))$ and t/T for $t \in [0, T)$, T being the switching period. The envelope of the function $S(x,t)$ coincides again with the graph of variable bandwidth hysteresis cycle Δh obtained from (12.39) and (12.38). Unfortunately, this method depends on the system parameters; thus, it is not suitable for practical implementation. The results achieved using a T-period Zero Order Hold in series with the equivalent control $d(k) = 0.5(1 + u_{eq}(kT))$ are depicted in Fig. 12.16. If this figure

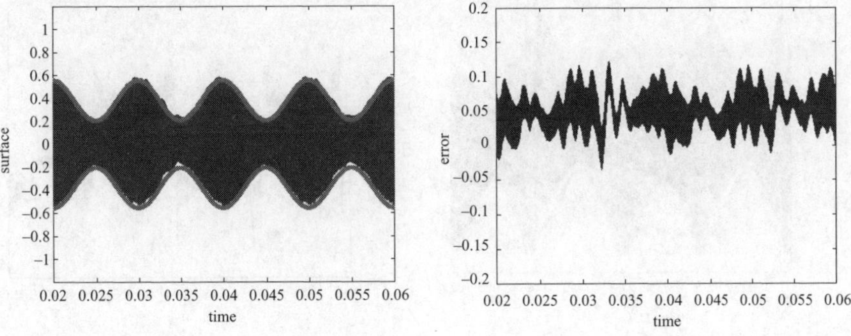

Figure 12.15 $S(x,t)$ *and voltage error using the equivalent control – without ZOH – as duty cycle (20 kHz)*

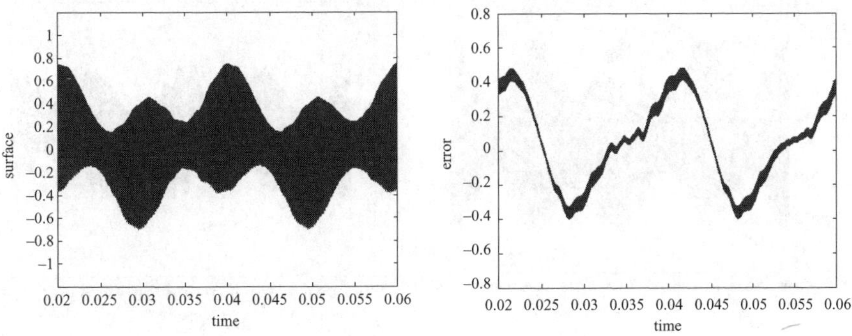

Figure 12.16 $S(x,t)$ *and voltage error using the equivalent control – with ZOH – as duty cycle (20 kHz)*

is compared with Fig. 12.15, the harmful effect of the sampling in the control action can be observed.

When evaluating the system at a frequency of 10kHz, the results worsen, (see Figs 12.17–12.19). As expected, this effect is stronger in those methods that require a Zero Order Hold. In spite of this, the ZAD method provides a maximum error of 2.5 per cent with respect to the output signal amplitude. This is the same error obtained with the strategy that uses the equivalent control with the Zero Order Hold. These results show that sampling with a Zero Order Hold prevents the envelope of the function $S(x,t)$ from coinciding with the value of the hysteresis cycle obtained if a technique based on a variable bandwidth hysteresis cycle and a 10 kHz fixed switching frequency were established.

12.4.3   Analogue electronic implementation

Quasi-sliding mode control can be implemented by analogue techniques as well as by programmable digital platforms. The advantages of analogue implementation lie

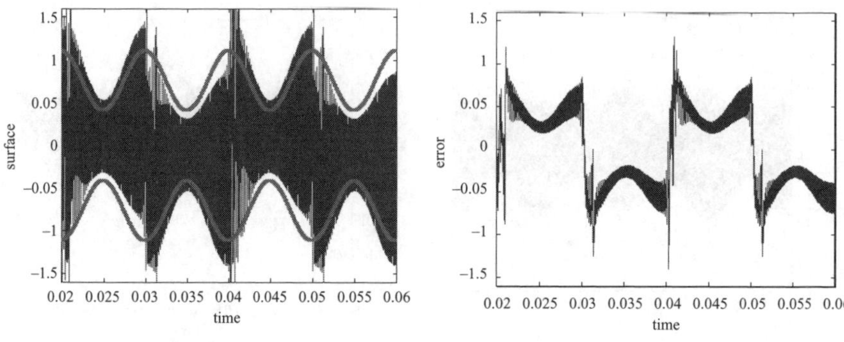

Figure 12.17 $S(x,t)$ and voltage error using ZAD (10 kHz)

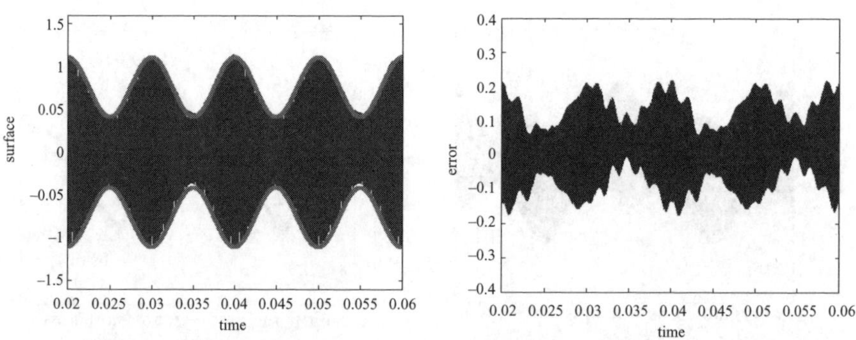

Figure 12.18 $S(x,t)$ and voltage error using the equivalent control – without ZOH – as duty cycle (10 kHz)

in the higher processing speed and in the capacity for integration. Analogue electronic platforms are based on circuits solving additions, derivatives, integrals and comparators.

As an illustrative example, Fig. 12.20 shows the implementation circuitry of a SMC with hysteresis for the regulation problem in the Buck converter. As can be seen in the diagram the voltage V_σ is the switching surface $\sigma = e + k(de/dt)$ appropriately scaled. The circuitry consists of

- A subtracting circuit that satisfies

$$V_{error} = \left(\frac{R_2}{R_1} + 1\right)\left(\frac{R_4}{R_3 + R_4}\right)V_{ref} - \left(\frac{R_2}{R_1}\right)V$$

  If $R_1 = R_2$ and $R_3 = R_4$, then $V_{error} = V_{ref} - V$.
- An inverter amplifier circuit that satisfies $V_p = -(R_6/R_5)V_{error}$.

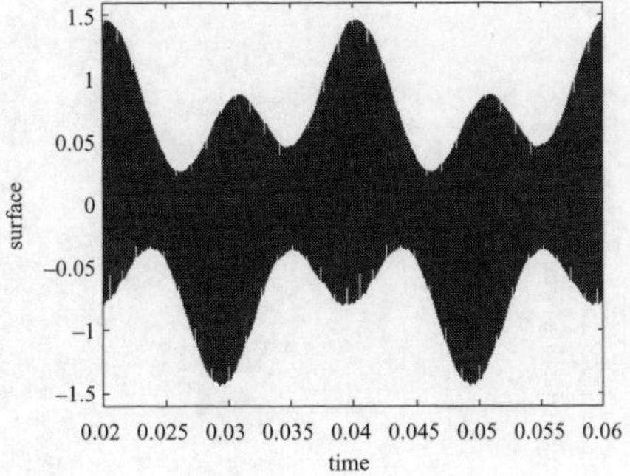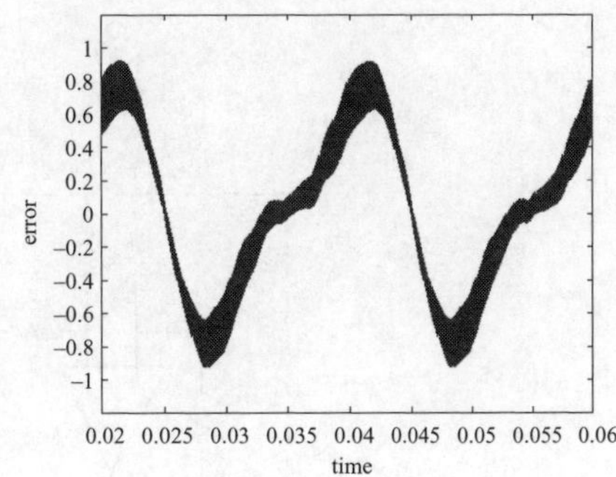

Figure 12.19   S(x,t) and voltage error using the equivalent control – with ZOH – as duty cycle (10 kHz)

- A derivative circuit that satisfies

$$V_d = -R_7 C_1 \frac{dV_{error}}{dt}$$

while $C_1 \gg C_2$. The capacitor C_1 must be used to reject switching noise.
- A summing circuit that satisfies

$$V_\sigma = -\left(\frac{R_{10}}{R_8}\right) V_p - \left(\frac{R_{10}}{R_9}\right) V_d$$

If $R_8 = R_9 = R_{10}$, then $V_\sigma = -\left(V_p + V_d\right)$.
- Finally, a hysteresis-cycle comparator circuit that satisfies

$$V_A = \left(\frac{R_{12}}{R_{13} + R_{12}}\right) V_{cc}$$

The hysteresis cycle is experimentally tuned by varying R_{12}.

Although there are circuits solving nonlinear operations, such as multipliers, exponentials, etc. tuning difficulties and sensitivity to noise and electronic interference lead designers to discard such implementations for complex control designs. There is ongoing research into analogue microelectronic implementations for quasi-sliding controllers. However, this has as yet not produced an automatic, easy-to-use procedure.

Some practical considerations have to be taken into account by designers. For instance, the current the diode and the transistor can manage is limited. In transient processes (start-up and step load changes) large discrepancies between the set point and measured values of the output voltage appear. This leads to a high current that can damage the semiconductors. A classical solution to this problem consists of defining two sliding surfaces (the original and a second one limiting the current) [7].

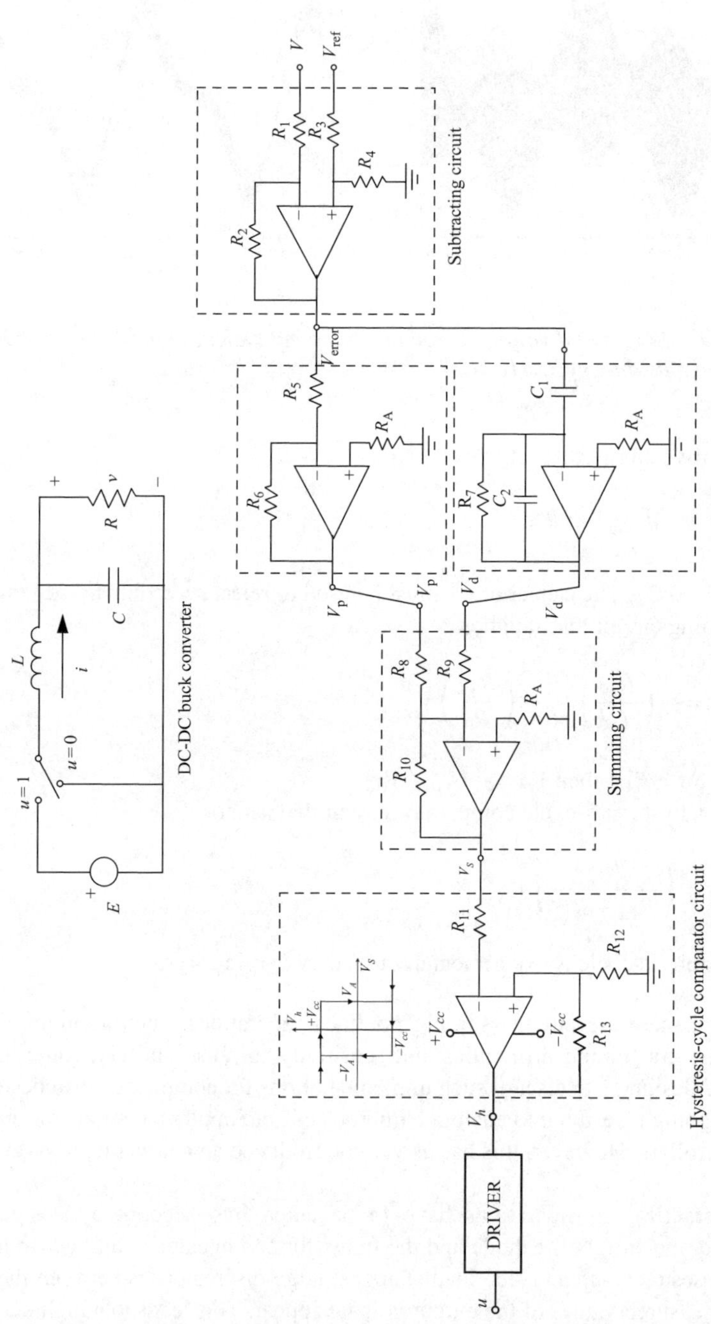

Figure 12.20   SMC analogue control circuitry

12.4.4 *Digital electronic implementations*

Digital implementations have some advantages over analogue implementations. For example, they are robust in the presence of electronic noise and interference, they are more flexible (changes only affect software and the corresponding hardware is not usually modified), and it is easier to manage complex functions using digital platforms rather than analogue ones. The first step in a digital implementation involves proper hardware selection. Several electronic devices can be considered, such as general-purpose microprocessors, Digital Signal Processors (DSP) and high-density programmable logic devices such as Field Programmable Gate Arrays (FPGA) or Complex Programmable Logic Devices (CPLD).

The selection procedure is based on several features such as processing speed, device capability, design environment and device price. The designer must consider the most restrictive requirement in order to select the right electronic device. For instance, a DSP is a good choice when the processor must solve complicated processing functions. The FPGA is the most appropriate when a high process speed is needed due to its implementation based on inner hardware. Process speed is one of the factors to consider in a digital implementation. This means that the control signal delay should be neglected with respect to the time characteristics of the process to be controlled (the switching period for power converters). In the control frame, particularly with sliding mode control, the processing speed is crucial and the designer should guarantee real-time control. Otherwise, delays must be included in the sliding mode control design procedure. Programming the controller directly on the hardware platform helps to reduce processing time.

Signal sampling is an important aspect for the implementation procedure. It can be managed through two parameters: the sampling rate, which must not restrict the system behaviour, and the number of A/D conversion bits, which must not add an important quantisation error. It is therefore necessary to choose appropriate A/D converters (minimum 12th bits).

12.5 Example: a ZAD inverter

This section is devoted to presenting some experimental results of the fixed switching frequency ZAD inverter. The ZAD algorithm, which was already introduced in the above section, has been programmed in a FPGA device. The algorithm is solved in less than 6 per cent of the switching period (42.67 microseconds), leading to a real-time controller implementation. The full-bridge buck inverter has been built with the following parameters:

- Buck converter: $E = 70$ V, $C = 80\,\mu$F, $L = 1.5$ mH, $R = 10\,\Omega$.
- Switching surface: $S := 1250(V_{ref} - v_0) + (V_{ref} - v_0)/dt$.
- Switching frequency $= 23$ kHz and the desired output voltage is $V_{ref}(t) = 45\sin(2\pi 50 t)$ V.

Figure 12.21 shows a block diagram of the XC4010E FPGA-based implementation of the ZAD quasi-sliding control algorithm. This block diagram includes an analogue signal conditioner, an analogue-to-digital converter (ADC) and an FPGA programmable logic device with its corresponding external clock and EEPROM memory to store the FPGA configuration.

The signal conditioner supplies the value of the switching surface to the ADC, and is designed by means of conventional OpAmp-based circuitry. The buck inverter output voltage is sensed by means of an AD215BY wideband isolation amplifier, whereas the capacitor current is acquired with an LA25-NP current sensor. The computational procedure [18] is based on a switching surface synchronous sampling at twice the desired switching frequency. Then, as can be seen in Fig. 12.22, during the Kth period the following samples are known:

$$S_1 = S(x(t_{K_0}), t_{K_0}); \quad S_2 = S\left(x\left(t_{K_0} + \frac{T}{2}\right), t_{K_0} + \frac{T}{2}\right);$$

$$S_3 = S(x(t_{(K+1)_0}), t_{(K+1)_0})$$

It should be pointed out that the values of S_1 and S_3 are obtained by sampling 826 ns prior to the end of the period to avoid switching noise.

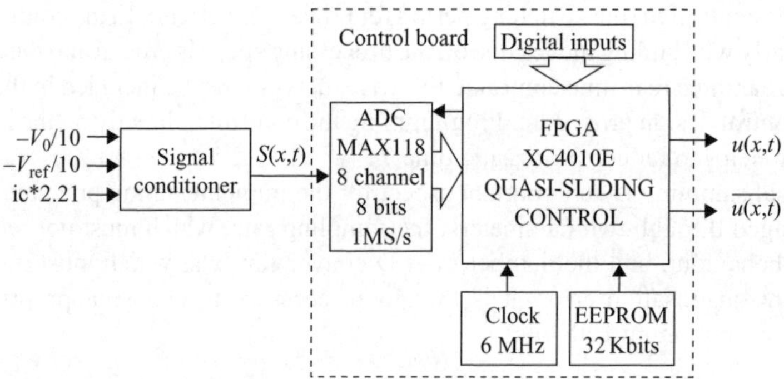

Figure 12.21   ZAD quasi-sliding control block diagram

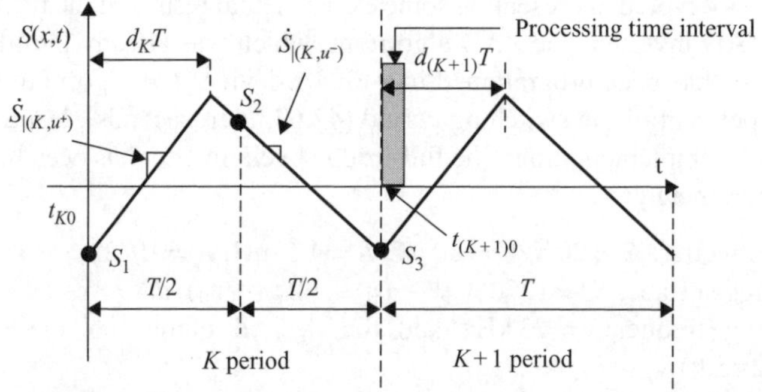

Figure 12.22   Algorithm procedure

Assuming that d_K, the duty cycle of the Kth period, is known, the first step is the computation of parameter D, defined as:

$$D = |\dot{S}_{|(K,u^+)}| + |\dot{S}_{|(K,u^-)}| \tag{12.44}$$

which corresponds to the denominator of the control law given in Table 12.2. This parameter can be easily calculated from S_1, S_2, S_3 and d_K. That is, if $S_1 < 0$ and $d_k \leq 0.5$ (as depicted in Fig. 12.22), the following relations hold:

$$\begin{cases} \dot{S}_{|(K,u^-)} = \dfrac{2(S_3 - S_2)}{T} \\ S_3 = S_1 + \dot{S}_{|(K,u^+)}d_K T + \dot{S}_{|(K,u^-)}(1 - d_K)T \end{cases} \tag{12.45}$$

hence

$$D = \frac{2S_2 - S_1 - S_3}{d_K T} \tag{12.46}$$

Similarly, the expressions for the parameter D depending on the sign of S_1 and the value of d_K can be easily derived. These expressions, normalised with respect to the switching period T, are summarised in Table 12.3.

It should be noted that these expressions may be applied provided that the two derivatives of (12.44) are defined during the Kth period, which implies that the control value switches during the period. However, in the transient state the ZAD algorithm holds the control action and the switching surface may remain positive (or negative) throughout the period. In this case, one of the two derivatives of (12.44) is not defined. Nevertheless, parameter D can be deduced as

$$D = \frac{\partial S}{\partial x} g(x)(u^+ - u^-) = \frac{2E}{LC} \tag{12.47}$$

The FPGA implementation algorithm can identify this fact and then assign the value given by (12.47) to D, which has been previously introduced and stored in the FPGA. Once the value of D is known, the next step is the computation of the switching surface derivatives $\dot{S}_{|(K,u^+)}$ and $\dot{S}_{|(K,u^-)}$. As shown in Table 12.4, these derivatives may also be easily computed from the values of S_1, S_2, S_3 and D. The value of $d_{(K+1)}$ is computed by assuming that the switching surface derivatives vary slowly with

Table 12.3   Expressions for parameter D

	$d_K > 0.5$	$d_K \leq 0.5$
$S_1 \geq 0$	$D = \dfrac{S_1 + S_3 - 2S_2}{1 - d_K}$	$D = \dfrac{S_1 + S_3 - 2S_2}{d_K}$
$S_1 < 0$	$D = \dfrac{2S_2 - S_1 - S_3}{1 - d_K}$	$D = \dfrac{2S_2 - S_1 - 2S_3}{d_K}$

Table 12.4   Switching surface derivatives in terms of S_1, S_2, S_3 and D

	$d_K > 0.5$	$d_K \leq 0.5$
$S_3 \geq 0$ and $S_1 \geq 0$	$\lvert\dot{S}_{\lvert(k,u+)}\rvert = 0 - 2(S_2 - S_1)$	$\lvert\dot{S}_{\lvert(k,u+)}\rvert = D - 2(S_3 - S_2)$
$S_3 \geq 0$ and $S_1 < 0$	$\lvert\dot{S}_{\lvert(k,u+)}\rvert = D - 2(S_2 - S_1)$	$\lvert\dot{S}_{\lvert(k,u+)}\rvert = 0 - 2(S_3 - S_2)$
$S_3 < 0$ and $S_1 \geq 0$	$\lvert\dot{S}_{\lvert(k,u-)}\rvert = D + 2(S_2 - S_1)$	$\lvert\dot{S}_{\lvert(k,u-)}\rvert = 0 + 2(S_3 - S_2)$
$S_3 < 0$ and $S_1 < 0$	$\lvert\dot{S}_{\lvert(k,u-)}\rvert = 0 + 2(S_2 - S_1)$	$\lvert\dot{S}_{\lvert(k,u-)}\rvert = D + 2(S_3 - S_2)$

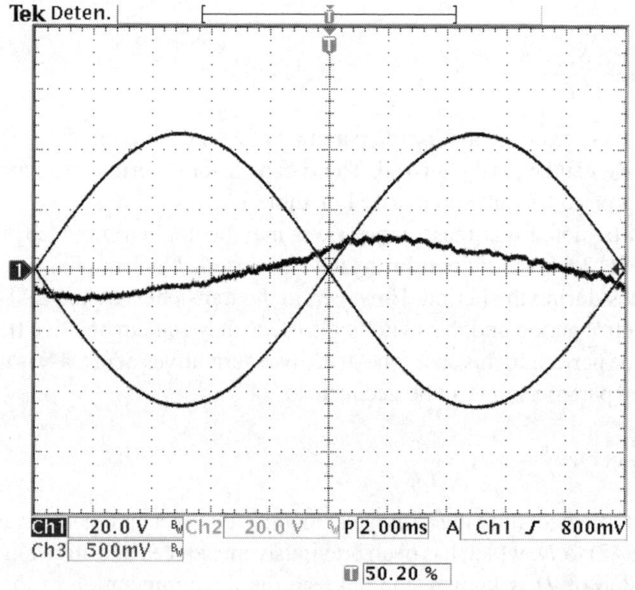

Figure 12.23   Measured output voltage (20 V/div), reference voltage (20V/div) and voltage error (0.5V/div)

respect to the switching period (this assumption is reasonable due to the low output voltage ripple), which enables the following approximation:

$$\dot{S}_{\lvert(K,u+)} \sim \dot{S}_{\lvert(K+1,u+)}; \quad \dot{S}_{\lvert(K,u-)} \sim \dot{S}_{\lvert(K+1,u-)} \tag{12.48}$$

For instance, in the case of the second row of Table 12.2, the duty cycle is finally computed as:

$$d_{k+1} \simeq 1 - \sqrt{\frac{\lvert\dot{S}_{\lvert(K,u+)}\rvert - 2(\lvert S(x(t_{(K+1)_0}), t_{(K+1)_0})\rvert/T)}{\lvert\dot{S}_{\lvert(K,u+)}\rvert + \lvert\dot{S}_{\lvert(K,u-)}\rvert}} \tag{12.49}$$

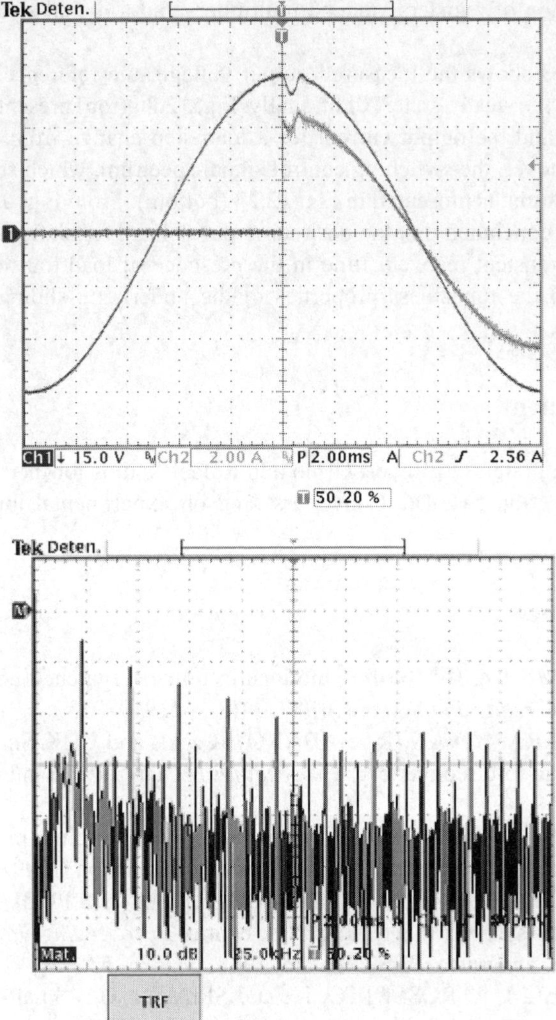

Figure 12.24   *Measured output voltage (15V/div) and output current (2A/div) for a load step change from open circuit to $R = 10\,\Omega$ (top). Switching control signal spectrum (bottom)*

which can be rewritten in terms of S_1, S_2, S_3 and, according to Tables 12.3 and 12.4, as:

$$d_{K+1} \simeq 1 - \sqrt{\frac{((S_1 + S_3 - 2S_2)/d_K) + 2S_2 - 4S_3}{(S_1 + S_3 - 2S_2)/d_K}} \qquad \text{for } d_K < 0.5 \qquad (12.50)$$

The FPGA algorithm both identifies the different cases of Tables 12.2–12.4 and computes the corresponding expressions. These tasks are carried out by means of the proper connection of registers, adders, multipliers and a digital square root circuit extractor.

Figure 12.23 shows the measured output voltage, the reference signal and the voltage error in the steady-state. Additionally, Fig. 12.24 (top) presents the measured output voltage and the output current for a load step change from open circuit to $R = 10\,\Omega$. Moreover, the switching control signal spectrum, which shows the fixed-frequency operation, is presented in Fig. 12.24 (bottom). From Figs 12.23 and 12.24 (top), it can be concluded that the desired output voltage regulation is achieved, as well as a fast transient recovery time in the presence of load transients. This may be attributed to the robustness properties of the underlying sliding mode control principles.

Acknowledgment

The authors are grateful to Robert Griñó and Rafael Ramos for their valuable comments in the section 'AC-DC Converters' and on experimental implementations, respectively.

12.6   References

1 SIRA-RAMÍREZ, H.: 'Sliding motions in bilinear switched networks', *IEEE Trans. on Circuits and Systems*, 1987, **34**(8), pp. 919–933

2 VENKATARAMANAN, R., SABANOVIĆ, A., and ĆUK S.: 'Sliding mode control of dc-to-dc converters'. *Proceedings IECON 1985*, San Francisco, 1985, pp. 251–258

3 UTKIN, V. I., GULDNER, J., and SHI., J.: 'Sliding mode control in electromechanical systems' (Taylor and Francis, London, 1999)

4 CARPITA, M., MARCHESIONI, M., OBERTI, M., and PUGUISI. L.: 'Power conditioning system using sliding mode control'. *Proceedings Power Electronic Specialist Conference* (PESC) Kyoto, 1988, pp. 623–633

5 MATTAVELLI, P., ROSSETTO, L., and SPIAZZI, G.: 'Small-signal analysis of dc-to-dc converters with sliding mode control', *IEEE Transactions on Power Electronics*, 1997, **12**(1), pp. 96–102

6 SIRA-RAMÍREZ, H. and ILIC, M.: 'A geometric approach to the feedback control of switch mode dc-to-dc power supplies', *IEEE Trans. on Circuits and Systems*, 1988, **35**, pp. 1291–1298

7 BILALOVIĆ, F., MUŠIĆ, O., and ŠABANOVIĆ, A.: 'Buck converter regulator operating in the sliding mode'. *Proceedings 7th International Power Conversion Conference (PCI)*, Orlando, 1983, pp 331–340

8 ESCOBAR, G., CHEVREAU, D., ORTEGA, R., and MENDES. E.: 'An adaptive passivity-based controller for a unity power factor rectifier', *IEEE Trans. on Control Systems Technology*, 2001, **9**(4), pp. 637–644

9 SIRA-RAMÍREZ, H.: 'Differential geometric methods in variable structure control', *Int. J. Control*, 1988, **48**, pp. 1359–1390

10   BÜHLER, H.: 'Réglage par mode de glissement' (Presses Polytechniques Romandes, Lausanne, 1986)

11   NICOLAS, B., FADEL, M., and CHÉRON, Y.: 'Sliding mode control of dc-to-dc converters with input filter based on the Lyapunov-function approach'. *Proceedings of European Power Electronics Conference (EPE)*, Seville, 1995, pp. 1338–1343

12   RUIZ, J. M., LORENZO, S., LOBO I., and AMIGO, J.: 'Minimal ups structure with sliding mode control and adaptive hysteresis band'. *Proceedings of International Conference on Industrial Electronics Control and Instrumentation (IECON)*, Asilomar, California, 1990, pp. 1063–1067

13   MALESANI, L., ROSSETTO, L., SPIAZZI, G., and ZUCCATO, A.: 'An ac power supply with sliding-mode control', *IEEE Industry Applications Magazine*, 1996, pp. 32–38

14   SILVA, J. F. and PAULO, S. S.: 'Fixed frequency sliding modulator for current mode pwm inverters'. *Proceedings of Power Electronic Specialist Conference (PESC)*, San Francisco, 1993, pp. 623–629

15   PINHEIRO, H., MARTINS, A. S., and PINHEIRO, J. R.: 'A sliding mode controller in single phase voltage source inverters'. *International Conference on Industrial Electronics Control and Instrumentation (IECON)*, Bologna, 1994, pp. 394–398

16   FOSSAS, E., GRIÑÓ, R., and BIEL. D.: 'Quasi-sliding control based on pulse width modulation, zero averaged dynamics and the L2 norm' (World Scientific, Singapore, 2001), pp. 335–344

17   FOSSAS, E., BIEL, D., RAMOS, R., and SUDRIÁ A.: 'Programmable logic device applied to the quasi-sliding control implementation based on zero averaged dynamic'. 40th IEEE Conference on *Decision and Control (CDC'01). Orlando, Florida, USA*, 2001, pp. 1825–1830

18   RAMOS, R., BIEL, D., FOSSAS, E., and GUINJOAN, F.: 'A fixed-frequency quasi-sliding control algorithm: application to power inverters design by means of fpga implementation', *IEEE Transactions on Power Electronics*, 2003, **18**(1), pp. 344–355

Chapter 13

Sliding modes in motion control systems

Asif Sabanovic and Karel Jezernik

13.1 Introduction

The complexity and nonlinear dynamics of motion control systems, along with the high-performance required during operation, require complex, often nonlinear control system design, to fully exploit system capabilities. Basic goals for motion control systems include trajectory tracking, velocity control and control of the force exerted by the system on the environment with torque or force as the control input. The torques or forces are on the other hand the outputs of actuators, often electrical motors, with their own complex nonlinear dynamics. In most approaches to motion control systems, the dynamics of torque or force is neglected and controllers are designed assuming perfect tracking in the torque or force control loop, which is not the case in many systems and such a design procedure may create some difficulties in systems with high demands. Direct application of discontinuous control in motion control systems with torque or force as the input may lead to chattering [1, 2] and some precaution must be taken in order to overcome chattering related problems. One of the ways to accomplish this is by smoothing the control in the ε-vicinity of the sliding mode manifold [1, 2] or in the framework of the discrete-time sliding mode design (due to the fact that in such systems the control action may be continuous [3]). Another solution for avoiding chattering is to augment the description of the system with torque or force dynamics – actuator dynamics.

In this chapter, the main problems in motion control systems such as position tracking, force (torque) control along with control and state estimation in induction electrical machines will be discussed. In the first part, a generalised approach to sliding mode control in motion control systems will be presented with some illustrative examples. We will then discuss the control of induction machines as one example of systems that include fast dynamics associated with the electromagnetic system and will present the latest results in the application of sliding mode methods for induction machine state and parameter estimation.

13.2 SMC in motion control system

For a 'fully actuated' mechanical system (number of actuators equal to the number of the primary masses) the mathematical model may be found from the Euler-Lagrange formulation in the following form:

$$\dot{\mathbf{q}}_1 = \mathbf{q}_2$$
$$\mathbf{M}(\mathbf{q}_1)\dot{\mathbf{q}}_2 + \mathbf{N}(\mathbf{q}_1, \mathbf{q}_2, t) = \boldsymbol{\tau} - \mathbf{G}_{ext} \tag{13.1}$$

where $\mathbf{q}_1 \in \Re^n$ stands for the vector of generalised positions, $\dot{\mathbf{q}}_1 = \mathbf{q}_2$ stands for the vector of generalised velocities, $\mathbf{M}(\mathbf{q}_1) \in \Re^{n \times n}$ is the generalised positive definite inertia matrix with bounded parameters hence $M^- \leq \|\mathbf{M}(\mathbf{q}_1)\| \leq M^+$, $\mathbf{N}(\mathbf{q}_1, \mathbf{q}_2, t) \in \Re^{n \times 1}$ represents the vector of coupling forces including gravity and friction and is bounded by $\|\mathbf{N}(\mathbf{q}_1, \mathbf{q}_2, t)\| \leq N^+$, $\boldsymbol{\tau} \in \Re^{n \times 1}$ with $\|\boldsymbol{\tau}\| \leq \tau_0$ as the vector of generalised input forces and $\mathbf{G}_{ext} \in \Re^{n \times 1}$ with $\|\mathbf{G}_{ext}\| \leq g_0$ as the vector of generalised external forces. M^-, M^+, N^+, τ_0, g_0 are the known scalars. Note that many different norms may be employed but the most common one is the 2-norm. The interested reader is referred to textbooks on robotics for a detailed treatment of the derivation of equations (13.1). In system (13.1), the vector $(\mathbf{N}(\mathbf{q}_1, \mathbf{q}_2, t) + \mathbf{G}_{ext})$, which contains most of the unknown parameters of the system, can be treated as a disturbance vector satisfying matching conditions [4]. The model (13.1) may be rewritten as n second order systems of the form

$$\dot{q}_1 = q_2$$
$$m_{ii}\dot{q}_{i2} + n_i = \tau_i - g_{exti} - \sum_{j=1, j \neq i}^{n} m_{ij}\dot{q}_{j2}, \qquad i = 1, \dots, n \tag{13.2}$$

where the elements of the inertia matrix are bounded $m_{ij}^- \leq |m_{ij}(t)| \leq m_{ij}^+$, the elements of the vector $\mathbf{N}(\mathbf{q}_1, \mathbf{q}_2, t)$ are bounded $n_i^- \leq |n_i(t)| \leq n_i^+$ and the elements of the external force vector are bounded by $g_{0i}^- \leq |g_{exti}(t)| \leq g_{0i}^+$ and the input generalised torques are bounded $\tau_{0i}^- \leq |\tau_i(t)| \leq \tau_{0i}^+$.

13.2.1 Control problem formulation

The configuration of a mechanical system is defined by vectors of generalised positions and generalised velocities, thus allowing motion control problems to be defined as a requirement to enforce certain dependence between generalised coordinates $\sigma(\mathbf{q}_1, \mathbf{q}_2, t) = \mathbf{0}$; $\sigma \in \Re^n$. In general, that dependence may be expressed by a nonlinear function. Without any loss of generality, in this chapter we will assume $\sigma(\mathbf{q}_1, \mathbf{q}_2, t) = \mathbf{0}$ is linear with respect to the generalised vectors as depicted in (13.3):

$$\sigma(\mathbf{q}_1, \mathbf{q}_2, t) = C\mathbf{q}_1 + \mathbf{q}_2 - \mathbf{f}(t) = \mathbf{0}, \qquad \sigma(\mathbf{q}_1, \mathbf{q}_2, t) \in \Re^{n \times 1}, \quad \mathbf{C} > 0,$$
$$\sigma = [\sigma_1, \sigma_2, \dots, \sigma_n]^T \tag{13.3}$$

where $\mathbf{f}(t) \in \Re^{n \times 1}$ is the known continuous and bounded function of time $\|\mathbf{f}(t)\| \leq f_0$ with a continuous and bounded first time derivative. Requirement (13.3) is equivalent to enforcing a sliding mode in the manifold $S_q = \{\mathbf{q}_1, \mathbf{q}_2 : \boldsymbol{\sigma}(\mathbf{q}_1, \mathbf{q}_2, t) = \mathbf{0}\}$, elements of $\boldsymbol{\sigma}(\mathbf{q}_1, \mathbf{q}_2, t)$ being $\sigma_i = c_i q_{i1} + q_{i2} - f_i(t)$, $i = 1, 2, \ldots, n$. If a sliding mode is established in the manifold (13.3), then the equivalent control, being the solution of $\dot{\boldsymbol{\sigma}}|_{\tau=\tau_{eq}} = \mathbf{C}\dot{\mathbf{q}}_1 + \dot{\mathbf{q}}_2 - \dot{\mathbf{f}}(t)|_{\tau=\tau_{eq}} = \mathbf{0}$, is determined as

$$\boldsymbol{\tau}_{eq} = \mathbf{M}(\dot{\mathbf{f}}(t) - \mathbf{C}\mathbf{q}_2) + \mathbf{N} + \mathbf{G}_{ext}, \tag{13.4}$$

and the equations of motion (13.1) with a sliding mode in manifold (13.3) are reduced to $\mathbf{q}_2 = \mathbf{f}(t) - \mathbf{C}\mathbf{q}_1$. Consequently, sliding mode control may be effectively applied in motion systems (13.1) to control problems that may be defined as depicted in (13.3).

For robotic systems, position tracking and force tracking are the two basic control problems. Selecting the reference trajectory as $\mathbf{q}_1^{ref}(t)$, the position-tracking problem can be specified as a requirement that a sliding mode is enforced in the manifold (13.5)

$$S_{q_1} = \{\mathbf{q}_1, \mathbf{q}_2 : \boldsymbol{\sigma}(\mathbf{q}_1, \mathbf{q}_2, t) = \mathbf{C}(\mathbf{q}_1^{ref} - \mathbf{q}_1) + (\dot{\mathbf{q}}_1^{ref} - \mathbf{q}_2) = \mathbf{0}, \mathbf{C} > 0\}$$

$$S_{q_1} = \{\mathbf{q}_1, \mathbf{q}_2 : \sigma(\mathbf{q}_1, \mathbf{q}_2, t) = \mathbf{f}(t) - (\mathbf{C}\mathbf{q}_1 + \mathbf{q}_2)\}, \mathbf{f}(t) = (\mathbf{C}\mathbf{q}_1^{ref} + \mathbf{q}_2^{ref}) \tag{13.5}$$

Assume that the contact force can be modelled as

$$\mathbf{F} = \mathbf{K}(\mathbf{q}_{e1} - \mathbf{q}_1) + (\dot{\mathbf{q}}_{e1} - \mathbf{q}_2) \tag{13.6}$$

where $\mathbf{q}_{e1}$ is the generalised coordinate of the contact point of the robot tip with the environment, $\mathbf{K} > 0$ is the spring coefficient matrix. The force control problem in which the contact force $\mathbf{F}$ should track its reference $\mathbf{F}^{ref}(t)$ can be specified as a requirement that a sliding mode is enforced in the manifold (13.7)

$$S_f = \{\mathbf{q}_1, \mathbf{q}_2, t : \mathbf{F}^{ref} - (\mathbf{K}(\mathbf{q}_{e1} - \mathbf{q}_1) + \mathbf{q}_{e2} - \mathbf{q}_2) = \mathbf{0}\}$$

$$S_f = \{\mathbf{q}_1, \mathbf{q}_2 : \boldsymbol{\sigma}(\mathbf{q}_1, \mathbf{q}_2, t) = -\mathbf{f}(t) + (\mathbf{K}\mathbf{q}_1 + \mathbf{q}_2)\}, \tag{13.7}$$

$$\mathbf{f}(t) = -(\mathbf{F}^{ref} - \mathbf{K}\mathbf{q}_{e1} - \mathbf{q}_{e2})$$

Both the trajectory tracking problem (13.5) and the force control problem (13.7) are mathematically defined in the same way as the general motion control problem (13.3) and thus both can be solved in the framework of sliding mode control systems by enforcing a sliding mode in selected manifolds. Moreover the combination of the two tasks is natural since it only requires a change of the siding mode manifold.

13.2.2   Selection of control input

A few different approaches may be used to design the control inputs for system (13.1), (13.2) with a sliding mode in manifold (13.3). Here we will discuss some of the possibilities in order to demonstrate the richness of the sliding mode design approach to motion control systems.

Discontinuous control. First we will demonstrate a straight forward sliding mode approach by selecting a discontinuous control input [5]. In this framework the control is selected in the following form (13.8)

$$\boldsymbol{\tau} = -\tau_0 \operatorname{sign}(\boldsymbol{\sigma}) \Rightarrow \tau_i = -\tau_{0i} \operatorname{sign}(\sigma_i), \qquad i = 1, \ldots, n \tag{13.8}$$

The existence of a sliding mode in manifold (13.3) can be proven by selecting, for each component σ_i of the sliding mode function, a Lyapunov function candidate as $v_i = \frac{1}{2}\sigma_i^2$ ($i = 1, \ldots, n$). Due to the fact that the matrix that multiplies the control is diagonal, such a selection of the Lyapunov function candidate is consistent with the results presented in Chapter 1. Time derivatives $\dot{v}_i = \sigma_i \dot{\sigma}_i$ along the trajectories of the system (13.2) with control (13.8) are

$$\dot{v}_i = \sigma_i \left(c_i \dot{q}_{i1} - f_i + \frac{1}{m_{ii}} \left(\tau_i - g_{exti} - n_i - \sum_{j=1,j\neq i}^{n} m_{ij}\dot{q}_{j2} \right) \right) \tag{13.9}$$

The derivative of functions $f_i(t)$ as well as the elements of inertia matrix, the elements n_i ($i = 1, 2, \ldots, n$) of vector $\mathbf{N}(\mathbf{q}_1, \mathbf{q}_2, t)$ and the elements of the external force vector are bounded. This assumption guarantees the existence of

$$\dot{v}_i \leq -\frac{\tau_{0i}}{m_{ii}^+}|\sigma_i| + |\sigma_i| \left(\dot{f}_i^+ + c_i|q_{i2}| + \frac{1}{m_{ii}^-} \left(g_{exti}^+ + n_i^+ + \sum_{j=1,j\neq i}^{n} m_{ij}^+ |\dot{q}_{j2}| \right) \right)$$

$$\tag{13.10}$$

With the amplitude of control

$$\tau_{0i} > m_{ii}^+ \left(\dot{f}_i^+ + c_i|q_{i2}| + \frac{1}{m_{ii}^-} \left(g_{exti}^+ + n_i^+ + \sum_{j=1,j\neq i}^{n} m_{ij}^+ |\dot{q}_{j2}| \right) \right)$$

the time derivative of the Lyapunov function candidate becomes

$$\dot{v}_i \leq -\mu|\sigma_i|, \qquad \mu > 0 \tag{13.11}$$

Consequently the convergence to the intersection of the manifolds $\sigma_i = 0$ is established. Each component of the control input undergoes discontinuity by taking values from the set $\{-\tau_{0i}, \tau_{0i}\}$. Direct implementation of algorithm (13.8) may result in chattering so it may not be suitable for direct application. An approach to reduce the effect of the discontinuous control is to implement (13.8) as $\tau_i = \hat{\tau}_{eq}^{est} - \tau_{0i}\,\text{sign}(\sigma_i)$ where $\hat{\tau}_{eq}^{est}$ is the estimated control torque that may be calculated either from the system's model using available measurements and estimated parameters or from disturbance estimation. In this case the discontinuous part of the control may be small depending on how close the estimate is to its real value. Another approach for chattering elimination suggested [1, 2] is a continuous approximation of the discontinuous control in a δ-vicinity of the sliding mode manifold. In many cases this is not a remedy for the problem. In References 6 and 7 it is shown that chattering caused by unmodelled dynamics may be eliminated in systems with asymptotic observers, where the observers serve as a bypass for the high frequency component. Another solution is to apply a discrete-time sliding mode design procedure, which results in a continuous control as discussed in Chapter 5 or a higher order sliding mode approach as discussed in Chapter 6.

Discrete-time sliding mode control. In contrast to continuous time SMC, in discrete-time SMC motion in the sliding mode manifold may occur if the control is continuous [3, 8, 9]. The discrete-time implementation of the sliding mode control is essentially the application of the equivalent control determined as a solution of $\sigma_{k+1}|_{u_k=u_k^{eq}} = 0$. Such implementation requires information on parameters, system states and external disturbances and may not be easy to apply in some motion control systems due to the nonlinearity of the system and large variations of parameters. Discrete time realisation of systems with calculation of the equivalent control may be greatly simplified using estimation techniques aimed at deriving a value of the equivalent control. In this framework, disturbance observers, which will be discussed later in this chapter, and neural networks, which are discussed in a separate chapter, may provide very useful results.

Here a realisation of discrete-time sliding mode control, which produces motion that, strictly speaking, does not ensure a finite reaching time but results in a smooth control assuring quasi-sliding mode, is discussed. The approach is based on enforcing a certain structure of the time derivative for the selected Lyapunov function candidate. For system (13.1), asymptotic stability of the solution $\sigma(\mathbf{q}_1, \mathbf{q}_2, t) = C\mathbf{q}_1 + \mathbf{q}_2 - \mathbf{f}(t) = \mathbf{0}$ can be assured if a control input is selected such that the Lyapunov function candidate $v_l = (\sigma^T \sigma)/2$ has time derivative $\dot{v}_l = -\sigma^T \mathbf{D}\sigma$, $\mathbf{D} > 0$, [7] (for simplicity in most of the cases $\mathbf{D} = \text{diag}\{d_{ii}\}$). After some algebra, one can obtain $\dot{v}_l = -\sigma^T \dot{\sigma} = -\sigma^T \mathbf{D}\sigma$, $\mathbf{D} > 0$, and $\sigma^T (\dot{\sigma} + \mathbf{D}\sigma) = \mathbf{0}$ which depends on the control due to the presence of the term $\dot{\sigma}$. The control can be selected to enforce $(\dot{\sigma} + \mathbf{D}\sigma)|_{\sigma \neq 0} = \mathbf{0}$. By applying the sample and hold process with the sampling interval T, the discrete-time control that satisfies the given requirements can be determined as

$$\tau(k) = \tau(k-1) + T^{-1}[(1 + \mathbf{D}T)\sigma(k) - \sigma(k-1)], \qquad \mathbf{D} > 0 \qquad (13.12)$$

Application of the approximated control (13.12) to system (13.1), (13.3) leads to

$$\sigma^T(k)\sigma(k-1) = \sigma^T(k)(\mathbf{I} - T\mathbf{D})\sigma(k) \qquad (13.13)$$

If $\mathbf{D}$ is a diagonal matrix, then for each of the components in (13.13), one can write $\sigma_i(k)\sigma_i(k-1) = \sigma_i^2(k)(1 - Td_{ii})$ and d_{ii} may be selected so that $0 < (1 - Td_{ii}) < 1$, which ensures existence of a quasi-sliding mode. This solution is similar to the so-called β-equivalent sliding mode control approach [10].

13.2.3   Sliding mode disturbance observer

Disturbance compensation is an established design approach in the framework of motion control systems. The main idea is very simple – the disturbance observer is constructed and the output of such an observer is fed to the control input of the system. As a result, an augmented system consisting of the original plant and disturbance observer appears linearised and then the control should be selected for such an augmented system. A PD controller in most cases satisfies the system's requirements. In such a framework, sliding mode methods can be applied for the design of a disturbance observer and the sliding mode controller of the augmented plant. Sliding mode application for disturbance estimation can be explained for the ith subsystem in (13.2). The system can be rewritten as

$$\hat{m}_{ii}\dot{q}_{i2} = \tau_i - n_i - g_{exti} - \sum_{j=1, j\neq i}^{n} m_{ij}\dot{q}_{j2} - (m_{ii} - \hat{m}_{ii})\dot{q}_{i2} = \tau_i - d_i \qquad (13.14)$$

where $\hat{m}_{ii}$ is the estimated value of the inertia and d_i stands for the total disturbance for which some components may be known or measured. Assuming that q_{i2} is measured and d_e represents the known part of the total disturbance, a model (observer) for the system (13.14) can be constructed in the form

$$\hat{m}_{ii}\dot{\hat{q}}_{i2} = \tau_i - d_e - u_i \qquad (13.15)$$

where u_i stands for the model control input. By selecting u_i, the sliding mode is enforced on $\sigma_i = q_{i2} - \hat{q}_{i2} = 0$ and one can find the value of equivalent control

$$u_{ieq} = d_i - d_e = \left(n_i + g_{exti} + \sum_{j=1, j\neq i}^{n} m_{ij}\dot{q}_{j2} + (m_{ii} - \hat{m}_{ii})\dot{q}_{i2} \right) - d_e \qquad (13.16)$$

The equivalent control represents the difference between the total disturbance and the known part of the system's disturbance. Selection of the control in (13.15) may follow all approaches discussed in Section 13.2.2. If the control enforcing a sliding mode in $\sigma_i = q_{i2} - \hat{q}_{i2} = 0$ is selected as $u_i = -d_{oi}\,\text{sign}(\sigma_i)$ with $|d_{oi}| > |d_i|$ then the average value of the control $\eta\dot{u}_{iav} + u_{iav} = u_i$ tends to the equivalent control $u_{iav} \to u_{ieq}$ if the filter time constant tends to zero ($\eta \to 0$) [11]. If the control input in the system (13.14) is selected as $\tau_i = u_{iav} + \hat{m}_{ii}v_i$, for $\eta \to 0$ the dynamics of the system reduces to $\dot{q}_{i1} = q_{i2}$, $\ddot{q}_{1i} = \dot{q}_{i2} = v_i$ representing a linear double integrator plant (in the motion control literature this is often called the nominal plant) with v_i as the control input representing the desired acceleration of the augmented motion system. Selection of $v_i = k_{pi}(q_{1i}^{ref} - q_{1i}) + k_{di}(\dot{q}_{1i}^{ref} - \dot{q}_{1i}) + \ddot{q}_{1i}^{ref}$ gives motion of the closed loop system as $k_{pi}(q_{1i}^{ref} - q_{1i}) + k_{di}(\dot{q}_{1i}^{ref} - \dot{q}_{1i}) + (\ddot{q}_{1i}^{ref} - \ddot{q}_{1i}) = 0$ representing a second order system with design parameters K_p and K_d. It is interesting to notice that the discrete-time sliding mode approach discussed in Section 13.2.2 leads to the same closed loop system motion with $\sigma_i = c_{ii}(q_{1i}^{ref} - q_{1i}) + (\dot{q}_{1i}^{ref} - \dot{q}_{1i})$ under the conditions $\dot{\sigma}_i + d_{ii}\sigma_i = 0$, with sliding mode motion being $c_{ii}d_{ii}(q_{1i}^{ref} - q_{1i}) + (c_{ii} + d_{ii})(\dot{q}_{1i}^{ref} - \dot{q}_{1i}) + (\ddot{q}_{1i}^{ref} - \ddot{q}_{1i}) = 0$ with $c_{ii}d_{ii} = k_{pi}$, $c_{ii} + d_{ii} = k_{di}$. The equivalency of the sliding mode control approach and the disturbance observer with PD controller and desired acceleration feed-forward term is obvious. A more general case is discussed in Reference 12.

The same idea may be applied for system (13.1) by constructing the model (13.17)

$$\dot{\hat{\mathbf{q}}}_2 = \hat{\mathbf{M}}^{-1}(\boldsymbol{\tau} - \mathbf{u}) \qquad (13.17)$$

where $\hat{\mathbf{M}}$, $\hat{\mathbf{q}}_2$ are estimates of the inertia matrix and the generalised velocity; $\mathbf{u}$ is the model control input, which should be selected to enforce a sliding mode in the manifold $\boldsymbol{\sigma}_{q_2} = \mathbf{q}_2 - \hat{\mathbf{q}}_2 = 0$. The equivalent control for the observer (13.17) in the manifold $\boldsymbol{\sigma}_{q_2} = \mathbf{0}$ can be calculated as $\mathbf{u}_{eq} = \mathbf{N} + \mathbf{G}_{ext} + (\mathbf{M} - \hat{\mathbf{M}})^{-1}\dot{\mathbf{q}}_2$ – which represents the total disturbance and parameter uncertainty in system (13.1). Following the same idea as in the scalar case and selecting the control input in (13.1) as $\tau = \mathbf{u}_{eq} + \hat{\mathbf{M}}\mathbf{v}$, motion of the augmented system can be written as $\dot{\mathbf{q}}_1 = \mathbf{q}_2$, $\dot{\mathbf{q}}_2 = \mathbf{v}$. The equivalency with sliding mode control may be established in the same way as in the previous case.

13.3   Timing-belt servosystem

In the following section we will demonstrate application of the earlier results to a timing-belt driven servosystem depicted in Fig. 13.1.

Forces F_1, F_2 and F_3 acting on the load depend on the stretch of the belt and its derivative – they thus depend on both motor and load position. The variables and parameters are: θ_1 angular position of the pulley driven by the servomotor; θ_2 angular position of the un-driven side pulley; $T = K_T i$ torque developed by the servomotor; $T_L(\theta, \omega)$ friction torque at the servomotor side; F_B belt elasticity force; F_D belt internal friction force; G gear ratio (if present in the system); $x_m = 2\pi\theta/G$ and v_m longitudinal position and velocity of the belt on the periphery of the pulley 1; x and v longitudinal position and velocity of the load; F_L friction force at the load side; m_{mot} equivalent mass on the motor side; m equivalent mass on the load side; r radius of the pulleys. A simplified description of the motor-belt-mass system can be modelled as a two-mass system with nonlinear spring. By combining dynamics of the servomotor and the dynamics of the load side one can develop a state space description of the overall system (13.18), with the total belt force given by (13.19) with the elasticity force $F_B(x_m, x)$ of the equivalent spring defined in (13.20) and damping force $F_D(v_m, v)$ due to the belt internal friction defined in (13.21):

$$
\begin{bmatrix} \dot{x} \\ \dot{v} \\ \dot{x}_m \\ \dot{v}_m \end{bmatrix} = \begin{bmatrix} 0 & 1 & 0 & 0 \\ -\dfrac{K}{m} & 0 & \dfrac{K}{m} & 0 \\ 0 & 0 & 0 & 1 \\ \dfrac{K}{m_{mot}} & 0 & -\dfrac{K}{m_{mot}} & 0 \end{bmatrix} \begin{bmatrix} x \\ v \\ x_m \\ v_m \end{bmatrix} + \begin{bmatrix} 0 \\ 0 \\ 0 \\ \dfrac{1}{m_{mot}} \end{bmatrix} [F_{mot}]
$$

$$
+ \begin{bmatrix} 0 & 0 \\ \dfrac{1}{m} & 0 \\ 0 & 0 \\ 0 & -\dfrac{1}{m_{mot}} \end{bmatrix} \begin{bmatrix} F_D - F_L \\ F_D + F_{Lmot} \end{bmatrix} \tag{13.18}
$$

$$
F_{tBelt} = F_B(\theta, x) + F_D(\omega, v), \quad F_{mot} = \frac{G K_T i}{r} \tag{13.19}
$$

$$
F_B = K(x)(x_m - x); \quad K(x) = \frac{1}{(1/C_0) + (1/(K_1 + K_2))};
$$

$$
K_1(x) = \frac{K}{l_{L0} - (x_0 + x)}; \quad K_2(x) = \frac{K}{x_0 + x} \tag{13.20}
$$

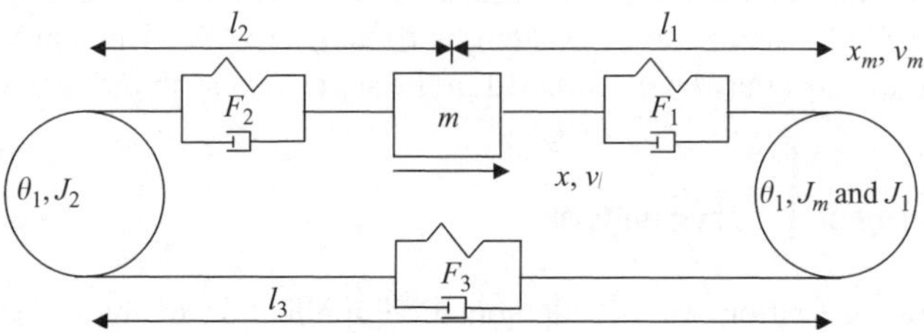

Figure 13.1   Timing-belt servosystem

$$F_D = K_D(x)(v_{mot} - v); \qquad K_D(x) = K_0 \left| \sqrt{\frac{m_{mot} m}{m_{mot} + m}} K(x) \right|; \qquad K_0 > 0$$

$$(13.21)$$

where C_0 stands for the elasticity coefficient of the gear and coupling; K_T stands for the motor torque constant; i stands for the motor current; $K_1(x)$ stands for the elasticity coefficient of the un-driven side of the belt; $K_2(x)$ stands for the elasticity coefficient of the driven side of the belt; l_{L0} the total length of the belt on the load side; x_0 the length of the belt when $x = 0$; $K_D(x)$ stands for the damping coefficient. In the above model the dynamics of the actuator with current (torque) controller are disregarded. The open loop motion of the experimental system with 6 kg payload is depicted in Fig. 13.2 for pulse changes in the motor reference current.

The oscillations in the system are shown in the motor velocity and belt stretch diagrams. The presence of the large friction force results in both motor and load not returning close to the initial position. The main problem of the system under investigation is oscillation of the load due to the belt elasticity, nonlinearity of the belt forces and large friction. The aim of the control system design is to achieve smooth motion with the mass of the load taking any value between (2–26) kg. It is desirable to have a simple controller structure with minimal possible tuning of any parameters. Control of the motor position can be designed in the framework of SMC control using the results presented in Section 13.2.2. The discontinuous control application is not suitable in this case due to the possibility of exciting belt oscillation so the discrete-time sliding mode or the disturbance rejection method could be used.

13.3.1   *Experimental verification*

Experimental verification is performed on a timing-belt driven linear drive DGEL25-1500-ZR-KF (FESTO) equipped with the electrical servomotor MTR-AC-70-3S-AA. The experimental set-up consists of the original motor driver attached to the dSPACE DS1103 module hosted in the PC with dSPACE software Control Desk v.2.0 and the MATLAB 6.0.0.88.R12. All experiment sampling in the controller loop is kept at $T_l = 10^{-3}$ s. The encoder reading loop has the sampling time $T_e = 250 \times 10^{-6}$ s. The observer has a sampling interval $T_o = 25 \times 10^{-6}$ s. The position and velocity of the motor are measured from an incremental encoder with 1024 ppr. Load position

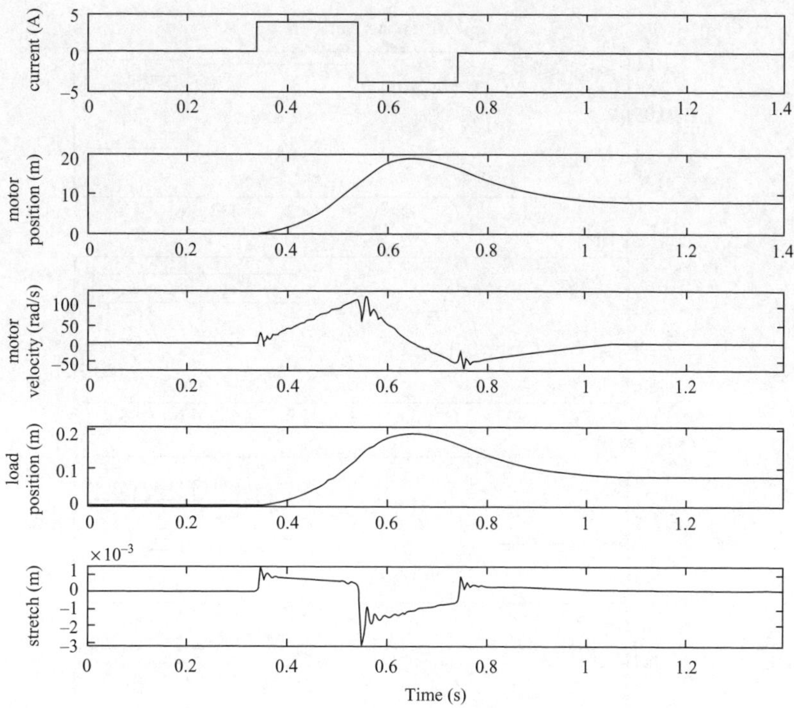

Figure 13.2   Transients in the open loop timing-belt servosystem with 6 kg
payload

is measured by the linear incremental encoder with a resolution of 3×10^{-6} m per pulse. With direct drive application and radius of pulley $r = 0.0205$ m, the ± 1 pulse of the motor incremental encoder gives motion of the timing-belt at the periphery of the driven pulley $\pm 15.71 \times 10^{-6}$ m. The experimental system was designed for point-to-point movement of the actuator shaft. A disturbance observer estimating the total disturbance on the motor shaft is designed and its output is fed to the motor reference current. Since motor angular velocity is measured and motor current is assumed to be equal to its reference value, the observer has the simple structure of the first order system $\dot{\hat{v}}_m = (F_{mot} - u)/m_{mot}$ and selecting $u = U_0 \operatorname{sign}(\varepsilon_{vm})$ with U_0 a large enough positive constant, sliding mode existence in $\varepsilon_{vm} = v_m - \hat{v}_m = 0$ is guaranteed and $u_{eq} = (F_B + F_D + GT_L/r + \Delta m_{mot}\dot{v}_m)$ thus representing the total disturbance on the motor shaft. Selecting the reference current $i^{ref}(k) = u_{eq}r/GK_T + sat(i^{ref}(k-1) + K_u((1+DT)\sigma(k) - \sigma(k-1)))$ sliding mode motion is guaranteed in the manifold $\sigma = C_F(x_m^{ref} - x_m) + (v_m^{ref} - v_m)$. In the experiments, the following parameters have been used: $K_u = 2 \times 10^{-5}$, $C_F = 450$, $D = 250$, where x_m^{ref}, v_m^{ref} are references of the belt position and velocity at the periphery of the driven pulley. In Fig. 13.3 transients for 10 cm motion with a load of 6 kg (Fig. 13.3a) and 26 kg

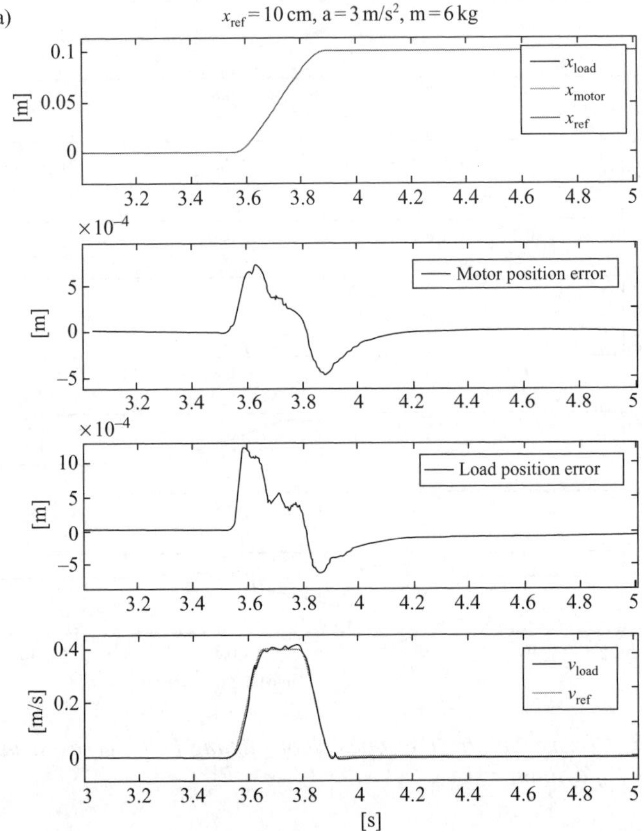

Figure 13.3   *Transients in the motor position change of 10 cm: a) with m = 6 kg and acceleration 3 m/s²*

(Fig. 13.3b) are depicted. Acceleration is $1\,\text{m/s}^{-2}$ for a system with a load of 26 kg and $3\,\text{m/s}^{-2}$ for a load of 6 kg.

Both experiments showed very small overshoot and high positioning accuracy. In Fig. 13.4, transients for 50 cm point-to-point movement are depicted. They show the same behaviour as the motion in Fig. 13.3 with positioning accuracy within one encoder pulse. In order to show such a behaviour the motion of 1 cm is depicted in Fig. 13.5. The pulsation of the motor position error is visible while the load position is not changing.

13.3.2   Belt stretch control

Since the force developed by the belt depends on the belt's stretch, it is natural to look at the system description that has $e_x = (x_m - x)$, $v_x = (v_m - v)$ as the systems coordinate. Then the structure of the system (13.18) could be represented taking the

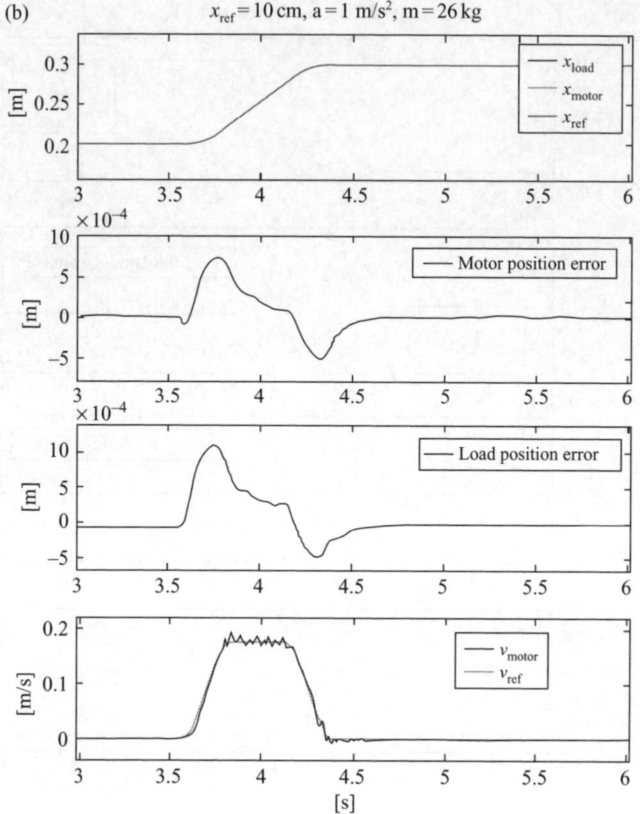

Figure 13.3   Continued. b) m = 26 kg and acceleration 1 m/s²

load side motion and the stretch of the system as coordinates

$$
\begin{bmatrix} \dot{x} \\ \dot{v} \\ \dot{e}_x \\ \dot{v}_x \end{bmatrix} =
\begin{bmatrix}
0 & 1 & 0 & 0 \\
0 & 0 & \dfrac{K}{m} & \dfrac{1}{m} \\
0 & 0 & 0 & 1 \\
0 & 0 & -\dfrac{K}{m_e} & -\dfrac{K_D}{m_e}
\end{bmatrix}
\begin{bmatrix} x \\ v \\ e_x \\ v_x \end{bmatrix} +
\begin{bmatrix} 0 \\ 0 \\ 0 \\ \dfrac{1}{m_{mot}} \end{bmatrix} [F_{mot}]
$$

$$
+ \begin{bmatrix}
0 & 0 \\
-\dfrac{1}{m} & 0 \\
0 & 0 \\
-\dfrac{1}{m} & -\dfrac{1}{m_{mot}}
\end{bmatrix}
\begin{bmatrix} F_L \\ F_{Lmot} \end{bmatrix}
\tag{13.22}
$$

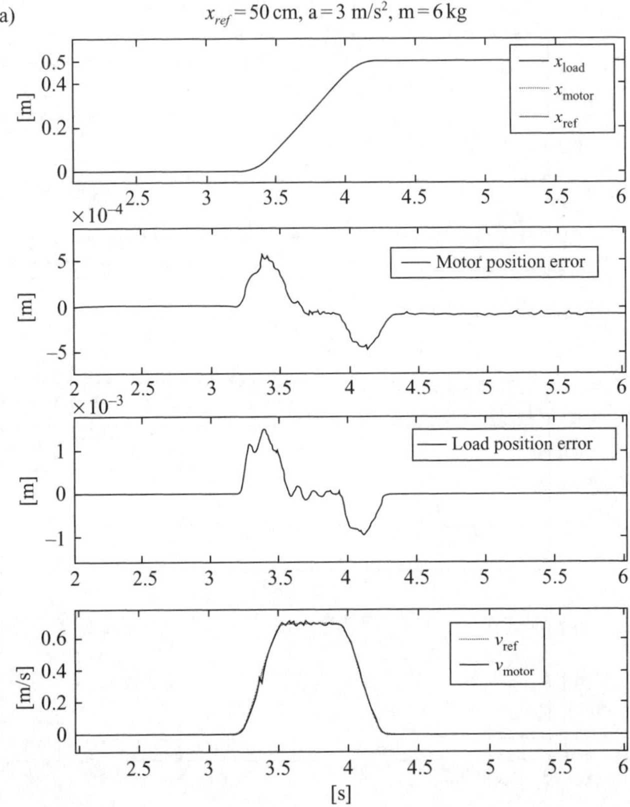

Figure 13.4   *Transients in the motor position change for 50 cm: a) with m = 6 kg and acceleration 1 m/s^2*

$$m_e = \frac{m_{mot} \cdot m}{m_{mot} + m} \tag{13.23}$$

The structure of the system (13.22) as depicted in Fig. 13.6 shows that the oscillatory part of the system is now confined to the block describing the belt stretch dynamics and the load side motion is described as a simple double-integrator plant.

Belt stretch control is based on the mathematical model (13.22) which, for the stretch dynamics, could be written as

$$\dot{e}_x = v_x$$

$$\dot{v}_x = -\frac{K_D}{m_e} v_x - \frac{K}{m_e} e_x - d + \frac{1}{m_{mot}} F_{mot}, \qquad d = \left(-\frac{1}{m} F_L - \frac{1}{m_{mot}} F_{Lmot} \right)$$

$$\tag{13.24}$$

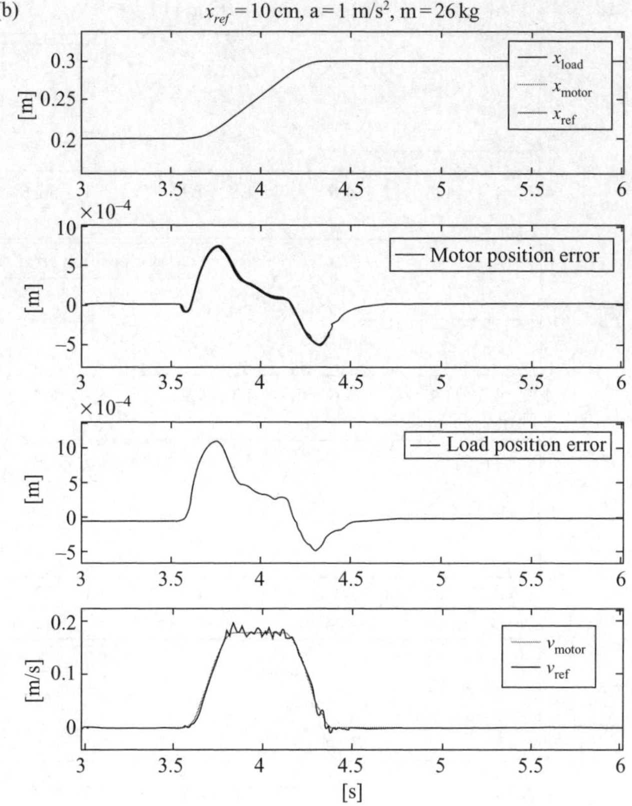

Figure 13.4   Continued. b) $m=26\,kg$ and acceleration $1\,m/s^2$

Assume that the position and velocity of the motor and load are measured, then stretch e_x and its derivative are available. System (13.24) has the same structure as system (13.2), and thus is suitable for application of the results presented in Section 13.2.2. Discontinuous control cannot be applied due to the elastic coupling in the system and the possibility of exciting timing-belt oscillations. Defining the sliding mode line as $\sigma_x = \dot{\varepsilon}_x + C_1\varepsilon_x$, the equivalent control can be determined as

$$u_{eq} = F_{mot}^{eq} = m_{mot}\left(\frac{K}{m_e}e_x + \left(\frac{K_D}{m_e}+C_1\right)\dot{e}_x + \ddot{e}_x^r + d\right)$$

and the control input $u = u_{eq} + D\sigma_x$ will guarantee $\dot{\sigma}_x + D\sigma_x = 0$ and consequently asymptotic stability of the solution $\sigma_x = 0$. The control input has the following form

$$u = F_{mot}^{ref} = m_{mot}(C_1 De_x + (C_1 + D)\dot{e}_x + \ddot{e}_x^r) + m_{mot}\left(\frac{K}{m_e}e_x + \frac{K_D}{m_e}\dot{e}_x + d\right)$$

$$(13.25)$$

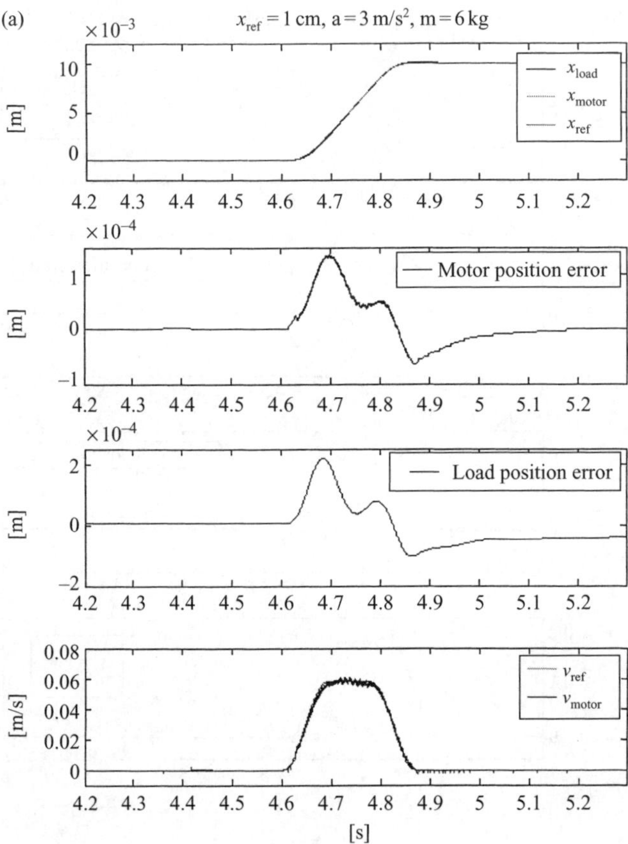

Figure 13.5 *Transients in the motor position change for 1 cm: a) with $m = 6$ kg and acceleration 1 m/s^2;*

For this implementation, the disturbance $((K/m_e)e_x + (K_D/m_e)\dot{e}_x + d)$ should be estimated. The disturbance observer may be designed in a relatively straightforward fashion based on the previously discussed results. In [11] a similar problem is discussed for shaft elasticity and a simple observer is proposed. As shown in Section 13.2.2, the discrete-time version of algorithm (13.25) could be implemented as $u(k) = (F_{mot}^{ref}(k)) = u(k) + K_F(D\sigma(k) - (\sigma(k) - \sigma(k-1))/T)$ where T is the sampling interval and K_F is the design parameter. From the systems structure, it is obvious that the belt force $F = F_B + F_D$ control can be implemented in the same way.

13.4   Control and state observers for induction machine

Control of the induction machine (IM) is still a challenging problem due to its nonlinear dynamics, limited possibility to measure or estimate necessary variables

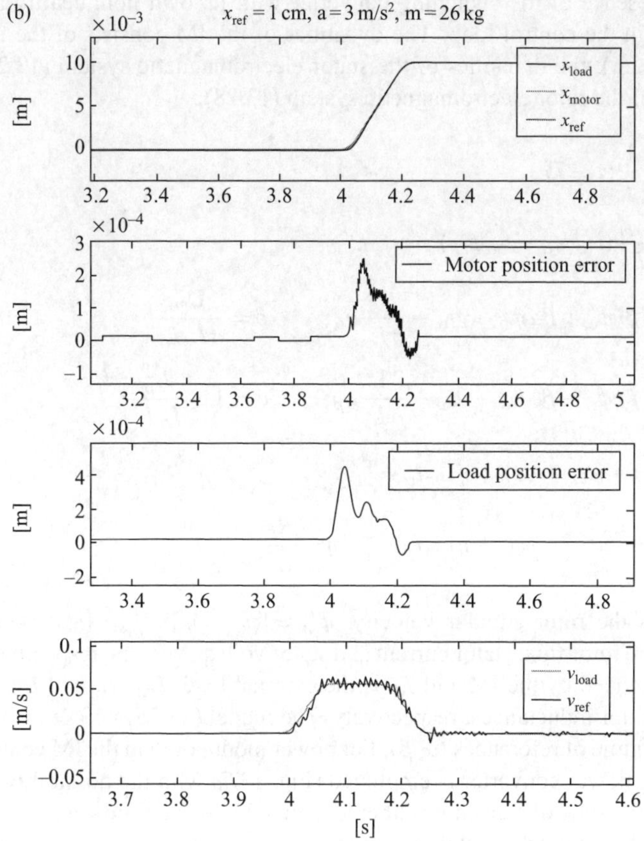

Figure 13.5   Continued. b) m = 26 kg and acceleration 1 m/s^2

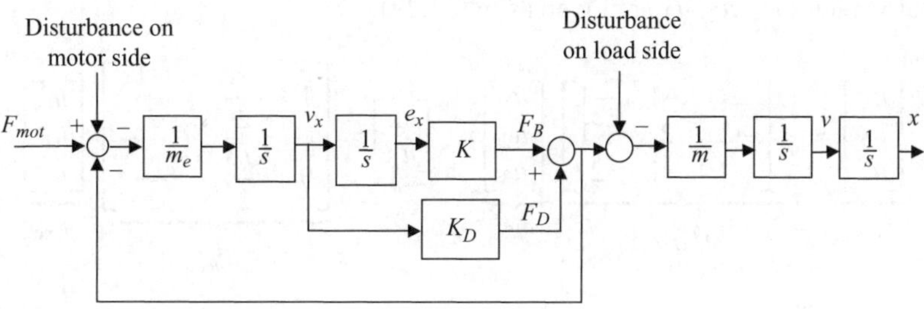

Figure 13.6   The structure of system (13.22)

and the presence of the switching converter with its own nonlinearity as a power modulator in the control loop. The dynamics of the IM consists of the mechanical motion (13.26), the dynamics of the stator electromagnetic system (13.27) and the dynamics of the rotor electromagnetic system (13.28).

$$\frac{d\omega}{dt} = \frac{1}{J}(\tau - T_L)$$

$$\tau = \frac{3L_m}{2L_r}(i_{s\beta}\phi_{r\alpha} - i_{s\alpha}\phi_{r\beta})$$

(13.26)

$$\frac{di_\alpha}{dt} = \beta\eta\phi_\alpha + \beta\omega\phi_\beta - \gamma i_\alpha + \frac{1}{\sigma L_s}u_\alpha; \qquad \beta = \frac{L_m}{\sigma L_s L_r}$$

$$\frac{di_\beta}{dt} = \beta\eta\phi_\beta - \beta\omega\phi_\alpha - \gamma i_\beta + \frac{1}{\sigma L_s}u_\beta; \qquad \sigma = 1 - \frac{L_m^2}{L_s L_r}$$

(13.27)

$$\frac{d\phi_\alpha}{dt} = -\eta\phi_\alpha - \omega\phi_\beta + \eta L_m i_\alpha$$

$$\frac{d\phi_\beta}{dt} = -\eta\phi_\beta + \omega\phi_\alpha + \eta L_m i_\beta; \qquad \eta = \frac{R_r}{L_r}$$

(13.28)

where ω is the rotor angular velocity, $\phi_{\alpha\beta}^T = [\phi_\alpha \quad \phi_\beta]$, $i_{\alpha\beta}^T = [i_\alpha \quad i_\beta]$ and $u_{\alpha\beta}^T = [u_\alpha \quad u_\beta]$ are rotor flux, stator current and stator voltage vectors, respectively; τ is the torque developed by the IM and T_L is the external load, L_m, L_s and L_r are mutual, stator and rotor inductances, respectively. The model (13.26)–(13.28) is written in a stationary frame of references (α, β). For power modulation in the IM control system, a switching power converter is employed (Fig. 13.7) with the possibility to connect each stator winding of a machine either to + or − bar of a DC power source. The converter switches may take eight distinct configurations S_i, $i = 1, 2, \ldots, 8$ thus defining eight distinct values $\mathbf{u}(S_i)$.

The converter's output voltages u_1, u_2, u_3 are taking values from the discrete set $\{0, V_0\}$. With motor stator windings in a star connection, the relationship between machine phase voltages u_a, u_b, u_c, stator voltage vector $u_{\alpha\beta}^T = [u_\alpha \quad u_\beta]$ and converter output voltages u_1, u_2, u_3 are given as in (13.29)

$$\underbrace{\begin{bmatrix} u_a \\ u_b \\ u_c \end{bmatrix}}_{u_{abc}} = \underbrace{\frac{1}{3}\begin{bmatrix} 2 & -1 & -1 \\ -1 & 2 & -1 \\ -1 & -1 & 2 \end{bmatrix}}_{T_{123}^{abc}} \underbrace{\begin{bmatrix} u_1 \\ u_2 \\ u_3 \end{bmatrix}}_{u_{123}}; \qquad \underbrace{\begin{bmatrix} u_\alpha \\ u_\beta \end{bmatrix}}_{u_{\alpha\beta}} = \underbrace{\begin{bmatrix} 1 & \dfrac{-1}{2} & \dfrac{-1}{2} \\ 0 & \dfrac{\sqrt{3}}{2} & \dfrac{-\sqrt{3}}{2} \end{bmatrix}}_{T_{abc}^{\alpha\beta}} \underbrace{\begin{bmatrix} u_a \\ u_b \\ u_c \end{bmatrix}}_{u_{abc}}$$

$$\mathbf{u}_{\alpha\beta} = \mathbf{T}_{abc}^{\alpha\beta}\mathbf{T}_{123}^{abc}\mathbf{u}_{123}$$

(13.29)

where $\mathbf{T}_{abc}^{\alpha\beta}$ stands for the transformation matrix from $(a, b, c) \to (\alpha, \beta)$ frame of reference; $\mathbf{T}_{123}^{abc}$ stands for the transformation matrix from $(1, 2, 3) \to (a, b, c)$ frame of

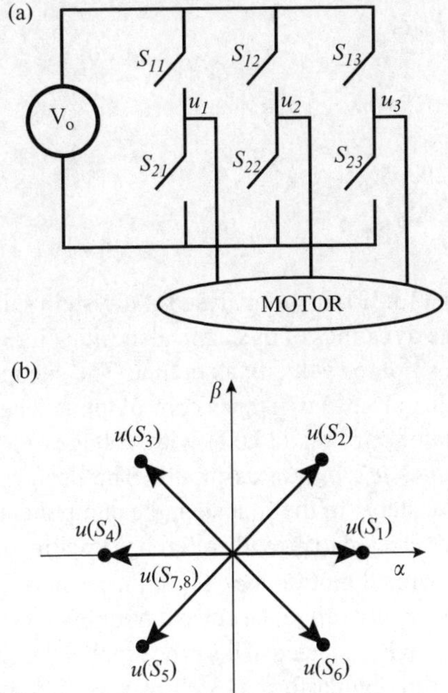

Figure 13.7   *a) Connection of the switching converter and IM machine, and b) control vectors corresponding to permissible switches configuration*

reference. For mechanical motion control, the system design model (13.26)–(13.28) is usually rewritten in the so-called field oriented frame of reference (d, q) in which the d-axis is collinear with, and the q-axis is orthogonal to, the vector of the rotor flux. The matrix $\mathbf{T}_{\alpha\beta}^{dq}$ describes the transformation from the (α, β) to the (d, q) frame of reference and the matrix $\mathbf{T}_{\alpha\beta}^{dq}\mathbf{T}_{abc}^{\alpha\beta}\mathbf{T}_{123}^{abc}$ describes transformation from converter output voltages $\mathbf{u}_{123}$ to $\mathbf{u}_{dq}$ voltages (13.30):

$$
\underbrace{\begin{bmatrix} x_d \\ x_q \end{bmatrix}}_{x_{dq}} = \underbrace{\begin{bmatrix} \cos\rho & \sin\rho \\ -\sin\rho & \cos\rho \end{bmatrix}}_{T_{\alpha\beta}^{dq}} \underbrace{\begin{bmatrix} x_\alpha \\ x_\beta \end{bmatrix}}_{x_{\alpha\beta}}, \quad \mathbf{u}_{dq} = \underbrace{\mathbf{T}_{\alpha\beta}^{dq}\mathbf{T}_{abc}^{\alpha\beta}\mathbf{T}_{123}^{abc}}_{T_{123}^{dq}}\mathbf{u}_{123} \tag{13.30}
$$

$$
\frac{di_d}{dt} = \beta\eta\phi_d + \omega i_q - \gamma i_d + \frac{1}{\sigma L_s} u_d
$$

$$
\tag{13.31}
$$

$$
\frac{di_\beta}{dt} = -\omega(\beta\phi_d + i_d) - \gamma i_q + \frac{1}{\sigma L_s} u_q
$$

$$\frac{d\phi_d}{dt} = -\eta\phi_d + \eta L_m i_d$$

$$\frac{d\rho}{dt} = \omega + \frac{\eta L_m}{\phi_d} i_q \tag{13.32}$$

$$\dot{\theta} = \omega$$

$$J\dot{\omega} = \tau - T_L = \frac{3L_m}{2L_r}\phi_d i_q - T_L \tag{13.33}$$

The system (13.31)–(13.33) can be analysed as a system split into three blocks: the first block represents the dynamics of mechanical motion (13.33) with i_q (or torque τ) as control input and position or velocity as output. The second block represents the dynamics of the rotor flux (13.32) with i_d as control input. The third block represents the dynamics of the stator currents (13.31) with voltages u_d, u_q and consequently (13.30) converter voltages u_1, u_2, u_3 as inputs. The design of IM motion control can be carried out in two steps. In the first step, the components of the current vector $\mathbf{i}_{dq}^T = [i_d \quad i_q]$ should be selected to provide reference tracking in the rotor flux control loop i_d, and in the mechanical motion loop i_q. In the second step the voltages u_d, u_q should be determined to ensure reference current tracking and then from (13.30) converter voltages $\mathbf{u}_{123}$ should be selected. This procedure is the same as used for sliding mode control of systems in regular form [13]. The rotor flux dynamics is a first order system with a scalar control and reference tracking can be achieved if a sliding mode is guaranteed in $S_d = \{\phi_d, i_d : \sigma_d = \phi_d^{ref} - \phi_d = 0\}$. The mechanical motion is of the same form as system (13.2) and position tracking requires the establishment of sliding mode motion in $S_q = \{\theta, \omega, i_q : \sigma_q = c(\theta^{ref} - \theta) + (\dot{\theta}^{ref} - \theta) = \sigma_q = 0\}$. Selection of a discontinuous control is not suitable here due to the fact that the determined components of the current vector will be set as references in the current control loop. One of the possible solutions for enforcing a quasi-sliding mode in the manifolds S_d and S_q is selection of $i_d(k) = i_d(k-1) + K_d((1 - Td_d)\sigma_d(k) - \sigma_d(k-1))$ and $i_q(k) = i_q(k-1) + K_q((1 - Td_q)\sigma_q(k) - \sigma_q(k-1))$ which results in a sliding mode motion $d_d(\phi_d^{ref} - \phi_d) + (\dot{\phi}_d^{ref} - \dot{\phi}_d) = 0$ and $cd_q(\theta^{ref} - \theta) + (c + d_q)(\dot{\theta}^{ref} - \theta) + (\ddot{\theta}^{ref} - \ddot{\theta}) = 0$, respectively. Thus determined values of the stator current should be treated as references $i_d^{ref} = i_d(k)$ and $i_q^{ref} = i_q(k)$, respectively, in the current control loop. A sliding mode in the intersection of manifolds $i_d^{ref} - i_d = \sigma_{di} = 0$ with $\sigma L_s((di_d^{ref}/dt) - \beta\eta\phi_d - \omega i_q + \gamma i_d) = u_{deq}$ and $i_q^{ref} - i_q = \sigma_{qi} = 0$ with $\sigma L_s((di_q^{ref}/dt) + \omega(\beta\phi_d + i_d) + \gamma i_q) = u_{qeq}$ can be enforced by selecting $u_d = U_0 \text{sign}(\sigma_{di})$ and $u_q = U_0 \text{sign}(\sigma_{qi})$ with $U_0 > \max(\sup_t|u_{deq}|, \sup_t|u_{qeq}|)$, to guarantee that components i_d, i_q of the stator current track their references.

In early contributions on the sliding mode control of IM [14], a solution based on the possibility of describing the motor mechanical motion and rotor flux dynamics was considered as follows:

$$\dot{\theta} = \omega$$

$$\dot{\omega} = \frac{\tau - T_L}{J}$$

$$\dot{\tau} = \frac{3L_m}{2L_r}\phi_d(-(\eta+\gamma)i_q + \omega(i_d + \beta\omega\phi_d)) + \frac{3L_m}{2L_r}\frac{\phi_d}{\sigma L_s}u_q = f_\tau + K_u u_q \quad (13.34)$$

$$f_\tau = \frac{3L_m}{2L_r}\phi_d(-(\eta+\gamma)i_q + \omega(i_d + \beta\omega\phi_d)), \quad K_u = \frac{3L_m}{2L_r}\frac{\phi_d}{\sigma L_s}$$

$$\dot{\phi}_d = v_d$$

$$\dot{v}_d = -\eta v_d + \eta L_m(\beta\eta\phi_d + \omega i_q - \gamma i_d) + \frac{\eta L_m}{\sigma L_s}u_d = f_\phi + K_\phi u_d \quad (13.35)$$

$$f_\phi = -\eta v_d + \eta L_m(\beta\eta\phi_d + \omega i_q - \gamma i_d), \quad K_\phi = \frac{\eta L_m}{\sigma L_s}$$

Then a sliding mode in manifolds $\sigma_{du} = d_d(\phi_d^{ref} - \phi_d) + (\dot{\phi}_d^{ref} - \dot{\phi}_d) = 0$ with $(1/K_\phi)(d_d(\dot{\phi}_d^{ref} - \dot{\phi}_d) + (\ddot{\phi}_d^{ref} - f_\phi)) = u_{deq}$ and $\sigma_{qu} = cd_q(\theta^{ref} - \theta) + (c + d_q)(\dot{\theta}^{ref} - \theta) + (\ddot{\theta}^{ref} - \ddot{\theta}) = 0$ with $(1/K_u)(cd_q(\dot{\theta}^{ref} - \dot{\theta}) + (c + d_q)(\ddot{\theta}^{ref} - \ddot{\theta}) + (\dddot{\theta}^{ref} + (T_L/J) - (f_\tau/J))) = u_{qeq}$ can be enforced if the component stator voltage is selected as $u_d = U_0\,\text{sign}(\sigma_{du})$ and $u_q = U_0\,\text{sign}(\sigma_{qu})$ with $U_0 > \max(\sup_t|u_{deq}|, \sup_t|u_{qeq}|)$. Equations of motion of the sliding mode in manifolds S_{du} and S_{qu} are $d_d(\phi_d^{ref} - \phi_d) + (\dot{\phi}_d^{ref} - \dot{\phi}_d) = 0$ and $dcd_q(\theta^{ref} - \theta) + (c + d_q)(\dot{\theta}^{ref} - \theta) + (\ddot{\theta}^{ref} - \ddot{\theta}) = 0$, respectively. These equations of motion have the same form as in the case where motor torque dynamics are neglected.

As the result of such a design procedure, the stator voltage vector in the (d, q) frame of reference is determined. Each of the control vector components is taking values from the set $\{-U_0, +U_0\}$. In order to complete the control system design a switching sequence for the converter switches defining the outputs u_1, u_2, u_3 should be determined. To determine which one of the eight configurations $S_i, i = 1, 2, \ldots, 8$ should be applied, one should map vector $\mathbf{u}_{dq}$ to vector $\mathbf{u}(S_i)$. The matrix $\mathbf{T}_{123}^{dq}$ is a 2×3 matrix, and thus different algorithms for mapping $\mathbf{u}_{dq}$ to $\mathbf{u}_{123}(S_i)$ can be used offering room for deriving different PWM strategies for the selection of the switching sequences. Indeed, many different solutions can be found in the literature [7, 15, 16]. Below a few published algorithms will be explained in more detail.

Algorithm 1. This algorithm is based on the projection of the (d, q) reference frame to an (a, b, c) reference frame thus using the pseudoinverse $[\mathbf{T}_{123}^{dq}]^+ = \frac{3}{2}[\mathbf{T}_{123}^{dq}]^T$

$$\begin{bmatrix} u_1 \\ u_2 \\ u_3 \end{bmatrix} = \left[\mathbf{T}_{123}^{dq}\right]^+ \begin{bmatrix} u_d \\ u_q \end{bmatrix} = \frac{3}{2}\left[\mathbf{T}_{123}^{dq}\right]^T \begin{bmatrix} u_d \\ u_q \end{bmatrix} \quad (13.36)$$

This algorithm (13.36) was first published in Reference 14 and is the basis for the so-called space vector PWM algorithms. Based on this idea, many algorithms are devised in the form of so-called transition tables [7, 16]. This algorithm provides a larger switching frequency when compared with all other solutions.

Algorithm 2.   The foundation of this algorithm is very simple: the control should be selected to give the minimum rate of change of the control error [16]. The controls should be selected from (13.37):

$$\mathbf{S}_i \Leftarrow \begin{cases} \min_{j} \|u_{eq}(t) - u_j(\mathbf{S}_j)\| \\ \mathrm{sign}\{[u_{deq} - u_d(\mathbf{S}_j)] \cdot \sigma_d(t)\} = -1 \\ \mathrm{sign}\{[u_{qeq} - u_q(\mathbf{S}_j)] \cdot \sigma_q(t)\} = -1 \end{cases} \tag{13.37}$$

This algorithm gives good results in steady state operation (when the change of the current is limited to a current ripple) but it does not perform very well when large transients in the current are needed (as produced for a sudden change in load).

Algorithm 3.   In algorithms (13.36) and (13.37), the switching pattern does not depend on the actual amplitude of the error. An interesting improvement of algorithm (13.37) has been proposed [15]. The idea is simple. The value of the switching function at the end of the switching interval T is minimised. The switching pattern is then selected as

$$\mathbf{S}_i = \left\{ \mathbf{S}_j \middle| \min_{j} \|\boldsymbol{\sigma}(0) + \dot{\boldsymbol{\sigma}}(\mathbf{S}_j)T\| \right\} \tag{13.38}$$

Further improvement is possible if two control vectors are allowed to be used in one switching interval. Assuming that from (13.38) a vector $\mathbf{S}_k$ is selected, by applying simple linear approximation the second vector is determined from (13.39):

$$\mathbf{S}_i = \left\{ \mathbf{S}_j \middle| \min_{j} \|\boldsymbol{\sigma}(0) + \dot{\boldsymbol{\sigma}}(\mathbf{S}_k)\alpha T + \dot{\boldsymbol{\sigma}}(\mathbf{S}_j)(1 - \alpha)T\| \right\} \tag{13.39}$$

Algorithm (13.39) allows optimisation of the switching pattern by selecting an optimal value for the coefficient $0 \leq \alpha \leq 1$. This algorithm shows very good behaviour in both steady state and transient conditions. The above algorithms can be applied for three phase voltage source converters (inverters and rectifiers) or for other types of three phase electrical machines, without any change.

13.5   Induction machine flux and velocity observer

The design of an observer that will give a good estimate of the rotor flux is the key to motor control. In so-called sensorless drives, estimation of rotor flux and rotor angular velocity is the key to successful design. In this section we explore the IM estimation issues in the framework of sliding mode control.

In Reference 18 – where the first ideas on IM identification in a sliding mode framework are presented – the rotor time constant η and angular velocity ω are treated as the control in a stator current model. This solution is further used in a closed loop torque control system [19].

In general, SMC based IM observers use stator current dynamics and selection of the additional control input in such a way that the estimated current tracks real currents. A stator current observer may be generalised in the following form:

$$\frac{d\hat{i}_\alpha}{dt} = E_\alpha + \frac{1}{\sigma L_s} u_\alpha + V_\alpha$$

$$\frac{d\hat{i}_\beta}{dt} = E_\beta + \frac{1}{\sigma L_s} u_\beta + V_\beta$$

(13.40)

where V_α and V_β are components of the observer's control vector. Then the estimation error dynamics become:

$$\frac{d\varepsilon_{i\alpha}}{dt} = \beta\eta\phi_\alpha + \beta\omega\phi_\beta - \gamma i_\alpha - E_\alpha - V_\alpha$$

$$\frac{d\varepsilon_{i\beta}}{dt} = \beta\eta\phi_\beta - \beta\omega\phi_\alpha - \gamma i_\beta - E_\beta - V_\beta$$

(13.41)

If components V_α and V_β of the control vector are selected such that a sliding mode exists in $\varepsilon_{i\alpha} = 0$, $\varepsilon_{i\beta} = 0$ then the following is true:

$$V_{\alpha eq} = \beta\eta\phi_\alpha + \beta\omega\phi_\beta - \gamma i_\alpha - E_\alpha = f_\alpha(\phi, i, \omega, \eta, \beta, \gamma)$$

$$V_{\beta eq} = \beta\eta\phi_\beta - \beta\omega\phi_\alpha - \gamma i_\beta - E_\beta = f_\beta(\phi, i, \omega, \eta, \beta, \gamma)$$

(13.42)

By selecting different structures of vector $\mathbf{E}^T = [E_\alpha \quad E_\beta]$, the equivalent control $\mathbf{V}^T_{eq} = [V_{\alpha eq} \quad V_{\beta eq}]$ will have different values. This offers a range of possibilities in determining $f_\alpha(\phi, i, \omega, \eta, \beta, \gamma)$, $f_\beta(\phi, i, \omega, \eta, \beta, \gamma)$ as functions of selected variables (rotor flux, rotor angular velocity, currents) and some of the machine parameters. By proper selection of functions (13.42) one is able to determine at least two of the unknown variables or parameters or combination of variables and parameters of the machine. This leads to a variety of structures that may be derived from this approach.

Selection of the observer control vector $\mathbf{V}^T = [V_\alpha \quad V_\beta]$, to enforce a sliding mode in $\varepsilon_{i\alpha} = 0$, $\varepsilon_{i\beta} = 0$, may follow different procedures of sliding mode control. In the discontinuous control framework, selecting $V_\alpha = V_0 \operatorname{sign}(\varepsilon_{i\alpha})$ and $V_\beta = V_0 \operatorname{sign}(\varepsilon_{i\beta})$ with $V_0 > \max(\sup_t |f_\alpha|, \sup_t |f_\beta|)$ will guarantee a sliding mode in $\varepsilon_\alpha = 0$, $\varepsilon_\beta = 0$ and the observer outputs are equal to the motor currents. With such a selection, the equivalent control $V_{\alpha eq}$, $V_{\beta eq}$ can be determined using simple first order filters. Discrete-time design also may be used in determining the structure of the controller in the motor current tracking loop. After determining the equivalent control and knowing the structure of f_α, f_β from (13.42) one can determine two unknowns – these being variables or parameters of the machine.

In References 20 and 21, relation (13.42) was used to determine the rotor flux vector assuming that parameters of the machine and the angular velocity are known. If $\mathbf{E}^T = [-\gamma i_\alpha \ -\gamma i_\beta]$ is selected then from (13.42) rotor flux can be determined as

$$\begin{bmatrix} \phi_\alpha \\ \phi_\beta \end{bmatrix} = \frac{1}{\beta} \begin{bmatrix} \eta & \omega \\ -\omega & \eta \end{bmatrix}^{-1} \begin{bmatrix} V_{\alpha eq} \\ V_{\beta eq} \end{bmatrix}$$

(13.43)

This work also describes an approach that enables angular velocity to be estimated. The idea uses the fact that in addition to the stator circuit observer (13.40) a rotor flux observer may be derived by substituting $V_{\alpha eq} = \beta\eta\phi_\alpha + \beta\omega\phi_\beta$ and $V_{\beta eq} = \beta\eta\phi_\beta - \beta\omega\phi_\alpha$ into (13.28) to obtain

$$\frac{d\hat{\phi}_\alpha}{dt} = -\frac{1}{\beta}V_{\alpha eq} + \eta L_m i_\alpha$$

$$\frac{d\hat{\phi}_\beta}{dt} = -\frac{1}{\beta}V_{\beta eq} + \eta L_m i_\beta \tag{13.44}$$

From (13.44) the rotor flux can be estimated thus providing additional information that can be used to determine rotor angular velocity and rotor time constant from (13.45)

$$\begin{bmatrix} \hat{\phi}_\alpha \\ \hat{\phi}_\beta \end{bmatrix} = \frac{1}{\beta} \begin{bmatrix} \hat{\eta} & \hat{\omega} \\ -\hat{\omega} & \hat{\eta} \end{bmatrix}^{-1} \begin{bmatrix} V_{\alpha eq} \\ V_{\beta eq} \end{bmatrix} \tag{13.45}$$

The estimated motor angular velocity and time constants can be found as

$$\begin{bmatrix} \hat{\eta} \\ \hat{\omega} \end{bmatrix} = \frac{1}{\sqrt{\hat{\phi}_\alpha^2 + \hat{\phi}_\beta^2}} \begin{bmatrix} \hat{\phi}_\alpha & \hat{\phi}_\beta \\ \hat{\phi}_\beta & -\hat{\phi}_\alpha \end{bmatrix} \begin{bmatrix} V_{\alpha eq} \\ V_{\beta eq} \end{bmatrix} \tag{13.46}$$

In both References 20 and 21 proof of convergence of the estimated rotor flux (13.44) to its real value is complicated and is not complete, but ample simulation and experimental results are presented to demonstrate the validity of the presented approach.

Further improvement of the above approach has been presented [19]. An observer that allows estimation of rotor flux, angular velocity and rotor time constant is discussed. The solution vector $\mathbf{E}$ in (13.40) is selected as $\mathbf{E}^T = [-\vartheta i_\alpha \ -\vartheta i_\beta]$; $\vartheta = R_s/\sigma L_s$, and then the components of the equivalent control in (13.42) are determined as $V_{\alpha eq} = \beta\eta\phi_\alpha + \beta\omega\phi_\beta - \beta L_m\eta i_\alpha$ and $V_{\beta eq} = \beta\eta\phi_\beta - \beta\omega\phi_\alpha - \beta L_m\eta i_\beta$. Assuming that the rate of change of the angular velocity ω and the rate of change of the rotor time constant η are small, and so $\dot{\omega} = 0$, $\dot{\eta} = 0$, one can design an observer of components of vector V_{eq}^T in the following form:

$$\begin{bmatrix} \dot{\hat{L}}_\alpha \\ \dot{\hat{L}}_\beta \end{bmatrix} = -\begin{bmatrix} \hat{\eta} & \hat{\omega} \\ -\hat{\omega} & \hat{\eta} \end{bmatrix}\begin{bmatrix} L_\alpha \\ L_\beta \end{bmatrix} - \beta L_m\hat{\eta}\begin{bmatrix} i_\alpha \\ i_\beta \end{bmatrix} - K\begin{bmatrix} \varepsilon_{L\alpha} \\ \varepsilon_{L\beta} \end{bmatrix};$$

$$\begin{bmatrix} \varepsilon_{L\alpha} \\ \varepsilon_{L\beta} \end{bmatrix} = \begin{bmatrix} V_{\alpha eq} - \hat{L}_\alpha \\ V_{\beta eq} - \hat{L}_\beta \end{bmatrix} \tag{13.47}$$

where adaptation of the rotor time constant and speed is governed by (13.48)

$$\begin{bmatrix} \dot{\hat{\eta}} \\ \dot{\hat{\omega}} \end{bmatrix} = \begin{bmatrix} V_{\alpha eq} + \beta L_m i_\alpha & V_{\beta eq} + \beta L_m i_\beta \\ V_{\beta eq} & -V_{\alpha eq} \end{bmatrix}\begin{bmatrix} V_{\alpha eq} - \hat{L}_\alpha \\ V_{\beta eq} - \hat{L}_\beta \end{bmatrix} \tag{13.48}$$

$$\begin{bmatrix} \phi_\alpha \\ \phi_\beta \end{bmatrix} = \frac{1}{\beta}\begin{bmatrix} \eta & \omega \\ -\omega & \eta \end{bmatrix}^{-1}\begin{bmatrix} V_{\alpha eq} + \beta L_m\eta i_\alpha \\ V_{\beta eq} + \beta L_m\eta i_\beta \end{bmatrix} \tag{13.49}$$

Convergence is assured since the derivative of the Lyapunov function $v_l = \frac{1}{2}[\varepsilon_{L\alpha}^2 + \varepsilon_{L\beta}^2 + \varepsilon_{\omega}^2 + \varepsilon_{\eta}^2]$ where $\varepsilon_{\omega} = \omega - \hat{\omega}$ and $\varepsilon_{\eta} = \eta - \hat{\eta}$ can be expressed as $v_l = -k[\varepsilon_{L\alpha}^2 + \varepsilon_{L\beta}^2] \leq 0$. This solution shows applicability of the SMC approach for design of nonlinear observers, and it represents a very good background for sensorless drive design. Possible limitations due to the assumption that the angular velocity is a slowly changing variable seem acceptable in most of the operational modes of the drive. The presented solution for the observer seems the most complete until now. Further work should be directed towards elimination of the assumption of constant angular speed and this can be achieved only if the mechanical motion and load torque of the drive are estimated.

13.6   Conclusions

In this chapter the sliding mode design method and its application to motion control systems are discussed. The general solution for motion control systems with generalised forces or torques as control inputs is derived and its application to the timing-belt servosystem as an illustrative example is shown. In this framework the dynamics of the subsystem that generates generalised force is neglected and the force control system is assumed ideal in the sense that it perfectly tracks reference values. The realisation of the control input in both a continuous time and discrete-time framework is discussed. IM induction machine motion control and state estimation is discussed with the aim to show the validity of the SMC approach in cases where the dynamics of the torque/force generation is taken into account. It was shown that the same motion dynamics as attained in the previous case could be achieved here too. The design of the IM rotor flux and velocity observer is discussed in the last part of the chapter. The usefulness of the SMC approach is demonstrated in this case too.

13.7   References

1 SLOTINE, J.-J.: 'Sliding mode controller design for nonlinear systems', *Int. J. Control*, 1983, **40**(2), pp. 421–434

2 WANG, W. J. and WU, G. H.: 'Variable structure control design on discrete-time systems – another point viewpoint', *Control-theory and Advanced Technology*, 1992, **8**(1), pp. 1–16

3 DRAKUNOV, S. V. and UTKIN, V. I.: 'On discrete-time sliding modes', *Proc. of Nonlinear Control System Design Conf.*, March 1989, Capri, Italy, pp. 273–78

4 DRAZENOVIC, B.: 'The invariance conditions in variable structure systems', *Automatica*, 1969, **5**, pp. 287–295

5 YOUNG, K.-K. D.: 'Controller design for a manipulator using theory of variable structure systems', *IEEE Transaction on Systems, Man and Cybernetics*, 1978, **8**, pp. 210–18

6   UTKIN, V. I.: 'Sliding mode control design principles and applications to electric drives', *IEEE Tran. Ind. Electr.*, 1993, **40**(1), pp. 421–434

7   SABANOVIC, A.: 'Sliding modes in power converters and motion control systems', *International Journal of Control Special Issue on Sliding Mode Control*, 1993, **57**(5), pp. 1237–59

8   FURUTA, K. 'Sliding mode control of a discrete system', *System and Control Letters*, 1990, **14**(2), pp. 145–52

9   UTKIN, V. I.: 'Sliding mode control in discrete time and difference systems', in ZINOBER, A. S. (Ed.): 'Variable structure and Lyapunov control' (Springer Verlag, London, 1993)

10   FURUTA, K.: 'VSS type self-tuning control-β equivalent control approach', in Proceedings of ACC, San Francisco, 1993, pp. 980–84

11   UTKIN, V., GULDNER, J., and SHI, J.: 'Sliding modes in electromechanical systems' (Taylor & Francis, London, 1999)

12   FRIDMAN, L.: 'Singularly perturbed analysis of chattering in relay control systems', *IEEE Transactions on Automatic Control*, 2002, **47**(12), pp. 2079–2084

13   LUK'YANOV, A. G. and UTKIN, V. I.: 'Methods of reducing equations of dynamic systems to a regular form', *Aut. Remote Control*, 1991, **42**(12), pp. 413–20

14   ŠABANOVIC, A. and IZOSIMOV, D. B.: 'Application of sliding mode to induction motor control', *IEEE Transaction on Industrial Applications*, 1981, **IA 17**(1)

15   HOLTZ, J., LOTZKAT, W., and KHAMBADKONE, A.: 'On continuous control of PWM inverters in the over modulation range including the Six-Step Mode', *Proceedings of IECON'92*, 1992, San Diego, USA, pp. 307–12

16   ŠABANOVIC, N., OHNISHI, K., and ŠABANOVIC, A.: 'Sliding modes control of three phase switching converters', *Proceedings of IECON'92*, 1992, San Diego, USA, pp. 319–25

17   CHEN, D., KOBAYASHI, H., OHNISHI, K., and SABANOVIC, A.: 'Direct instantaneous distortion minimization control for three phase converters', *Transaction IEE of Japan*, 1997, **117**-D(7) (in Japanese)

18   IZOSIMOV, D.: 'Multivariable nonlinear induction motor state identifier using sliding modes', (in Russian), in MEEROV, M. V. (Ed): 'Problems of multivariable systems control' (Nauka, Moscow, 1983)

19   YAN, Z. and UTKIN, V.: 'Sliding mode observers for electrical machines – an overview', *Proceedings of IECON 2002*, vol. 3, pp. 1842–47

20   SAHIN, C., SABANOVIC, A., and GOKASAN, M.: 'Robust position control based on chattering free sliding modes for induction motors', *Proceedings of IEEE IECON 95*, Florida, 1995, pp. 512–17

21   DERDIYOK A., GÜVEN, M., RAHMAN, H., INANÇ, I., and XU, L.: 'Design and implementation of a new sliding-mode observer for speed-sensorless control of induction motor', *IEEE Transaction on IE*, 2002, **49**(5), pp. 1177–82

Chapter 14

Sliding mode control for automobile applications

Vadim I. Utkin and Hao-Chi Chang

14.1 Introduction

The items under study in this chapter embrace four issues: automotive alternators, combustion engines AFR (air-fuel ratio), diesel engines NOx and ABS (anti-lock brake system) control. First, the automotive alternator is governed by nonlinear equations and back EMF (Electromotive Force) should be estimated to optimise rectifier performance. For the sake of cost reduction, it should be done only with readily available battery current instead of generator angular position measurement.

Next, the desired AFR in a combustion engine depends on fuelling rate, air flow rate and the mass of fuel in the fuel film. The mass can be measured under no conditions. Information from the AFR sensor, governed by a 1st-order equation, is obtained with certain delay. A sliding mode observer is designed to get the quantity of mass in the fuel film based on this information. The estimation result then may be utilised for fuel injector diagnosis.

The third issue covers diesel engine NOx control. New technology applied to contemporary diesel engines is to append the VG (Variable Geometry) turbocharger. It enables us to control exhaust gas recirculation and compressor air flow rate simultaneously. The system exhibits unstable zero dynamics so that the feedback control is designed to stabilise the system. A regular form approach from the sliding mode control methodology [1] is employed here for controller design.

Finally, an important issue, automobile ABS, is discussed. The tyre traction force depends on the road conditions and is always an unknown function. There exists the need to design an optimisation scheme that is capable of finding the maximum point from an unknown function. A sliding mode self-optimisation method, without measuring the gradient of the unknown tyre traction force function, is utilised to maximise tyre traction forces, or, in other words, vehicle deceleration.

14.2   Estimator for automotive alternator

One of the modern approaches used in the automobile industry for optimising the operating point conditions of today's three-phase alternator is to employ a controllable rectifier through the six-step switching algorithm [2]. In the following section, a sliding mode observer is designed to supply the necessary information to the switching algorithm.

Basically, the dynamics of a three-phase generator may be described by the following four equations:

$$\frac{di_1}{dt} = -\frac{R_\omega}{L}i_1 - \frac{v_o}{6L}(2u_1 - u_2 - u_3) + \frac{1}{L}e_1(t)$$

$$\frac{di_2}{dt} = -\frac{R_\omega}{L}i_2 - \frac{v_o}{6L}(2u_2 - u_1 - u_3) + \frac{1}{L}e_2(t)$$

$$\frac{di_3}{dt} = -\frac{R_\omega}{L}i_3 - \frac{v_o}{6L}(2u_3 - u_1 - u_2) + \frac{1}{L}e_3(t)$$

$$\left(\frac{R_L + R_b}{R_L(t)}\right)\frac{du_c}{dt} = -\frac{1}{R_L(t)C}u_c + \frac{1}{2C}\sum_{i=1}^{3}u_k i_k$$

$$(14.1)$$

See Table 14.1 for the nomenclature of the model parameters.

More details of this model may be found in Reference 3. In addition, we assume that the following ramp function is an acceptable model describing the time-varying engine speed:

$$\omega = \alpha t + \beta \tag{14.2}$$

$$e_1 = -A_{amp}(\alpha t + \beta)\sin\left(\frac{\alpha}{2}t^2 + \beta t + \gamma\right) \tag{14.3}$$

where α is acceleration and β, γ are constants.

The observer design will be separated into two parts: first, the design of the estimator for the sum of the load current and its derivative is considered, and second, a nonlinear asymptotic observer (NAO) is designed.

To estimate the load current, we define a new variable:

$$s_1 = u_1 i_{battery} - \hat{i}_1 \tag{14.4}$$

Table 14.1   *Nomenclature of a three-phase generator model (14.1)*

Parameters	Explanation	Parameter	Explanation
$i_{k=1,2,3}$	Phase currents	$u_{m=1,2,3}$	Switching signals (+1 or −1)
R_ω	Winding resistance	u_c	Voltage of capacitor
R_L	Load resistance	v_o	DC output voltage
R_b	Battery resistance	C	Battery resistor
L	Winding inductance	$e_{m=1,2,3}$	Back EMFs

Since $u_1 i_{pink} = i_1$ within the observation window when $u_1 \neq u_2 = u_3$ and $i_{battery} = i_{link} - i_{load}$, the derivative of s_1 is of the form

$$\dot{s}_1 = \frac{d}{dt} i_1 - u_1 \frac{d}{dt} i_{load} - \frac{d}{dt} \hat{i}_1$$

$$= -\frac{1}{L} \left[R_\omega s_1 - e_1 + M_1 \operatorname{sgn}(s_1) + R_\omega u_1 i_{load} + u_1 L \frac{d}{dt} i_{load} \right] \tag{14.5}$$

where i_{pink} is the current in the output of the rectifier. Enforcing sliding mode on the surface $s_1 = u_1 i_{battery} - \hat{i}_1$ such that the average value of $M_1 \operatorname{sgn}(s_1)$ is of the form

$$z = [M_1 \operatorname{sgn}(s_1)]_{eq} = \left[e_1 - R_\omega u_1 i_{load} - u_1 L \frac{d}{dt} i_{load} \right]$$

$$= \left[e_1 - u_1 \frac{1}{R_L} \left(R_\omega v_o + L \frac{d}{dt} v_o \right) \right] = e_1 - u_1 q h(t) \tag{14.6}$$

where the voltage v_o is available and its derivative may be found as

$$\frac{d}{dt} v_o = \frac{1}{C} i_{battery} + R_{battery} \frac{d}{dt} i_{battery} \tag{14.7}$$

and the function $h(t)$ is the equivalent sum of load current and its derivative. Note that the battery current is readily available and its derivative may be found from

$$\frac{d}{dt} i_{battery} = -\frac{R_\omega}{L} i_{battery} - \frac{v_o}{6L} (2u_1 - u_2 - u_3) + \frac{1}{L} M_1 \operatorname{sgn}(s_1) \tag{14.8}$$

Therefore, the function $h(t)$ may be found from (14.6). Note that the parameter $q = 1/R_L$ is considered as unknown constant here to embrace the fact that it is a variable parameter. It will be estimated by a 5th-order nonlinear asymptotic observer is proposed:

$$\dot{\hat{e}}_1 = \hat{e}'_1 - L_{11}[\hat{e}_1 - (z + u_1 \hat{q} h(t))]$$

$$\dot{\hat{e}}'_1 = -\left(\hat{\omega}^2 + \frac{3\hat{\alpha}^2}{\hat{\omega}^2} \right) \hat{e}_1 - \frac{3\hat{\alpha}^2}{\hat{\omega}^2} \hat{e}'_1 - L_{21}[\hat{e}_1 - (z + u_1 \hat{q} h(t))]$$

$$\dot{\hat{\omega}} = \hat{\alpha} - L_{31}(\hat{\omega} - \omega) \tag{14.9}$$

$$\dot{\hat{\alpha}} = -L_{41}(\hat{\omega} - \omega)$$

$$\dot{\hat{q}} = L_{51} u_1 ([\hat{e}_1 - (z + u_1 \hat{q} h(t))])$$

Results of simulation may be found in Fig. 14.1 which evidently demonstrates the performance of the proposed cascaded observers.

14.3   Estimation of fuelling rate and AFR using UEGO

The engine fuelling rate $\dot{m}_{fi}$ as a control action should be selected such that the AFR in the cylinder is equal to the desired value. To illustrate the fuel film and AFR (denoted as ϕ_c which is equal to $\dot{m}_{ac}/\dot{m}_{fi}$ where $\dot{m}_{ac}$ is the air flow rate in the

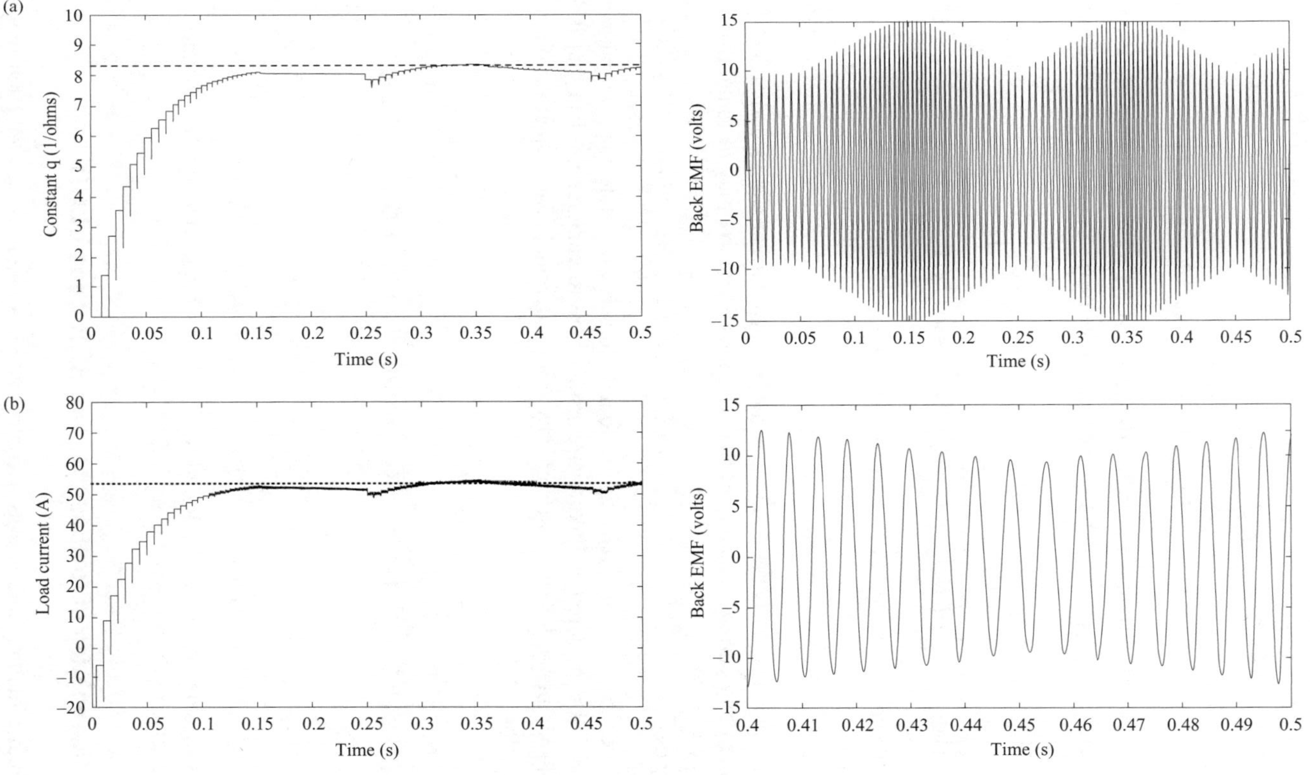

Figure 14.1   *Estimation results of a) unknown constant q, b) load current, and back EMF*

cylinder) estimation approach, we first make use of the proportional measurement of air-fuel ratio provided by UEGO, the proportional oxygen sensor [4]. The reading of the UEGO sensor, ϕ_{UEGO}, is governed by the following equation in terms of crank angle θ

$$\frac{d\phi_{UEGO}(\theta)}{d\theta} = -\frac{1}{\tau_{UEGO}\omega}\phi_{UEGO}(\theta) + \frac{1}{\tau_{UEGO}\omega}\phi_c(\theta - \theta_d) \tag{14.10}$$

where ω is engine speed and τ_{UEGO} is the sensor time constant. The estimation problem is approached by first estimating the in-cylinder *AFR* with delay, $\hat{\phi}_c(\theta - \theta_d)$. Then, the estimate of fuel film mass $\hat{m}_{ff}(\theta - \theta_d)$ can be obtained knowing $\dot{m}_{fi}(\theta - \theta_d)$. The observer is constructed as follows

$$\frac{d\hat{\phi}_{UEGO}(\theta)}{d\theta} = -\frac{1}{\tau_{UEGO}\omega}\hat{\phi}_{UEGO}(\theta) + \frac{1}{\tau_{UEGO}\omega}\Psi \tag{14.11}$$

$$\Psi = \Psi_o \, \mathrm{sign}(\phi_{UEGO} - \hat{\phi}_{UEGO}) \tag{14.12}$$

After finite time, sliding occurs and the equivalent value of Ψ approaches $\phi_c(\theta - \theta_d)$. The in-cylinder AFR estimate can be obtained through low pass filtering of the discontinuous function Ψ:

$$\phi_c(\theta - \theta_d) = \Psi_{eq} = z \tag{14.13}$$

Similarly, the value of the in-cylinder airflow rate may be estimated. Please refer to Reference 5 for details. Next, the value of $m_{ff}(\theta)$, the fuel mass in the fuel film, may be found from

$$\frac{\dot{m}_{ac}}{\phi_c} = \dot{m}_{fc} = \frac{1}{\tau}m_{ff} + (1 - X)\dot{m}_{fi} \tag{14.14}$$

where X is a known parameter. Next, using $\phi_c(\theta - \theta_d)$ and $\dot{m}_{fi}(\theta - \theta_d)$, $m_{ff}(\theta)$ may be found by the following convolution of the fuel film dynamics from $\theta - \theta_d$ to θ in the form of a predictor:

$$\hat{m}_{ff}(\theta) = e^{-(1/\tau_f\omega)\theta}m_{ff}(\theta - \theta_d) + \int_{\theta - \theta_d}^{\theta} e^{-(1/\tau_f\omega)(\theta - \tau)}\frac{X}{\omega}\dot{m}_{fi}(\tau)\,d\tau \tag{14.15}$$

Finally, the desired value or the engine control action, $\dot{m}_{fi}(\theta)$ may be found from (14.14) with $\phi_c = \phi_{c,desired}$, the desired AFR.

Figure 14.2 depicts the estimation result based on this method. The fuel flow rate through the fuel injector is estimated by augmenting an additional state variable for $\dot{m}_{fi}$ using the UEGO sensor measurement. Since the commanded fuel flow rate for the injector is known, just additive (possibly injector fault) fuel flow rate is estimated and then added to the commanded fuel flow rate. Without the presence of an injector fault, the additional state variable has zero value and constant disturbance can be obtained with a constant injector fault [6, 7].

This estimation can be used for diagnosis for the fuel injector. Basically, the essence of the diagnostic approach is to compare the difference between real and

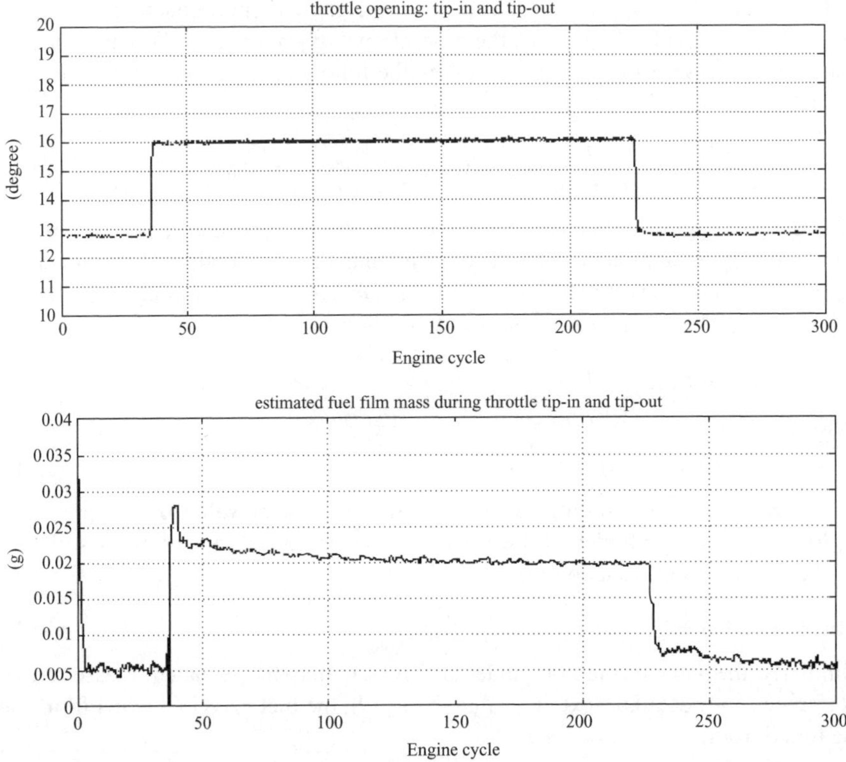

Figure 14.2   Fuel film mass estimation result during throttle tip-in and tip-out

estimated fuel injector output. The fuel injector may have failed if the difference between them exceeds a certain threshold. Estimation results are demonstrated in the Fig. 14.3.

14.4   NOx control for EGR-VGT diesel engine

Exhaust gas recirculation (EGR) combined with the variable geometry turbocharging provides an important avenue for NOx emission reduction. In this chapter we study the problem of EGR-VGT control from the sliding mode design perspective. The departure point for our work is the reduced order model [8] that we use for the control design:

$$\dot{p}_1 = k_1 \left[\frac{\eta_C}{T_a C_p} \frac{P_C}{((p_1/p_a)^\mu - 1)} - k_e p_1 + W_{egr} \right] \tag{14.16}$$

$$\dot{p}_2 = k_2 (k_e p_1 + W_f^d - W_{egr} - u_{2t}) \tag{14.17}$$

$$\dot{P}_C = -\frac{1}{\tau} \left[P_C - \eta_m \eta_t T_2 C_p \left(1 - \left(\frac{p_a}{p_2} \right)^\mu \right) u_{2t} \right] \tag{14.18}$$

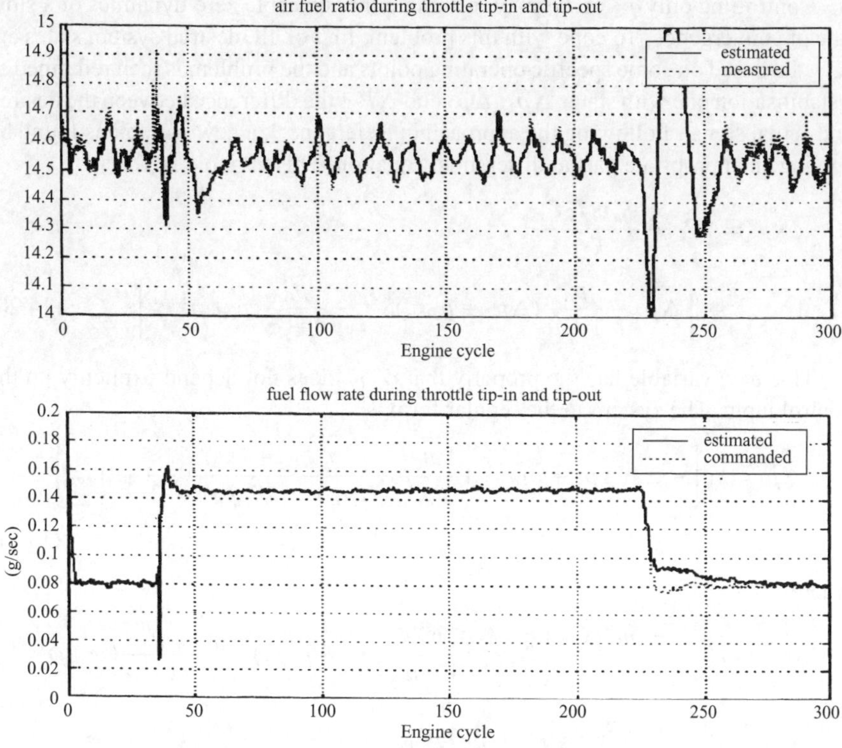

Figure 14.3   *Injected fuel flow rate estimation result during throttle tip-in and tip-out*

where:

p_1/p_2:	intake/exhaust manifold pressure.
P_C:	compressor power.
k_e:	pumping rate constant.
W_{egr}:	EGR flow rate which is equal to u_{12} or u_{21}.
η_C, η_t:	compressor and turbine isentropic efficiencies.
η_m:	turbocharger mechanical efficiency.
τ:	turbine-to-compressor power transfer time constant.
C_p, C_V:	specific heats at constant pressure, volume.

To account for the fact that the turbine cannot immediately realise the commanded flow because of the actuator dynamics, a first order linear system is introduced as

$$\dot{u}_{2t} = -\frac{1}{\tau_A}u_{2t} + \frac{1}{\tau_A}v_{2t} \tag{14.19}$$

where τ_A is the time constant, and v_{2t} is the commanded VG turbine flow rate, that is achieved in steady-state.

Controlling only p_2 or P_C may result in either unstable zero dynamics or a slow rate of convergence. To cope with this problem, first of all, desired system states are selected based on some specific operating points and the problem is then redefined as a stabilisation one with states Δp_1, Δp_2 and ΔP_C, the difference between the desired and actual signal. Following the approach in Reference 9, a new variable is found by solving an appropriate partial differential equation. This variable is

$$\phi_{\Delta P_C} = \Delta P_C + \frac{\eta_m \eta_t T_2 C_p}{k_2 \tau}$$

$$\times \left[\Delta p_2 - \frac{p_a^\mu}{1-\mu}(\Delta p_2 + p_{2des})^{1-\mu} + \frac{p_a^\mu}{1-\mu}(p_{2des})^{1-\mu} \right] \qquad (14.20)$$

This new variable has the property that $\dot{\phi}_{\Delta P_C}$ does not depend explicitly on the control input. The system in the regular form is

$$\Delta \dot{p}_1 = k_1 \left[-k_e \cdot (\Delta p_1 + p_{1des}) + \frac{\eta_c}{T_a C_p} \frac{\phi_{\Delta P_C} + (**)}{(\Delta p_1 + p_{1des}/p_a)^\mu - 1} + W_{egr} \right]$$

$$(14.21)$$

where

$$(**) = P_C^d - \frac{\eta_m \eta_t T_2 C_p}{k_2 \tau} \left[\Delta p_2 - \frac{p_a^\mu}{1-\mu}(\Delta p_2 + p_{2des})^{1-\mu} + \frac{p_a^\mu}{1-\mu}(p_{2des})^{1-\mu} \right]$$

$$\dot{\phi}_{\Delta P_C} = -\frac{1}{\tau}\phi_{\Delta P_C} - \frac{1}{\tau}P_C^d + \frac{\eta_m \eta_t T_2 C_p}{k_2 \tau^2}$$

$$\times \left\{ \Delta p_2 + \frac{p_a^\mu}{\mu - 1}\left[(\Delta p_2 + p_{2des})^{1-\mu} - (p_{2des})^{1-\mu}\right] \right.$$

$$\left. + k_2 \tau \left[1 - \left(\frac{p_a}{\Delta p_2 + p_{2des}} \right)^\mu \right] [k_e(\Delta p_1 + p_{1des}) + W_f^d - W_{egr}] \right\}$$

$$(14.22)$$

$$\Delta \dot{p}_2 = k_2[k_e(\Delta p_1 + p_{1des}) + W_f^d - W_{egr}] - k_2(u_2 + u_{2des}) \qquad (14.23)$$

Note that the control u_2 explicitly affects only (14.23). The EGR flow rate, W_{egr}, is treated as an external input to these equations as it is controlled by a separate controller.

The state variable, Δp_2, involved in (14.22) may be treated as a fictitious control input,

$$\Delta p_2 = -K_S \cdot \phi_{\Delta P_C}, \qquad K_S > 0 \qquad (14.24)$$

to speed up the dynamics of the state, $\phi_{\Delta P_C}$. This relationship can be held by enforcing the sliding mode on the surface

$$s = \Delta p_2 + K_S \cdot \phi_{\Delta P_C} \qquad (14.25)$$

With the actuator dynamics (14.19) included, the relative degree of the system from u_2 to s increases by one and the proposed control algorithm is not directly applicable. What we can do, however, is to modify the sliding surface as

$$S = \dot{s} + \alpha \cdot s, \qquad \alpha > 0 \tag{14.26}$$

where

$$\dot{s} = \dot{p}_2 \left[1 + K_S \frac{T_2 \eta_{tm} \eta_{tuis} c_p}{\tau_{tc} k_2} \left(1 - \left(\frac{p_{atm}}{p_2} \right)^{\mu} \right) \right] - \frac{K_S P_C}{\tau_{tc}}$$

$$+ K_S \frac{T_2 \eta_{tm} \eta_{tuis} C_p}{\tau_{tc} k_2} \left(1 - \left(\frac{p_{atm}}{p_2} \right)^{\mu} \right) (u_2 + u_{2des}) \tag{14.27}$$

and where α controls the rate of decay (after the sliding mode is enforced). The controller for v_{2t} is then developed to enforce the sliding mode on the surface (14.26). For the sake of brevity, the corresponding observer design is not included in this chapter. The information may be found in Reference 10.

The performance of the proposed SM control is examined by simulation. The controller parameters (K_S, α and M), were determined from simulations of the reduced order model augmented with VGT actuator dynamics. We assumed that W_{egr} is generated by a first order lag with the steady-state value of W_{egr}^d. The closed loop responses are illustrated in Fig. 14.4.

The same controller with the same set-points was applied to the more accurate, higher order model and acceptable performance has been observed in Fig. 14.5. Note

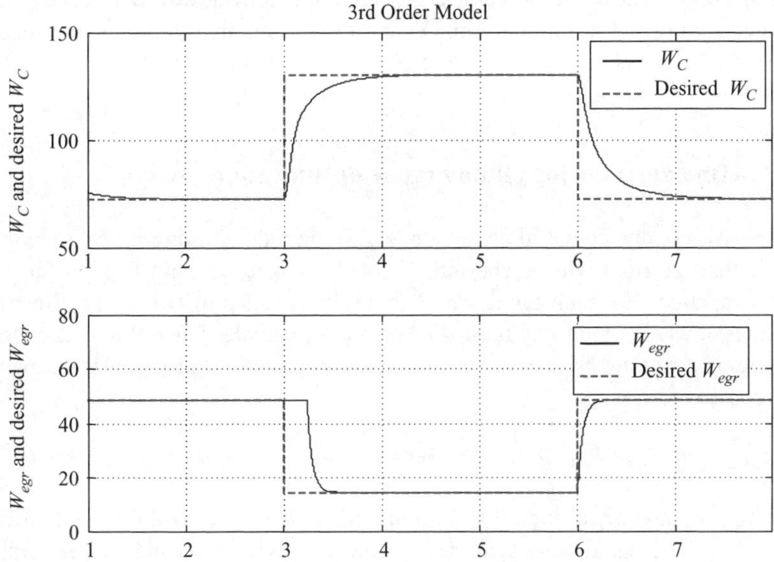

Figure 14.4   Desired versus regulated values of compressor mass flow rate and EGR mass flow rate

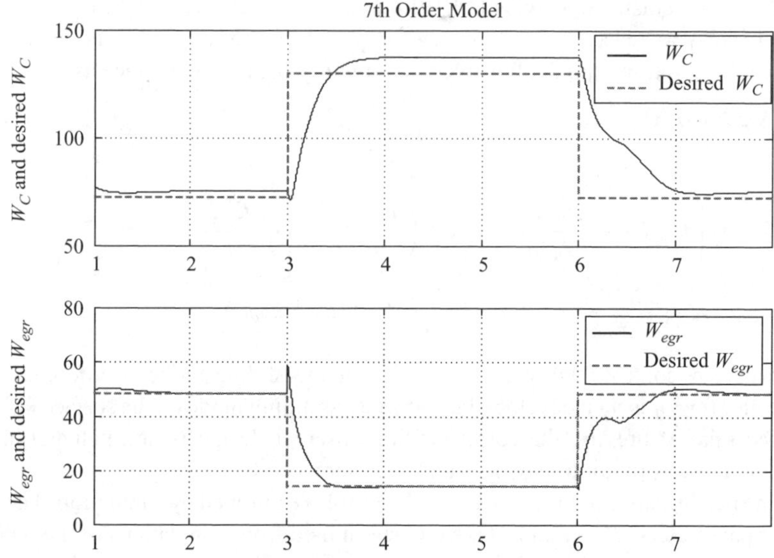

Figure 14.5   *Closed loop responses on the higher order (7th order) model*

that, for the scenario of the higher order model, the EGR flow was generated by assuming that the EGR valve position is driven by a first order lag towards the desired position. The non-monotonic character of the EGR flow response is due to the dependence of this flow on intake and exhaust manifold pressures. The discrepancy of the models results in the steady-state error observed in Fig. 14.5.

14.5   ABS control using sliding mode optimisation

This study is conducted based on the model of longitudinal motion. The assumption is made that the right wheels and left wheels have different slip-friction functions and the functions are both unknown. The model consists of two parts: Rotational Dynamics of wheels and Longitudinal Dynamics of vehicle. The following equations of rotational dynamics are derived under the assumption that the engine is disengaged in the course of braking:

$$J_w \dot{\omega}_i = R_w F_{ti} - B_t \omega_i - T_{bi}, \qquad i = 1, 2, 3, 4 \tag{14.28}$$

where J_w, R_w are wheel inertia and radius, respectively, B_t is the viscous friction of the wheel bearing, F_{ti}s are tyre tractive forces with the road and T_{bi}s are braking forces. The governing equations for longitudinal motion are of the form

$$(M_{car} + 4 * m_w) \cdot \dot{V}_L = -F_{t1} - F_{t2} - F_{t3} - F_{t4} - F_a \tag{14.29}$$

where the nonlinear terms of (14.29) may be found below:

$$F_a = C_d A_f V_L^2 \qquad \text{(aerodynamic drag force)}$$

$$F_{t1} = \frac{N_{front}\mu(\lambda_1)}{F_{t2}} = \frac{N_{front}\mu(\lambda_2)}{F_{t3}} = \frac{N_{rear}\mu(\lambda_3)}{F_{t4}} = N_{rear}\mu(\lambda_4)$$

$$N_{front} = \frac{1}{2}\left[\left(1 - \frac{a}{\ell}\right)M_{car}g - \frac{M_{car}h_{cg}}{\ell}\dot{V}_L\right] \qquad \text{(front wheel normal forces)}$$

$$N_{rear} = \frac{1}{2}\left[\left(\frac{a}{\ell}\right)M_{car}g + \frac{M_{car}h_{cg}}{\ell}\dot{V}_L\right] \qquad \text{(rear wheel normal forces)}$$

$$\lambda_i = \frac{V_L - R_w\omega_i}{V_L} \qquad \text{(slip rate)}$$

The slip-friction functions, $\mu(\lambda_i)$, are unknown and depend only on the road surfaces/conditions. In Fig. 14.6, typical plots of the functions are presented. Usually, the maximum value of friction is for relative slip $\lambda = 0.2$ for many road conditions.

As the slip-friction is an unknown function, the brake controller should be able to maximise the coefficient of friction using the fact that the local gains, which depend on the locations of real and optimal slips, of the plant are not available. To find the maximum tractive force which is, in reality, an unknown function, the self-optimisation algorithm based on sliding mode control is employed [11–13].

Assume that $f(\lambda)$ is the function to be maximised and this function has only one extremum point. The basic idea of the theory is to design sliding mode control to force the system to track an introduced monotonously increasing function $\dot{g}(t) = \rho_o v > 0$.

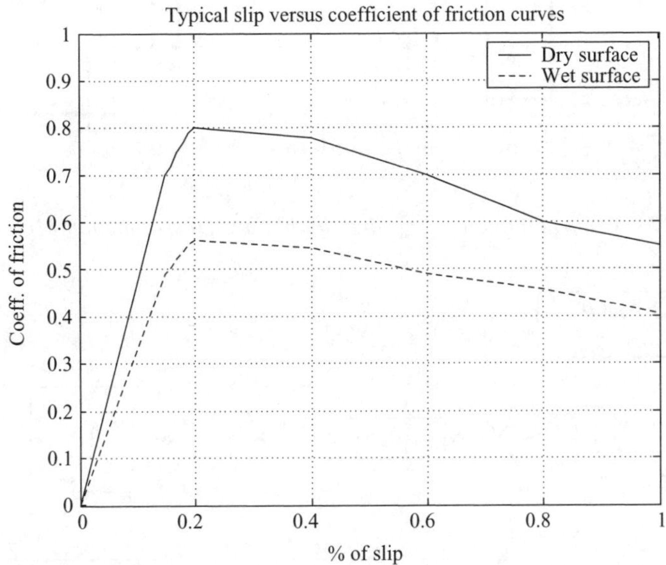

Figure 14.6   Slip versus coefficient of friction curves for surfaces

Let the system output y be the tyre tractive force. Define the tracking error as $\varepsilon = g - y$ and let $s_1 = \varepsilon$ and $s_2 = \varepsilon + \delta$ where δ is the design parameter to be selected. Note that the braking torque, T_{b1}, is handled as a control action which may have two different values, $(-F_{t1} R_w)$ or $(T_{b1} - F_{t1} R_w)$. Then the motion equation w.r.t. angular velocity of one wheel, for example, the first one associated with the control algorithm, is of the form

$$J_w \dot{\omega}_1 = -B_t \omega_1 - u \tag{14.30}$$

where u is the control input from sliding mode optimiser

$$u = \begin{cases} T_{b1} - F_{t1} R_w, & s_1 s_2 > 0 \\ -F_{t1} R_w, & s_1 s_2 < 0 \end{cases} \tag{14.31}$$

Modify the reference function by adding a double-hysteresis component (as shown in Fig. 14.7). By enforcing a sliding mode on either the surface $s_1 = 0$ or $s_2 = 0$, the system output can be maximised by tracking the reference function $g(t)$, where $\dot{g} = \rho_o + Mv(s_1, s_2)$. As for the rest of the wheels, their control signals follow the first wheel (or the master wheel) where the proposed sliding mode optimiser is installed. A schematic diagram of the self-optimiser is shown in Fig. 14.8 which reveals the whole control design.

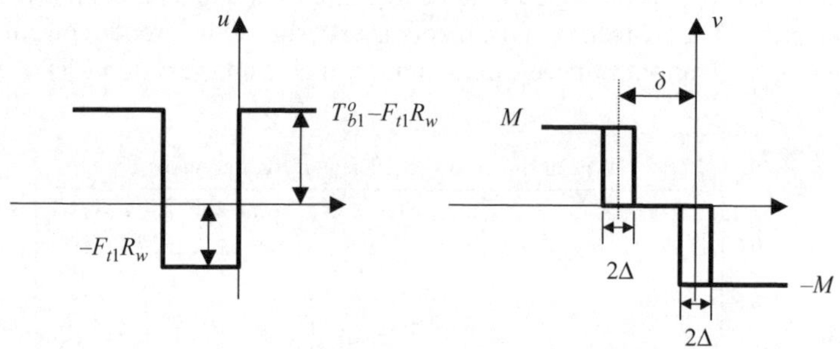

Figure 14.7   Control input $u(s_1, s_2)$ and double-hysteresis function $Mv(s_1, s_2)$

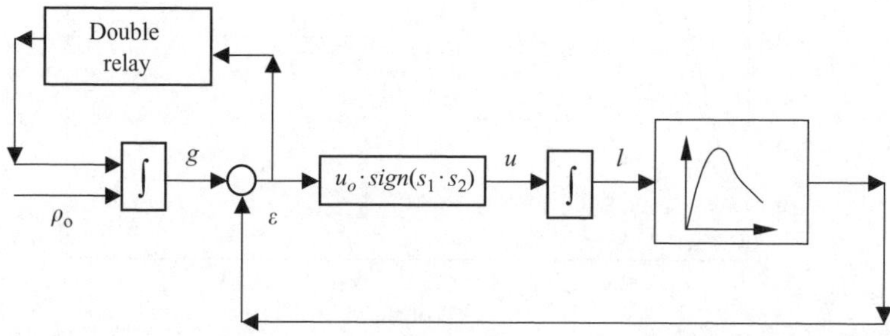

Figure 14.8   Sliding mode self-optimiser

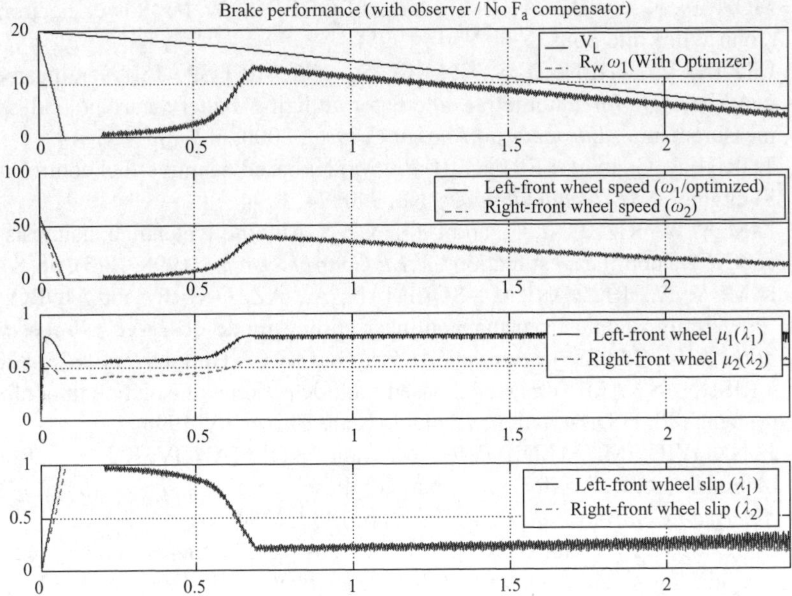

Figure 14.9   Braking performance after wheels are locked up

The following simulation is performed under the scenario that no on-board accelerometer is available and, initially, constant brake torque is applied at the very beginning to totally lock up all the wheels. The controllers take over after 0.2 s to regulate the brake torque. As can be seen from Fig. 14.9, the controllers are capable of reaching optimal slip after 0.5 s and the optimal level is maintained during whole braking process.

14.6   Conclusions

Automobile-related control and estimation issues usually involve either high-order, high levels of nonlinearities or limited system state information. Evidently, from the reported results, sliding mode control theory demonstrates its capability of dealing with various problems, either control or estimation ones, with limited information. The results are promising and worthy of further modification or extension.

14.7   References

1   UTKIN, V.: 'Sliding modes in control and optimization' (Springer Verlag, Berlin, 1991)

2   MOHAN, N., UNDELAND, T. M., and ROBBINS, W. P.: 'Power electronics' (John Wiley and Sons, New York, 1995)

3   UTKIN, V., CHEN, D. S., ZAREI, S., and MILLER, J.: 'Nonlinear estimator design for automotive alternator utilizing battery current and speed measurements', *European Journal of Control*, 2000, **6**(2), pp. 135–149

4   TURIN, R. C. and GEERING, H. P.: 'Model based adaptive fuel control in an SI engine', *SAE Technical paper*, No. 940374, 1994

5   KIM, Y. W., RIZZONI, G., and UTKIN, V.: 'Automotive engine diagnosis and control via nonlinear estimation', *IEEE Control Systems*, 1998, **18**(5), pp. 84–99

6   KIM, Y. W., RIZZONI, G., SOLIMAN, A., AZZONI, P., and MORO, D.: 'Powertrain diagnostic using nonlinear sliding mode observer', *Proceedings of the IFAC/IMACS symposium SAFEPROCESS'97*, Hull, UK, pp. 825–830

7   KRISHNASWAMI, V.: 'Model based fault detection and isolation in nonlinear systems', Ph.D Dissertation, The Ohio State University, 1996

8   JANKOVIC, M., JANKOVIC, M., and KOLMANOVSKY, I.: 'Robust nonlinear controller for turbocharged diesel engines', *Proceedings of the American Control Conference*, 1998, pp. 1389–1394

9   LUK'YANOV, A. and UTKIN, V.: 'Methods of reducing dynamic systems to a regular form', *Automation and Remote Control*, 1981, **42**(1), pp. 5–13

10   UTKIN, V. I., CHANG, H.-C., KOLMANOVSKY, I., and COOK, J.: 'Sliding mode control for variable geometry turbocharged diesel engine', *ACC2000*, Chicago, U.S.A., 2000

11   KOROVIN, S.: 'Sliding mode control in static optimization', *Automation and Remote Control*, 1972, **33**(4), pp. 50–60 (in Russian)

12   KOROVIN, S. and UTKIN, V.: 'Using sliding modes in static optimization and nonlinear programming', *Automatica*, Journal of IFAC, 1974, (10), pp. 525–532

13   UTKIN, V.: 'Sliding modes and their applications in variable structure systems' (Moscow, Nauka, 1974) (in Russian, English translation by MIR, 1978)

Chapter 15

The application of sliding mode control algorithms to a diesel generator set

Keng Boon Goh, Sarah K. Spurgeon and N. Barrie Jones

15.1 Introduction

This chapter demonstrates the application of both standard, first order sliding mode control and higher order sliding mode control techniques to a specific diesel power generator. The specific first order methods considered are sliding mode integral tracking (SMIT) control and sliding mode model-following (SMMF) control. A second order sliding mode (SOSM) control is employed to demonstrate the higher order sliding mode technique. The SMIT and SMMF control algorithms are modified from the methods in References 1, 2 and 3, respectively. The SOSM control algorithm is modified from References 4 and 5.

The specific problem under consideration involves idle speed control of a diesel power generator set. Real-time microprocessor based sliding mode controllers are applied for idle speed control of the generator. The idle speed control problem is formulated as a speed tracking and disturbance (i.e. electrical load) rejection problem. The tracking requirement ensures that the engine speed follows a reference speed set point. The disturbance rejection requirement ensures that the engine speed does not deviate too much from the set point in the presence of electrical load disturbances and also that the engine speed regains the reference speed as soon as possible so that the frequency of the electricity generated is maintained.

This chapter briefly formulates the three mentioned control algorithms and describes the diesel generator, hardware setup and data acquisition system. The control design process is described and implementation results are presented. Finally, some concluding remarks on the relative performance of the different sliding mode control algorithms are given.

15.2   Sliding mode integral tracking (SMIT) control system

The SMIT control technique introduces an additional state (an integral error state) into the system. The controller synthesis is then to minimise the tracking error. The control law requires all the internal states of the system to be available. Thus, a sliding mode observer is designed to estimate these states for the controller. Consider the system

$$\dot{x}(t) = Ax(t) + Bu(t)$$
$$y(t) = Cx(t) \tag{15.1}$$

where $A \in \mathfrak{R}^{n \times n}$, $B \in \mathfrak{R}^{n \times m}$, $C \in \mathfrak{R}^{p \times n}$ and $m = p$, i.e. the system is square. The matrices A, B and C are the system, input and output distribution matrices respectively. The variables $u(t)$ and $y(t)$ are the input and output respectively. The pair (A, B) is assumed to be controllable, $\det(CB) \neq 0$ and the invariant zeros of (A, B, C) stable. The nominal linear system is first converted into regular form:

$$\dot{x}_1(t) = A_{11}x_1(t) + A_{12}x_2(t) \tag{15.2}$$

$$\dot{x}_2(t) = A_{21}x_1(t) + A_{22}x_2(t) + B_2 u(t) \tag{15.3}$$

$$y(t) = C_1 x_1(t) + C_2 x_2(t)$$

where $A_{11} \in \mathfrak{R}^{(n-m) \times (n-m)}$, $B_2 \in \mathfrak{R}^{m \times m}$ and $C_2 \in \mathfrak{R}^{p \times p}$. The square matrix C_2 is nonsingular because $C_2 B_2 = CB$ which is nonsingular by assumption and B_2 is nonsingular by construction. The tracking control technique uses an integral action methodology where $x_r \in \mathfrak{R}^p$ is given by:

$$\dot{x}_r(t) = r(t) - y(t) \tag{15.4}$$

where the differentiable signal $r(t)$ satisfies

$$\dot{r}(t) = \Gamma(r(t) - R) \tag{15.5}$$

where $\mathfrak{R} \in \mathfrak{R}^p$ is a reference demand vector and $\Gamma \in^{p \times p}$ is a stable design matrix. Equation (15.5) represents an ideal reference model where $r(t)$ defines a dynamic profile that ultimately converges to the demand vector, R. The matrix Γ will determine the dynamic response of the system output. Combine both the integral action state and the previous system state to yield the new state vector as follows:

$$x_{tol} = \begin{bmatrix} x_r \\ x \end{bmatrix} \tag{15.6}$$

Assume a partition of the state vector, x_{tol}, to isolate the input channel. The new augmented system can be re-partitioned in the form

$$\dot{z}_1(t) = \tilde{A}_{11}z_1(t) + \tilde{A}_{12}z_2(t) + B_r r(t)$$
$$\dot{z}_2(t) = \tilde{A}_{21}z_1(t) + \tilde{A}_{22}z_2(t) + B_2 u(t) \tag{15.7}$$

where $z_1 \in \mathfrak{R}^n$ and $z_2 \in \mathfrak{R}^m$. The system input now only affects the states in z_2. $B_r = [I_p \quad 0]^T$ is the input distribution matrix for the demand signal $r(t)$. The newly partitioned state matrix is given as

$$\tilde{A} = \begin{bmatrix} \tilde{A}_{11} & \tilde{A}_{12} \\ \tilde{A}_{21} & \tilde{A}_{22} \end{bmatrix} = \left[\begin{array}{c|c} \begin{matrix} 0 & -C_1 \\ 0 & A_{11} \\ 0 & A_{21} \end{matrix} & \begin{matrix} -C_2 \\ A_{12} \\ A_{22} \end{matrix} \end{array} \right] \tag{15.8}$$

The sliding mode controller seeks to induce a sliding motion on a sliding surface. A possible sliding surface can be proposed as

$$\ell_{SMIT} = \{(z_1, z_2, r) \in \Re^{n+p} : s(z_1, z_2, r) = 0\}$$
$$s(z_1, z_2, r) = S_1 z_1 + S_2 z_2 - S_r r \tag{15.9}$$

where $S_1 \in \Re^{m \times n}$, $S_2 \in \Re^{m \times m}$ and $S_r \in \Re^{p \times p}$ are design parameters that govern the reduced order motion. When ideal sliding motion occurs, $s(z_1, z_2) = 0$ and (15.9) can be rewritten as

$$z_2(t) = -S_2^{-1} S_1 z_1(t) + S_2^{-1} S_r r(t) \tag{15.10}$$

It is possible to choose S_2 so that it is invertible. Let $M = S_2^{-1} S_1$. The system dynamic during the ideal sliding motion on ℓ_{SMIT} can be obtained by substituting (15.10) into the first equation in (15.7):

$$\dot{z}_1(t) = (\tilde{A}_{11} - \tilde{A}_{12} M) z_1(t) + (\tilde{A}_{12} S_2^{-1} S_r + B_r) r(t) \tag{15.11}$$

The matrix M is seen to have the role of a state feedback controller for the z_1 subsystem and this matrix can be determined via a state feedback method if the pair $(\tilde{A}_{11}, \tilde{A}_{12})$ is completely controllable. The following reachability condition is used to determine the control

$$\dot{s} = \Phi s - \rho \, \text{sign}(s) \tag{15.12}$$

where Φ is a stable design matrix and the eigenvalues of this matrix determine the speed at which the system goes into sliding motion. Differentiating (15.9) and substituting from (15.7) yields

$$\dot{s} = S \tilde{A} x_{tol} + S_1 B_r r + S_2 B_2 u - S_r \dot{r} \tag{15.13}$$

where $S = [S_1 \quad S_2]$. The proposed control law is given as

$$u = u_L(x_{tol}, r) + u_N \tag{15.14}$$

where u_L and u_N represent the linear and nonlinear components of the control law. With the reachability condition (15.12), the linear part of the control is derived from (15.12) and (15.13) as follows:

$$u_L(x_{tol}, r) = L_{x_{tol}} x_{tol} + L_r r + L_{\dot{r}} \dot{r} \tag{15.15}$$

where

$$L_{x_{tol}} = -\Lambda^{-1} S\tilde{A} + \Lambda^{-1} \Phi S$$
$$L_r = -\Lambda^{-1} S_1 B_r - \Lambda^{-1} \Phi S_r \qquad (15.16)$$
$$L_{\dot{r}} = \Lambda^{-1} S_r$$

where $\Lambda = S_2 B_2$. The linear control law will drive the system into sliding motion asymptotically.

However, in order to make the controller robust against matched uncertainties and to achieve sliding in finite time, a nonlinear control component is required. The nonlinear control component, u_N is approximated to reduce the chattering effect and it is given as

$$u_N = \begin{cases} -\rho_N \Lambda^{-1} \dfrac{Ps}{\|Ps\| + \delta_c} & \text{if } s \neq 0 \\ 0 & \text{otherwise} \end{cases} \qquad (15.17)$$

where P is a symmetric positive definite matrix satisfying the equation

$$P^{-1} \Phi^T + \Phi P^{-1} = -Q \qquad (15.18)$$

for some positive definite matrix Q. δ_c is a small positive constant known as the smoothing factor. It is used to eliminate the chattering in the otherwise discontinuous control action. This positive constant can be tuned during implementation. Here, ρ_N is defined as the positive scalar function:

$$\rho_N(e_y) = \rho_{c1} + \rho_{c2} \|e_y\| + \rho_{c3} \|e_y\|^2 \qquad (15.19)$$

where ρ_{c1}, ρ_{c2} and ρ_{c3} are positive design scalars and e_y is the speed error. The second and third terms in (15.19) will only be introduced in the application to improve the speed recovery. Note that $\delta_c = 0$, yields finite time convergence to the sliding manifold. For smooth, positive δ_c, convergence to a boundary layer of the sliding manifold is assured in finite time. The above control law requires all the system states to be available and thus, an observer is required. The system is considered in regular form. The observer has the following form:

$$\dot{\hat{x}} = A\hat{x} + Bu - GCe + v \qquad (15.20)$$

where e is the state error and $G \in \Re^{n \times p}$, the observer linear gain, is defined as

$$G = \begin{bmatrix} A_{12} C_2^{-1} \\ A_{22} C_2^{-1} - C_2^{-1} A_{22}^{\phi} \end{bmatrix} \qquad (15.21)$$

where A_{22}^{ϕ} is a stable design matrix. The gain G is chosen such that the closed-loop observer matrix, $A_{closed} = A - GC$, has stable eigenvalues and satisfies the Lyapunov equation:

$$\hat{P} A_{closed} + A_{closed}^T \hat{P}^T = -\hat{Q} \qquad (15.22)$$

where $\hat{Q}$ is some positive definite matrix and $\hat{P}$ is a Lyapunov matrix that satisfies the structural constraint

$$C^T F^T = \hat{P}B \tag{15.23}$$

where $F \in \Re^{m \times m}$ is a nonsingular design matrix and is given as $[\hat{P}C_2 B_2]^T$. The sliding surface for the observer is:

$$\ell_{observer} = \{e \in \Re^n : s(e) = 0\}, \qquad s(e) = Fe \tag{15.24}$$

The nonlinear part of the observer is defined as

$$v = \begin{cases} -\rho_o \dfrac{Fe}{\|Fe\| + \delta_o} & \text{if } Fe \neq 0 \\ 0 & \text{otherwise} \end{cases} \tag{15.25}$$

where ρ_o is a positive design scalar and δ_o is a small positive constant, the smoothing factor. This nonlinear component can reject parameter uncertainties and maintains a sliding motion.

15.3   Sliding mode model-following (SMMF) control system

The SMMF control system design requires a reference model that can be obtained via linear feedback design on the nominal plant model whereby appropriate desirable nominal dynamics are prescribed. The sliding mode control is used to provide robustness. The sliding mode control term consists of a continuous term and a discontinuous term. The objective of the control is to minimise the error between the model and controlled plant. The control law needs all the internal states of the system to be available and so an observer must be used. Consider (15.1) as the plant model where here a subscript 'p' is used to indicate a plant matrix:

$$\begin{aligned} \dot{x}(t) &= A_p x(t) + B_p u(t) \\ y(t) &= C_p x(t) \end{aligned} \tag{15.26}$$

The corresponding ideal model is defined as

$$\begin{aligned} \dot{w}(t) &= A_m w(t) + B_m r(t) \\ y_m(t) &= C_m w(t) \end{aligned} \tag{15.27}$$

where $w \in \Re^n$ is the state vector of the ideal model, $r \in \Re^r$ is a reference input and A_p, B_p, A_m and B_m are compatibly dimensioned matrices. It is assumed that the pair (A_p, B_p) is controllable and the ideal model is stable. Let the tracking error state, $e(t)$, be defined by

$$e(t) = x(t) - w(t) \tag{15.28}$$

and the derivative be

$$\dot{e}(t) = \dot{x}(t) - \dot{w}(t) \tag{15.29}$$

From (15.26) and (15.27), the dynamic of the model-following error system (15.29) becomes

$$\dot{e}(t) = A_p x(t) - A_m w(t) + B_p u(t) - B_m r(t) \tag{15.30}$$

Adding and subtracting a term $A_m x$ in (15.30) yields

$$\dot{e}(t) = A_m e(t) + (A_p - A_m)x(t) + B_p u(t) - B_m r(t) \tag{15.31}$$

For perfect model following, the following condition holds [6] for time, t.

$$\begin{aligned}
\dot{x}(t) - \dot{w}(t) = 0 &\quad \Rightarrow \quad \dot{x}(t) = \dot{w}(t) \\
x(t) - w(t) = 0 &\quad \Rightarrow \quad x(t) = w(t)
\end{aligned} \tag{15.32}$$

Assume some arbitrary term, feeding forward from the model states, is added to the control action and (15.26) becomes

$$\dot{x}(t) = A_p x(t) + B_p(u(t) + G_{fwd} w(t)) \tag{15.33}$$

where the matrix G_{fwd} is some arbitrary gain. From the perfect model following conditions in (15.32), equations (15.27) and (15.33) become

$$A_p x(t) + B_p u(t) + B_p G_{fwd} w(t) = A_m w(t) + B_m r(t) \tag{15.34}$$

Rearranging (15.34) to obtain an expression for the control gives

$$u(t) = B_p^+ (A_m w(t) + B_m r(t) - A_p x(t) - BG_{fwd} w(t)) \tag{15.35}$$

where B_p^+ denotes the Moore-Penrose pseudo-inverse of the matrix B_p. Substituting this control expression into (15.34) yields

$$A_p x(t) + B_p B_p^+ (A_m w(t) + B_m r(t) - A_p x(t) - BG_{fwd} w(t))$$

$$+ BG_{fwd} w(t) - A_m w(t) - B_m r(t) = 0$$

Since $B_p B_p^+ B_p = B_p$ by definition, the above equation can be simplified as follows:

$$(B_p B_p^+ - I)(A_m - A_p)x(t) + (B_p B_p^+ - I)B_m r(t) = 0 \tag{15.36}$$

It can be seen that the direct use of the model states w in the control loop has no effect on the condition for model following. Based on (15.32), equation (15.36) is satisfied for all $x(t), w(t)$ and r if

$$(B_p B_p^+ - I)(A_p - A_m) = 0 \tag{15.37}$$

$$(B_p B_p^+ - I)B_m = 0 \tag{15.38}$$

If (15.37) and (15.38) hold, consider a control law with the following structure:

$$u(t) = u_1(t) + u_2(t) \tag{15.39}$$

where

$$u_1(t) = B_p^+ (A_m - A_p)x(t) + B_p^+ B_m r(t) \tag{15.40}$$

$$u_2(t) = -Me(t) \tag{15.41}$$

Equations (15.37) and (15.38) are the conditions for perfect model following and (15.39) is a control law for implementing it. The generation of (15.39) requires the model states. By substituting the control law (15.39) into (15.31) gives

$$\dot{e}(t) = (A_m - B_p M)e(t) \tag{15.42}$$

Thus, the dynamics of the model-following error system reduces to the closed-loop matrix $(A_m - B_p M)$. With proper selection of eigenvalues for M, the errors' settling rate can be controlled. A well-known theorem from linear algebra states that for the system of simultaneous equations denoted by $HD = F$, a solution for D exists if and only if rank $[H \quad F] = \text{rank}[H]$. Assume the following rank conditions hold

$$\text{rank}[B_p \quad A_p - A_m] = \text{rank}[B_p]$$
$$\text{rank}[B_p \quad B_m] = \text{rank}[B_p] \tag{15.43}$$

According to the theory, there thus exist matrices G_{mfwd} and L of suitable dimensions such that

$$B_p L = A_m - A_p \quad \Rightarrow \quad A_m = B_p L + A_p$$
$$B_p G_{mfwd} = B_m \quad \Rightarrow \quad B_m = B_p G_{mfwd} \tag{15.44}$$

These effectively define the model dynamics. Hence (15.40) for u_1 can now be expressed as

$$u_1(t) = Lx(t) + G_{mfwd}r(t) \tag{15.45}$$

Consider (15.27) at steady state, whereby

$$w_{ss} = -A_m^{-1} B_m r \tag{15.46}$$

At steady state, the reference model output equation becomes

$$y_{mss} = C_m w_{ss}$$

Substituting from (15.46) and (15.44),

$$y_{mss} = -C_m A_m^{-1} B_m r$$
$$= -C_m A_m^{-1} B G_{mfwd} r$$

If the model is to attain the reference, it follows that

$$G_{mfwd} = \text{inv}(-C_m A_m^{-1} B) \tag{15.47}$$

The term L in (15.44) can be obtained as a linear feedback matrix by selecting appropriate eigenvalues for the reference model. Thus, the reference model matrix can be written as

$$A_m = A_p + B_p L \tag{15.48}$$

u_2 in (15.39) is augmented with a discontinuous element in the control law to provide robustness. By considering the error states, (15.28), an error dependent switching function is defined as

$$\ell_{SMMF} = \{e(t) \in \Re^n : s(e(t)) = 0\}, \qquad s(e(t)) = Se(t) \tag{15.49}$$

When a sliding motion takes place

$$Se(t) = 0 \tag{15.50}$$

By differentiating (15.50), substituting (15.31) into it and rearranging the equation gives

$$\dot{s}(e(t)) = S\dot{e}(t) = S(A_m e(t) + (A_p - A_m)x(t) + B_p u(t) - B_m r(t)) \tag{15.51}$$

Applying the control component (15.4), (15.51) reduces to

$$\dot{s}(e(t)) = S(A_m e(t) + B_p u_2(t)) \tag{15.52}$$

The reachability condition for the controller is given as in (15.12). By considering both (15.52) and (15.12), the control law u_2 is expressed as

$$u_2 = u_l + u_n \tag{15.53}$$

where

$$u_l(t) = -(SB_p)^{-1}(SA_m - \Phi S)e(t)$$

$$u_n(t) = -\rho_N(t, e)(SB_p)^{-1}\frac{P_2 Se(t)}{\|P_2 Se(t)\| + \delta} \tag{15.54}$$

where δ is a small positive constant used to smooth the discontinuous control action and $P_2 \in \Re^{m \times m}$ is a symmetric positive definite matrix satisfying the Lyapunov equation

$$P_2 \Phi + \Phi^T P_2 = -I \tag{15.55}$$

ρ_N is defined as

$$\rho_N(e_y) = \rho_{c1} + \rho_{c2}\|e_y\| + \rho_{c3}\|e_y\|^{Pow} \tag{15.56}$$

where ρ_{c1}, ρ_{c2} and ρ_{c3} are positive design scalars and e_y is the system output error. It is assumed that SB_p is chosen to be nonsingular. The complete model-following variable structure control scheme has the form of

$$u(t) = u_1(t) + u_l(t) + u_n(t) \tag{15.57}$$

The above control law requires all the system states to be available and thus, a similar observer used for the SMIT control implementation is employed here.

15.4   Second order sliding mode (SOSM) control system

A 'super-twisting' 2-sliding controller [4], which needs only measurement of the sliding variable, is used. It is assumed that upper bounds on the system dynamics are known. Consider a system of the form

$$\dot{x}(t) = \phi(t, x) + \gamma(t, x)u(t) \tag{15.58}$$

where $x = s$ and $\phi(t, x)$, $\gamma(t, x)$ are smooth uncertain functions with $|\phi| \leq \Phi_{SOSM} > 0$, $0 < \Gamma_m \leq \gamma \leq \Gamma_M$. The super-twisting algorithm converges to the 2-sliding set ($s = \dot{s} = 0$) in finite time. The trajectories of the super-twisting algorithm are characterised by twisting around the origin on the phase portrait of the sliding variable. The super-twisting algorithm defines the control law, $u(t)$ as a combination of two terms:

$$u(t) = u_1(t) + u_2(t) \tag{15.59}$$

$$\dot{u}_1(t) = \begin{cases} -u & |u| > 1 \\ -W_{SOSM}\,\mathrm{sign}(s) & |u| \leq 1 \end{cases} \tag{15.60}$$

$$u_2(t) = \begin{cases} -\lambda_{SOSM}|s_0|^{\rho_{SOSM}}\,\mathrm{sign}(s) & |s| > s_0 \\ -\lambda_{SOSM}|s|^{\rho_{SOSM}}\,\mathrm{sign}(s) & |s| \leq s_0 \end{cases} \tag{15.61}$$

and sufficient conditions for finite time convergence are

$$W_{SOSM} > \frac{\Phi_{SOSM}}{\Gamma_m} > 0$$

$$\lambda_{SOSM}^2 \geq \frac{4\Phi\Gamma_M(W_{SOSM} + \Phi_{SOSM})}{\Gamma_m^3(W_{SOSM} - \Phi_{SOSM})} \tag{15.62}$$

$$0 < \rho_{SOSM} \leq 0.5$$

The choice of $\rho_{SOSM} = 0.5$ assures that sliding order 2 is achieved. The following section describes the implementation findings and reports on the modifications required to improve the system performance.

15.5 Diesel generator system

The diesel generator set under consideration is the Perkins 1000 Series, four-litre, four-cylinder, turbo-charged diesel engine. The engine runs at a nominal speed of 1500 rpm and can generate maximum power of 65 kW. The generated power is dissipated via an electrical resistor load bank. The permissible load value varies from 1 kW up to 65 kW. A change of load facility is used to assess the designed controllers performance.

The test system comprises of a dSPACE™ DS1103 real-time processor board and a computer. The real-time processor board comprises a Texas Instrument TMS 320F240 DSP microcontroller with the input/output expanded via a number of different socket connections. dSPACE™ integrated software, the ControlDesk, allows real-time implementation, parameter setting, display of the input/output in graphical format, on-board control and data acquisition. The hardware configuration is shown in Fig. 15.1.

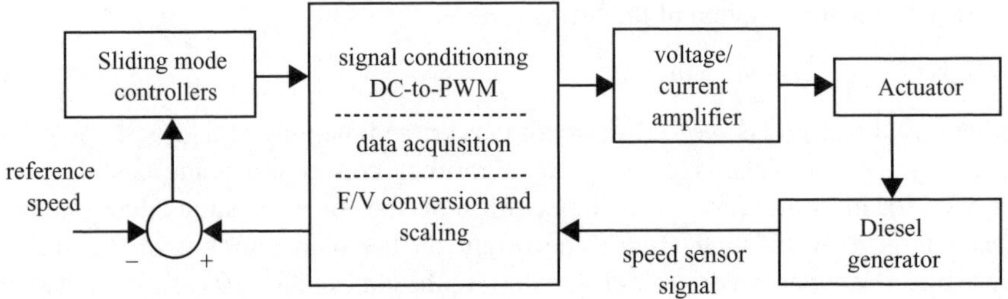

Figure 15.1   Hardware set-up for the controllers

15.6   Control systems setting and simulation

A diesel generator model developed from system identification is used for both the SMIT and SMMF control system designs. The model equations are:

$$\dot{x}(t) = \begin{bmatrix} -0.0055 & -0.0026 \\ 0.0026 & -0.7170 \end{bmatrix} x(t) + \begin{bmatrix} 4.9617 \\ -1.1836 \end{bmatrix} u(t)$$

$$y(t) = [4.9617 \quad 1.1836]x(t)$$

(15.63)

The designed controllers are written as Matlab™ C-Mex S-functions for dSPACE™ implementation. The S-functions are converted into C code using a specific function in the Matlab™/RealTime toolbox. This C code was further cross-compiled into an executable file to be run on the DSP micro-controller.

The sliding mode designs are carried out using the available toolbox [7]. For the SMIT design, the poles are set to $[-1.0; -1.2]$ and the resulting switching function has the value of $S = [0.0346 \quad 0.0337 \quad 0.1000]$. Other controller design parameters have the following settings $S_r = -0.018$, $P2 = 0.02$, $\Phi = -1.0$, $\delta_c = 0.1$, $L_r = 0.0679$, $L_{\dot{r}} = 0.0353$ and $L_{x_{tol}} = [0.0679 \quad -0.1661 \quad 0.4857]$. The observer parameters are given by

$$\delta_o = 0.1, \quad F = [0.5804], \quad G = \begin{bmatrix} 4.2730 \\ -0.9824 \end{bmatrix}, \quad A_{22}^{\phi} = -20$$

Three control gains are introduced to the respective linear control component, i.e. $g\_L_{x_{tol}}$, g_L_r and $g\_L_{\dot{r}}$. Direct substitution into the reachability condition shows that this is a valid parameterisation for effective online tuning. The ρ_{c1}, ρ_{c2}, ρ_{c3} and ρ_o design parameters are also open for tuning. The control package is designed to allow the user to tune the gains online while testing the actual plant performance in real-time. The gains $g\_L_{x_{tol}}$ and ρ_{c1} are tuned in such a way that the controller performance could cope with large load variations, as well as small step loads. Both the g_L_r and $g\_L_{\dot{r}}$ are tuned to obtain speed tracking at difference dynamic profile of Γ. Three different Γ values are considered in the test, i.e. $0.1, 0.5$ and 1.0. The higher the value, the faster the tracking is required. The following gain settings are found to give better load disturbance rejection and speed tracking capability: $g\_L_{x_{tol}} = 0.00765$, $\rho_{c1} = 0.1$, $\rho_{c2} = 0$, $\rho_{c3} = 0$, $\rho_o = 0.1$, $g\_L_r = 0.001$ and $g\_L_{\dot{r}} = 0.001$.

For the SMMF control system design, the reference model is chosen to have poles of $p = -0.6 \pm 0.001\,j$ and the resulting model is

$$A_m = \begin{bmatrix} -0.15015 & -0.0802 \\ 0.1209 & -0.6985 \end{bmatrix}$$

$$B_m = \begin{bmatrix} 0.1011 \\ -0.0241 \end{bmatrix} \tag{15.64}$$

The desired poles of the sliding motion are set to $[-0.35]$ and the resulting switching function has the value of $S = [-0.5054 \quad 2.1911]$. The design parameters for the SMMF control system have the following setting: $L = [-0.10 \quad -0.0156]$, $\Phi = -0.35$, $G_{mfwd} = [0.0204]$, $S = [-0.5054 \quad 2.1911]$, $P_2 = 1.4286$ and $\Phi = -0.35$. The corresponding observer settings are chosen as

$$A_{22}^{\phi} = -20, \quad F = 0.5804 \quad \text{and} \quad G = \begin{bmatrix} 4.2730 \\ -0.9824 \end{bmatrix}$$

As before, two control gains are introduced, i.e. g_u_l and g_u_1, to tune u_l and u_1. The ρ_{c1}, ρ_{c2}, ρ_{c3} and ρ_o parameters are also open for online tuning. The initial simulation test results show that the engine speed dipped slightly for each added load. The speed dropped 270 rpm at a load of 60 kW. The results consistently show a steady state error in the speed signal. One way to achieve zero steady state error is by introducing integral action in the controller. The resulting control law becomes

$$u(t) = u_1(t) + u_2(t) + K_{int} u_3 \tag{15.65}$$

where $\dot{u}_3 = e(t)$ and K_{int} is the gain of the integral control. The sliding motion is insensitive to the matched uncertainties. With the inclusion of the integral action, this property of a sliding motion is assumed to hold. The tuning parameters have the following values: $g\_u_l = 1$, $g\_u_1 = 1.13$, $\rho_{c1} = 0.01$, $\rho_{c2} = 0.0005$, $\rho_{c3} = 0.0005$, $\rho_o = 150$, $pow = 1.5$, $\rho_o = 0.1$ and $K_{int} = 0.002$.

In general, any r-sliding controller that keeps $s = 0$ needs $s, \dot{s}, \ddot{s}, \ldots, s^{(r-1)}$ to be made available. In the case of engine speed control, this implies that acceleration should either be measured or else an observer constructed to estimate it. Because the 'super-twisting' 2-sliding algorithm is used, no knowledge of the engine acceleration is required and hence speed control of the diesel engine is based on speed error measurement alone. A sliding variable, s, is defined as the difference between the measured engine speed, N_{mea}, and the desired nominal speed, N_{nom}:

$$s = N_{mea} - N_{nom} \tag{15.66}$$

The control algorithm from (15.60) and (15.61) for systems linear in the control where $s_0 = \infty$ can be simplified as follows:

$$u(t) = -\lambda_{SOSM} |s|^{\rho_{SOSM}} \operatorname{sign}(s) + u_1 \tag{15.67}$$

$$\dot{u}_1 = -W_{SOSM} \operatorname{sign}(s) \tag{15.68}$$

Effectively the controller can be tuned via three parameters, ρ_{SOSM}, λ_{SOSM} and W_{SOSM}. In the SOSM control system, an additional gain, K_p, which is applied to

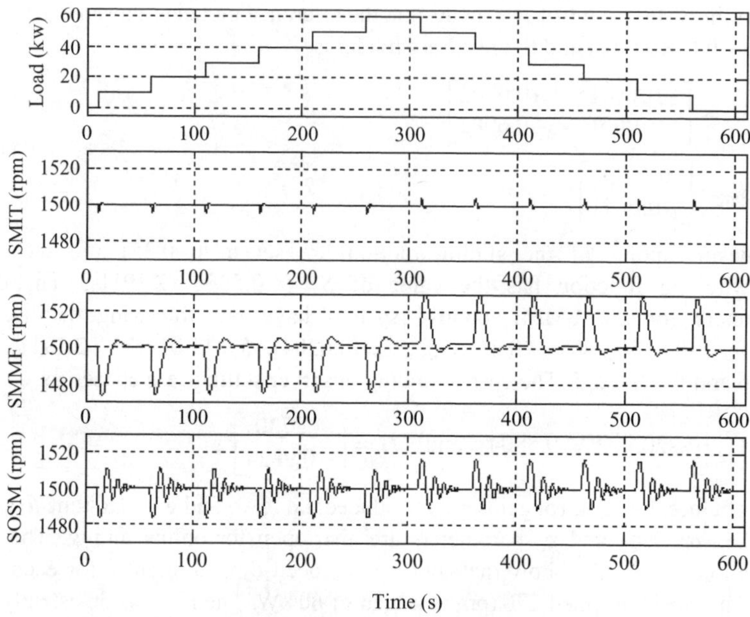

Figure 15.2   Simulation results for the designed controllers

the feedback speed error, is introduced to improve the control system performance. The effective control algorithm from (15.67) has become

$$u(t) = -\lambda_{SOSM} |s|^{\rho_{SOSM}} \operatorname{sign}(s) + u_1 + K_p s \tag{15.69}$$

These gains are set to $W_{SOSM} = 0.1$, $\lambda_{SOSM} = 0.01$ and $K_p = -0.001$.

Since the engine model is a SISO system, no load input is available for simulation testing of the controllers. To simulate a load disturbance, the identified model is approximated by a pole-zero pair and an approximate integrator. The pole-zero component can be considered to be an actuator and the integrator the fundamental dynamics of the generator. The load is assumed to act between the two transfer functions. Figure 15.2 shows the simulated load and engine speed signal for the respective controllers during the loading condition. The load increases by 10 kW at each step up to 60 kW and then decreases back to 0 kW. It is noticed that the designed controllers are able to maintain the engine speed. The SMIT controller gives the shortest settling time.

15.7   Control systems implementation results

The designed controllers are assessed with respect to three test criteria. First, the performance is assessed at the nominal speed of 1500 rpm. The dynamic behaviour of interest is the speed change with load. The generator is subjected to small step

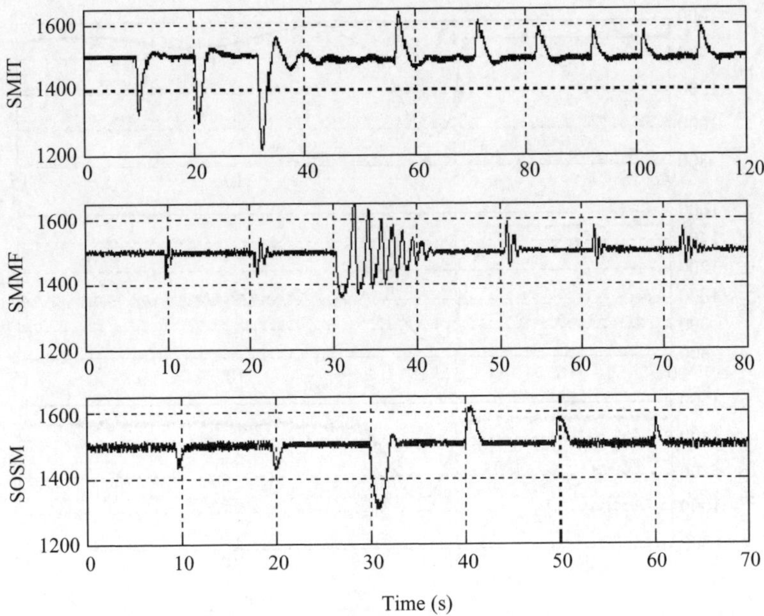

Figure 15.3   Speed response at a step load of 20 kW at 1500 rpm reference speed

changes of load and a single large step load. The speed transient responses are studied. The first test simulates the situation of small varying demands by the consumer. This can be achieved by applying a small load (i.e. 20 kW) to the engine. The second test assesses the controllers in the presence of a large step change in load (i.e. 60 kW). This corresponds to a sudden high power demand by the consumer. The resulting performances are shown in Figs 15.3 and 15.4 respectively. Figure 15.3 shows the engine speed response when step loads of 20 kW (i.e. from 0 kW to 60 kW and then in steps of 20 kW) are applied to the diesel generator. The speed recovers within 3 s for a step load of 20 kW for the SOSM controller. The SMIT controller shows the ability to cope with step loads up to 60 kW. However, it exhibits a longer settling time and larger speed dips. During the step off load (i.e. from 60 kW to 40 kW), the engine speed hit the speed safety level (i.e. about 1650 rpm) and the engine fuel is cut off to shutdown the engine automatically. Thus, the load decrement was reduced to 10 kW when moving down from 60 kW to 10 kW. The subplot 1 in Fig. 15.3 shows the resulting effect. The SMMF controller copes reasonably well with the step load. The controller performance shows a smaller speed dip but oscillates when the load is changed from 40 kW to 60 kW. The engine speed settles down after about 12 s. During the large load test, the SOSM controller settles down in about 2.5 s. A large speed drop is observed due to the large change in load. The settling time for the SOSM controller is faster than the other controllers. The SMMF controller struggles with such a large step load change and does not regain the set point; the speed settles down to about 1000 rpm. The SMMF controller can handle a maximum step load of up to

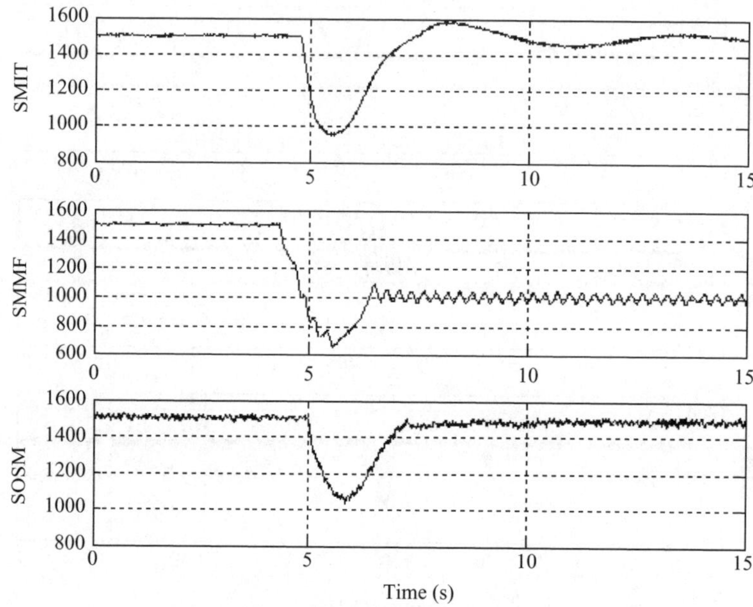

Figure 15.4   Speed response at a step load of 60 kW at 1500 rpm reference speed

45 kW. Overall, the SOSM shows a faster settling time to large step changes in load especially at the nominal generator speed, 1500 rpm. The SMMF controller cannot tolerate large step changes in load at any speed. The SMIT controller copes well in both load situations but with a longer settling time. It is noted that the engine test performance is slightly better than the simulation performance. This is because the model used to simulate the load disturbance condition is an approximate model and does not represent the engine performance in the steady state as well as the identified model; the latter was inappropriate for demonstrating the response under load.

The second test evaluates the robustness of the designed controllers. The objective of this test is to show the robustness of the controller performance at different operating conditions to investigate the wide envelope performance. The reference speed of the generator is set to 1350 rpm and 1200 rpm. These changes are done online during the test. Similar test procedures to those used in the preceding paragraph are employed here: the generator is subjected to a small steps load of 20 kW and a large speed load of 60 kW; the speed response is studied. Figures 15.5, 15.6, 15.7 and 15.8 show the speed response for the mentioned tests respectively. The first observation of the test is that all controllers maintain the engine reference speed setting at zero loading condition. The SOSM shows a faster speed recovery than the other controllers at 1350 rpm. The SMIT controller shows similar performance and copes reasonably well with electrical load but with a longer settling time compared to the SOSM controller. The SMMF controller tracks the speed well at the reference setting. However, the controller cannot cope with a large step load of 60 kW at 1350 rpm. This test shows the

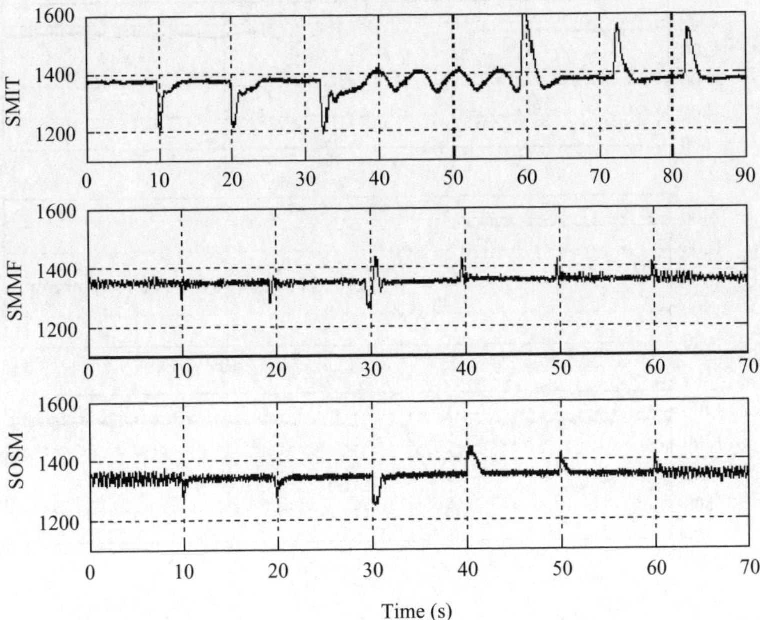

Figure 15.5   *Speed response at a step load of 20 kW at 1350 rpm reference speed*

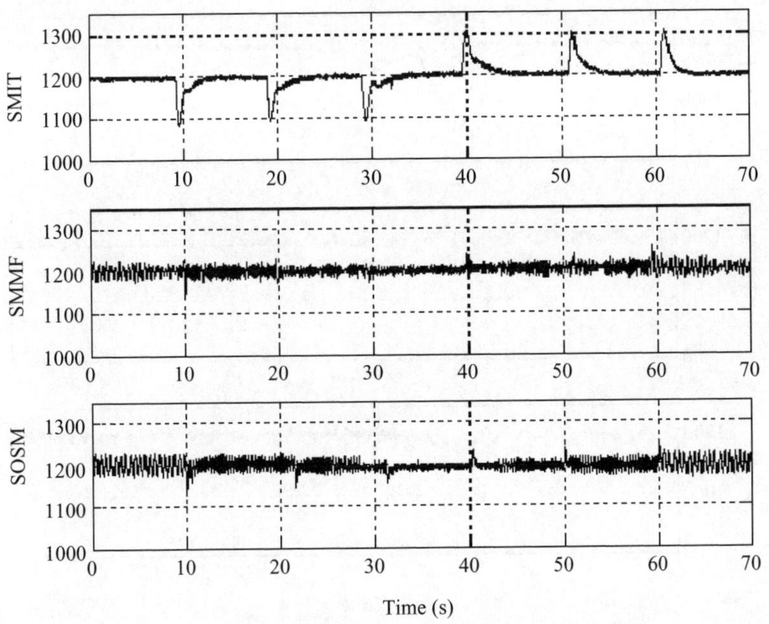

Figure 15.6   *Speed response at a step load of 20 kW at 1200 rpm reference speed*

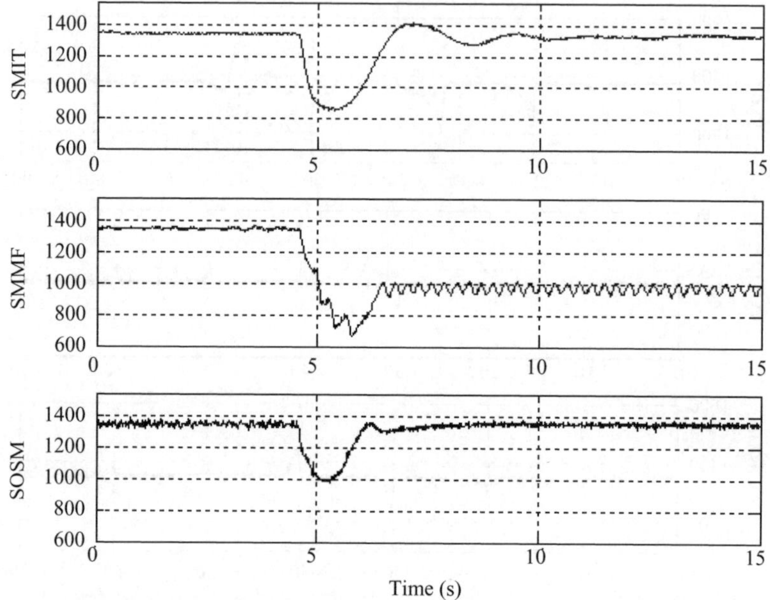

Figure 15.7   Speed response at a step load of 60 kW at 1350 rpm reference speed

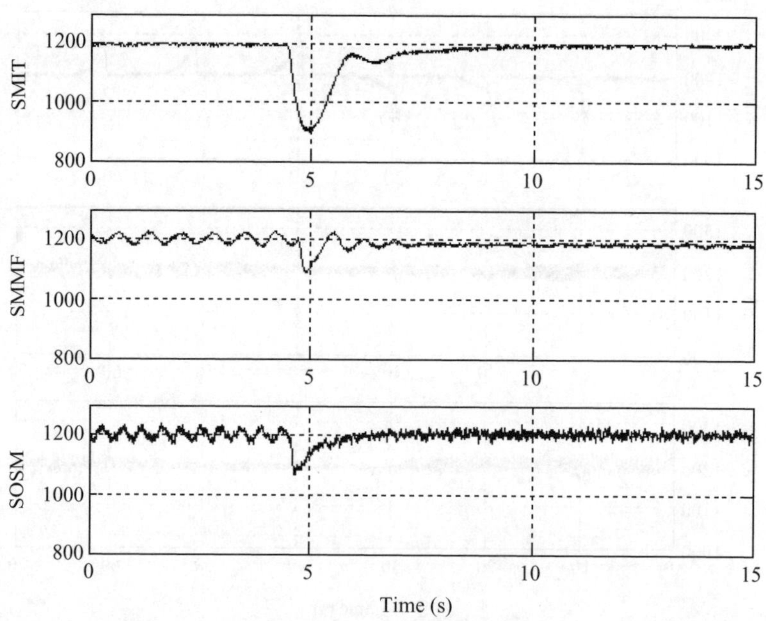

Figure 15.8   Speed response at a step load of 60 kW at 1200 rpm reference speed

SOSM controller is robust and capable of controlling diesel engine speed over a wide operating envelope. The analysis shows the SOSM performs better at a reference speed of 1350 rpm. During large loading conditions, the SOSM and SMIT controllers perform better than the SMMF controller.

The final test examines the speed tracking capability of both the SMIT and SMMF controllers. For the SMIT controller, the dynamic profile of speed tracking depends on the term Γ. The demand vector R is set to change in the fashion of step-changes from 0 rpm to -50 rpm to 0 rpm to 50 rpm. The resulting performance is shown in Fig. 15.9. Three different settings of Γ were set at 0.1, 0.5 and 1.0. The higher the Γ value, the faster the speed tracking rate. At Γ equal to 0.1 and 0.5, the controller maintained the engine speed closely to the demand signal R at a slower rate. At $\Gamma = 1.0$, the controller tracked very rapidly and closely to the demand value. The tracking performance was determined by both the g_L_r and $g\_L_{\dot{r}}$ settings. For validation of the controller performance, the practical gain settings were applied to the simulation model. A similar demand vector R is applied to the simulation system. The dynamic

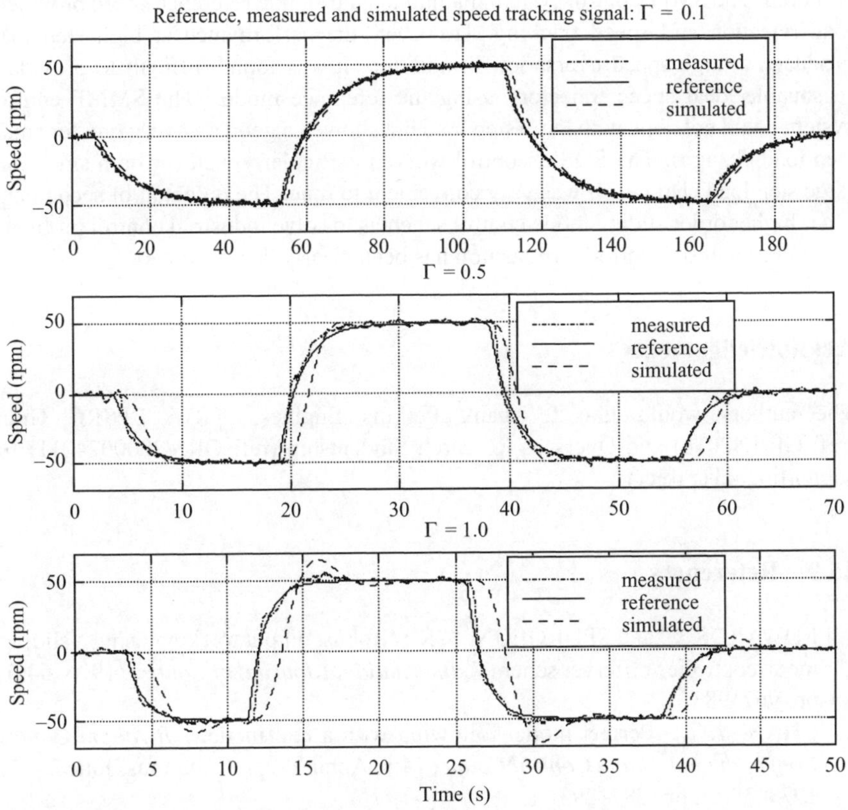

Figure 15.9   Speed tracking response for the SMIT controller

response of the speed tracking was determined by the three different settings of Γ. The resulting performance is shown in Fig. 15.9. It can be seen that the simulated output tracks in the same fashion as the demand signal. However, at a faster tracking rate, the simulated generator speed shows a small overshoot and slight delay in the response. The SMMF controller produced reasonably good speed tracking but the desired reference produced by the model is not adequate for real application.

15.8   Conclusion

Three different control system designs for generator idle speed control have been considered. Two techniques use standard sliding mode control methods. A higher order sliding mode method has also been considered. The associated design procedures have been described. The control techniques were then applied. Controller performance tests have been carried out. First, different loadings at nominal engine speed have been applied. Second, controller robustness has been investigated by setting different reference speed settings. Finally, speed tracking capability has been explored. The SMIT control system maintained the engine reference speed provided, load rejection and speed tracking. However, the performance at high step load produced a large speed overshoot. In addition, it was found difficult to provide a reasonable ideal speed trajectory using the reference model. The SMMF control system could not cope with large step load but showed a good recovery rate for small step load changes. The SOSM control worked particularly well for both small and large step load changes. It was very convenient to tune. The potential of second and other higher order sliding mode control schemes to solve industrial control problems in a robust and straightforward fashion has been clearly demonstrated.

Acknowledgements

The authors would like to thank Perkins Engines, TRW, EPSRC Grant (ref: GR/L42018) and Overseas Research Studentship (ref: ORS/2000024011) for supporting this project.

15.9   References

1   EDWARDS, C. and SPURGEON, S. K.: 'Robust output tracking using a sliding-mode controller/observer scheme', *International Journal of Control*, 1996, **64**(5), pp. 967–983

2   CHAN, Y. T.: 'Perfect model following with a real model', *Jt Autom Control Conf of the Am Autom Control Counc*, 14th, Annu, Prepr, Columbus, June 20–22, 1973, **10**(5), pp. 287–293

3   SPURGEON, S. K., YEW, M. K., ZINOBER, A. S. I., and PATTON, R. J.: 'Model-following control of time-varying and nonlinear avionics systems',

in ZINOBER, A. S. I. (Ed.): 'Deterministic control of uncertain systems' (Peter Peregrinus, London, 1990), pp. 96–114
4   LEVANT, A.: 'Sliding order and sliding accuracy in sliding model control', *International Journal of Control*, 1993, **58**(6), pp. 1247–1263
5   KHAN, M. K., SPURGEON, S. K., and PULESTON, P. F.: 'Robust speed control of an automotive engine using second order sliding modes', *Proceedings of the European Control Conference*, 2001, pp. 974–978
6   ERZBERGER, H.: 'On the use of algebraic methods in the analysis and design of model following control systems', TN D-4663, NASA, 1968
7   EDWARDS, C. and SPURGEON, S. K.: 'Sliding mode control: theory and applications' (Taylor & Francis, London, 1998)

Chapter 16

Motion control of underwater objects by using second order sliding mode techniques

Giorgio Bartolini, Alessandro Pisano,
Elisabetta Punta and Elio Usai

16.1   Introduction

It is well known that, disregarding implementation aspects, the sliding mode approach constitutes the most simple and effective control method to deal with the control of complex mechanical systems, in particular robotic systems. Indeed any joint velocity profile can be tracked by a completely decoupled discontinuous control law, exploiting some good properties of the inertia matrix, the complexity of which is relatively independent from the complexity of the model, rapidly increasing with the number of degrees of freedom.

This control approach has not been considered suitable for real implementation due to the so-called chattering phenomenon. This phenomenon, which is tied to the discontinuous nature of the control strategy, is dangerous, not because the control signals chatter, but because they chatter at a frequency that is not sufficiently high due to non-idealities (finite bandwidth of the actuators, time delays, etc.). In this sense, often the use of a continuous approximation of the discontinuous control signals [1], worsens this phenomenon.

Another way to avoid chattering is based on the use of an observer [2] controlled so as to generate a sliding mode on a manifold in the observer state space which tends to coincide with the ideal sliding manifold. The discontinuous high-frequency control is filtered by the high-gain unmodelled dynamics giving rise to a continuous control which turns out to be close to the so-called equivalent control [2].

Another way to avoid chattering without using observers consists of enforcing a second-order sliding mode on a manifold in the original state space, i.e., a sliding regime on a surface $s[x(t)] = 0$ in the system state space, with $\dot{s}[x(t)]$ identically equal to zero, a regime enforced by a control signal depending on $s[x(t)]$, but directly acting only on $\ddot{s}[x(t)]$. A list of possible second order sliding mode algorithms has been presented [3].

The first two sections of this chapter are devoted to recalling briefly the basic results of the SOSM control method for single-input and multi-input systems. The rest of the chapter is aimed at tailoring this approach to an important robotic application domain, that is the control of underwater objects.

It is important to stress that the chattering phenomenon, considered as the main drawback of the VSS approach, belongs to the class of implementation aspects. This means that, while nobody can find theoretical 'faults' in the sliding mode theory, based on rigorous mathematical tools, equipped with good regularity properties [2, 4], there are strong doubts about the existence of practical devices able to implement the relevant control laws. This fact explains the strong effort made by the researchers involved in this area to analyse every aspect related to the practical implementation of the SM approach. A fundamental contribution in this direction is Reference 5.

In this spirit, this chapter reports the results of accurate experiments made on a simple laboratory prototype of a jet based propulsion system which will be used in the design of more complex underwater vehicles with the twofold aim of validating the effectiveness of the control strategy and analysing the performance of an innovative actuator.

16.2   Nonlinear output-feedback control via 2-SM controllers and 2-SM differentiators

Consider the nonlinear SISO system

$$\dot{x} = f(x) + g(x)u$$
$$y = h(x) \tag{16.1}$$

with unavailable state vector $x \in R^n$, scalar control variable $u \in R$ and measurable output $y \in R$. Let f, g and h be unknown, sufficiently smooth, vector-fields of appropriate dimension satisfying proper growth constraints to be specified. The heavy uncertainty of the system prevents immediate reduction of (16.1) to any normal form by means of standard approaches based on the knowledge of f, g and h.

If conditions $L_g L_f h(x) = L_g L_f^2 h(x) = \cdots = L_g L_f^{r-2} h(x) = 0$ and $L_g L_f^{r-1} h(x) \neq 0$ hold globally, then system (16.1) possesses a globally-defined relative degree r [6] (L_g, L_f denote the Lie derivatives), and the input-output dynamics can be expressed correspondingly as

$$y^{(r)} = L_f^r h(x) + L_g L_f^{r-1} h(x)u \tag{16.2}$$

Let $\zeta = [y, \dot{y}, \ldots, y^{(r-1)}]^T$, then it is always possible [6] to define a vector $\eta \in R^{n-r}$ such that the map

$$x = \Phi(\zeta, \eta) \tag{16.3}$$

is a diffeomorphism on R^n and the η dynamics, which are referred to as the 'internal dynamics' [6], can be expressed in the following form

$$\dot{\eta} = q(\zeta, \eta) \tag{16.4}$$

If $r = n$, there are no internal dynamics and the system is said to be 'fully linearisable' [6].

The following assumptions are made.

Assumption 1.   *The internal dynamics (16.4) are input-to-state stable (ISS).*

Assumption 2.   *The drift term, $L_f^r h(x)$, and the control gain, $L_g L_f^{r-1} h(x)$, of the input-output dynamics (16.2) are globally bounded and Lipschitz.*

Let y_R be a desired smooth output response, and consider the error dynamics

$$e^{(r)} = L_f^r h(x) - y_R^{(r)} + L_g L_f^{r-1} h(x) u \tag{16.5}$$

where $e = y - y_R$.

The output-feedback control problem with higher-order sliding modes has been dealt with in recent work [7, 8] and can be characterised by a three-step procedure.

Step 1 – Sliding manifold design

The sliding variable is usually expressed as follows:

$$\sigma = e^{(r-1)} + \sum_{i=0}^{r-2} c_i e^{(i)}, \qquad e^{(0)} = e \tag{16.6}$$

where the coefficients c_i $(i = 0, 2, \ldots, r-2)$ are chosen such that the polynomial $P(\lambda) = \lambda^{(r-1)} + \sum_{i=0}^{r-2} c_i \lambda^{(i)}$ is Hurwitz.

Step 2 – Estimation of the sliding variable

The error derivatives are not available and must be evaluated by means of some accurate real-time device robust against measurement noise and possibly finite-time converging. Recently an arbitrary-order differentiator has been proposed [8, 9] based on higher-order sliding modes which appears to be an effective, yet robust, solution.

If the input/output relative degree is $r = 2$, then only the first derivative of e needs to be estimated, and the arbitrary-order differentiator [8, 9] reduces to the first-order differentiator [10]:

$$\varepsilon(t) = q_0(t) - e(t)$$

$$\dot{q}_0(t) = q_1(t) - \kappa_0 |\varepsilon(t)|^{1/2} \operatorname{sign}(\varepsilon(t)) \tag{16.7}$$

$$\dot{q}_1(t) = -\kappa_1 \operatorname{sign}(\varepsilon(t))$$

The tuning conditions are

$$\begin{cases} \kappa_1 > C_2 \\ \kappa_0^2 > 4C_2 \dfrac{\kappa_1 + C_2}{\kappa_1 - C_2} \end{cases} \tag{16.8}$$

where C_2 is a Lipschitz constant of the error derivative $\dot{e}$.

If $r > 2$ one could implement a cascade series of differentiators of the type (16.7) to estimate the higher-order derivatives. However, by performing noise propagation analysis it is seen that a higher-order differentiator [9], specifically designed for multiple differentiation, is more effective.

Step 3 – Stabilisation of the sliding variable

Consider the nonlinear uncertain second-order sliding variable dynamics

$$\ddot{\sigma} = \varphi(\zeta, \zeta_R, \eta, u) + \gamma(\zeta, \eta)\dot{u} \tag{16.9}$$

where $\zeta_R = [y_R, \dot{y}_R, \ldots, y_R^{(r-1)}]^T$.

From Assumption 2, in view of the fact that vector ζ_R is norm-bounded, it can be concluded that the so-called 'equivalent control' [2] is bounded. Thus, as long as the closed-loop system evolves within a (possibly large) compact domain containing the 2-sliding manifold $\sigma = \dot{\sigma} = 0$, the following additional assumption can be met.

Assumption 3. *Three positive constants F, Γ_1, Γ_2 can be found such that the uncertainties φ and γ satisfy the following boundedness conditions*

$$\begin{aligned} |\varphi| &\leq F \\ 0 &< \Gamma_1 \leq \gamma \leq \Gamma_2 \end{aligned} \tag{16.10}$$

Conditions (16.10), whose local validity is to be stressed, constitute a particular case of more general state-dependent bounds for the uncertain sliding variable dynamics [11], which allows for a particularly simple constant-parameter realisation of the so-called '*Sub-optimal*' 2-SMC algorithm [3]:

$$u(0) = 0$$

$$\dot{u}(t) = \begin{cases} -U_M \text{sign}(\sigma(t) - \frac{1}{2}\sigma(0)), & 0 \leq t < t_{M_1} \\ -\alpha(t)U_M \text{sign}(\sigma(t) - \frac{1}{2}\sigma(t_{M_i})) & t_{M_i} \leq t < t_{M_{i+1}}, \qquad i = 1, 2, \ldots \end{cases} \tag{16.11}$$

$$\alpha(t) = \begin{cases} \alpha^*, & \sigma(t_{M_i})(\sigma(t) - \frac{1}{2}\sigma(t_{M_i})) \leq 0 \\ 1, & \sigma(t_{M_i})(\sigma(t) - \frac{1}{2}\sigma(t_{M_i})) > 0 \end{cases} \tag{16.12}$$

where t_{M_i} $(i = 1, 2, \ldots)$ is the sequence of the time instants at which $\dot{\sigma} = 0$, and parameters U_M and α^* are chosen according to the tuning rules

$$\alpha^* > \frac{\Gamma_2}{3\Gamma_1} \tag{16.13}$$

$$U_M > \max \left\{ \frac{4F}{3\alpha^*\Gamma_1 - \Gamma_2}, \frac{F}{\Gamma_1} \right\} \tag{16.14}$$

The above control law steers to zero both σ and $\dot{\sigma}$ in a finite time [3, 11]. The reader is referred to [8, 12] for details regarding the separation principle establishing the closed loop stability of combined 2-SMC/2-SMD schemes.

Remark 1. Since $\dot{\sigma}$ is not known, the sequence t_{M_i} is unavailable, but can be approximately detected using only sampled measurements of σ carried out at the time instants $t_k = k\tau$ [11]. It has been shown that the resulting approximate implementation of the controller guarantees the reaching of an $O(\tau^2)$ boundary layer of the sliding manifold $\sigma = 0$ [11, 13].

16.3   A multi-input version of the control problem

A possible generalisation to multi-input cases of the previously considered problem has been presented [14 and 15].

In contrast with the standard multi-input first order sliding mode control, a Lyapunov-like approach is not readily available for second order sliding mode control of uncertain systems. We limit ourselves to considering some particular cases that can be used to deal with the control problem in the presence of a sufficiently wide class of uncertainties. Consider a system characterised by the following equation:

$$\dot{x} = A(x) + B(x)u \tag{16.15}$$

where $x \in \Re^n$ and $u \in \Re^n$.

The vector u collects the control signals devoted to steering x to zero. The system is assumed to be uncertain with some structural assumptions whose reasonableness will be discussed in the next sections. The procedure follows the same steps previously performed for the single input case.

Differentiate (16.15) to obtain

$$\ddot{x} = F(x, \dot{x}, u) + B(x)\dot{u} \tag{16.16}$$

where $F(x, \dot{x}, u) = (d/dt)[A(x)] + (d/dt)[B(x)]u$, x is assumed measurable, while $\dot{x}$ is not available due to uncertainties in system dynamics. The vector field $F^T(x, \dot{x}, u) = [F_1(x, \dot{x}, u), \ldots, F_n(x, \dot{x}, u)]$ is uncertain and for simplicity sake we assume that, in a sufficiently large open set, the following holds $F_i(x, \dot{x}, u) < \bar{F}_i$, where $\bar{F}_i$ are known constants.

The matrix $B(x)$ is assumed uncertain, but with known bounds of the modulus of its components, and positive definite.

Let $B(x)$ be not only positive definite but also sufficiently diagonally dominant [14], equation (16.16) can be rewritten as

$$\ddot{x}_i = F_i(x, \dot{x}, u) + \sum_{j=1, j\neq i}^{n} b_{ij}(x)\dot{u}_j + b_{ii}(x)\dot{u}_i$$

Assume that any control signal $\dot{u}_i$ has the form

$$\dot{u}_i = -U_{Max}\, \text{sign}\left(x_i - \tfrac{1}{2}x_{i\,Max}\right)$$

$$\ddot{x}_i = \bar{F}_i - \sum_{j=1, j\neq i}^{n} b_{ij}(x)U_{Max}\, \text{sign}\left(x_j - \frac{1}{2}x_{j\,Max}\right)$$

$$- b_{ii}(x)U_{Max}\, \text{sign}\left(x_i - \tfrac{1}{2}x_{i\,Max}\right)$$

By considering the worst case, that is the one which gives the highest value of U_{Max}

$$\ddot{x}_i = \bar{F}_i - g_i(x)U_{Max}\, \text{sign}\left(x_i - \tfrac{1}{2}x_{i\,Max}\right)$$

$$g_{1i}(x) < g_i(x) < g_{2i}(x)$$

$$g_{1i}(x) = b_{ii}(x) - \sum_{j=1, j\neq i}^{n} |b_{ij}(x)|$$

$$g_{2i}(x) = b_{ii}(x) + \sum_{j=1, j\neq i}^{n} |b_{ij}(x)|$$

a value of U_{Max} can be derived

$$U_{Max} \geq \bar{U}_{Max} \tag{16.17}$$

$$\bar{F}_{Max} = \max_{1 \leq i \leq n} \bar{F}_i$$

$$\bar{U}_{Max} = \max\left(\frac{\bar{F}_{Max}}{g_1^*}, \frac{4\bar{F}_{Max}}{3g_1^* - g_2^*}\right)$$

$$g_1^* = \min_{1 \leq i \leq n} g_{1i}, \qquad g_2^* = \max_{1 \leq i \leq n} g_{2i}, \qquad 3g_1^* > g_2^*$$

The control law $\dot{u}_i = -U_{Max}\text{sign}(x_i - \tfrac{1}{2}x_{i\,Max})$ with U_{Max} satisfying inequality (16.17) is sufficient to steer to zero the vector x and $\dot{x}$.

It is possible to deal with the case in which $B(x)$ is positive definite but not necessarily diagonally dominant (like the inertia matrix of a Lagrangian system). In order to explain the proposed procedure, consider again (16.16) and an observer

$$\ddot{y} = w \tag{16.18}$$

Define the observation error as $e = x - y$. Its dynamics are described by

$$\ddot{e} = F(x, \dot{x}, u) + B(x)\dot{u} - w \tag{16.19}$$

Equation (16.19) can be rewritten as

$$e = e_1$$
$$\dot{e}_1 = e_2 \tag{16.20}$$
$$\dot{e}_2 = [F(x,\dot{x},u) + B(x)\dot{u}] - w$$

with e_2 not available. Now, for this equation, the control matrix is diagonal dominant and a second order sliding mode control algorithm can be found as above, capable of steering e_1 and e_2 to zero in finite time. By simultaneously applying the equivalent control method, on $e_1 = e_2 = 0$, w is equivalent to

$$w_{eq} = F(x,\dot{x},u) + B(x)\dot{u}$$

This expression can be substituted in the observer equation (16.18)

$$\ddot{y} = F(x,\dot{x},u) + B(x)\dot{u}$$

The vector $\dot{y}$, contrarily to $\dot{x}$, is now available. The control matrix $B(x)$ is positive definite, so well known Lyapunov-like procedures can be adopted, as in References 1 and 2, to design the control $\dot{u}$ to establish a first order sliding motion on $s = \dot{y} + cy = 0$, that is to steer y and consequently x to zero asymptotically.

16.4   Mathematical model

An underwater vehicle (UV) can be considered as a rigid body with 6 degrees of freedom (d.o.f.).

Translational and rotational motions of the UV are described by nonlinear dynamic equations. Moreover the motions of the vehicle can be expressed referring to either an earth-fixed or body-fixed frame of reference [16–19].

Considering the I-frame position of the vehicle $x \in R^3$, the unit quaternion representing the attitude $q = [\eta, \epsilon^T]^T$, with $\eta \in R$ and $\epsilon \in R^3$, the linear and angular velocities of the UV in the B-frame $v \in R^3$ and $\omega \in R^3$, the body-fixed reference frame (B-frame) and the inertial reference frame (I-frame) are related by the transformation matrix $J(q)$

$$\begin{bmatrix} \dot{x} \\ \dot{q} \end{bmatrix} = \begin{bmatrix} R(q) & O_{3\times3} \\ O_{4\times3} & \frac{1}{2}U(q) \end{bmatrix} \begin{bmatrix} v \\ \omega \end{bmatrix} \Leftrightarrow \dot{\xi} = J(q)v$$

The rotation matrix $R \in SO(3)$ from the I to B frame can be written in terms of the Euler parameters as

$$R(q) = I_{3\times3} + 2\eta S(\epsilon) + 2S^2(\epsilon)$$

The coordinate transformation matrix $U(q)$ can be written

$$U(q) = \begin{bmatrix} -\epsilon^T \\ \eta I_{3\times3} + S(\epsilon) \end{bmatrix} = \begin{bmatrix} -\epsilon^T \\ T(q) \end{bmatrix}$$

where $S(a)$ is a skew-symmetric matrix defined such that $S(a)b = a \times b$.

The transformation matrix $J(q)$ has full rank, i.e. $\text{rank}[J(q)] = 6 \ \forall q$ therefore the kinematic equations contain no singular points.

The Newton's equations of motion for the UV regarded as a rigid body with constant mass m can be written as [16–17]

$$M_{RB}\dot{v} + C_{RB}(v) = \tau_{RB}$$

$$M_{RB} = \begin{bmatrix} mI_{3\times3} & -mS(r_G) \\ mS(r_G) & I_0 \end{bmatrix}$$

$$C_{RB}(v) = \begin{bmatrix} O_{3\times3} & -mS(v) - mS(S(\omega)r_G) \\ -mS(v) - mS(S(\omega)r_G) & -S(I_0\omega) + mS(S(v)r_G) \end{bmatrix}$$

where $r_G = [x_G, y_G, z_G]^T$ is the centre of gravity, I_0 is the inertia matrix (positive definite) of the vehicle with respect to the B-frame origin and τ_{RB} is the vector of the external applied forces and moments.

In order to obtain a complete dynamic model for the vehicle, the added inertia, hydrodynamic damping, and restoring forces and moments need to be considered. The added inertia matrix M_A is assumed to be positive definite and constant. Consideration of the added mass introduces Coriolis and centrifugal terms which can be represented by the product $C_A(v)v$ where the matrix $C_A(v)$ is skew-symmetric. The dissipative forces and moments due to hydrodynamic damping are collected in the vector $D(v)v$, and the matrix $D(v)$ is positive. Finally the gravitational and buoyancy forces, W and B respectively, give rise to the restoring forces and moments that are collected in the vector $g(q)$.

16.4.1   Vehicle dynamics in the B-frame

The dynamic model for the UV in the B-frame can be expressed as [16–18]

$$M\dot{v} + C(v)v + D(v)v + g(q) = \tau \tag{16.21}$$

where $M = M_{RB} + M_A$, $C(v) = C_{RB}(v) + C_A(v)$, and M is assumed to be constant and positive definite, while $C(v)$ is skew-symmetric.

16.4.2   Thruster dynamics

As for the control signal τ, the following dynamic equation must be added

$$\tau = Hf \tag{16.22}$$

where f is the vector of the thrusts developed by the thrusters and H is a known matrix relating the external forces with the total forces and torques in the B-frame. A simplified model of the dynamics of the actuators is

$$f_i = -K_i\omega_{ti}|\omega_{ti}| \tag{16.23}$$

$$I_i\dot{\omega}_{ti} = (v_{ai} - K_i\omega_{ti})\frac{R}{K_i} + \Delta_i \tag{16.24}$$

where Δ_i is a term that collects all the disturbances.

In the complete dynamic model, the control actions are related in a nonlinear uncertain dynamical way to the true inputs, which are the voltages of the DC drives.

16.4.3 The position and attitude control

In order to apply the proposed methodology, differentiate (16.21) and (16.22) to obtain

$$M\ddot{v} = \psi(v, q, \dot{v}, \dot{q}, \tau) + \dot{\tau}$$

$$\dot{\tau} = H\dot{f}$$

From (16.23), by standard non smooth differentiation

$$\dot{f}_i = -2K_i |\omega_i| \dot{\omega}_i$$

and substituting (16.24), it can be obtained that

$$\dot{f}_i = -\frac{|\omega_{ti}|}{I_i} \left[(v_{ai} - K_i \omega_{ti}) \frac{R}{K_i} + \Delta_i \right] \tag{16.25}$$

Therefore, as for the control, the considered system dynamics can be expressed as

$$M\ddot{v} = \Psi(v, q, \dot{v}, \dot{q}, \omega_t, \Delta) + v$$

$$v = -H \left[\operatorname{diag}\left\{ \frac{|\omega_{ti}|}{I_i} \right\} \right] v_a$$

Choose $s_i = v_i - v_{id}(t)$

$$\dot{x}_1 = x_2$$

$$\dot{x}_2 = M^{-1}\Phi(v, q, \dot{v}, \dot{q}, \omega_t, \Delta, t) + M^{-1}v$$

$$x_1 = [s_1, \ldots, s_6]^T$$

and introduce an observer

$$\dot{y}_1 = y_2$$

$$\dot{y}_2 = w$$

$$y_1 = [y_{1_1}, \ldots, y_{1_6}]^T$$

The control laws, according to the previous sections, must be chosen as

$$w_i = -W_{Max} \operatorname{sign}\left(e_i - \tfrac{1}{2}e_{iMax}\right)$$

$$v_j = -V_M \operatorname{sign}(y_{2j} + \lambda\, y_{1j})$$

where e_i is the observation error $e_i = x_{1_i} - y_{1_i}$, and W_{Max} and V_M are chosen in order to dominate the system uncertainties, and in order to achieve in a finite time a second order sliding motion on $e_1 = 0$ first, and a first order sliding motion on $y_2 + \lambda\, y_1 = 0$ subsequently.

The above methodology can be applied only when the angular velocities of the motors are different from zero since any control is v_{ai} multiplied by $|\omega_{ti}|$. This means that as the position reaches the equilibrium point, the robustness of the control system is questionable and only a position inside a small neighbourhood of the desired one is guaranteed.

The fact that at zero velocity the system, controlled by thrusters, turns out to be locally uncontrollable suggests the idea of using – instead of a bidirectional thruster – two opposite mono-directional actuators.

Indeed let $\omega_1 = \omega_0 + \delta\omega_1$ and $\omega_2 = \omega_0 + \delta\omega_2$ be the angular velocity of the two opposite thrusters. The total thrust is given by

$$f_i = k(\omega_1^2 - \omega_2^2)$$

Let $\delta\omega = (\delta\omega_1 - \delta\omega_2)$, then

$$f_i = k(\omega_0\delta\omega + \Delta)$$

with $\Delta = \delta\omega_1^2 - \delta\omega_2^2$, which is linear in $\delta\omega$. In this way, the loss of controllability at low velocity can be overcome.

Similar results, but for a different reason, can be achieved by using another type of actuator, which is intrinsically mono-directional, that is a pair of opposite jets. The jet thrust can be expressed as $f_j = k(x_j)v_j^2$, v_j is the constant jet velocity and x_j is the position of a valve controlled by an external motor.

The advantage of this actuator is that the thrust is approximatively linear in the valve position which is considered to be the control. It makes sense to investigate, through experimental systems, these different approaches.

In the next section, as a first step in this comparison and as an experimental validation of the proposed approach, the control of a simple UV with two opposite jets is considered.

16.4.4   Simulation example

The proposed control law has been applied to an underwater vehicle [19] with 6 d.o.f. the behaviour of which has been simulated. The vehicle is assumed to be neutrally buoyant, and environmental disturbances were not incorporated in the simulation model.

The desired position x_d is generated by a 2nd order filter $\ddot{x}_d + 2\lambda_x\dot{x}_d + \lambda_x^2 x_d = \lambda_x^2 x_r$, where $\lambda_x = 0.75$ and x_r has the shape of a square wave between 0 and 5 m with period 20 s.

The desired attitude signal is generated in order to obtain the desired axis–angle parameterisation of $SO(3)$, that is, $R_d = R(n_d, \alpha_d)$. The chosen axis is constant and equal to $n_d = (1/\sqrt{3})[1, 1, -1]^T$, and the desired angle α_d is generated by a 2nd order filter $\ddot{\alpha}_d + 2\lambda_\alpha\dot{\alpha}_d + \lambda_\alpha^2\alpha_d = \lambda_\alpha^2\alpha_r$, where $\lambda_\alpha = 1$ and α_r has the shape of a square wave between 0 and $2\pi/3$ rad with period 20 s. Considering the holding relationships, the desired attitudes can be computed according to $\eta_d = \cos(\alpha_d/2)$, and $\epsilon_d = \sin(\alpha_d/2)n_d$. The vectors $\dot{\xi}_d$ and v_d are calculated from the kinematic equations and utilising $\omega_d = \dot{\alpha}_d n_d$. Simulation results are shown in Figs 16.1–16.3.

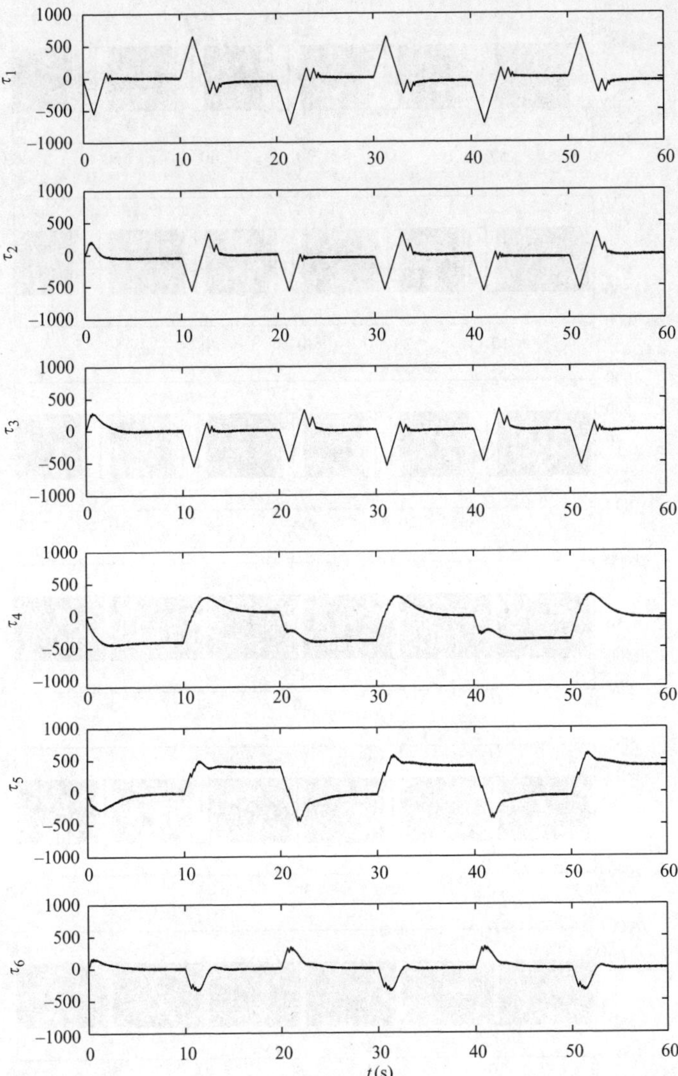

Figure 16.1   The continuous signals τ_i

16.5   Test results: motion control for an underwater vehicle prototype

An UV prototype has been recently built at the DIEE-University of Cagliari as a preliminary test-bed of a novel water-jet based propulsion system for underwater vehicles. The vehicle is about 150 cm long and 80 cm high. It contains a centrifugal pump feeding a hydraulic circuit and two variable-section nozzles actuated

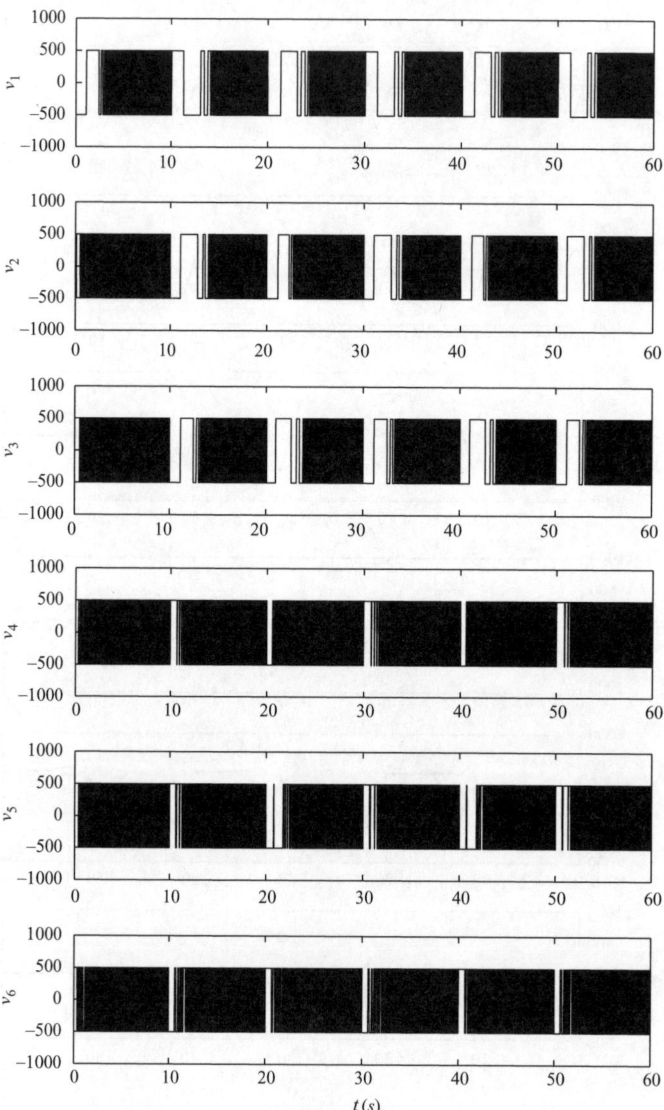

Figure 16.2   The discontinuous control signals v_i

by means of linear motor drives located at the opposite edges of the hydraulic circuit (Fig. 16.4).

The prototype is rigidly connected with a wheeled trolley that 'suspends' the UV into a water channel. This configuration allows the UV to move freely along the channel under the reaction force exerted by the water flow through the nozzles.

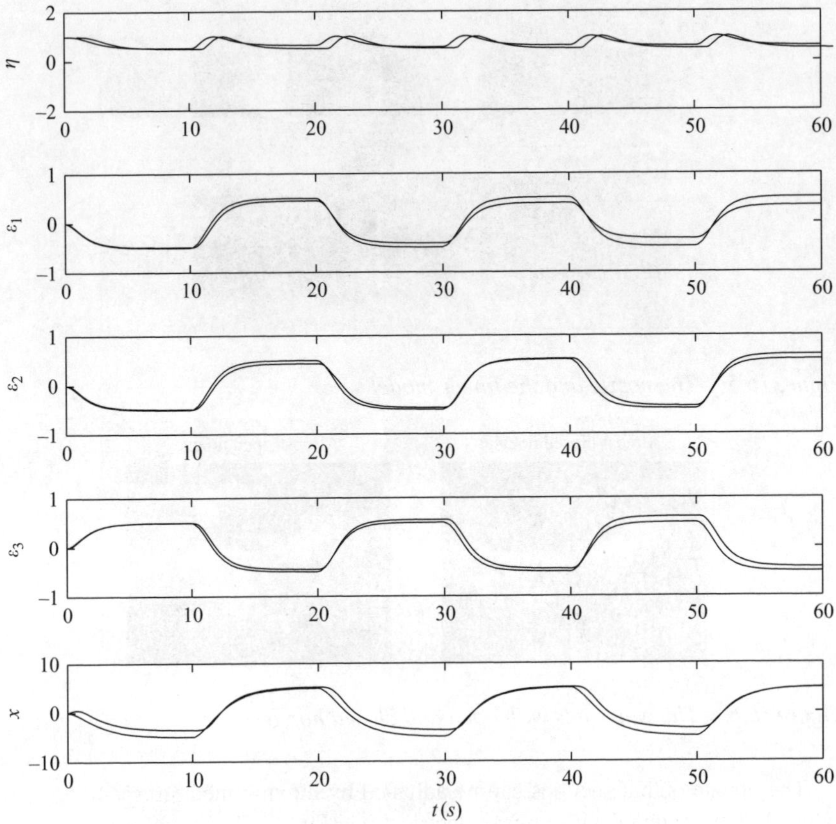

Figure axis labels (Figure 16.3): η, ε_1, ε_2, ε_3, x plotted against $t(s)$.

Figure 16.3   Orientation and position tracking results

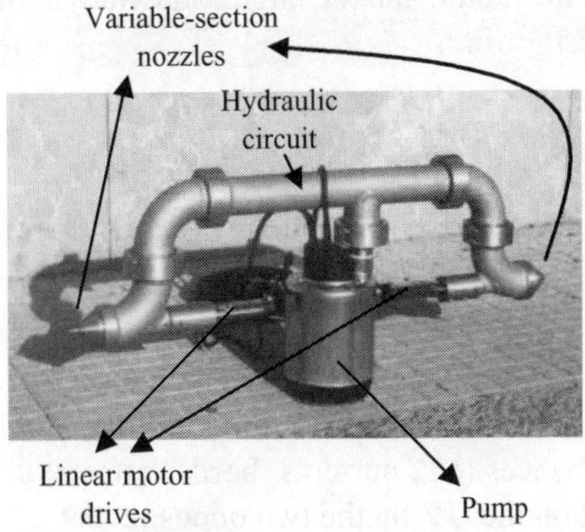

Figure 16.4   The UV prototype

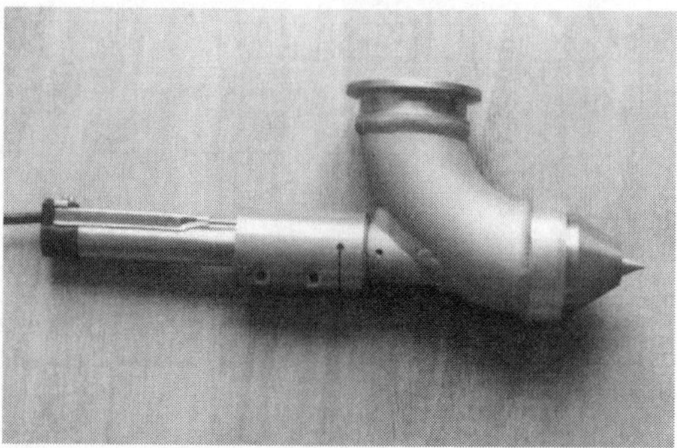

Figure 16.5   The nozzle and the linear model

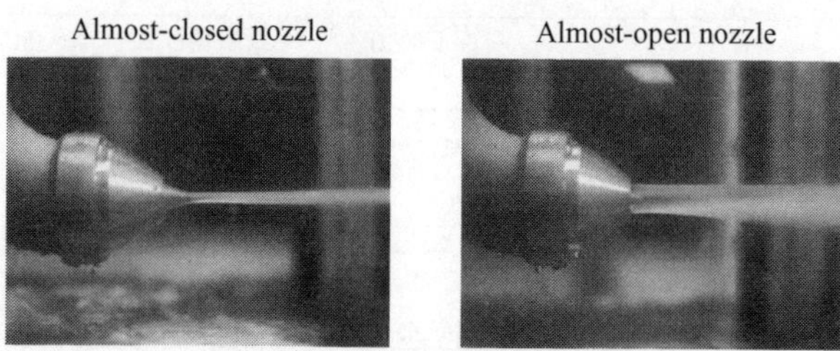

Figure 16.6   The water jets with open and closed nozzle

The nozzle output sections can be adjusted by moving the corresponding spear valve, directly-coupled with a linear electric drive (Fig. 16.5). The spear valve profile is shaped similarly to that of a Pelton turbine, so that the generated thrust is turbulence-free and almost-linearly dependent on the valve position. Figure 16.5 shows the details of the linear motor coupled with the spear valve, and Fig. 16.6 shows the hydro-jets in two different opening conditions of the nozzle. The direct mechanical coupling between the valve and the motor, and the large bandwidth of the latter, allow for very fast control of the thrust profile.

16.5.1   The UV model

The dynamics of the considered jet-propelled UV can be formulated as follows [17]:

$$(M_v + M_a)\ddot{y} + k_1 \dot{y}|\dot{y}| + k_2 \dot{y} + d(t) = F_1(t) - F_2(t) \tag{16.26}$$

where $y(t)$ is the vehicle position, M_v is the vehicle mass, M_a represents the added mass effect, k_1 and k_2 are the viscous friction and drag coefficients, $d(t)$ accounts for the external disturbances (e.g. currents, border effects) and $F_1(t)$, $F_2(t)$ are the control thrusts exerted on the UV by the two opposite jets.

With good approximation, it can be assumed that $F_1(t)$ and $F_2(t)$ depend instantaneously on the positions, z_1 and z_2, of the spear valves (Fig. 16.7) via some nonlinear

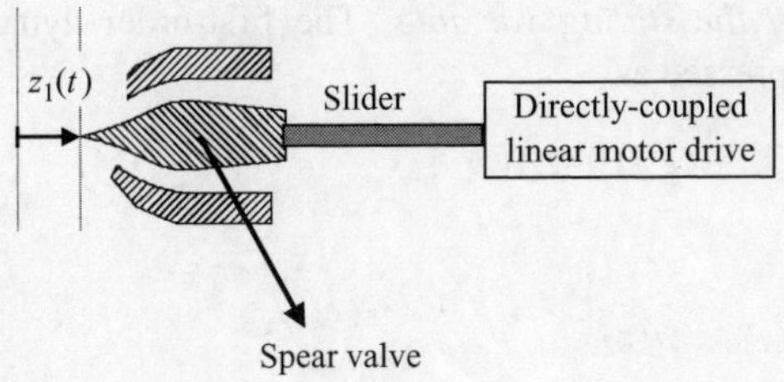

Figure 16.7   Nozzle position notation

function h

$$F_1 = h(z_1), \quad F_2 = h(z_2) \tag{16.27}$$

The positions of the two spear valves are defined according to the notation represented in Fig. 16.7: z_1 (z_2) is zero when the nozzle is closed and increases when the spear valve is opening. Thus, z_1 and z_2 remain always non-negative.

The function h is very difficult to determine, and therefore it is considered uncertain in the present treatment. Obviously h is strictly positive, monotonically increasing and zero when its argument is zero.

The system parameters M_v, M_a, k_1, k_2 as well as the disturbance $d(t)$ are unknown. Collecting together the uncertainties affecting the UV dynamics, system (16.26) can be rearranged as follows

$$\ddot{y} = f(\dot{y}, t) + g[h(z_1) - h(z_2)] \tag{16.28}$$

where $g = 1/(M_a + M_v)$ with implicit definition of the function $f(\dot{y}, t)$.

16.5.2   Controller design

Define the position tracking error and its derivative by

$$e = y - y^*, \quad \dot{e} = \dot{y} - \dot{y}^* \tag{16.29}$$

By (16.28), the relative degree between the position error e and the control variables z_1 and z_2 is $r = 2$, and the above outlined design procedure yields the following steps.

Sliding manifold design: according to (16.6), the sliding variable is defined as

$$\sigma(t) = \dot{e}(t) + ce(t), \qquad c > 0 \tag{16.30}$$

Estimation of the sliding variable: the actual vehicle velocity error $\dot{e}$ is estimated in real-time by means of the differentiator (16.7), with the parameters κ_0 and κ_1 set on the basis of (16.8) where C_2 is sufficiently large so that the following condition is met:

$$|\ddot{e}| \leq C_2 \tag{16.31}$$

Stabilisation of the sliding variable: The first-order dynamics of the sliding variable can be expressed as

$$\dot{\sigma} = f(\dot{y}, t) - \ddot{y}^* + c\dot{e}(t) + gA(z_1, z_2) \tag{16.32}$$

where

$$A(z_1, z_2) = h(z_1) - h(z_2) \tag{16.33}$$

To minimise the energy consumption, the two nozzles should not be both opened at the same time. This corresponds to keeping either z_1 or z_2 zero at each time instant.

Define the dummy control variable

$$\delta_z = z_1 - z_2 \tag{16.34}$$

subjected to the aforementioned constraints

$$z_1 \geq 0, \quad z_2 \geq 0, \quad z_1 \cdot z_2 = 0 \tag{16.35}$$

By combining (16.33) and (16.34)–(16.35), it can be written that

$$A \equiv A(\delta_z) = \begin{cases} h(\delta_z), & \delta_z \geq 0 \\ -h(-\delta_z), & \delta_z < 0 \end{cases} \tag{16.36}$$

Differentiating further (16.32) and considering (16.33)–(16.36) one obtains

$$\ddot{\sigma} = \varphi(y, \dot{y}, \delta_z, t) + g\frac{dA}{d\delta_z}\dot{\delta}_z \tag{16.37}$$

The dynamics (16.37) are formally equivalent to (16.9) with $u = \delta_z$. Since by physical arguments it can be asserted that the function φ is bounded and $A(\delta_z)$ is strictly increasing, it follows that conditions (16.10) hold for some constants F, Γ_1 and Γ_2.

A cascade compensation scheme is employed for the control system (Fig. 16.8): the 'high-level' vehicle controller drives each linear motor (LM) controller with a reference profile for the LM position.

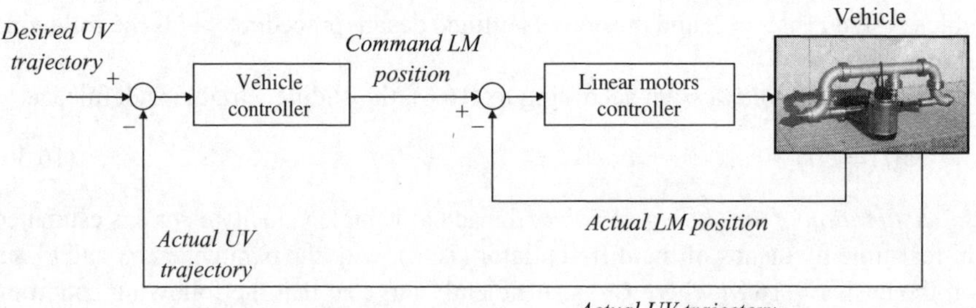

Figure 16.8   A schematic representation of the control architecture

Given a command profile δ_z^* for δ_z, the desired profiles z_1^* and z_2^* for the spear valve positions are obtained by inverting (16.34) under conditions (16.35):

$$\begin{cases} z_1^* = \max\{\delta_z^*, 0\} \\ z_2^* = -\min\{\delta_z^*, 0\} \end{cases} \tag{16.38}$$

The inner loop in Fig. 16.8 might cause the actual profiles of z_1 and z_2 to be largely different from the prescribed ones z_1^* and z_2^*. Nevertheless, considering the large bandwidth of the LM drive control system (over 15 Hz) and the considerable inertia of the UV, it is possible to regard the LM dynamics as a singular perturbation sufficiently fast to preserve the sliding mode stability [20]. The robustness against fast unmodelled dynamics is indeed one of the most important features of the SMC approach as far as practical implementation is concerned.

Let us summarise the overall controller. The sliding manifold is

$$\hat{\sigma}(t) = \hat{\dot{e}} + ce, \qquad c > 0 \tag{16.39}$$

where $e = y - y^*$ and $\hat{\dot{e}}$ is computed by using differentiator (16.7). The reference position profiles for the linear motors are set according to (16.38), where $\dot{\delta}_z^*$ is a discontinuous signal defined according to the sub-optimal 2-SMC algorithm (16.11)–(16.14).

By relying on the separation principle demonstrated [12], it can be asserted that the following conditions are ideally simultaneously fulfilled after a finite time

$$|\hat{\dot{e}} - \dot{e}| = 0 \tag{16.40}$$

$$|\hat{\sigma}| = 0 \tag{16.41}$$

which means that $\sigma = \dot{e} + ce$ vanishes in finite-time. The exponential convergence to zero of e follows from trivial arguments.

Remark 2.   The combined 2-SMC/2-SMD scheme [12] was based on a proper online adjustment of both the controller and differentiator parameters. Nevertheless, the adaptive gain tuning procedure [12] is sufficient but not necessary for the effectiveness of the algorithm, which is often attained using suitably-tuned constant values for several controller and differentiator parameters.

Remark 3.   Due to actual implementation effects (e.g. noise [8] and discretisation [13]), conditions (16.40) and (16.41) are guaranteed only approximately

$$|\hat{\dot{e}} - \dot{e}| \leq \varepsilon_1 \tag{16.42}$$

$$|\hat{\sigma}| \leq \varepsilon_2 \tag{16.43}$$

with $\varepsilon_1, \varepsilon_2 \approx 0$. Under mild smoothness requirements, the system behaviour is 'regular', and the tracking error e exponentially converges toward a small neighbourhood of zero.

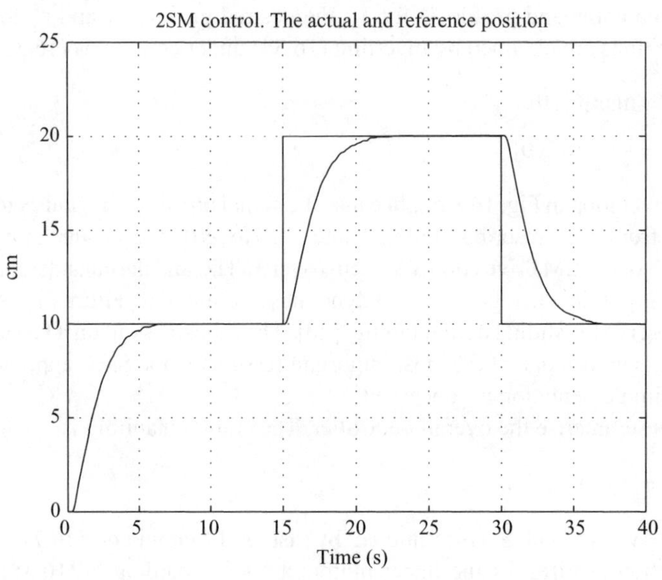

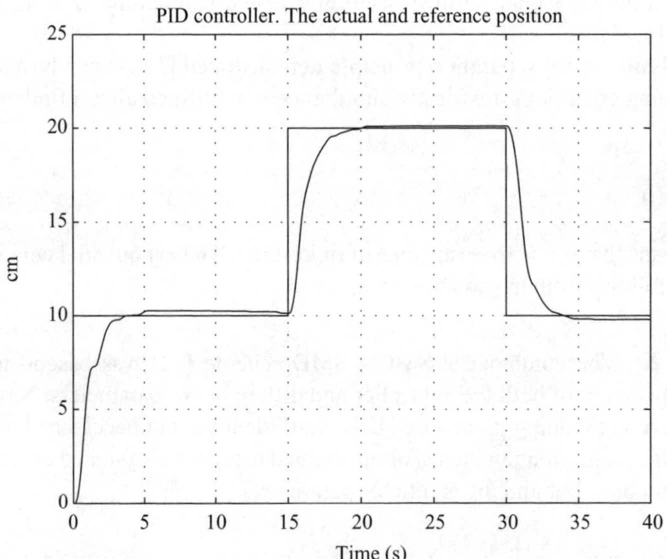

Figure 16.9   *Regulation test with 2-SMC (top) and PID-control (bottom): the actual and desired position profile*

16.5.3   The experimental setup: implementation issues and test results

In this section the experimental setup is described in some detail.

The linear motor drives (factored by *Linmot*^{TM}) have a rated bandwidth between 15 and 20 Hz. A dedicated driver module allows the force w applied to the slider to be

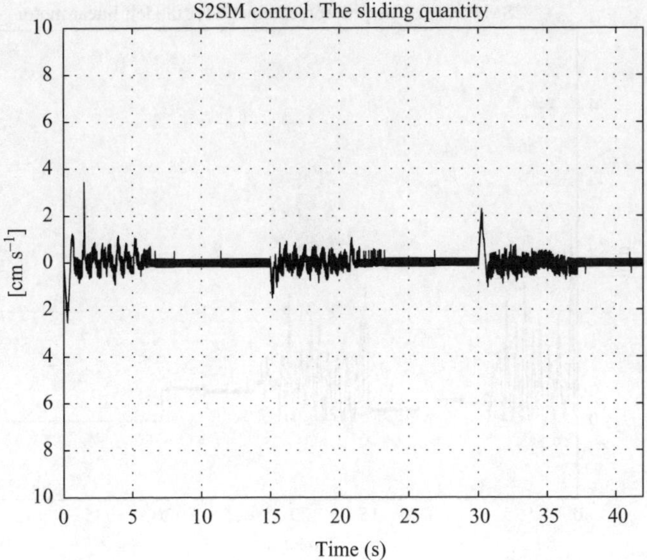

Figure 16.10   Regulation test with 2-SMC. The sliding variable time history

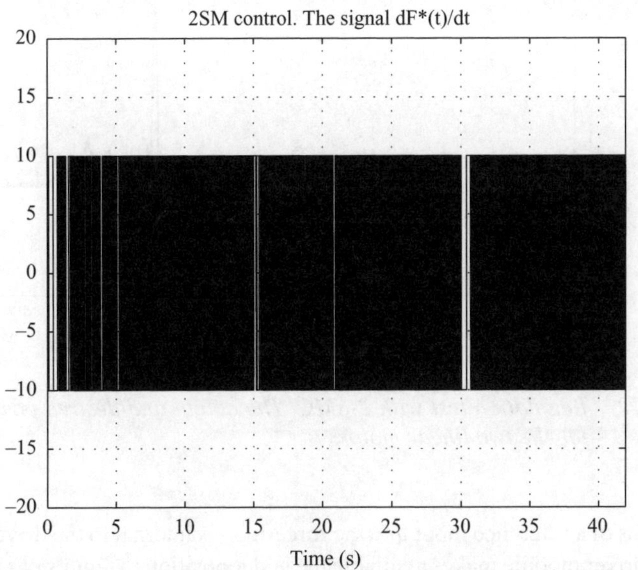

Figure 16.11   Regulation test with 2-SMC. The discontinuous signal $\dot{\delta}_y^*(t)$

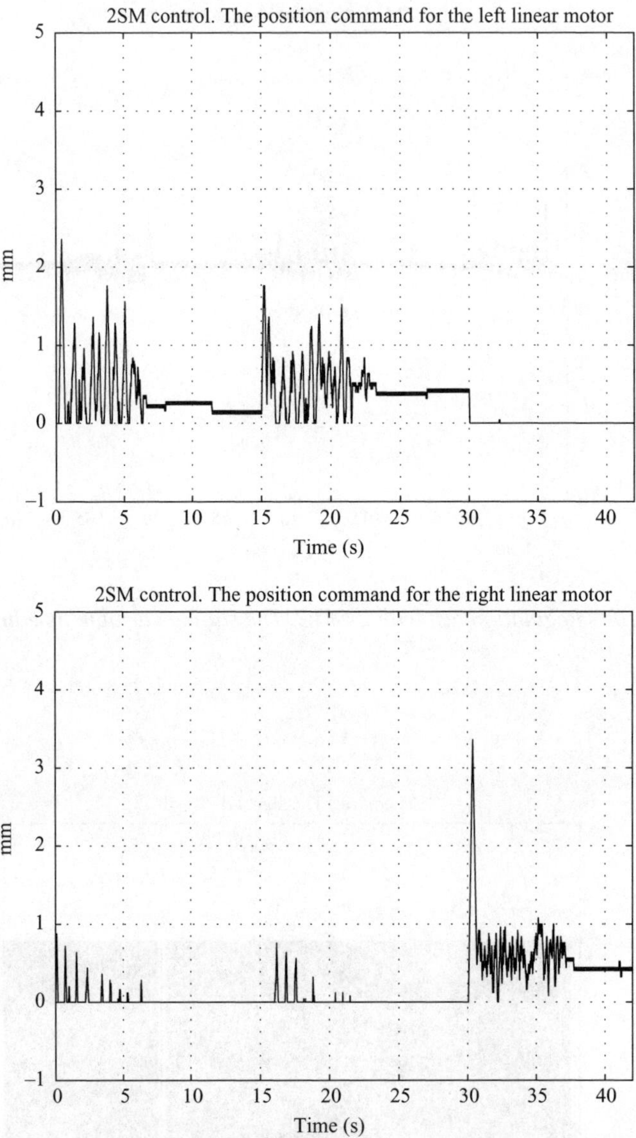

*Figure 16.12   Regulation test with 2-SMC. The actual and desired position profile
for the two linear motors*

set by means of a reference input w^* (the force-loop is internal to the driver). Further-more, the driver module makes available the slider positions y_1 and y_2 as incremental encoder-like signals. A PID position-force loop has been closed externally to the LM controller.

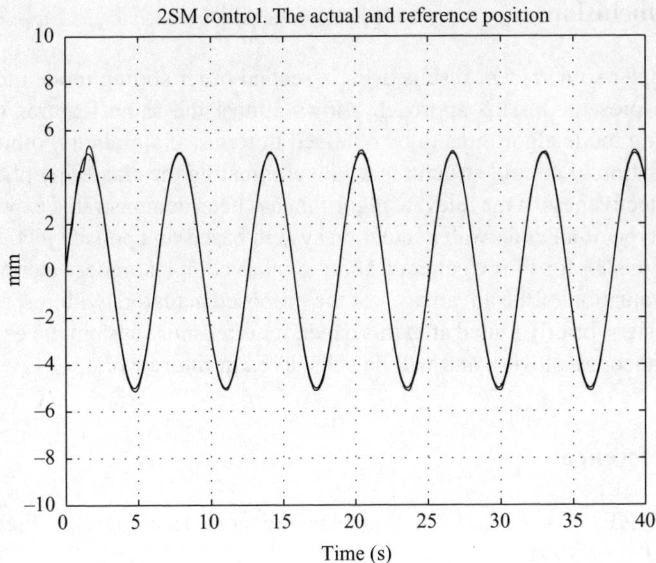

Figure 16.13   Tracking test with 2-SMC. The actual and desired position profiles

The control system has been implemented on a PC-based platform (Pentium 2 processor at 350 MHz). The computational burden of the control system is limited, and much less computing power would be sufficient. The controller and the differentiator have been discretised by the classical backward-difference method with a sampling step of 2 ms (the sample-and-hold effect has been analysed [13]).

The parameters of the controller and of the differentiator are set as follows:

Sub-optimal 2-SMC: $U_M = 10, \quad \alpha^* = 1$

2-SMD: $\kappa_0 = 12, \quad \kappa_1 = 20$

The proposed 2-SM controller/observer scheme has been implemented using the above parameter set, and its performance has been compared with that of a classical PID controller with gains $K_P = 2$, $K_I = K_D = 1$.

In the first test, a piecewise-constant reference position was used. Figure 16.9 reports the actual and desired trajectory obtained using the two different approaches, and provides evidence that the VSC is more accurate. Figure 16.10 shows the time evolution of the sliding variable, while Fig. 16.11 reports the discontinuous signal δ_z^*. Figure 16.12 reports the actual and desired position of the two linear motors. It can be seen that the two nozzles are never both opened at the same time instant.

Also a tracking test using a sinusoidal reference profile has been carried out. Figure 16.13 shows that the actual trajectory converges to the desired one after a very short transient.

16.6   Conclusions

In this chapter some recent results using a second order sliding mode methodology have been presented. This approach allows almost the same features of the first order sliding mode algorithms to be obtained in terms of simplicity, robustness and decentralisation of control structures, while eliminating the chattering phenomenon.

The effectiveness of the relevant algorithm has been demonstrated experimentally on a prototype of an underwater actuation system based on opposite jets. The results appear promising for future generalisation to more complex underwater objects.

The evaluation of the advantages of the proposed actuators with respect to traditional thrusters, briefly hinted at in this paper, requires more systematic experimental as well as theoretical work and will be considered in future work.

16.7   References

1   SLOTINE, J. J. E. and LI, W.: 'Applied nonlinear control' (Prentice Hall, New Jersey, 1991)

2   UTKIN, V. I.: 'Sliding modes in control and optimization' (Springer Verlag, Berlin, 1992)

3   BARTOLINI, G., FERRARA, A., LEVANT, A., and USAI, E.: 'On second order sliding mode controllers', in YOUNG, K. D. and OZGUNER, U. (Eds): 'Variable structure systems, sliding mode and nonlinear control', *Lecture Notes in Control and Information Sciences*, **24** (Springer Verlag, Berlin, 1999)

4   BARTOLINI, G. and ZOLEZZI, T.: 'Control of nonlinear variable structure systems', *J. Math. Anal. Appl.*, 1986, **118**, pp. 42–62

5   UTKIN, V. I., GULDNER, J., and SHI, J.: 'Sliding modes in control in electromechanical systems' (Taylor & Francis, London, 1999)

6   ISIDORI, A.: 'Nonlinear control systems' (Springer Verlag, Berlin, 1995, 3rd edn)

7   BARTOLINI, G., LEVANT, A., PISANO, A., and USAI, E.: 'Higher-order sliding modes for output-feedback control of nonlinear uncertain systems', in YU, X. and XU, J. X. (Eds): 'Variable structure systems: towards the 21st century', Volume 274 of *Lecture Notes in Control and Information Sciences* (Springer Verlag, Berlin, 2002) pp. 83–108

8   LEVANT, A.: 'Higher order sliding modes, differentiation and output-feedback control', *Int. J. Control*, 2003, **76**(9–10), pp. 924–941

9   LEVANT, A.: 'Higher order sliding modes and arbitrary-order exact robust differentiation', Proceedings of the 6th European Control Conference, Porto, Portugal, 2001

10   LEVANT, A.: 'Robust exact differentiation via sliding mode technique', *Automatica*, 1998, **34**, pp. 379–384

11   BARTOLINI, G., FERRARA, A., PISANO, A., and USAI, E.: 'On the convergence properties of a 2-sliding control algorithm for nonlinear uncertain systems', *Int. J. Contr.*, 2001, **74**, pp. 718–731

12   BARTOLINI, G., LEVANT, A., PISANO, A., and USAI, E.: 'On the robust stabilization of nonlinear uncertain systems with incomplete state availability', *ASME Journal of Dynamic Systems, Measurement, and Control*, 2000, **122**, pp. 738–745

13   BARTOLINI, G., PISANO, A., and USAI, E.: 'Digital second order sliding mode control for uncertain nonlinear systems', *Automatica*, 2001, **37**(9), pp. 1371–1377

14   BARTOLINI, G., FERRARA, A., USAI, E., and UTKIN, V. I.: 'On multi-input chattering-free second-order sliding mode control', *IEEE Trans. on Automatic Control*, 2000, **45**(9), pp. 1711–1717: 'Chattering elimination in the hybrid control of constrained manipulators via first/second order sliding modes control', *Dynamics and Control*, 1999, **9**(2), pp. 99–124

15   BARTOLINI, G., FERRARA, A., and PUNTA, E.: 'Multi-input second-order sliding mode hybrid control of constrained manipulators', *Dynamics and Control*, 2000, **10**, pp. 277–296

16   FOSSEN, T. I.: 'Nonlinear modelling and control of underwater vehicles' PhD Thesis, Department of Engineering Cybernetics, The Norwegian Institute of Technology, June 1991

17   FOSSEN, T. I.: 'Guidance and control of ocean vehicles' (John Wiley & Sons, Chichester, 1994)

18   HEALEY, A. J. and LIENARD, D.: 'Multivariable sliding mode control for autonomous diving and steering of unmanned underwater vehicles', *IEEE Journal of Oceanic Engineering*, 1993, **18**(3), pp. 327–339

19   FJELLSTAD, O. E. and FOSSEN, T. I.: 'Position and attitude tracking of auv's: a quaternion feedback approach', *IEEE Journal of Oceanic Engineering*, 1994, **19**(4), pp. 512–518

20   FRIDMAN, L.: 'Chattering analysis in sliding mode systems with inertial sensors', *Int. J. Control*, 2003, **76**(9–10), pp. 906–912

Chapter 17

Semiglobal stabilisation of linear uncertain system via delayed relay control

Leonid Fridman, Vadim Strygin and Andrei Polyakov

17.1 Introduction

The time delays that usually occur in relay and sliding mode control systems must be considered in system analysis and design [1]. On the other hand, the presence of time delay does not allow the sliding mode control to be designed in the space of state variables. Even in the simplest one-dimensional delayed relay control system only oscillatory solutions can occur [2]. That is why the following directions in relay delayed control require investigation:

- research into time delay compensation;
- control of the amplitude of any oscillations.

A Pade approximation of the delay reduces the relay delay output tracking problem to a sliding mode control problem for a nonminimum-phase system. This approach was suggested in Reference 3. The sliding mode control [4] was designed in the space of predictor variables (see [5] also). This approach allowed the eigenvalue assignment problem to be solved without any restriction on the time delay and spectral properties of the open loop system. However, the sliding mode control design in the space of predictor variables [6, 7]:

- cannot compensate even matched uncertainties;
- in the simplest case of square systems, if the dimensions of the state space and control vector are the same, the sliding mode design in the space of predictor variables can remove uncertainties in the space of predictor variables but cannot guarantee that the effects of uncertainties in the space of state variables will be compensated.

Robustness properties of Smith predictors with respect to uncertainties in the time delay were studied [7, 8]. The conditions for robustness of Smith predictors with respect to uncertainty in the time delay were formulated [8] in terms of stability margins.

In this chapter we propose an algorithm for relay control gain adaptation ensuring semiglobal stabilisation of unstable systems.

17.1.1   *Oscillatory nature of relay delayed systems*

17.1.1.1   Example of steady modes

The equation

$$\dot{x}(t) = -\operatorname{sign} x(t-1) \tag{17.1}$$

has a 4-periodic solution

$$g_0(t) = \begin{cases} t & \text{for } -1 \le t \le 1 \\ 2-t & \text{for } 1 \le t \le 3 \end{cases}$$

$$g_0(t+4k) = g_0(t), \qquad k \in \mathbf{Z}$$

Since

$$\dot{g}_0(t) = -\operatorname{sign}[g_0(t-1-4n)]$$

and transforming t to $(4n+1)t$ we obtain

$$\frac{1}{4n+1}[g_0((4n+1)t)]' = -\operatorname{sign} \frac{1}{4n+1} g_0((4n+1)(t-1))$$

This means that there exists a countable set of periodic solution, *a steady mode* (see Fig. 17.1), or more briefly, a SM. Namely, it is easy to verify that the $4/(4n+1)$-periodic function

$$g_n(t) = \frac{1}{4n+1} g_0((4n+1)t), \qquad t \in \mathbf{R}$$

is a solution of (17.1) for each integer $n \ge 1$. It is necessary to remark that the initial function φ_n, which generates the corresponding steady mode g_n, has $2n$ zeros on the time interval $(-1, 0)$.

Remark 1.   *In Reference 2 it was shown, that each solution $x(t) \not\equiv 0$ of (17.1), after finite time, coincides with one of the $g_n(t+\alpha)$ for some $n \ge 0$, $\alpha \in \mathbf{R}$. Consequently, in the simplest scalar relay delayed control system, only oscillatory solutions can occur. Moreover, a solution $g_n(t)$ is stable for $n = 0$, and unstable for $n \ge 1$.*

17.1.1.2   Stabilisation of oscillations in the simplest unstable systems

Consider the stabilisation problem for the simplest unstable system

$$\dot{x} = \lambda x, \qquad x \in \mathbf{R}, \quad \lambda > 0 \tag{17.2}$$

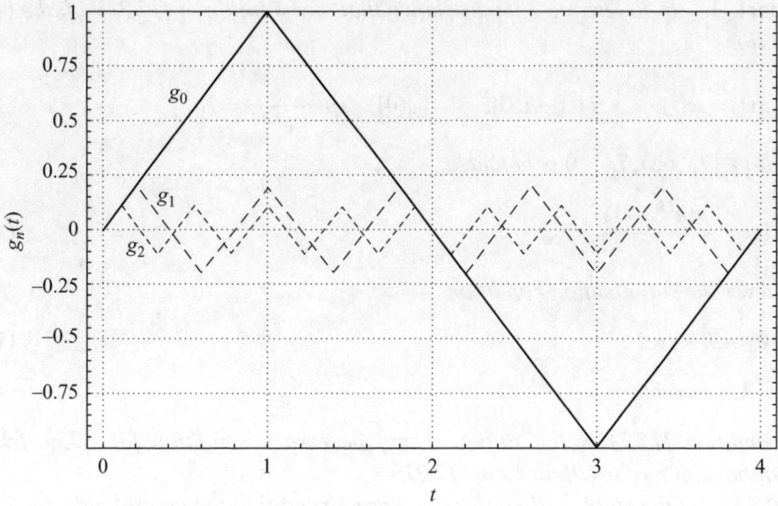

Figure 17.1   The set of steady modes

via delay relay control law of the form $u = -p \cdot \operatorname{sign} x(t - h)$, where $h \geq 0$ is a time delay and $p > 0$ is a control gain. In this case the equation of the controlled system has the form:

$$\dot{x}(t) = \lambda x - p \cdot \operatorname{sign} x(t - h) \tag{17.3}$$

Let us compute the constant $R > 0$ for which the system (17.3) with the initial function

$$\varphi(t) = \begin{cases} -1 & t \in [-h, 0) \\ R & t = 0 \end{cases} \tag{17.4}$$

has a stable periodic solution for $t > 0$.

It is obvious that $\operatorname{sign}[x(t - h)] = 1$ for $t \in [0, h)$. Therefore, at the switching moment $t = h$ we will have

$$x(h) = \left(R + \frac{p}{\lambda} \right) e^{\lambda h} - \frac{p}{\lambda}$$

If the condition $\dot{x}(h) = \lambda x(h) - p < 0$ holds, the function $x(t)$ could decrease from $t = h$ until the next switching instant. Substituting $x(h)$ into the last inequality, we have

$$(\lambda R + p)e^{\lambda h} - p < 0$$

$$R < \frac{p(2 - e^{\lambda h})}{\lambda e^{\lambda h}}$$

Since $R > 0$, then the right hand side of the last inequality should be positive too. Hence $\lambda h < \ln 2$.

Remark 2.   *In Reference 2 it was shown that any solution of (17.3) with the initial conditions*

$$x(t) = \varphi(t), \qquad t \in [-h, 0], \qquad |\varphi(0)| < p \frac{2 - e^{\lambda h}}{\lambda e^{\lambda h}} = r_0 \tag{17.5}$$

for all $t \in [T_0, \infty), T_0 > 0$ is bounded

$$|x| < \frac{p(e^{\lambda h} - 1)}{\lambda} = r_\infty \tag{17.6}$$

whenever the stabilisation condition

$$0 < \alpha h < \ln 2 \tag{17.7}$$

holds. Moreover:

- *condition (17.7) is the sufficient and necessary condition for relay delayed stabilisation of oscillations in (17.3);*
- *the size of the stabilisation domain is proportional to the control gain p.*

Remark 3.   *This means that to ensure a semiglobal stabilisation via relay delayed control an adaptation of the relay delay control gain is necessary.*

17.1.2   Problem formulation

Consider a linear system with delayed control of the form:

$$\frac{dx}{dt} = Ax + Bu(x(t - h(t))) \tag{17.8}$$

where $x \in R^n$ is the state vector, A, B are real matrices, $h(t)$ ($0 < h(t) \leq h_0$, $\forall t \in [0, \infty)$) is a continuous function describing an uncertainty in the time delay, $u \in R^m$ is a relay control vector bounded in every bounded domain $\|x\| \leq D$, $x \in R^n$.

Let us denote as $x(t)$ the solution to the system (17.8) with the initial condition:

$$x(t) = \varphi(t), \quad (-h_0 \leq t \leq 0), \qquad \varphi(t) \in \mathbf{C}[-h_0, 0]\,^1 \tag{17.9}$$

The size of the stabilisation domain is proportional to the control gain. That is why to achieve *nonlocal stabilisation* for system (17.8) we need to use a sufficiently large initial relay control gain in order to stabilise the solutions of (17.8) with sufficiently large initial conditions. On the other hand, due to the oscillatory properties of relay delay systems, one can conclude that it is impossible to achieve asymptotic stability for the solutions to the system (17.8) via relay control with a finite number of gain's switches.

---

[1] The values of relay delayed control for $t \in [0, h_0]$ will be defined below through the value of the initial function $\varphi(t), t \in [-h_0, 0]$ but for the state variable x only the restriction at the initial point $x(0) = \varphi(0)$ is necessary.

Definition 1 [10].   *The zero solution to the system (17.8) is said to be semiglobally stabilisable, if for any $R > 0$ there exists a control $u(t - h(t))$, such that the inequality $\|\varphi(0)\| < R$ implies*

$$\|x(t)\| \to 0 \qquad \text{for } t \to \infty$$

P.I. control algorithms for amplitude control of a one-dimensional relay system with delay in the input were suggested in Reference 11. The following algorithm for control of the amplitudes of the oscillations was proposed in Reference 2: since, after finite time, all solutions coincide with the periodic solution, one can extrapolate the next zero for the periodic solution, and reduce the control gain near to zero of the periodic solution. This algorithm needs only the knowledge of the sign of the state variable with delay but requires a *stabilisation condition* (17.7). This algorithm is valid for any constant delay satisfying condition (17.7) and does not depend on the value of the delay. *Stabilisation condition* (17.7) and the algorithm for stabilisation was generalised [12] for *linear second order* relay delay systems.

In Reference 13, the *stabilisation condition* (17.7) was generalised for MIMO systems and a delayed relay control algorithm proposed, allowing *local stabilisation* of oscillation amplitude to be achieved for controllable systems.

In this chapter, a relay control law and semiglobal stability conditions are proposed ensuring semiglobal stabilisation for the zero solution of the system (17.8). The proposed control law requires the knowledge of:

- the amplitude of the solutions at the delayed time instant;
- an upper bound for the time delay;
- an upper bound on the initial conditions.

17.2   Two simple cases

17.2.1   Scalar system

Consider the scalar control system with the continuous uncertain time delay $h(t)$, $(0 < h(t) < h_0)$

$$\dot{x} = \alpha x + u(x(t - h(t))) \tag{17.10}$$

$$x(t) = \phi(t), \qquad t \in [-h_0, 0] \tag{17.11}$$

where $\alpha > 0$ and the initial function $\phi(t) \in C^1_{[-h_0, 0]}$ satisfies the following condition

$$|\phi(0)| < R \tag{17.12}$$

where R is a positive constant (see Definition 1).

Our aim is to design a relay control that will ensure semiglobal stabilisation of the system (17.10). As can be seen from Section 17.1.2, it cannot be achieved by means of the classical relay control $u(t - h(t)) = -p \cdot \text{sign}[x(t - h(t))]$ with the constant control gain $p > 0$, since the system (17.10) under such a control has only oscillatory solutions. Therefore the control gain should be changed in compliance

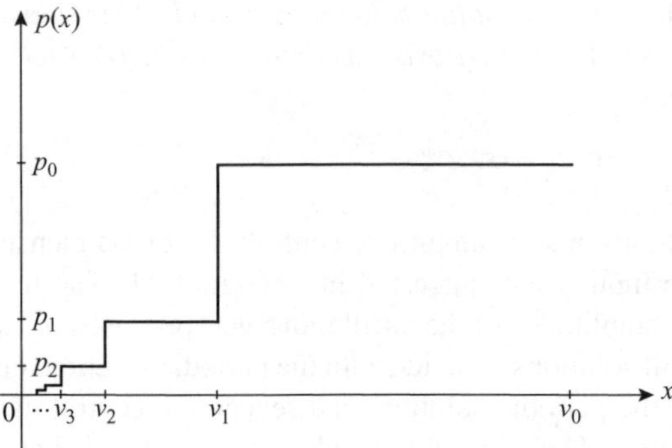

Figure 17.2   Stepwise function $p(\cdot)$

with some law, for example, proportional to the radius of the neighbourhood of stable oscillations.

The idea of the control algorithm is the following:

- consider the amplitude of stabilisation domain for each value of relay delay control gain as the amplitude of the initial conditions for the next step;
- by decreasing the control gain, enter into a smaller neighbourhood of zero.

Now the desired control can be found in the form

$$u(t - h(t)) = -p(|x(t - h(t))|)\ \text{sign}\ [x(t - h(t))]$$

where $p(\cdot)$ is a stepwise function, i.e. $p(\cdot)$ can take only the following values $p_0 > p_1 > p_2 > \cdots > p_k > \cdots > 0$. Each step in $p(\cdot)$ corresponds to some neighbourhood of zero $U_k = \{x : |x| < v_k\}$ in which $p(\cdot) \leq p_k$ (see Fig. 17.2). To find the form of $p(\cdot)$ and system of neighbourhoods $\{U_k\}$, let us return to equation (17.3) and conditions (17.5) and (17.6). First of all let us compare the conditions (17.5) and (17.12). The inequality (17.5) is the necessary condition for the existence of an oscillatory solution to the system (17.3), consequently we should choose p_0 satisfying the condition (17.5), i.e.

$$R\frac{\alpha e^{\alpha h_0}}{2 - e^{\alpha h_0}} < p_0$$

On the other hand, the radius of the stabilisation domain r_∞ should be less than R, otherwise decreasing the control gain leads to system instability. In other words we have

$$p_0 < R\frac{\alpha}{e^{\alpha h_0} - 1}$$

Thus we can define p_0 in the form $p_0 = \alpha' R$, where

$$\alpha' \in I_{\alpha, h_0} = \left(\frac{\alpha e^{\alpha h_0}}{2 - e^{\alpha h_0}}, \frac{\alpha}{e^{\alpha h_0} - 1} \right) \tag{17.13}$$

At the same time, it is necessary to take into account that the interval $I_{\alpha h_0}$ could not be empty. Therefore the following condition should hold

$$\frac{\alpha e^{\alpha h_0}}{2 - e^{\alpha h_0}} < \frac{\alpha}{e^{\alpha h_0} - 1}$$

or equivalently

$$\alpha h_0 < L = \tfrac{1}{2}\ln 2 \tag{17.14}$$

Condition (17.14) means that the neighbourhood of initial conditions (17.5) is more than the amplitude of steady oscillations in (17.6).

Now we need to find the neighbourhood U_0. The radius v_0 defines the maximum possible value of $|x(t)|$. As it was shown in Section 17.1.2, this value can be easily found

$$v_0 = \left(R + \frac{p_0}{\alpha}\right) e^{\alpha h_0} - \frac{p_0}{\alpha}$$

Denoting

$$\gamma = \left(1 + \frac{\alpha'}{\alpha}\right) e^{\alpha h_0} - \frac{\alpha'}{\alpha} \tag{17.15}$$

we will have $v_0 = R\gamma$.

So we have the condition (17.14) and the parameter p_0 to design the relay control that will ensure stabilisation of the solution $x(t)$ of the system (17.10) with initial condition $|\phi(0)| < R$ in the neighbourhood with radius $r < R$. Now we should design the algorithm decreasing the control gain and radius of the neighbourhood. It is reasonable to find these parameters in the form

$$p_k = \alpha' R d^{-k} \quad \text{and} \quad v_k = R\gamma d^{-k}$$

where $d > 1$ is some number that should be found. Since the radius of stable oscillations r is less than R then v_1 should be less than R too, i.e. $R\gamma d^{-1} < R$, hence $d > \gamma$. The parameter γ depends on α and α', so we can find the supremum of γ

$$\sup_{\alpha \in (0, 1/2 \ln 2), \alpha' \in I_{\alpha, h_0}} \gamma(\alpha, \alpha') = \sup_{\alpha \in (0, 1/2 \ln 2), \alpha' \in I_{\alpha, h_0}} e^{\alpha h_0} + \alpha' \frac{e^{\alpha h_0} - 1}{\alpha} = \sqrt{2} + 1 > 2$$

Let us define d as a natural number $d = 3 > \gamma$, and the indicator function

$$H_{v_k}(|x|) = \begin{cases} 1, & \text{for } |x| > v_k \\ 0, & \text{for } |x| \le v_k \end{cases} \tag{17.16}$$

whose zero value indicates that $x \in U_k$. In this case, the desired control law takes the form

$$u(t - h(t)) = -2\alpha' R \sum_{n=1}^{\infty} 3^{-n} H_{v_n}(|x(t - h(t))|) \, \text{sign}\, [x(t - h(t))] \tag{17.17}$$

The relay control law (17.17) has the following properties:

1. If $v_{k+1} < x(t - h(t)) \leq v_k$, then $u(x(t - h(t))) = -\alpha' R 3^{-k}$.
2. If $-v_k \leq x(t - h(t)) < -v_{k+1}$, then $u(x(t - h(t))) = \alpha' R 3^{-k}$.
3. If $|x(t - h(t))| \leq v_k$, then $|u(x(t - h(t)))| \leq \alpha' 3^{-k} R$.
4. $|u(x(t - h(t)))| \leq \alpha' R$, $(\forall t \geq 0)$.
5. $\gamma < 3$.
6. $v_k - 3^{-k} R(\alpha'/\alpha) = 3^{-k} \varepsilon(\gamma - (\alpha'/\alpha)) < 0$.

Properties 1–4 arise from convergence of the following series

$$\sum_{n=l}^{\infty} 3^{-n} = \frac{3^{-l+1}}{2}$$

Property 5 is realised because of choosing $d = 3 > \gamma$ and Property 6 can be easily proved

$$\gamma - \frac{\alpha'}{\alpha} = e^{\alpha h_0} - \alpha' \frac{2 - e^{\alpha h_0}}{\alpha} < e^{\alpha h_0} - \frac{\alpha e^{\alpha h_0}}{2 - e^{\alpha h_0}} \frac{2 - e^{\alpha h_0}}{\alpha} = 0$$

17.2.2   System stability

Theorem 1.   *The zero solution of the system (17.10) under control (17.17) is semiglobally stable.*

Full analytical evidence of this theorem is proposed in the appendix to this chapter, so we produce only the outline of the proof. It is obvious that convergence of $x(t)$ to zero can be proved using the following statements

- any solution $x(t)$ of the system (17.10) under the control (17.17) is situated in v_0 – neighbourhood of zero, i.e. $|x(t)| < v_0 = R\gamma$ for all $t > 0$;
- if $|x(t)| < v_k$ for all $t > T_k$, then there exists $T_{k+1} > 0$ such that $|x(t)| < v_{k+1}$ for all $t > T_{k+1}$.

The first statement can be proved following the strategy that we have used in Section 17.1.2. The proof of the second statement needs more precise study of the system dynamics. First of all we will find the sufficient condition for staying in the neighbourhood U_k: *if the solution $x(t)$ is situated in the neighbourhood U_k during the time interval $[T_k - h_0, T_k]$ and $|x(T_k)| < v_k/\gamma$, then $x(t)$ will never leave this neighbourhood in future.* However, existence of such a time instant T_k should be proved. Then we will prove that *for any $t > 0$ and any $\varepsilon > 0$ there exists such a time instant $T_\varepsilon > t : |x(T_\varepsilon)| < \varepsilon$.* This means that there exists an arbitrarily small value of the considered solution. It turns out that existence of such a time instant T_ε guarantees the existence of T_k.

17.2.3 Stabilisation of a second order system with unstable complex conjugate eigenvalues

Consider the system

$$\begin{pmatrix} \dot{x} \\ \dot{y} \end{pmatrix} = \begin{pmatrix} \alpha & -\beta \\ \beta & \alpha \end{pmatrix} \begin{pmatrix} x \\ y \end{pmatrix} + \begin{pmatrix} u_1(x(t-h(t)), y(t-h(t))) \\ u_2(x(t-h(t)), y(t-h(t))) \end{pmatrix} \tag{17.18}$$

where $(x(t), y(t))^T \in R^2$ is the state vector, $\alpha > 0$ and $\beta > 0$, $(u_1, u_2)^T \in R^2$ is the control vector and $h(t)$ $(0 < h(t) \le h_0)$. Let us find a control law of the form:

$$\begin{pmatrix} u_1(x(t-h(t)), y(t-h(t))) \\ u_2(x(t-h(t)), y(t-h(t))) \end{pmatrix} = S \begin{pmatrix} u_R(x(t-h(t))) \\ u_R(y(t-h(t))) \end{pmatrix} \tag{17.19}$$

where S is a 2×2 square matrix and $u_R(\cdot)$ has already been considered in the previous section

$$u_R(\cdot) = -2\alpha' R \sum_{i=1}^{\infty} 3^{-i} H_{v_i}(|\cdot|) \, \text{sign}[\cdot]$$

where $v_i = 3^{-i+1} R$. It is obvious that for the case $S = I$, the control law (17.19) defines the vector field shown on Fig. 17.3. Since the solution of the open-loop system has a

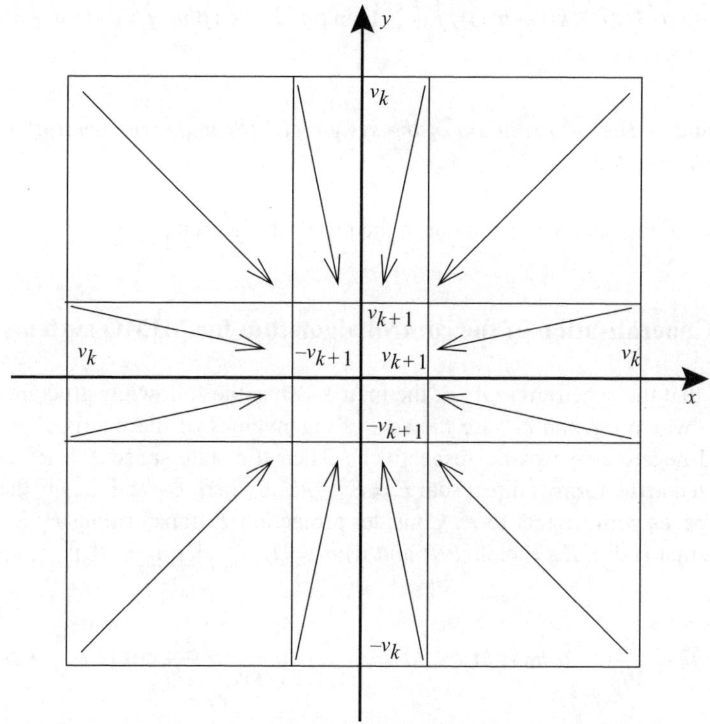

Figure 17.3   Vectorial field in the complex case

spiral form, then supposing that α is sufficiently small we can suppose that

$$\begin{pmatrix} x(t) \\ y(t) \end{pmatrix} \approx \begin{pmatrix} \cos\beta h(t) & -\sin\beta h(t) \\ \sin\beta h(t) & \cos\beta h(t) \end{pmatrix} \begin{pmatrix} x(t-h(t)) \\ y(t-h(t)) \end{pmatrix}$$

Now we can choose the matrix S in the form:

$$S = \frac{1}{2} \begin{pmatrix} \cos\beta h_0 & -\sin\beta h_0 \\ \sin\beta h_0 & \cos\beta h_0 \end{pmatrix}$$

It can be proved that the proposed control guarantees semiglobal stabilisation of the system (17.18) with parameters $0 < \beta h_0 < \pi/4$ and $\alpha h_0 < M$, where

$$M = \max_{t \in [0,(\pi/4)-\beta h_0]} \frac{t}{6\sqrt{2}} \cos\left(t + \frac{\pi}{4} + \beta h_0\right)$$

and

$$\alpha' \in \left(0, \frac{\pi}{4h_0} - \beta\right) : \alpha < \frac{\alpha'}{6\sqrt{2}} \cos\left(\alpha' h_0 + \frac{\pi}{4} + \beta h_0\right)$$

Thus the desired control has the form

$$\begin{pmatrix} u_1(x(t-h(t)), y(t-h(t))) \\ u_2(x(t-h(t)), y(t-h(t))) \end{pmatrix} = \frac{1}{2} \begin{pmatrix} \cos\beta h_0 & -\sin\beta h_0 \\ \sin\beta h_0 & \cos\beta h_0 \end{pmatrix} \begin{pmatrix} u_R(x(t-h(t))) \\ u_R(y(t-h(t))) \end{pmatrix}$$

$$(17.20)$$

Theorem 2.   *The zero solution of the system (17.18) under the control (17.19) is semiglobally stable.*

The proof of this theorem is similar to the proof of Theorem 1 [14].

17.3   Generalisation of the control algorithm for MIMO systems

Assume that the spectrum $\sigma(A)$ of the matrix A has the following structure $\sigma(A) = \sigma_+ \cup \sigma_-$, where σ_+ and σ_- are the sets of eigenvalues of the matrix A with positive and negative real parts, respectively. Then the state space $E = R^n$ could be represented in the form of direct sum $E = E_+ \oplus E_-$, where E_+ and E_- are the invariant subspaces with respect to A. Consider projection P transforming $P : R^n \to E_+$. Suppose that i) $\dim E_+ = \text{rank}(PB)$ and ii) $\sigma_+ = \{\lambda_i\}_{i=1}^{l} \bigcup \{\alpha_j \pm i\beta_j\}_{j=1}^{\nu}$, $\lambda_i h_0 < L$, $L = \frac{1}{2} \ln 2$,

$$0 < \beta_i < \frac{\pi}{4h_0}, \quad \alpha_i h_0 < M_j, \quad M_j = \max_{t \in [0,(\pi/4)-\beta_j h_0]} \frac{t}{6\sqrt{2}} \cos\left(t + \frac{\pi}{4} + \beta_i h_0\right)$$

$$(17.21)$$

and all the eigenvalues from σ_+ are simple.

In this case, there exists a nonsingular coordinate transformation G such that the system (17.8), after substituting $z = (z_1, z_2)^T = G^{-1}x$, can be rewritten as follows

$$
\begin{aligned}
\dot{z}_1 &= A^+ z_1 + B^+ u \\
\dot{z}_2 &= A^- z_2 + B^- u
\end{aligned}
\tag{17.22}
$$

where $A^+, B^+ \in R^{m \times m}$, $A^- \in R^{(n-m) \times (n-m)}$, $B^- \in R^{(n-m) \times m}$, $\det(B^+) \neq 0$ and

$$
A^+ = \begin{pmatrix}
\lambda_1 & 0 & \cdots & \cdots & \cdots & \cdots & \cdots & 0 \\
\cdots & \cdots & \cdots & \cdots & \cdots & \cdots & \cdots & \cdots \\
0 & \cdots & \lambda_l & 0 & \cdots & \cdots & \cdots & 0 \\
0 & \cdots & 0 & \alpha_1 & -\beta_1 & 0 & \cdots & 0 \\
0 & \cdots & 0 & \beta_1 & \alpha_1 & 0 & \cdots & 0 \\
\cdots & \cdots & \cdots & \cdots & \cdots & \cdots & \cdots & \cdots \\
0 & \cdots & \cdots & \cdots & \cdots & 0 & \alpha_v & -\beta_v \\
0 & \cdots & \cdots & \cdots & \cdots & 0 & \beta_v & \alpha_v
\end{pmatrix}
$$

Thus we can choose a control law of the form

$$
u = [B^+]^{-1} u_0
$$

and reduce the control problem for the system (17.8) to the $l + v$ simple control problem, which was already considered in Sections 17.2.1 and 17.2.3.

Theorem 3. *If the conditions i) and ii) hold, then the system (17.8) is semiglobally stabilisable.*

17.4 Semiglobal stabilisation of a mechanical system via relay delayed control

17.4.1 *Stabilisation of linearised mechanical systems via relay delayed control*

Let us consider the linearised mechanical system

$$
H\ddot{q} - P\dot{q} + Wq = Bu(t - h(t))
\tag{17.23}
$$

where $q \in R^n$ is a vector of generalised coordinates, $H > 0$, P, W are constant $n \times n$ matrices, $u \in R^m$, $(m \leq n)$ is a control vector, B is a $n \times m$ gain matrix, $0 \leq h(t) \leq h_0$ is an uncertain continuous time delay. Our aim is to apply the proposed control algorithm for the stabilisation of the system (17.23).

Assumption 1. *Suppose that the spectrum σ_+ of the system (17.23) satisfies condition ii) and $l + v = m$.*

17.4.1.1 Matrix Riccati equation

For initial system decoupling, consider the following matrix Riccati equation

$$
X^2 + BX + C = 0
\tag{17.24}
$$

where $X, B, C \in R^{n \times n}$.

Let us find the solution of the equation (17.24) following [9]. Consider the operator

$$M = \begin{pmatrix} 0 & I_n \\ -C & -B \end{pmatrix} \tag{17.25}$$

Assume the operator M has an n-dimensional invariant subspace E_n. Denote as

$$U = \begin{pmatrix} U_1 \\ U_2 \end{pmatrix} = (e_1, e_2, \ldots, e_n)$$

where $e_1, e_2, \ldots, e_n$ are the basis in E_n, and U_1, $U_2 - n \times n$ square matrices. If $\det(U_1) \neq 0$ then the solution of the equation (17.24) can be written in the following form:

$$X = U_2 U_1^{-1} \tag{17.26}$$

Indeed, since the vectors $\{e_i\}$ are a basis in E_n, then

$$MU = UM_{E_n}$$

where the matrix $M_{E_n}(n \times n)$ is the restriction of operator M to the invariant subspace E_n. Taking into account the structure of the matrix M we have

$$U_2 = U_1 M_{E_n}$$
$$-CU_1 - BU_2 = U_2 M_{E_n}$$

Let us multiply each equality by U_1^{-1}

$$U_2 U_1^{-1} = U_1 M_{E_n} U_1^{-1}$$
$$-C - BU_2 U_1^{-1} = U_2 U_1^{-1} U_1 M_{E_n} U_1^{-1}$$

Denoting $X = U_2 U_1^{-1}$ and substituting the first equality into the second one produces (17.24). This means that X is the solution of the Riccati equation.

17.4.1.2   System factorisation

To reduce the system (17.23) to the system that was already considered in Section 17.3, let us factorise the initial system, for example

$$H \left(\frac{d}{dt} - C^+ \right) \left(\frac{d}{dt} - C^- \right) q = Bu(t - h(t)) \tag{17.27}$$

where C^+ and C^- are some constant square matrices and the spectrum of C^- is located in the left-half complex plane. It is obvious that the matrices C^+ and C^- satisfy equations

$$\begin{aligned} C^+ + C^- &= H^{-1} P \\ C^+ C^- &= H^{-1} W \end{aligned} \tag{17.28}$$

Hence, the matrix C^- satisfies the matrix equation

$$(C^-)^2 - H^{-1} P C^- + H^{-1} W = 0 \tag{17.29}$$

Thus, to factorise the system (17.23), we should find the solution of the matrix Riccati equation (17.29). The matrix of the solution of this Riccati equation should have eigenvalues with negative real parts.

According to the proposed algorithm, we consider the matrix

$$M = \begin{pmatrix} 0 & I_n \\ -H^{-1}W & H^{-1}P \end{pmatrix} \tag{17.30}$$

It can be proved that $\text{spec}(M) = \text{spec}(C^+) \cup \text{spec}(C^-)$.

$$\begin{aligned}
\det(M - \lambda I_{2n}) &= \det\begin{pmatrix} -\lambda I_n & I_n \\ -H^{-1}W & H^{-1}P - \lambda I_n \end{pmatrix} \\
&= \det(\lambda I_n(-H^{-1}P + \lambda I_n) - (-\lambda I_n)(-H^{-1}W)(-\lambda I_n)^{-1}I_n) \\
&= \det(\lambda^2 I_n - \lambda H^{-1}P + H^{-1}W) \\
&= \det((\lambda I_n - C^+)(\lambda I_n - C^-)) \\
&= \det(\lambda I_n - C^+)\det(\lambda I_n - C^-)
\end{aligned}$$

Since the spectrum of the system (17.23) coincides with the spectrum M, then $R^{2n} = E_+ \oplus E_-$, where E_+, E_- are the invariant subspaces of M corresponding to the eigenvalues in the right and left-half complex plane, respectively. From Assumption 1 and condition $m \leq n$ it follows that $\dim(E_+) \leq n$. In this case the matrix M can have an invariant subspace $E_n : E_n \subset E_-$ and matrix C^- can be found by formula (17.26).

Denoting $z = \dot{q} - C^-q$, we have

$$\dot{z}(t) = H^{-1}C^+z(t) + H^{-1}Bu(t - h(t)) \tag{17.31}$$

On the one hand, applying the control law proposed in Section 17.3, we ensure the asymptotic decrease of $z(t)$ to zero. On the other hand, $\dot{q} = C^-q + z(t)$ and the eigenvalues of C^- have negative real parts. Thus $q(t)$ will tend to zero too.

17.4.2 *Generalisation of control algorithm to the case of nonlinear mechanical system*

Let us consider the following mechanical system

$$H(q)\ddot{q} + P(q,\dot{q})\dot{q} + W(q) = Bu(t - h(t)) \tag{17.32}$$

where $q \in R^n$ is a vector of generalised coordinates, $H(q) > 0$, $P(q,\dot{q})$ are square matrices, $W(q)$ is a vector-function, $u \in R^m$, $(m \leq n)$ is a control vector, B is a gain matrix $(n \times m)$ and $0 \leq h(t) \leq h_0$ is an uncertain time delay. Let $q = 0$ be the unstable equilibrium point of the system (17.32). The decoupling problem for the system (17.32) can be reduced to the matrix Riccati equation by the following.

Define $X = C$ as the solution of the Riccati equation

$$X^2 + P(0,0)H^{-1}(0)X + \frac{\partial W(0)}{\partial q}H^{-1}(0) = 0$$

and the spectrum of the matrix C is situated in the left-half complex plane. This means that the matrix

$$M = \begin{pmatrix} 0 & I_n \\ -\dfrac{\partial W(0)}{\partial q} H^{-1}(0) & -P(0,0)H^{-1}(0) \end{pmatrix}$$

should have at least n eigenvalues with negative real parts and the linear operator M should have an invariant subspace corresponding to these eigenvalues. Denote

$$v = H(q)\dot{q} - CH(q)q \tag{17.33}$$

It is easy to see that $\dot{q} = H^{-1}(q)CH(q)q + H^{-1}(q)v$ and, for any q, the eigenvalues of the matrix $H^{-1}(q)CH(q)$ have negative real parts. This means that the last system is stable to a first approximation at least. Consequently, to ensure semiglobal stabilisation of the system (17.32), we have to design the control law ensuring $\|v(t)\| \to 0$.

Let us find the time derivative of $v(t)$

$$\dot{v} = H(q)\ddot{q} - CH(q)\dot{q} + f_1(q,\dot{q})$$

where $\|f_1(q,\dot{q})\| \le K_1^1 \|q\|^2 + K_2^1 \|q\| \|\dot{q}\| + K_3^1 \|\dot{q}\|^2$. Using (17.32) we have

$$\dot{v} = (-P(q,\dot{q}) - CH(q))\dot{q} - W(q) + Bu(t - h(t)) + f_1(q,\dot{q})$$
$$= (-P(q,\dot{q})H^{-1}(q) - C)(CH(q)q + v) - W(q) + Bu(t - h(t)) + f_1(q,\dot{q})$$
$$= D_1 H(q)q + D_2 v + Bu(t - h(t)) + f_2(q,\dot{q})$$

where

$$D_1 = -P(0,0)H^{-1}(0)C - C^2 - \frac{\partial W(0)}{\partial q} H^{-1}(0)$$

$$D_2 = P(0,0)H^{-1}(0) - C$$

$$\|f_2(q,\dot{q})\| \le K_1^2 \|q\|^2 + K_2^2 \|q\| \|\dot{q}\| + K_3^2 \|\dot{q}\|^2$$

Since C is the solution of a matrix Riccati equation, then $D_1 = 0$. Finally one can conclude that

$$\dot{v}(t) = D_2 v(t) + Bu(t - h(t)) + f_2(q,\dot{q})$$

Usually the function $f_2(q,\dot{q})$ can be disregarded for the control design, so we have the system that was already considered. It is necessary to remark that system (17.32) can be reduced to the system (17.23) by means of linearisation. However, in this case we will have nonlinearity $f(q,\dot{q},\ddot{q})$ on the right hand side of (17.23).

17.5   Numerical examples

17.5.1   Stabilisation of inverted pendulum

Consider the stabilisation problem of an inverted pendulum. With linear approximation, the equation of oscillation of the inverted pendulum with unit mass under relay delay control has the form

$$\ddot{\theta} + k\dot{\theta} - p\theta = u(t - h(t)) \tag{17.34}$$

where $k > 0$ is a friction coefficient, $p = g/l > 0$, $h(t)$ is an uncertain time delay $(0 \leq h(t) \leq h_0)$. It is easy to see that the spectrum of the open-loop system consists of two real eigenvalues: $\lambda_1 = \frac{1}{2}(\sqrt{k^2 + 4p} - k) > 0$, and $\lambda_2 = -\frac{1}{2}(\sqrt{k^2 + 4p} + k) < 0$. Equation (17.34) can be rewritten in the form:

$$\left(\frac{d}{dt} - \lambda_1\right)\left(\frac{d}{dt} - \lambda_2\right)\theta = u(t - h(t))$$

Denoting $z(t) = \dot{\theta}(t) - \lambda_2\theta(t)$, we will have

$$\dot{z} = \lambda_1 z + u(t - h(t))$$

As was shown in Section 17.2.1, the last equation is semiglobally stabilisable under the condition $\lambda_1 h_0 < \frac{1}{2}\ln 2$ by the control law

$$u(z(t - h(t))) = -2\alpha' R \sum_{n=1}^{\infty} 3^{-n} H_{v_n}(|z(t - h(t))|)\, \text{sign}[z(t - h(t))] \tag{17.35}$$

Since $\dot{\theta} = \lambda_2\theta + z(t)$ and $\lambda_2 < 0$ then $|z(t)| \to 0$ implies $|\theta(t)| \to 0$.

Consider the following model

$$\ddot{\theta} + 0.3\dot{\theta} - 0.04\theta = u(t - h(t)) \tag{17.36}$$

$$\theta(t) = \sin 3t \tag{17.37}$$

$$\dot{\theta}(t) = 3\cos 3t \qquad \text{for } t \in [-h_0, 0] \tag{17.38}$$

where $h(t) = 0.75 + 0.25\sin(30t)$, $h_0 = 1$, control has the form (17.35) with parameters $\alpha' = 0.5$, $R = 3$, $v_n = R3^{-n}\gamma$, $\gamma = 1.841\,37$. Figure 17.4 presents the simulation results for this model.

17.5.2   Double inverted pendulum

Consider a more complicated mechanical system consisting of the two inverted pendulums as shown in Fig. 17.5. The dynamics of the pendulums are described by the following equation [15]:

$$H(q)\begin{bmatrix} \ddot{q}_1 \\ \ddot{q}_2 \end{bmatrix} + P(q, \dot{q})\begin{bmatrix} \dot{q}_1 \\ \dot{q}_2 \end{bmatrix} + W(q) = \begin{bmatrix} F_1 \\ F_2 \end{bmatrix} \tag{17.39}$$

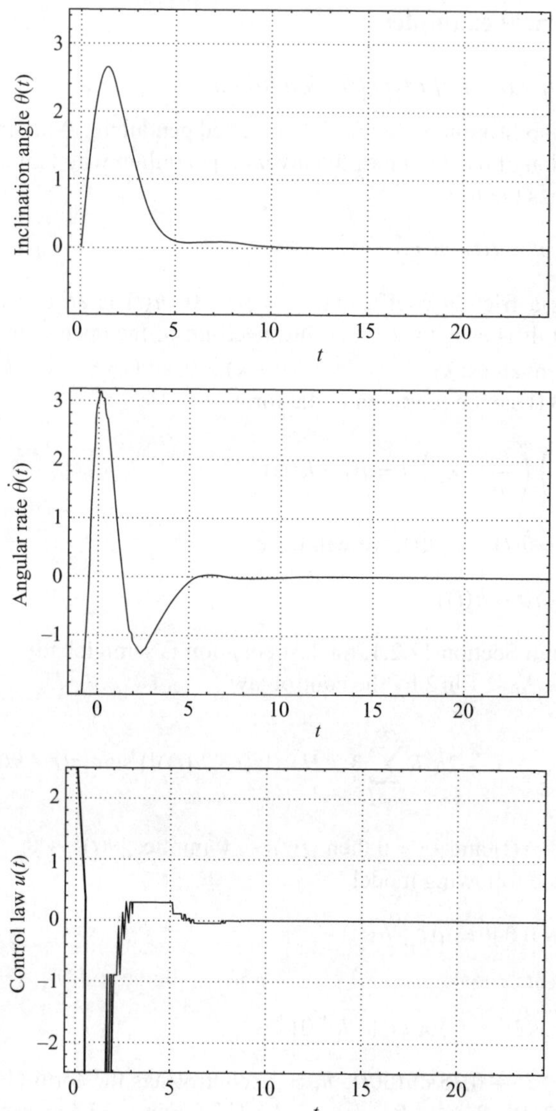

Figure 17.4   Simulation results for the inverted pendulum model

where

$$H = \begin{bmatrix} J_0 + I_1 + m_1 l_1^2 + m_2 L_1^2 & m_2 L_1 l_2 \cos(q_1 - q_2) \\ m_2 L_1 l_2 \cos(q_1 - q_2) & m_2 l_2^2 + I_2 \end{bmatrix}$$

$$W = \begin{bmatrix} -g(m_1 l_1 + m_2 L_1) \sin q_1 \\ -m_2 g l_2 \sin q_2 \end{bmatrix}$$

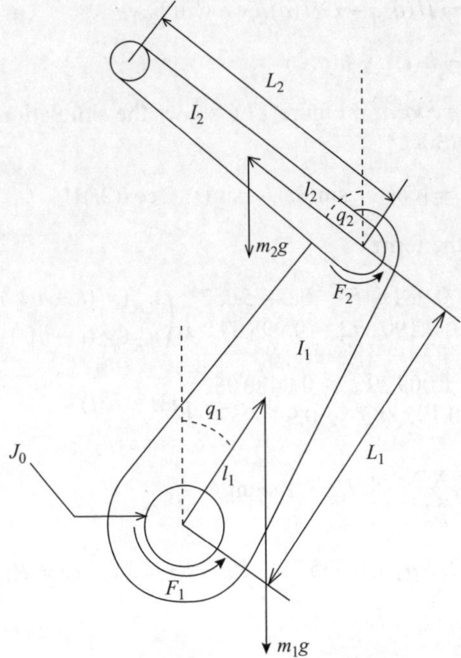

Figure 17.5   Double inverted pendulum

and

$$P = \begin{bmatrix} 0 & V\dot{q}_2 \\ -V\dot{q}_1 & 0 \end{bmatrix}, \qquad V = m_2 L_1 l_2 \sin(\theta_1 - \theta_2)$$

where m_i, I_i are the mass and inertia of each link, L_i is the total length of the link and l_i is the distance from the centre of gravity of each link to its pivot point. It is assumed that both control forces F_1 and F_2 are available and $[F_1, F_2]^T = u(t - h(t))$.

We will consider the following parameters of the double inverted pendulum: $m_1 = 0.132$, $m_2 = 0.088$, $L_1 = 0.3032$, $L_2 = 0.3545$, $l_1 = 0.1274$, $l_2 = 0.1209$, $I_1 = 0.0562$, $I_2 = 0.0314$, $J_0 = 0.000\,006$, $g = 9.8$. According to the control algorithm proposed in Section 17.4.2, we should find a matrix $C < 0$ that satisfies the following equation

$$C^2 + P(0,0)H^{-1}(0)C + \frac{\partial W(0)}{\partial q}H^{-1}(0) = 0$$

For the problem under consideration, the last equation has the form

$$C^2 + \begin{pmatrix} -6.4492 & 0.6330 \\ 0.1548 & -3.2115 \end{pmatrix} = 0$$

Hence

$$C = \begin{pmatrix} -2.5385 & 0.1462 \\ 0.0357 & -1.7906 \end{pmatrix}$$

After substituting $v = H(q)\dot{q} - CH(q)q$ we will have

$$\dot{v} = -Cv + u(t - h(t)) + f(q, \dot{q})$$

where $f(q, \dot{q}) = O(q^2, \dot{q}q, \dot{q}^2)$. Figure 17.6 shows the simulation results for the case $h(t) = 0.1 + 0.03\sin(30t)$,

$$[q_1, q_2, \dot{q}_1, \dot{q}_2]^T = [\cos(t), \sin(2t), -\sin(t), 2\cos(2t)]^T, \qquad t \in [-h_0, 0]$$

with control law of the form

$$u(t - h(t)) = \begin{pmatrix} 0.981\,615 & 0.0475\,057 \\ -0.190\,872 & 0.998\,871 \end{pmatrix} \begin{pmatrix} u_R^1(y_1(t - h(t))) \\ u_R^2(y_2(t - h(t))) \end{pmatrix}$$

$$y(t - h(t)) = \begin{pmatrix} 1.009\,39 & -0.0480\,062 \\ 0.192\,883 & 0.991\,957 \end{pmatrix} v(t - h(t))$$

$$u_R^i(\cdot) = -2p_i R_i \sum_{n=1}^{\infty} 3^{-n} H_{v_n^i}(|\cdot|)\mathrm{sign}[\cdot]$$

where $p_1 = 6.489\,82$, $p_2 = 6.835\,75$, $R_1 = R_2 = 3\pi$, $v_n^1 = R_1 3^{-n} 2.392\,16$, $v_n^2 = R_2 3^{-n} 2.260\,66$.

17.6 Appendix

17.6.1 Staying in the neighbourhood

Lemma 1. *If there exists $T_k \geq 0$ such that $|x(t)| \leq v_k$ for all $t \in [T_k - h_0, T_k]$ and $|x(T_k)| \leq v_k/\gamma$, then $|x(t)| \leq v_k$ for all $t \geq T_k$.*

Proof. Suppose by contradiction, that there exists $T' \geq T_k$ such that $|x(T')| > v_k$. Then, from the condition $|x(T)| \leq v_k/\gamma$ it follows that there exists $t^* \geq T : |x(t^*)| = v_k/\gamma$ and $|x(t)| > v_k/\gamma, \forall t \in (t^*, T']$, and moreover there exists $T^* > t^* : |x(T^*)| = v_k$ and $|x(t)| \leq v_k, \forall t \in [t^*, T^*]$.

Let us show that $T^* - t^* \geq h_0$. Taking into account that $|x(t)| \leq v_k$ for $t \in [T - h_0, T]$ and $|x(t)| \leq v_k$ for $t \in [T, T^*]$, from Property 3 one has $|u(t - h(t))| \leq \alpha' R3^{-k}$ for $t \in [T, T^*]$.

Let us consider the case $x(t^*) > 0$ and $x(T^*) > 0$ (other cases can be proved analogously).

$$\dot{x} \leq \alpha x + \alpha' 3^{-k} R$$

$$x(t^*) = \frac{v_k}{\gamma}$$

and from the Bellman lemma

$$x(t) \leq \left(\frac{v_k}{\gamma} + \frac{\alpha'}{\alpha} R3^{-k} \right) e^{\alpha(t - t^*)} - \frac{\alpha'}{\alpha} R3^{-k}$$

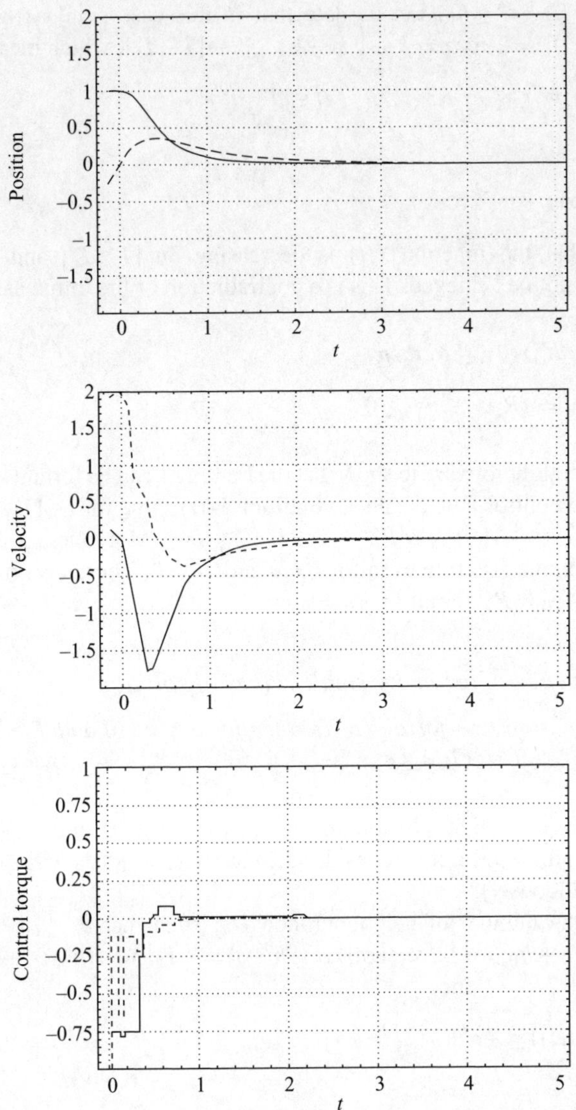

Figure 17.6   Simulation results for the double inverted pendulum model

For $t = T^*$ one has

$$x(T^*) = v_{k+1} = \gamma R 3^{-k} = \left(\left(1 + \frac{\alpha'}{\alpha}\right) e^{\alpha h_0} - \frac{\alpha'}{\alpha}\right) 3^{-k} R$$

$$\leq 3^{-k} R \left(\left(1 + \frac{\alpha'}{\alpha}\right) e^{\alpha(T^* - t^*)} - \frac{\alpha'}{\alpha}\right)$$

which yields $T^* - t^* \geq h_0$. Let us note that in this case $\text{sign}[x(t - h(t))] = 1$ and $H_{v_n}(|x(t - h(t))|) = 1$ for $n = k+1, k+2, \ldots, t \in [T^*, T']$, which means

$$u(x(t - h(t))) \geq -\alpha' R3^{-k}, \qquad \forall t \in [T^*, T'] \tag{17.40}$$

Now

$$\dot{x}(T^*) \leq \alpha v_k - \alpha' 3^k R < 0 \tag{17.41}$$

This means that the function $x(t)$ is decreasing on $[T^*, T']$ and the inequality $x(T') > v_k$ cannot be achieved. This is a contradiction of the initial assumption.

Corollary 1.   *If* $|x(0)| \leq R$, *then*

$$|x(t)| \leq v_0 = \gamma R\gamma, \qquad \forall t \geq 0$$

Proof.   It is straight forward to show that the Lemma 1 can be formulated and proved in terms of the control law. Namely, condition $|x(t)| \leq v_k$ for $t \in [T - h_0, T]$ can be replaced to $|u(x(t - h(t))| \leq \alpha' R3^{-k}, \forall t \in [T, T + h_0]$. Moreover, it is necessary to remark that Lemma 1 is true even for $k = 0$ and $T_0 = 0$. Therefore, from Property 4 we have $x(t) \leq v_0 = R\gamma$ for all $t > T_0 = 0$.

17.6.2   *Existence of arbitrary small values of solution*

Lemma 2.   *If* $|x(t)| < v_k$ *for all* $t > T_k$, *then for any* $\varepsilon > 0$ *and* $T \geq T_k$ *there exists a time instant* $T_\varepsilon \geq T : |x(T_\varepsilon)| \leq \varepsilon$.

Proof.   Suppose by contradiction that there exist $\varepsilon_0 > 0$ and $T_0 \geq T_k$ such that $|x(t)| > \varepsilon_0$ for all $t \geq T_0$. Let us consider the case $x(t) > \varepsilon_0$ (the case $x(t) < -\varepsilon_0$ can be proved analogously).

Let n be the number of neighbourhood U_n, such that $v_n \leq \varepsilon_0 < v_{n-1}$. In this case, for $t \geq T_0 + h_0$ we have $\text{sign}[x(t - h(t))] = -1$ and $H_i(|x(t - h(t))|) = 1$ for $i = n, n+1, n+2, \ldots$. Hence

$$u(x(t - h(t))) \leq -\alpha' R3^{-n+1}$$

and

$$\dot{x}(t) \leq \alpha x(t) - \alpha' R3^{-n+1}$$

Let us show that in this case we will have $x(t) > v_{n-1}$ for all $t > T_0 + h_0$. Suppose by contradiction that there exists $t^1 > T_0 + h_0$ such that $x(t^1) \leq v_{n-1}$ then

$$\dot{x}(t^1) \leq \alpha x(t^1) - \alpha' R3^{-n+1} \leq \alpha \gamma R3^{-n+1} - \alpha' R3^{-n+1} < 0 \quad \text{(see 6)}$$

Moreover, from the Bellman lemma we have

$$x(t) \leq \left(\gamma R3^{-n+1} - \frac{\alpha'}{\alpha} R3^{-n+1} \right) e^{\alpha(t - t^1)} + \frac{\alpha'}{\alpha} R3^{-n+1} \equiv v(t)$$

Hence, there exists a time instant $t^2 : v(t^2) = v_n$, and

$$x(t^2) \leq v_n$$

This contradicts the initial assumption. Thus $x(t) > v_{n-1}$ for all $t > T_0 + h_0$.

Analogously we can prove that $x(t) > v_{n-2}$ for all $t > T_0 + h_0$, etc. Finally we will have that $x(t) > v_k$, which contradicts with the condition $|x(t)| \leq v_k$ for all $t > T_k$.

17.6.3   Proof of Theorem 1

From Lemma 1 we have that $|x(t)| < v_0$ for all $t > 0$. Therefore to prove the theorem it is enough to show the following.

Proposition 1.   *If $|x(t)| \leq v_k$ for all $t \geq T_k$, then there exists such T_{k+1} that $|x(t)| \leq v_{k+1}, \forall t \geq T_{k+1}$.*

Proof.   According to Lemma 2, there exists a time instant $T_\varepsilon > T_k + 2h_0$ such that $|x(T_\varepsilon)| \leq \varepsilon = \delta 3^{-k} R$, where $\delta = 1 - (2(\gamma + 1))/(3(e^{\alpha h_0} + 1))$, $0 < \delta < \frac{1}{3}$. Let us show that $|x(t)| < v_{k+1}$ for all $t > T_\varepsilon$.

Suppose, by contradiction, there exists a time instant $T' > T_\varepsilon : x(T') > v_{k+1}$ (the case $x(T') < -v_{k+1}$ can be proved analogously). Let $T^* < T'$, and $x(T^*) = v_{k+1}$.

In this case two alternative cases are possible:

1) $|x(t)| < v_{k+1}$ for all $t \in [T_\varepsilon - h_0, T_\varepsilon]$;
2)   there exists $t^* \in [T_\varepsilon - h_0, T_\varepsilon] : |x(t^*)| = v_{k+1}$ and $|x(t)| < v_{k+1}$ for all $t \in (t^*, T_\varepsilon]$.

Initially, let us consider case 1. It is easy to see that in this case $|x(t - h(t))| < v_{k+1}$ for all $t \in [T_\varepsilon, T']$.

Then from Property 3 we have

$$\dot{x}(t) \leq \alpha x(t) + \alpha' R 3^{-k-1}$$

$$x(T_\varepsilon) \leq \varepsilon$$

and

$$x(t) \leq \left(\varepsilon + \frac{\alpha'}{\alpha} R 3^{-k-1} \right) e^{\alpha(t - T_\varepsilon)} - \frac{\alpha'}{\alpha} R 3^{-k-1}$$

Hence

$$x(T^*) = v_{k+1} = \gamma R 3^{-k-1} \leq \left(\varepsilon + \frac{\alpha'}{\alpha} R 3^{-k-1} \right) e^{\alpha(T^* - T_\varepsilon)} - \frac{\alpha'}{\alpha} R 3^{-k-1}$$

and

$$T^* - T_\varepsilon \geq \frac{1}{\alpha} \ln \left(\frac{\gamma + (\alpha'/\alpha)}{3\delta + (\alpha'/\alpha)} \right)$$

Let us show that $T^* - T_\varepsilon \geq h_0$.

$$\frac{1}{\alpha} \ln \left(\frac{\gamma + (\alpha'/\alpha)}{3\delta + (\alpha'/\alpha)} \right) \geq h_0$$

$$\frac{\gamma + (\alpha'/\alpha)}{3\delta + (\alpha'/\alpha)} \geq e^{\alpha h_0}$$

$$\gamma + \frac{\alpha'}{\alpha} \geq (3\delta + (\alpha'/\alpha)) e^{\alpha h_0}$$

$$\left(1 + \frac{\alpha'}{\alpha} \right) e^{\alpha h_0} - \frac{\alpha'}{\alpha} + \frac{\alpha'}{\alpha} \geq \left(3\delta + \frac{\alpha'}{\alpha} \right) e^{\alpha h_0}$$

$$e^{\alpha h_0} \geq e^{\alpha h_0} 3\delta$$

Hence, for $t \in [T^*, T']$ we have $\text{sign}[x(t - h(t))] = 1$ and $H_{v_n}(|x(t - h(t))|) = 1$ for $n = k+1, k+2, \ldots$.

$$\dot{x}(T^*) \leq \alpha x(T^*) - \alpha' R3^{-k-1} = \alpha \gamma R3^{-k-1} - \alpha' R3^{-k-1} < 0$$

Then $x(t)$ is a decreasing function on the interval $[T^*, T']$ and the inequality $x(T') > v_{k+1}$ cannot be achieved.

Consider Case 2.

Since $x(T_\varepsilon) \leq \varepsilon = \delta 3^{-k} R < 3^{-k-1} R = v_{k+1}/\gamma$ and $x(T^*) = v_{k+1} \geq v_{k+1}/\gamma$ then there exists a time instant $T \in [T_\varepsilon, T^*] : x(T) = v_{k+1}/\gamma$.

Let us estimate the size of the time interval $[t^*, T]$. From the condition $|x(t)| < v_k$ and Property 3 one has $|u(t - h(t))| < \alpha' R3^{-k}$. Then from the differential inequalities

$$\dot{x}(t) \geq \alpha x(t) - \alpha' R3^{-k}$$

$$\dot{x}(t) \leq \alpha x(t) + \alpha' R3^{-k}$$

and initial conditions

$$x(t^*) = v_{k+1}$$

$$x(T_\varepsilon) = \varepsilon$$

Using the Bellman lemma we conclude

$$x(t) \geq \left(v_{k+1} - \frac{\alpha'}{\alpha} R3^{-k} \right) e^{\alpha(t - t^*)} + \frac{\alpha'}{\alpha} R3^{-k} \qquad (t \in [T^*, T_\varepsilon])$$

$$x(t) \leq \left(\varepsilon + \frac{\alpha'}{\alpha} R3^{-k} \right) e^{\alpha(t - T_\varepsilon)} - \frac{\alpha'}{\alpha} R3^{-k} \qquad (t \in [T_\varepsilon, T])$$

Let us consider the last inequalities at the time instants $t = T_\varepsilon$ and $t = T$, respectively

$$\varepsilon \geq x(T_\varepsilon) \geq \left(v_{k+1} - \frac{\alpha'}{\alpha} R3^{-k} \right) e^{\alpha(T_\varepsilon - t^*)} + \frac{\alpha'}{\alpha} R3^{-k}$$

$$x(T) = \frac{v_{k+1}}{\gamma} \leq \left(\varepsilon + \frac{\alpha'}{\alpha} R3^{-k} \right) e^{\alpha(T^* - T_\varepsilon)} - \frac{\alpha'}{\alpha} R3^{-k}$$

Hence,

$$t^* - T_\varepsilon \geq \frac{1}{\alpha} \ln \frac{\delta - (\alpha'/\alpha)}{(\gamma/3) - (\alpha'/\alpha)}$$

$$T - T_\varepsilon \geq \frac{1}{\alpha} \ln \frac{(1/3) + (\alpha'/\alpha)}{\delta + (\alpha'/\alpha)}$$

Then

$$T - t^* = T - T_\varepsilon + T_\varepsilon - t^* \geq \frac{1}{\alpha} \ln \left(\frac{(\alpha'/\alpha) - \delta}{(\alpha'/\alpha) - (\gamma/3)} \cdot \frac{(1/3) + (\alpha'/\alpha)}{\delta + (\alpha'/\alpha)} \right) \geq h_0$$

To prove that the last inequality is true:

$$\frac{(\alpha'/\alpha) - \delta}{(\alpha'/\alpha) - (\gamma/3)} \cdot \frac{(1/3) + (\alpha'/\alpha)}{\delta + (\alpha'/\alpha)} \geq e^{\alpha h_0}$$

$$\frac{\alpha'}{3\alpha} - \frac{\alpha'\delta}{\alpha} + \left(\frac{\alpha'}{\alpha}\right)^2 - \frac{\delta}{3} \geq e^{\alpha h_0} \left(\left(\frac{\alpha'}{\alpha}\right)^2 + \delta\frac{\alpha'}{\alpha} - \frac{\alpha'}{\alpha}\frac{\gamma}{3} - \frac{\delta\gamma}{3} \right)$$

$$\frac{\alpha'}{\alpha}\frac{\gamma e^{\alpha h_0} + 1}{3} + \frac{\delta}{3}(e^{\alpha h_0} - 1) \geq \left(\frac{\alpha'}{\alpha}\right)^2 (e^{\alpha h_0} - 1) + \frac{\alpha'}{\alpha}\delta(e^{\alpha h_0} + 1)$$

$$\frac{\gamma e^{\alpha h_0} + 1}{3} + \frac{\delta}{3}\frac{\alpha}{\alpha'}(e^{\alpha h_0} - 1) \geq \frac{\alpha'}{\alpha}(e^{\alpha h_0} - 1) + \delta(e^{\alpha h_0} + 1)$$

$$\frac{\gamma e^{\alpha h_0} + 1}{3} + \frac{\delta}{3}\frac{\alpha}{\alpha'}(e^{\alpha h_0} - 1) \geq \frac{\alpha'}{\alpha}(e^{\alpha h_0} - 1) + e^{\alpha h_0} + 1 - \frac{2(\gamma + 1)}{3}$$

$$\frac{\gamma(e^{\alpha h_0} + 2)}{3} + \frac{\delta}{3}\frac{\alpha}{\alpha'}(e^{\alpha h_0} - 1) \geq \frac{\alpha'}{\alpha}(e^{\alpha h_0} - 1) + e^{\alpha h_0} = \gamma$$

$$\frac{\gamma(e^{\alpha h_0} - 1)}{3} + \frac{\delta}{3}\frac{\alpha}{\alpha'}(e^{\alpha h_0} - 1) \geq 0$$

Thus we have $T - t^* \geq h_0$. Now from Lemma 1 it follows $|x(t)| \leq v_{k+1}$ for all $t \geq T$, which contradicts with the inequality $x(T') > v_{k+1}$.

17.7 References

1   UTKIN, V., GULDNER, J., and SHI, J.: 'Sliding modes in electromechanical systems' (Taylor & Francis, London, 1999)

2   FRIDMAN, L., FRIDMAN, E., and SHUSTIN, E.: 'Steady modes and sliding modes in relay control systems with delay', in BARBOT, J. P. and PERRUQUETTI, W. (Eds): 'Sliding mode control in engineering' (Marcel Dekker, New York, 2002), pp. 263–293

3   SHTESSEL, Y. B., ZINOBER, A., and SHKOLNIKOV, I. A.: 'Sliding mode control for nonlinear systems with output delay via method of stable system center', *ASME Journal of Dynamic Systems, Measurement, and Control*, 2003, **125**(2), pp. 253–257

4   ROH, Y. H. and OH, J. H.: 'Robust stabilization of uncertain input delay systems by sliding mode control with delay compensation', *Automatica*, 1999, **35**(11), pp. 1861–1865

5   RICHARD, J.-P., GOUIASBOUT, F., and PERRUQUETTI, W.: 'Sliding mode control in the presence of delay', *Kybernetica*, 2001, **37**(3), pp. 277–294

6   SING KIONG NGUANG: 'Comments on "Robust stabilization of uncertain input delay systems by sliding mode control with delay compensation"', *Automatica*, 2001, **37**(10), p. 1677

7   FRIDMAN, L., ACOSTA, P., and POLYAKOV, A.: 'Robust eigenvalue assignment for uncertain delay control systems', Proceedings of 3rd IFAC Workshop on *Time Delay Systems*, Santa Fe, NM, 2001

8   PALMOR, Z.: 'Stability properties of Smith dead time compensator controller', *International Journal Control*, 1980, **32**(8), pp. 937–949

9   IKRAMOV, K. D.: 'Numerical solution of matrix equation' (Nauka, Moscow, 1984)

10   ISIDORI, A.: 'Nonlinear control systems' (Springer Verlag, New York, 1989)

11   AKIAN, M., BLIMAN, P.-A., and SORINE, M.: 'Control of delay systems with relay', *IMA Journal Mathematical Control and Information*, 2002, **19**(1–2), pp. 133–155

12   SHUSTIN, E., FRIDMAN, E., and FRIDMAN, L.: 'Oscillations in a second order discontinuous systems with delay', *Discrete and Continuous Dynamical Systems*, 2003, **9**(2), pp. 339–357

13   FRIDMAN, L., STRYGIN, V., and POLYAKOV, A.: 'Stabilization of oscillations amplitudes via relay delay control', *International Journal of Control*, 2003, **76**(8), pp. 770–780

14   FRIDMAN, L., STRYGIN, V., and POLYAKOV, A.: 'Nonlocal stabilization via delayed relay control rejecting uncertainty in a time delay', *International Journal of Robust and Nonlinear Control*, 2004, **14**(1), pp. 15–37

15   LEDGERWOOD, T. and MISAWA, E.: 'Controllability and nonlinear control of rotational inverted pendulum', *ASME Journal of Dynamic Systems, Measurement, and Control*, 1992, **43**(1), pp. 81–88

Index